Electronic Sensor Design Principles

Get up to speed with the fundamentals of electronic sensor design with this comprehensive guide and discover powerful techniques to reduce the overall design timeline for your specific applications.

It includes:

- A step-by-step introduction to a generalized information-centric approach for designing electronic sensors, demonstrating universally applicable practical approaches to speed up the design process.
- Detailed coverage of all the tools necessary for effective characterization and organization of the design process, improving overall process efficiency.
- A coherent and rigorous theoretical framework for understanding the fundamentals of sensor design, to encourage an intuitive understanding of sensor design requirements.

Emphasizing an integrated interdisciplinary approach throughout, this is an essential tool for professional engineers and graduate students keen to improve their understanding of cutting-edge electronic sensor design.

Marco Tartagni is Professor of Electrical Engineering at the Alma Mater Studiorum, University of Bologna. He has more than twenty-five years of experience in microelectronic design, with an emphasis on applied optical, biochemical, aerospace, and nanotechnology sensor design.

Electronic Sensor Design Principles

MARCO TARTAGNI

Alma Mater Studiorum, University of Bologna

CAMBRIDGE
UNIVERSITY PRESS

CAMBRIDGE
UNIVERSITY PRESS

University Printing House, Cambridge CB2 8BS, United Kingdom

One Liberty Plaza, 20th Floor, New York, NY 10006, USA

477 Williamstown Road, Port Melbourne, VIC 3207, Australia

314–321, 3rd Floor, Plot 3, Splendor Forum, Jasola District Centre,
New Delhi – 110025, India

103 Penang Road, #05–06/07, Visioncrest Commercial, Singapore 238467

Cambridge University Press is part of the University of Cambridge.

It furthers the University's mission by disseminating knowledge in the pursuit of
education, learning, and research at the highest international levels of excellence.

www.cambridge.org
Information on this title: www.cambridge.org/9781107040663
DOI: 10.1017/9781139629225

First published 2021

Printed in the United Kingdom by TJ Books Limited, Padstow Cornwall

A catalogue record for this publication is available from the British Library.

Library of Congress Cataloging-in-Publication Data
Names: Tartagni, Marco, 1962– author.
Title: Electronic sensor design principles / Marco Tartagni, University of Bologna.
Description: Cambridge, United Kingdom ; New York, NY, USA : Cambridge
University Press, 2021. | Includes index.
Identifiers: LCCN 2020035625 | ISBN 9781107040663 (hardback) |
ISBN 9781139629225 (ebook)
Subjects: LCSH: Detectors – Design and construction.
Classification: LCC TK7872.D48 T37 2021 | DDC 681/.2–dc23
LC record available at https://lccn.loc.gov/2020035625

ISBN 978-1-107-04066-3 Hardback

To Marianna my grandmother

and my daughter Ottavia,

for teaching me how to feel,

what artificial sensors do not, love

Contents

Part IV   Problems and Solutions

Preface

One question I have constantly pondered in the course of my professional life is whether there are general principles in the sensing process that can be applied to the designing of electronic sensors. The aim of this book is to guide the reader along a train of argument leading to an affirmative answer to that question, which provided the title.

Textbooks outlining techniques of sensor design often follow one of two quite contrasting approaches. The first is to focus on one highly specific context linked to a single application. The second hangs the treatment on a broad classification of transduction and architecture techniques. The latter approach leads to the false idea that sensor design boils down to a series of separate cases classified according to their specific application. Such a view may be erroneously encapsulated in the idea that the highly interdisciplinary nature of this field consists in a reasoned assembly of various points geared only to enhance the design efficiency. In actual fact, interdisciplinarity in this domain (which calls for highly different techniques and models ranging from the mathematics of random processes to the science of measurements, signal conversion and processing, information theory, and transduction physics) reveals that the various different subjects are nothing but contextualizations of a few general principles. This book thus sets out to define a general methodology behind the designing of electronic sensors, regardless of the individual application: this will markedly reduce design time and enable the essential design variables to be swiftly identified.

One basic line of inquiry that the book pursues is the role of *information* in the sensing process. This led to a broader definition of *sensor*, focusing on its ability to extract information from the environment. Whereas information theory stemmed from the area of electric communications in which codes may be optimized so as to maximize the amount of data transmitted on the channel, with sensing, the perspective has to be reversed. Here, since the source of information pertains to observed nature, it cannot, as a rule, be altered, and design optimization can be seen as the maximizing of information conveyed by the process. It is in this context that certain basic definitions of sensor physics have been revised. By way of example, we will describe the concept of resolution not just in its original accepted meaning, but highlighting its connections to the amount of information conveyed in the sensing process: an aspect that will enable us to find effective methods of optimizing electronic sensor design.

Writing a textbook is a complex process in which the author is often torn between opposing courses of action. On the one hand, he or she will tend to go for formal rigor of presentation vis-à-vis the scientific community above all. On the other, the desire is to outline the subject as clearly as possible for the beginner, if possible, favoring an intuitive approach based on solving certain conceptual problems, sometimes at the expense of formal rigor. Depending on the target audience, the author will juggle between these two opposing approaches.

Yet if I consider my own teaching experience more closely, it has shown some surprising features. In many (though not all) cases, students to begin with prefer a formal approach to the discipline since that confers an apparent security; they are suspicious about the kind of treatment that starts from concrete examples and ends up in abstract concepts. On the other hand, those already familiar with the discipline will tend to focus more on the principles of the subject, picking out points that had previously eluded them. This accounts for the reaction to certain seminal scientific textbooks that favor a bottom-up approach to the subject. For all their conceptual rigor, they are unlikely to be popular as student course books, since by definition the abstracting process calls for a long period of maturation in the mind of a student. With this book, I have hence adopted a style that mixes the two: the intuitive approach is designed not around abstract hypotheses, but concrete numerical examples, concluding with a due degree of formalism suitable for the task of summing up.

The text is based on two courses I have taught for several years at the Cesena Campus of Bologna University. I must thank my past and present undergraduate, graduate, and Ph.D. students, first and foremost, for their responsiveness and dialogue that enabled me to correct and hone this experience of tuition. I find it hard to imagine writing a textbook without the critical backing of a course dealing with the subject. In drafting the book, I adopted a number of different approaches: trying them out on the course helped me choose among them. My thanks go to many people with whom it has been a pleasure to exchange notes. First of all, Alessandro Piovaccari for his friendship, constant support and encouragement, and above all for firing my imagination about basic principles of the subject. I must also thank Victor Zhirnov for points to do with the limits of sensing, Marco Chiani for information theory, Davide Dardari for the theory of estimation, Aldo Romani for piezoelectric transduction, Alberto Corigliano for mechanical transduction, Hywel Morgan for ionic transduction and biosensor physics, and Marco Crescentini for our many discussions on electronic noise and measurement science. Also, a special thanks to my colleagues Alessandra Costanzo, Alessandro Talamelli, Emanuele Giordano, Luigi Ragni, Elena Babini, Annachiara Berardinelli, Mauro Ursino, and Enrico Sangiorgi for scientific collaboration, friendship, and constant and effective support in this work. My thanks, too, to Roberto Trolli for various points to do with the philosophy of science, raised by him over many years of fraternal friendship and endless and engaging discussions. A special thanks to my older brother Flavio Tartagni for intriguing me into experimental science when I was a child. Last, and definitely not least, my sincere thanks to my publishers, Cambridge University Press, and in particular to Julie Lancashire, for the unreserved confidence they passed on to me from the outset. My thanks to all the editorial staff for their patience, constant assistance, and suggestions over these past years.

One important restriction must be noted in the range of subjects covered by the book. It analyzes sensors whose resolution is constrained by thermal noise – the area often referred to as "thermal noise limited sensors" - and by sensors limited by shot noise . Such an approach is suited to most sensors used at room temperature and for general applications. Analysis of the origin of the noise is thus briefly introduced from the angle of classical statistical mechanics. Of course, it is evident that in sensing and measurement science, the extreme outer limits are dictated by the uncertainty principle, the restraints of which are felt in far less usual cases than those here described. For this reason, sensors based on quantum mechanics principles will not figure in here.

The book falls into three parts. The first one contextualizes general concepts by definitions of design variables, characterization parameters, signals, and errors. This framework will discuss the design tradeoffs under the information theory viewpoint.

The second part focuses on the physical origin of the noise and its role in electronic interfaces design. Here, general design optimization approaches, techniques, and architectures are analyzed, also covering the time-domain sensing techniques.

Finally, the third part deals with selected topics on three different aspects of transduction physics: optical, ionic, and mechanical. The section makes no claim to cover all sides to transduction but only gives examples of typical applications. However, the latter part is not a bare addition to this book since physical transduction is an essential step of the overall sensor acquisition chain, and it should be considered a processing block like all the others in the design process. Therefore, I considered it necessary to show transduction examples to practice the overall sensor design optimization.

Chapter 1 introduces various concepts concerning information and signals, which will be dealt with in later chapters but meanwhile yield a definition of the artificial sensor. Chapter 2 introduces a number of basic parameters characterizing sensor physics, including precision/accuracy and respective tradeoffs, along with relevant figures of merits. Chapter 3 analyzes the main sensor design tradeoffs under the information theory perspective. Chapter 4 lists certain important features of the mathematical methods used in analyzing and synthesizing sensor systems. Chapter 5, by M. Chiani (Alma Mater, University of Bologna), introduces the compressive sensing approach.

Chapter 6 deals with the origins and models of noise in various contexts, including the time domain, such as phase noise. Chapter 7 provides models for calculating and optimizing noise in the case of devices and simple electronic circuits. Chapter 8 covers more complex electronic systems of extracting information that operates in signal and time-space.

Chapter 9 introduces a number of concepts of photonic transduction with particular reference to the photodiode and its configurations in area sensors. Chapter 10 deals with the ionic-electronic transduction that underlies biosensing. That chapter also provides an opportunity to cover some noise issues in biosensors, especially in biopotential sensing. Lastly, Chapter 11 tackles mechanical transduction with particular attention to piezoresistivity and piezoelectricity. The chapter also considers some examples in which resistive mechanical sensors are applied. Finally, Chapter 12 by M. Crescentini (Alma Mater, University of Bologna) offers a collection of electronic sensor design problems and related solutions.

The book is dedicated to the memory of Professors Silvio Cavalcanti and Claudio Canali.

Part I

Fundamentals

1 Introduction

The advancement of electronic devices went through the silent yet pivotal revolutions such as solid-state technology and integrated circuit miniaturization, among many others. However, through merging with the information and communication technologies (ICT), the increasing electronic integration was able to open new perspectives and markets in modern society. The ability to compute and to communicate is among the basic needs of the human community, and the progress of the ICT technologies showed unforeseen scenarios where in some cases – surprisingly – machines greatly surpassed our natural skills. However, in the enduring efforts to make artificial systems behave like humans, few areas are still lagging behind, and one of those is the ability to sense the environment as we do. In this framework, the pervasive implementation of sensing devices in consumer electronics is a highly pursued electronics industry paradigm. The number and variety of sensors implemented in consumer and industrial devices increased dramatically in the past decades: handled devices, automotive, robotics, healthcare, and living assistance.

Unfortunately, artificial sensing requires an approach beyond the borders of information science with the necessity to cope with an incredible number of physical and chemical transduction processes. The interaction with the environment of synthetic systems still shows open issues both at the transduction and at the information processing levels.

Designing sensing devices is always an exciting and challenging task. Very often, the ultimate question is: "Will we be able to detect it?" The answer is hidden in both the technology capabilities and the environment status, and it is not clear, at first sight, where the limitation of the approach is and the reason for the failure. Therefore, the arguments should not be treated as a collection of individual cases but with a general approach using the appropriate tools of abstraction and formalization, with a constant look to the biomimetic inspiration of sensing. In doing so, adopting a strategic and theoretical perspective, we can foresee the integration of massive computation with the sensing capabilities as one of the next incoming revolutions of information engineering.

This first chapter aims to set up the framework on which the book will be shaped, and it is intentionally based on informal descriptions of concepts. This is a nonrigorous approach but is a fundamental step toward an abstraction process about artificial sensing: the ideas behind the general definition of sensors, their main performance-limiting processes, and essential tradeoffs. Using this inductive approach, we will first define concepts, leaving the formalization to the following chapters of the book. If the reader is facing this field for the first time, the arguments could appear vague and fuzzy; thus, this chapter should be eventually reread as the *last* one.

1.1   Sensing as a Cognitive Process

The concept of a sensor would not exist without life. To grow, reproduce, and survive, any organic entity should perceive external signals to evaluate them, either as an opportunity or a danger. *Sensing* is not a mathematical or physical abstraction derived by inorganic matter, but it is a biological process since any living being should perceive, measure, and evaluate external stimuli to take *actions*. As with many other engineering concepts, this *feedback* model is taken from nature, and sensing is the primary input of such a loop mechanism.

Focusing on human beings, the word *sensor* is derived from the verb *to sense*, referring to the capabilities of human beings to perceive reality by means of sight, hearing, taste, smell, and touch. Sensing is a fundamental part of what is referred to as cognitive sciences, an interdisciplinary field aimed at studying the human mind and knowledge processes.

It has been common practice to analyze the sensing process in interdependent stages such as *sensation, perception*, and *consciousness*. The definitions and the borders between these domains significantly differ among the scientific (in a broader sense) communities. However, there is a general agreement on this sensorial experience segmentation. This partition is also reflected in artificial sensing systems.

Sensation is the primary process of receiving, converting, and transmitting information resulting from the stimulation of sensory receptors. Sensory stimuli are taken from the environment by means of physical transduction processes such as those operated by the eyes, where the photons coming from the scene are focused onto the retina in the same manner as a photo camera. The cones and rods of the retina work as transductors, detecting external energy from every single photon and sending the information by means of electrical messages to the brain. On the other hand, perception is the process of selecting, identifying, organizing, and interpreting sensory information. It is not a passive reception of stimuli but early processing: the information is collected, organized, and transmitted by nerves to the brain. Edge detection of objects in seeing and touch is an example of perception. Finally, consciousness is the more elaborated knowledge process: it is the brain's deepest interpretation of neural responses to sensory stimuli. It involves the capacity to sense or perceive and active use of those abilities, depending on previous experience. Humans can experience conscious and unconscious perception. If we relate it to machines, definition, context setting, learning, and adaptation could be possible processes ascribing to a sort of artificial "consciousness." Here, the concept of consciousness is restricted to a functional/phenomenal process, distinguished from the problem of self-awareness where implications are highly speculated in philosophy, with open issues.

In the past centuries, when there were weak boundaries between scientific and philosophical studies, the sensing process was highly conjectured, especially when it was considered a fundamental step for human perception and knowledge. The connection of sensory stimuli with the brain was observed and studied since the time of the

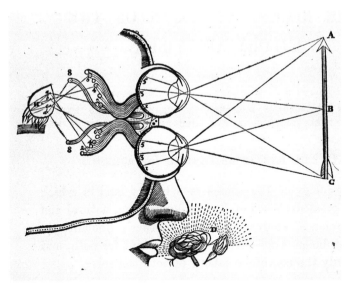

Figure 1.1 Picture taken from R. Descartes, *Tractatus de homine et de formatione foetus*, 1677 (edited posthumously), showing the process of seeing and smelling.

ancient Greeks and in Leonardo da Vinci's monumental work. Among others, it is interesting to note how Descartes remarkably analyzed in greater detail the sensing process in some of his writings, which were very useful in contributing to a general framework of cognitive sciences. As shown in Fig. 1.1, sensorial stimuli (sight and smell) are conveyed into an inner part of the brain, where they are interpreted. Even if some physiological aspects were not correct and Descartes's speculations were far beyond the pure phenomenological aspects of the matter (still unresolved and on debate today), the organization of the sensing process in several steps was profoundly analyzed, introducing modern concepts.

Helmholtz, maybe one of the last polymaths, gave another example of deep analysis of the sensing process in the nineteenth century. In his works (see excerpt illustrated in Fig. 1.2), he gave seminal contributions in the field of visual and auditory perceptions, envisioning a profound relationship between sensing and cognitive sciences. He claimed that human perception should be studied by the *process's physical, physiological, and psychological character*s. He even attempted to justify the perception of beauty related to the sensing process in some of his works.

It is no coincidence that the Greek word *aisthanesthai*, meaning "to perceive by senses and by the mind, to feel" is at the root of the word *aesthetics*: a branch of philosophy dealing with the perception and appreciation of art, taste, and beauty.

To summarize:

- Sensing is a biomimetic concept. Sensors engineering has frequently borrowed functionality models from life sciences psychophysiology and cognitive studies.
- A sensor should not be considered a pure transducer but part of an artificial cognitive process to grab as much information as possible from the environment.

THE RECENT PROGRESS OF THE
THEORY OF VISION.

---·▸◦◂·---

A COURSE OF LECTURES DELIVERED IN FRANKFORT AND HEIDEL-
BERG, AND REPUBLISHED IN THE PREUSSISCHE JAHRBUCHER, 1868.

I. THE EYE AS AN OPTICAL INSTRUMENT.

THE physiology of the senses is a border land in which
the two great divisions of human knowledge, natural and
mental science, encroach on one another's domain; in
which problems arise which are important for both, and
which only the combined labour of both can solve.

Figure 1.2 Excerpt from the *Popular Lectures on Scientific Subjects*, by H. von Helmholtz, 1873
(translated from German) envisioning the need for a strong relationship between the sensing and
the cognitive processes.

1.2   Aiming at a General Definition of Electronic Sensors

In electronic engineering, the word "sensor" embraces a broad class of systems
designed for highly different applications. A "sensor" could be roughly referred to as
"a system that transduces physical stimuli into data." However, this definition is too
vague and does not take the essence of artificial sensing; thus, a closer look should be
taken to understand the standard framework better.

Figure 1.3 shows four examples of systems referred to as "sensors": a weight scale,
a microphone, a heart rate monitor, and a machine vision system. They all collect
stimuli from the physical environment and convert them into data; however, they are
dealing with increasing complexity to achieve the related tasks.

A scale operates a static force measurement. We do not care about the variation of
weight within the measurement timeframe. On the other hand, the microphone needs to
follow the pressure variation (sound) on a surface versus time, and its time-domain
properties are a fundamental aspect of its design. Next, a heartbeat sensor uses patterns
in ECG signals associated with heartbeat events. Finally, a machine vision system
deals with many images to detect/count defective objects. The idea is that any
application identifies specific conditions of the signal to be identified and measured
by a custom sensing system. However, the idea of classifying sensors according to the
kind of signal could be misleading.

We are looking for not the stimulus itself, but something more complex hidden in
primary stimuli and is referred to as *information*. In simple words, the information
content is the essence of what we are looking for in the sensing process. The concept of

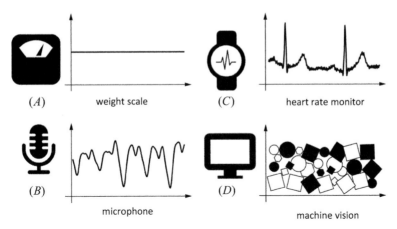

Figure 1.3 Input signals of systems commonly referred to as sensors. (A) Weight scale readout. (B) 1 ms of microphone recording from Etude no. 21 for piano by F. Chopin. (C) 2 s of heart rate biopotential recording. (D) A 2-D machine vision image processor.

information has been extensively treated and formalized in other disciplines; at the moment, it corresponds informally to the amount of knowledge that we gain during the sensing process aiming at the specific task application. We will use a more formal approach to the issue in Chapter 3.

1.2.1   Signals and Information

To illustrate the role of the information in the sensing process, we will use examples. In Fig. 1.4A is shown a heartbeat detector. The sensor's main task is to detect the number of beats in a given period of an ECG signal using a decision threshold. We informally link this to the "information" necessary for our application to understand the concept. The three signal examples of Fig. 1.4A are taken from a set of all possible ECG waveforms in the same time period, and we refer this to as *samples* in the *signal space*. In the first two cases, the system counts 8 beats, while in the last one, only 7. Therefore, we associate the result in a measurable space, referred to as *information space*. In other words, we say that samples in the signal space could be mapped in points in the information space. The acquisition process of a sensor is a correspondence between these two spaces. We will always refer to discrete information space.

In the second example of Fig. 1.4B, the sensing system should detect the number of circles/squares in images. Even in this case, the four sampled images belong to a very large signal space, for example, composed of all possible images of N × M black-and-white pixels. However, the "information" is relatively smaller than the signal space and could be organized in a two-dimensional space where the variables are the number of circles and the number of squares, respectively.

In these two examples, it is easy for human perception to identify the information in the signal space at first sight and check if the sensor system has correctly detected our task. However, there are other cases in which the information is more hidden than

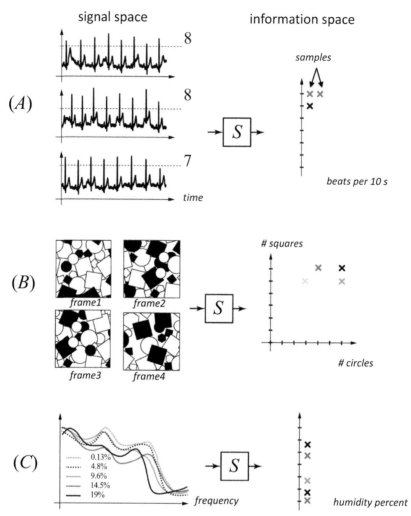

Figure 1.4 Signal and information spaces. (A) Biopotential heartbeat detector using a threshold. (B) Machine vision object counter. (C) Impedance spectrum humidity detector. Note that the first two examples have a discrete information space while the last is a continuous space.

previous examples, and machines could outperform human perception. For example, in the case of Fig. 1.4C, the signal is composed of five measured microwave impedance spectra related to a material having different water content (humidity). The idea is to use these spectra to implement a microwave humidity sensor, where the information is the percent humidity. It is hard to see any regular or monotonic behavior in spectra or in parts of them with respect to the stimulus (humidity). Our intuition concludes that there is no clear relationship between the humidity of the material and the spectra. In other words, it is not easy to see any significant information in the signal itself. However, suppose the signal is treated by suitable mathematical processing. In that case, we can set up a linear predictive model to detect the humidity based on microwave spectra so that signals can be mapped into distinguishable and ordered levels in the information space. The latter example shows that the information could be very hidden in signals, even beyond human capabilities to distinguish them in raw data. For this reason, in these cases, the information to be extracted is often referred to as *latent variables*.

The preceding examples are related to cases of different *complexity* of the task and require different processing resources to extract the information.

To summarize:

- The sensing process should be defined by a *task*, which qualifies the kind of *information* that should be measured. Thus, the application (task) determines the characteristics of the information space.
- *Signals* are functions representing states of the sensed environment carrying information. All the possible configurations of signals define the signal space.
- The information space has smaller dimensions of the signal space, and it is discrete. This means that the multiple elements of the signal space may have the same element in the information space.
- The sensing process is a function implying that each *sample* of the signal space has a correspondence in the information space.

1.2.2 The Simplest Case of an Analog-to-Digital Interface

The previous section identified a distinction between signals and their information content, mapped into the information space. However, if we refer to the simple analog-to-digital (A/D) conversion of a signal in the "analog domain," we can match more easily the two spaces since the analog value itself encodes the information. We can better understand this with the cases illustrated in Fig. 1.5. In Fig. 1.5A is shown a time-varying biopotential signal that is monitored by an A/D interface. Our task is to know the biopotential value evolution with respect to time, and thus it is precisely the information that we need. The A/D converter associates a specific analog value of the signal full scale with a binary-encoded discrete value. Therefore, the discrete values of the A/D converter are easily represented in the information space. The correspondence is made by associating an analog value with the converter's closest discrete value. The case of Fig. 1.5B is even more straightforward: the information is the static analog value of a weight sensor. Therefore, each measure (sample) is directly mapped in the information space. As before, multiple analog values may be mapped into the same coded value by the converter.

In summary:

- In the simple case of an A/D interface, the association between *information* and *signal* is closer because the signal value itself *represents* the information that we need to detect.
- The correspondence is made by associating an analog value with the closest discrete level of the A/D converter.

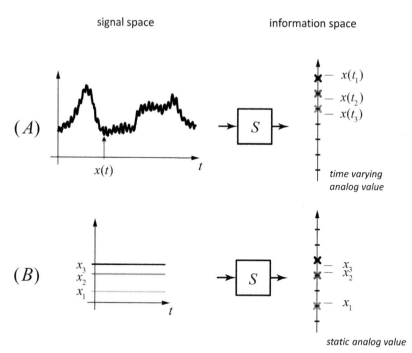

Figure 1.5 Sensing process in time-varying (A) and static (B) analog signals. The analog value of the signal space is associated with the closest discrete value of the information space.

1.2.3   The Role of Errors

Unfortunately, the sensing process's physical implementation is necessarily affected by *errors* due to the stochastic nature of random processes and nonidealities. Errors arise from either the environment or the sensing system itself and its nonperfect detection capabilities. Let us look at Fig. 1.6, where a biopotential is used to detect heartbeats utilizing a threshold as in Fig. 1.4A. In the absence of noise, during the time-lapse, we measure 8 beats, as shown in Fig. 1.6A. Now, assume that the sensing process is noisy and the same waveform as in Fig. 1.6A with added noise is shown in Fig. 1.6B. If we use the same detection approach, the count is no longer 8 but rather 10. The random process of noise changes the threshold crossing cases: there are some points (e.g., point M) that were not crossing the threshold before (without noise), while now they do, thanks to noise contribution. Conversely, there are other points (e.g., point N) that were crossing the threshold in the previous case, and now they do not pass the level because of the perturbation of noise. If we repeat the same procedure on a signal containing 8 beats in the presence of the noise, we may at one time count 7, another time 9, another time again 8, and so on. This means that we cannot say that the count is certain but, in the presence of noise, we can say that the "estimation of the count is given by 8 ± 2." Therefore, the presence of noise determines an *uncertainty* (Chapter 2) of the measure of ± 2 counts. The uncertainty due to noise could be visualized in the gray area across the tick equal to 8 in Fig. 1.6.

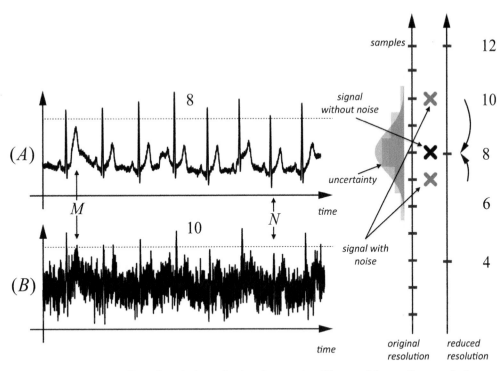

Figure 1.6 Effect of resolution reduction due to noise. Biopotential recording to calculate heartbeat (A). Same biopotential signal but with noise added (B). The addition of noise determines the increase of uncertainty of detection and thus a decrease of resolution.

In this example, the presence of errors (or noise) changes the information space situation. If we could have counted single beats without noise before, now we have an uncertainty of ± 2 counts. Therefore, previous levels are no longer truly distinguishable from each other because the *same* signal could give counts in the interval 8±2 due to noise. This fact reveals that the information space subdivision is not appropriate because the sentence "count = 8" has information similar to that one of "count = 10." This results in a *misclassification* because we classify 8 counts, although, in reality, they might be 10 or vice versa. This means that we have a high probability that the affirmations mentioned above reflect the same signal condition due to errors. This could be seen pictorially showing that the same uncertainty area covers samples. Therefore, it could be better to reduce the number of subdivisions (e.g., by grouping 4 levels) so that "count is between 6 and 10" and "count is between 2 and 6" has more significance from the information point of view because there is a lower probability that the two sentences correspond to the same signal. In this case, any sample giving a value in the uncertainty zone, identified by 8 ± 2, will be associated with the center of the interval, whose value is 8. In other words, by enlarging the classification zones by considering uncertainty, we reduce possible misclassification errors.

We can thus refer to the subdivisions of information space as *resolution levels*. In the presence of noise, we might set the resolution level of the order of the uncertainty so that it preserves significance from the information point of view, avoiding misclassification. As shown in Fig. 1.6, the higher the noise, the lower the resolution.

Turning back to the simplest case of the analog domain where information and signal are strictly related, we can observe that we could have infinitely small resolution levels in the absence of noise. It is thus the presence of noise that determines a finite resolution. This is congruent with the observation that any real sensing system (in which necessarily errors such as noise are present) does have a finite resolution and reinforces the already cited statement that the information space is discrete.

However, any application task always identifies a maximum in the needed information. For example, in the case of Fig. 1.4A, we do not need to detect more than 30 beats per 10 seconds, or in the one in Fig. 1.5A, we know that the biopotential will never surpass a value of hundreds of millivolts. The maximum achievable value in the information space is referred to as *full scale*. Therefore, the scale defines a *fixed number of resolution levels* under which the information space can be divided. The number of resolution levels expressed in terms of energy is referred to as *dynamic range* (Chapter 2). The number of information (resolution) levels achievable by a sensing system is a measure of the process's information.

To summarize:

- The physical sensing process is always affected by *errors* (e.g., noise) deriving from both the environment and from the sensing system itself.
- Errors define *uncertainty* in the measuring process, meaning that we cannot be sure of the result given by the system, but we can estimate the information with some degrees of confidence.
- The uncertainty sets the *resolution* of the sensing process to make the levels distinguishable from each other at the information space and reduce *misclassification*.
- Real applications always define a maximum level of the coded information or *full scale* in the information environment. This boundary identifies a limited number of *resolution levels* in the information space or *dynamic range* if expressed in energy terms.

One useful option is to express the number of discrete levels at the information space in terms of *bits*. This creates a significant link on the one hand within the context of A/D converters (Chapter 2) and, on the other hand, with information theory (Chapter 3). For example, referring to Fig. 1.6, if the number of resolution levels is coded in N bits and the noise reduces that number by grouping 4 adjacent levels, the resolution is degraded by 2 bits of information. We will return to this later.

Another important characteristic of the sensing process is the *minimum detectable signal* (MDS), which is the minimum variation of the signal to induce a significant variation in the information space (i.e., at least one resolution level). If we take the simpler example of Fig. 1.5A, where signal and information spaces overlap, the MDS is the amount of signal variation that is "equal" to that of noise. In other words, the signal to be detected should have a "strength" surpassing that one of noise. Thus the MDS is usually set to the noise "strength" and thus to the uncertainty of the process. Of course, we should define a metric for making a comparison (Chapter 2). In any case, if there were no errors (e.g., noise) at all, the MDS could be infinitely small, resulting in an infinite detection capability.

Figure 1.7 shows a conceptual view of the sensing process as a mapping between signal space and information space. As shown in Fig. 1.7A, each cross identifies a sample in the signal domain corresponding to another point in the information space. Any point of a subset of the signal space is thus mapped into another point in the information space within the uncertainty area due to errors. Again, the number of elements in the signal space is, in general, higher than that of the information space (e.g., the number of all possible images containing the same number of objects is mapped in the same point in the information space.). Following the discussion of Fig. 1.6, we can set the resolution's size to encompass most of the uncertainty so that the discrete resolution levels have a higher degree of confidence to be distinguishable from each other. In doing so, most of the points within a discrete level are associated with one point of the information space, shown in the figure with the crossing point between lines. Another way to see this is that by enlarging the resolution levels, we reduce the overlap between the uncertainty zone and thus the misclassification. The higher the noise, the lower the resolution and the lower the resolution levels since the full scale is fixed.

Following the illustrations of Fig. 1.7, the transition of samples between two adjacent points of the information space corresponds to the transition between two

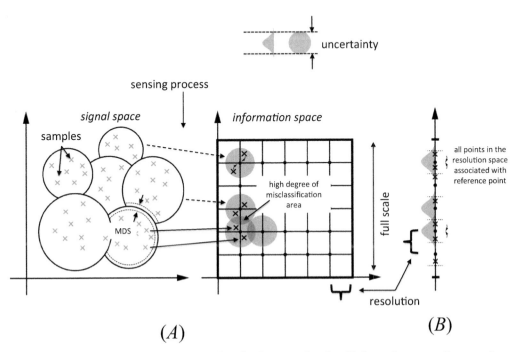

Figure 1.7 Sensing process as a function between signal and information space. Conceptual (A) multidimensional (B) one-dimensional representation.

signal space subsets. During the information space transition, the sample points cross overlapping zones of uncertainties. If it is in the middle of the overlapping, we are unsure whether the sample belongs to one discrete level or the adjacent one; thus, we have a high degree of misclassification. The overlapping misclassification area could be mapped in the signal space by an area across the subsets' boundaries. The thickness of this area identifies the MDS because it is the minimum variation of the signal required to have a distinguishable change in the information space.

The concept could be applied to the simplest case of a noisy analog to digital conversion shown in Fig. 1.7B. This is the simplest case where we can superpose the signal and the information space because the signal value itself encodes the information. Each sample in the analog signal space identifies a binary-encoded level in the information space; however, the noise implies the possibility of an error of assignment. To reduce that error, we can enlarge the resolution levels, but this implies that we reduce the conveyed information. As an example, let's take a noisy 8b converter. If the noise is so high that it covers 4 discrete levels (2 bits), this means that it is convenient to reduce the number of levels by adjacent groups of 4 levels to lower misclassification. Therefore, noise reduces the system's resolution from 8 bits into 6 equivalent bits $(8b - 2b = 6b)$ to make information (resolution) levels distinguishable. In summary, the number of distinguishable resolution levels expressed in bits is a measure of the information gained by the sensing process.

The corruption of the information by errors is conceptually exemplified in Fig. 1.8. The signal coming from the environment is physically affected by noise. Therefore, the amount of information gained at the input is limited by this effect arising from the source's random physical processes. However, in sensor system design, we have to deal with other sources of errors that further restrict the amount of information conveyed. For example, analog interfaces have intrinsic noises due to thermal noise given by electronic devices. Furthermore, the transduction process might have nonoptimal characteristics (e.g., presenting nonlinear or saturation effects), or the detection algorithm could induce assignment errors due to poor characterization of the model. All these sources of errors reduce the amount of information conveyed to the output. From the resolution point of view, the sensor design is aimed at reducing to as low as possible the information corruption, leveraging on design constraints and tradeoffs.

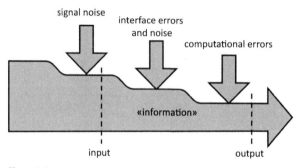

Figure 1.8 Degradation of the information in the sensing process due to errors/noise.

Another important point of the sensing process is the role of energy and time (Chapters 2 and 3). The corruption of information induced by errors could be accounted for by representing them in terms of energy/power whose effects set the detection limits. In other words, the amount of information conveyed by a sensing system could be referred to as the relationship between signal and errors (noise) energies. The simplest case of this tradeoff is the signal-to-noise-ratio.

As far as time is concerned, any measurement process requires some kind of energy/ power to be taken from the signal itself to be classified. As in biological organisms, where sensing is useful to organize an action, the measurement process should be performed in a determined amount of time. As such, when the signal change versus time is limited, it is said that signals have a characteristic *bandwidth*. In the example of the microphone, we can observe that the change of states versus time is limited in the audio bandwidth. We have to implement an interface that follows signal changes in the whole audio bandwidth in an optimal design perspective. Therefore, a sensing system's capability to perform classification in a determined amount of time is a primary constraint.

Thus, we could make a broader definition of an "electronic sensor" as follows from previous arguments.

> **Electronic Sensor** An electronic system that extracts the information required by the application from observed signals in a determined amount of time, namely an *information classifier*.

Turning back to our examples, we can see that this definition embraces the entirely different functions of the systems used as examples and gives some hints about the essence of the sensing process, based on the role of the information.

In summary:

- The number of distinguishable resolution levels expressed in equivalent bits at the readout is the measure of the information gained by the sensing process.
- One key aspect of sensor design is maximizing the number of resolution levels, which is equivalent to maximizing the information conveyed by the sensor system.
- Another key issue of sensor design is related to the time required by the sensor to resolve the task, which is linked to the bandwidth characterization of the electronic interface.
- The aforementioned characteristics allow us to define any kind of sensors as information classifiers performing the task in a limited amount of time.

1.3   Essential Building Blocks of Electronic Sensors

A preliminary and necessary step of the electronic sensing process is the transduction of the physical stimulus into electronic signals. This is performed by a *transducer*

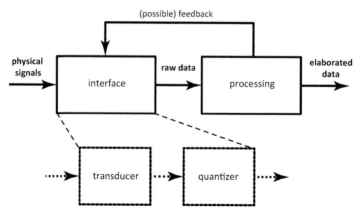

Figure 1.9 Building block structure of an electronic sensor.

interface, which converts the energy of physical signals into electronic states. Then, data should be organized and elaborated to extract the information depending on the specific task. This process could be performed by a system that follows the transducer and computes elaborations on the previous stage's raw data. The higher the complexity of information, the higher the "intelligence" (computational complexity) required by this block to achieve the result.

We can draw a general scheme of an electronic sensor structure illustrated in Fig. 1.9. Two parts could describe a generic sensor: the interface and the processing machine. The first one is devoted to the transduction and digitization of the signal, composed of two subblocks: the transducer and the quantizer. The transducer is the block that directly connects with the physical environment. It could be a simple amplifier or a more complex structure operating in the time or frequency domain. The quantizer is a block that is necessary to operate in the binary domain on which the elaboration block will work. A typical quantizer is an A/D converter, even if other digitizers might be used (Chapter 8).

A processing machine implements the second part to gain the maximum information from raw data. The main differences between the two parts can be summarized. The transducer converts the energy/power levels of physical signal states into raw data. Therefore, it acts as an energy detector, and the design should be mainly focused on optimization with respect to the energetic content of signals. One of the references of this optimization is the *signal-to-noise ratio*. Conversely, the processor searches the raw data's information to achieve the final task required by the sensor application. It makes complex elaborations of raw data that earlier stages cannot treat. For example, we can implement analog filters in the first blocks to optimize the signal-to-noise ratio, but we need data elaboration to implement complex algorithms, for example, Kalman's filters or machine learning classifiers.

From previous arguments, it is clear that the signal-to-noise-ratio (which is an energetic ratio) should *not* be considered as the reference for the ultimate limit of the sensor performance but just a first step to optimizing the overall chain. For example,

Table 1.1 Sensing approaches classified according to the information complexity (from lower to upper lines in the table).

Signal characteristics	Information extraction strategy
Unknown pattern/event classes	Unsupervised classification (learning)
Known pattern/event classes	Supervised classification (learning)
Pattern/event counting	Pattern matching
DC and AC measurement	Transducer and quantizer

there are radar or ECG signals whose energy is much lower than that of noise. However, it is possible to discriminate useful information using raw data processing techniques.

Looking back to the examples, we can see that the weight scale and the microphone are sensors that could be modeled by just the first block of the structure illustrated in Fig. 1.9. This is because the relationship between signal and information is closer to each other. On the other hand, the heartbeat detector needs to extract information at an upper level, for example, using pattern recognition or other advanced filtering techniques. These techniques can be implemented only at higher processing levels, on raw data. *Intelligent filtering, pattern recognition, compressive sensing* (Chapter 5 by M. Chiani), and *machine learning* are just a few examples of possible functionalities of this block.

Based on the role of the structure of Fig. 1.9, we can draw Table 1.1 showing different strategies of information detection according to the complexity of the information to be extracted. The bottom line is related to the transducer case, where the A/D conversion of the stimulus readily captures the information. This is the case of a direct measure where the transduction directly converts the intensity of a stimulus into information content. In the pattern/event line of the table, the information extraction is based on a priori knowledge of the pattern so that suitable algorithmic strategies could be conceived to get the right information content. The last upper two lines of the table represent a borderline case of sensing in which not only a measurement should be performed but also learning is involved. In supervised classification, the computational machine should learn the rule by way of predefined examples (training set). In the top line of the unsupervised learning case, the computational architecture should be able to identify classes of events/objects based on their inherent characteristics.

To summarize:

- Electronic sensors should be segmented into different processing stages, similar to the biological paradigm of the cognitive process aimed at extracting information from the environment.
- Depending on the information content's complexity, electronic sensors implement higher degrees of computation in the final operational blocks to increase the information classification.

A problematic and still debated argument is quantifying the complexity of information extraction. This is generally unknown; however, we can relate it to the minimum number of computational resources required to solve it.

1.4   At the Origin of Uncertainty: Thermal Agitation

Errors are differences from observed results from what we expect. One of the primary sources of error is *noise*, arising from random physical processes (Chapter 6), so that it is one of the main limiting factors of the sensing process. As discussed, noise limits the resolution of the electronic interfaces (Chapter 7); thus, the amount of information conveyed into the sensor.

If we look at Fig. 1.10, we can see a simple mechanical force transducer (Chapter 11). One end of a cantilever is anchored to a firm reference while a variable force is exerted on the other free end. On the cantilever's upper side, a laser beam is reflected toward a surface or a position-sensitive optical sensor (Chapter 9). Therefore, the position of the cantilever beam is proportional to the input force, realizing a force meter.

What is the limit of discrimination of this "sensor"? In principle, we can sense any slight variation of the input. If it is hard to distinguish variations on the screen, we can move the screen to a greater distance than the original one to look at variations clearer. In principle, this sensor has an "infinite" capability of discrimination (down to fundamental physical limits).

Unfortunately, nature made things a more complex way, and we know that any mechanical system is at a microscopic level subject to molecular agitation. In thermal equilibrium, any atom of the cantilever and any molecule of the surrounding gas is subject to a natural thermal agitation where the mean kinetic energy of any particle of the system is a microscopic expression of the temperature.

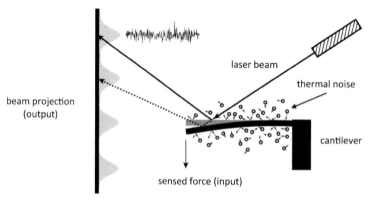

Figure 1.10 A mechanical sensor as a paradigm of electronic transducers. The discrimination of the sensor is limited by thermal noise.

Therefore, the cantilever is subject to natural and random displacements that determine an erratic movement of the beam projection; thus, the measure is affected by *uncertainty* so that we no longer can have an infinite capability to sense the input source in a limited amount of time. It is said that the discriminability of this sensor is thermal noise limited. We could average samples to increase the resolution of the system; however, this takes time and is limited by how fast the force signal is varying (signal bandwidth).

Understanding the balance between the signal (force) and the noise (thermal agitation) is necessary for sensor design. We will see that the above sensor mechanical paradigm is very much similar to electronic sensing systems. Even if fundamental physical limits are at the bottom of any sensor's resolution, thermal noise is one of the most common limitations in devices operating at room temperature and at the microscale.

1.5   Basic Constraints of Electronic Sensor Design

Looking back to all the examples of this chapter, we can devise several basic interdependences and constraints to be taken into account in sensor design.

- Resolution-bandwidth tradeoff. We can increase the resolution (i.e., reduce the uncertainty) by averaging the readouts by following the law of large numbers, as in the force sensor example. However, we must assume that the input force shall be stable during the averaging. This means that averaging is limited by signal bandwidth, and we cannot follow input signals that are faster than the averaging time. Therefore, the higher the resolution, the lower the bandwidth.
- Resolution–power consumption tradeoff. Depending on the complexity of the information to be extracted, we need to use higher amounts of power due to the larger computation requirements to gain higher information (resolution).
- Bandwidth–power consumption tradeoff. Since the computation requires energy consumption in real systems, the elaboration of higher information in a shorter time implies higher power consumption.

A block diagram of the foregoing relationships is shown in Fig. 1.11. The diagram shows the basic constraints of the sensor design divided into three main areas (Chapters 2 and 3). The first is related to the *information constraint* related to the amount of information conveyed by the sensing system, which is represented by the dynamic range. The *dynamic range* is in turn determined by the *operating range* and the input-referred *resolution* of the system. A second area is related to the *system's time constraint*, that is, the bandwidth. A third area is related to the *energy constraint*, which is represented by the power consumed by the sensing system.

A given electronic technology or architecture allows us to determine *figures of merit* (Chapter 3) relating and trading off the three areas mentioned above. Therefore, once we have two out of the three constraints set, we can determine the remaining. For example, once we have the required bandwidth and dynamic range for a given figure of

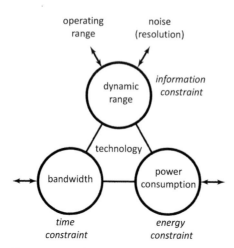

Figure 1.11 Basic design constraints of electronic sensor design.

merit, we can determine the minimum required power consumption for a given technology. Alternatively, we can determine the dynamic range for a given power budget and bandwidth, that is, its maximum achievable resolution.

Further Reading

Brillouin, L., *Science and Information Theory*, 2nd ed. Mineola, NY: Dover Publications, 1962.
Pierce, J. R., *An Introduction to Information Theory: Symbols, Signals & Noise*. Mineola, NY: Dover Publications, 1961.
von Helmholtz, H., *Popular Lectures on Scientific Subjects*, Harvard Un. D. Appleton, 1873. L.

2   Sensor Modeling and Characterization

This chapter will present a general overview of sensor characterization from a system perspective without any specific implementation detail. The systems will be defined based on input and output signal description, and the overall architecture will be discussed, showing how the information is transduced, limited, and corrupted by errors. One of the main points of this chapter is the characterization of the error model and how it could evaluate the measurement's uncertainty and its relationship to resolution, precision, and accuracy of the overall system. Finally, the quantization process, which is at the base of any digital data elaboration, will be illustrated, interpreted, and included in the error model.

2.1   Signals

Signals are representations of states of the environment modeled by mathematical functions carrying *information*. As discussed in Chapter 1, *signals and information are different entities*. Signals could be described by sequences, one- or multidimensional functions in space and time. For the moment, we will refer to "information" as the *measurable task of the sensing process*, being aware that it is a vague concept. More formal definition of information will be addressed in Chapter 3; however, this is a good starting point, to begin with. Take the example of searching for the number of objects (information) in an image (signal). In that case, the information is associated with the object's number, while the image is only the mean where the information is conveyed. Therefore, we can have an extremely high number of different images encoding the same information, and the task of the sensing process is to extract as much information as possible (the correct number of objects) from the raw data of a movie stream. This process could be easy or difficult based on the complexity of the task.

Since sensing is a measurement process, we can define two important classes:

- *DC measurements*, where the *information* is stable during the observation time. This means that we could perform a measurement in a reasonable amount of time that is much shorter than any change of the requested information.
- *AC measurements*, where the *information* changes versus time.

Therefore, AC measurements are associated with bandwidth, while DC measurements are not. AC and DC are acronyms historically derived from electric power transmission ("direct current" and "alternate current"); however, in modern contexts, the original significance has been extrapolated to refer to the role of time in signals.

A weight scale, a caliber, and a laser meter are examples of DC measurements. However, even a sinusoidal-based impedance meter is a DC measurement if the information "impedance" does not change over our interest observation time. Examples of AC measurements are detecting sound by a microphone or the acceleration data logging with inertial sensors. In these cases, the information "pressure of acoustic waves" and "acceleration" should be recorded in the time domain.

We have just defined a general concept in the measurement process. However, we will start by restricting signals to single-variable, time-dependent functions $x(t)$ where the information is related to the analog value of the signal itself. The above is the simplest case to start with, where *the analog value of the signal is strictly associated with the information*. In other terms, the analog value *is* somehow the information.

More specifically, we will refer to AC signals as those in which the information is associated with the displacement with respect to a reference such as the mean value in the observation time as illustrated in Fig. 2.1A with $\Delta x(t)$. Therefore, the signal is $x(t)$ but the component carrying information is $\Delta x(t)$. We will see the implication of this from the energetic point of view.

We can describe a DC signal as a constant time function shown in Fig. 2.1B to converge with the previous definition. Therefore, again, the information is in the signal itself, that is, in the distance of the signal value from a reference such as the zero input. In this case, we can say that the signal (and thus the information) is $\Delta x = x$ shown in Fig. 2.1B. It should be pointed out that the signals class defined in the preceding are just a subset of possible signals but cover a large part of "electronic sensors" systems.

Now, referring to *signals* (and implicitly to the information for our restricted definition that we started with), given the maximum and the minimum values of the allowed signal x_{max} and x_{min}, we can define some common characteristics:

- *Full scale(s)* is the maximum extension of the allowed signals x_{max}. If the signal is also defined for negative values, x_{min} could be provided as additional negative full scale.

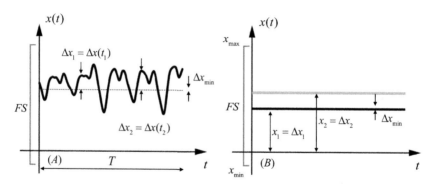

Figure 2.1   Example of AC (A) and DC (B) signals.

- *Full scale span FS* $= x_{max} - x_{min}$ is the algebraic difference between the maximum and minimum values of the signal excursion.
- *Operating range* is the set of values within the full-scale span $[x_{min}, x_{max}]$ (also written as $x_{min} \div x_{max}$).

It is common in the literature to find overlapping definitions of *full scale* and *full scale span*; in this book, we will often use the term "full scale" to refer to the "full scale span" for simplicity. The preceding definitions could refer to either signals or interfaces. In the case of signals, these definitions could be used to specify the requirements for a particular application. As far as interfaces are concerned, these definitions might result from a design. The full scale should not be confused with the *dynamic range* discussed in detail later in this book.

Another important characteristic of a sensor interface is the *minimum detectable signal* (MDS). The MDS could be defined in terms of information as:

- The *minimum detectable signal* Δx_{min} is the minimum variation of the signal state that induces a distinguishable change of the extracted information.

However, in our simplest case of analog-to-digital conversion, we can associate MDS with a minimum variation of the input signal Δx_{min} to change the least noiseless significant bit of the coded output.

The MDS is a fundamental concept in sensor design and could be applied to both DC and AC measurements. For example, in the DC case, we may want to measure pressure between 1 kPa and 50 kPa, having the possibility to distinguish variations of 10 Pa, which is the MDS. Alternatively, in the AC case, we require a biopotential amplifier having the capability to amplify signals as small as 500 nV.

The concept of "strength" of signals Δx will be associated with either the *amplitude* or the *power/energy* of stimulus variation. We will see that the latter is the more appropriate from both the physical and the information points of view.

A final consideration is again about DC/AC measurements. Assume that we need to measure the *amplitude* or *frequency* of a sinusoidal signal. In this case, the information is associated with "amplitude" and "frequency," and if they are stable in time, it is a DC measurement, while if they are changing in time (and so there is a signal bandwidth), this is an AC measurement. Therefore, the MDS refers to the minimum change of $x(t)$ (which is acting as a "carrier" as in the communication context) that is sufficient to detect a change of the frequency or amplitude by the sensing system.

2.2   The Sensor Interface: The Deterministic Model

A fundamental achievement in electronic sensor design is to characterize a *model* to predict its behavior. We can use different models according to the level of description we want to assume, for example, deterministic or stochastic, as will be discussed in Sections 2.3 and 2.8. We will start with the deterministic one as the first and simplest

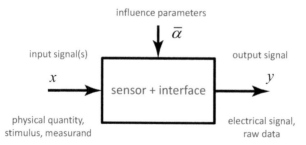

Figure 2.2   Behavioral representation of the deterministic model of the sensor interface.

step of sensor system description, given by a function F mapping a quasistatic input stimulus x with an output y

$$y = F(x) \text{ or simply: } y = y(x). \tag{2.1}$$

Thus, a sensor interface, as shown in Fig. 2.2, could be described as a black box where input signals are *physical quantities* or *stimuli*, and outputs are electronic signals transducing the physical information into raw data encoded in either analog or digital format.

The deterministic function F is referred to as the *quasistatic characteristic* of the sensor interface. The name "static" stems from the fact that the relationship (2.1) is valid, assuming extremely slow input variations. Of course, time is a fundamental parameter for sensor processing, and we will see how to treat time variations correctly in the following sections. From the experimental point of view, we can reconstruct the sensor's transfer function by recording the output with respect to a quasistationary variation of the input, that is, using variations of input signals much slower than any time constant of the interface.

For a better characterization of sensors, we have to consider the case in which the output depends not only on the input but also on other variables not carrying information that could be considered *influence parameters*. Influence quantities or parameters are unwanted dependencies of the measurement on secondary physical entities or sensor nonidealities. For instance, a resistance strain sensor is designed to measure a force; however, owing to the principle of transduction, the output also depends on the temperature. In other cases, an influence parameter could be a generic technology variation. For example, for the same known input, each production lot device gives different outputs under identical ambient conditions. This variation could be a simple amplifier offset variation or a combination of multiple technology variation effects. In other terms, influence parameters are variables that reduce the amount of information possible to acquire from the input. Therefore, we can add the dependence of the output on influence parameters to the sensor function as

$$y = F(x, \underbrace{\alpha_1 \ldots \alpha_K}_{\text{parameters}}) = F(x, \overline{\alpha}), \tag{2.2}$$

where $\overline{\alpha} = [\alpha_1, \alpha_2, \ldots, \alpha_K]$ is the set of influence parameters.

2.3 Quasistatic Ideal Characteristic and Sensitivity

In the simplest concept, a sensor interface should operate a measurement of an analog value, that is, a ratio of the measured data with a reference. In this case, to preserve the ratio for any value of the input span, the linear characteristic is the best one[1]. Therefore, the linear characteristic of the sensor is often referred to as the *ideal quasistatic characteristic*, as shown in Fig. 2.3 and characterized by a linear relationship

$$y = F(x) = S \cdot x + y_{\text{off}}, \tag{2.3}$$

where S is the slope of the characteristic and y_{off} is the output offset, and $x_{\text{off}} = -y_{\text{off}}/S$ is the offset on the abscissa.

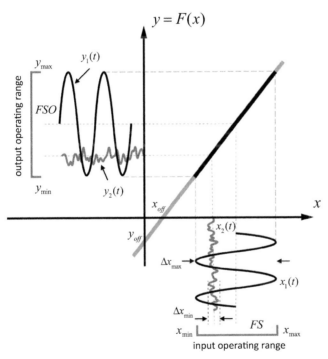

Figure 2.3 Quasistatic ideal characteristic of the interface with the application of two input signals, $x_1(t)$ and $x_2(t)$.

---

[1] It can also be shown that in the case in which the input could take with equal probability any value of the range (uniform probability distribution), the linear characteristic maximizes the mutual information conveyed by the system. However, nonlinear characteristics could be more effective in the information extraction for nonuniform distributions.

> **Notation** We will use $x \equiv [\xi]$ and $y \equiv [\eta]$ to indicate that the measurement of the physical input signal is expressed in ξ units (e.g., pascal, lumen, newton, etc.) and that of the electronic output signal is expressed in η units (e.g., volt, coulomb, least significant bit, etc.). We will refer to either *base* or *derived* International System of Units (SI) measurements. The notation $\equiv [\cdot]$ means that the expression is a dimensionless quantity. Units such as percent, permille, and parts-per notation (ppm, ppb, etc.) are derived dimensionless quantities.

Figure 2.3 shows how the values x_{min} and x_{max} are mapped into y_{min} and y_{max} and how *FS* is reflected into the full-scale output span $FSO = y_{max} - y_{min}$ by means of the characteristic. The figure illustrates two possible AC signals $x_1(t)$ $x_2(t)$ that could be applied to the system: the first one is a sinusoid and covers the whole operating range values, whilst the second one is a small stochastic signal comparable with the MDS.

Example An integrated pressure sensor transduces an input from -15 kPa to 25 kPa into an output from 0.2 V to 4.7 V. It is a DC measurement sensor. The input *full scales* are 25 kPa and -15 kPa (since the input is bidirectional), and 4.7 V is the full scale for the output. The input and output *full-scale spans* are $FS = (25 - (-15)) = 40$ kPa, $FSO = 4.7 - 0.2 = 4.5$ V, respectively. The input and output *operating ranges* are $[-15, 25]$ kPa and $[0.2, 4.7]$ V, respectively. Furthermore, the sensor is able to detect variations of the input pressure as low as 0.76 Pa, which is its MDS.

Example A electrophysiology interface senses biopotential variations up to ± 1 mV in a bandwidth of 30 Hz. It performs AC sensing. The full scales are -1 mV and 1 mV, and the input *full scale (span)* of the interface is $FS = 2$ mV. The input *operating range* is $[-1$ mV, 1 mV$]$. The MDS is given as $\Delta x_{min} = 1$ μV, meaning that we have a "distinguishable" output readout only if the input variation is greater than that value.

We will discuss primarily AC signals for generality; thus, *input* and *output signals* should be referred to as *variations* with respect to a *reference* or *bias point* (x_0, y_0) where $y_0 = y(x_0) = F(x_0)$. Therefore, the relationship between input and output becomes

$$F(x_0 + \Delta x) = y(x_0 + \Delta x) = y_0 + \Delta y, \tag{2.4}$$

where Δx and Δy are the input and output signals, respectively. The input/output full scales give the maximum amplitude of the input/output variations.

The sensor interface *gain* also referred to as *sensitivity*, is defined as the ratio of the input and output variations

$$S = \frac{\Delta y}{\Delta x} \equiv \begin{bmatrix} \eta \\ \xi \end{bmatrix}. \tag{2.5}$$

The dimension expression means that the sensitivity is measured by the ratio between the output and the input units, for example, [V/Pa] for an analog pressure sensor. We can also see the ideal characteristic expression as

$$y(x) = y_0 + \Delta y = y_0 \left(1 + \frac{\Delta y}{y_0}\right) = y_0[1 + S'\Delta x], \tag{2.6}$$

where $\Delta y / y_0$ is referred to as *relative output variation* and where

$$S' = \frac{1}{y_0}\frac{\Delta y}{\Delta x} = \frac{S}{y_0} \equiv \begin{bmatrix} 1 \\ \xi \end{bmatrix} \tag{2.7}$$

is the *relative sensitivity* of the sensor.

Example A platinum resistance temperature sensor is implemented by a metallic wire whose conductance is dependent on the temperature. Thus $y \leftrightarrow R \equiv [\Omega]$; $x \leftrightarrow T \equiv [°C]$. A typical first-order relationship for this kind of sensor is $R(T) = R_0[1 + \alpha_0(T - T_0)]$; $R_0 = 100\,\Omega$; $\alpha_0 = 0.00385$. Its sensitivity is

$$S = \frac{\Delta R}{\Delta T} = 0.385 \begin{bmatrix} \Omega \\ °C \end{bmatrix} \text{ and since}$$

$$R(T) = R_0[1 + \alpha_0(T - T_0)] = R_0\left[1 + \frac{\Delta R}{R_0}\right] = R_0[1 + S'\Delta T] \text{ its relative}$$

sensitivity is $S'\Delta T = \dfrac{\Delta R}{R_0} \rightarrow S' = \dfrac{1}{R_0}\dfrac{dR}{dT}\bigg|_{T_0} = \alpha_0 \equiv \begin{bmatrix} 1 \\ °C \end{bmatrix}.$

Example A strain gauge is a resistance sensor whose value is dependent on the applied force: $y \leftrightarrow R \equiv [\Omega]$; $x \leftrightarrow F \equiv [N]$. A first-order relationship is given by

$$R(F) = R_0\left(1 + \frac{\Delta R}{R_0}\right) = R_0\left(1 + \frac{G}{AE} \cdot F\right) = R_0[1 + S'\Delta F] \text{ where } E \text{ and } A \text{ are}$$

Young's module and section area of the rod on which it is applied, and G is a parameter called *gauge factor*. Using the values, $R_0 = 210\,\Omega$; $E = 73.0$ GPa; $A = 144$ mm^2; $G = 2.1$ we have the sensitivity and relative sensitivity of the weight sensor

$$S = \frac{\Delta R}{\Delta F} = R_0\frac{G}{AE} = 210 \cdot \frac{2.1}{144 \cdot 10^{-6} \cdot 73 \cdot 10^9} = 41.9 \cdot 10^{-6} \begin{bmatrix} \Omega \\ N \end{bmatrix} \text{ and}$$

$$S' = \frac{G}{AE} = 199.7 \cdot 10^{-9} \begin{bmatrix} 1 \\ N \end{bmatrix}, \text{ respectively.}$$

From the input/output operating ranges values, we can determine the parameters of the *ideal* characteristic function (2.3) as

$$S = \frac{FSO}{FS} = \frac{y_{\max} - y_{\min}}{x_{\max} - x_{\min}};$$

$$y_{\text{off}} = y_{\min} - x_{\min} \cdot S; \quad x_{\text{off}} = -\frac{y_{\text{off}}}{S} = x_{\min} - \frac{y_{\min}}{S}. \tag{2.8}$$

Example An analog pressure sensor transduces from -25 kPa to 25 kPa into an output from 0.1 V to 4.5 V. Its static characteristic is $y = S \cdot x + y_{\text{off}}$ where $S = (4.5 - 0.1)/(50 \times 10^3) = 88\mu\text{V/Pa}$ and $y_{\text{off}} = -25 \times 10^3 - 0.1 \cdot 88 \times 10^{-6} = 2.3V$ and $x_{\text{off}} = 0.1 + 25 \times 10^3/88 \times 10^{-6} = -261$ Pa.

An alternative graphical representation of the ideal sensor input/output relationship is illustrated in Fig. 2.4A. The plot shows two opposite-oriented axes connecting input and output values. The two signals are going from minimum to maximum values within the full scale, and a given input perturbation Δx is mapped into an output perturbation Δy by means of a central *focal point*. The graphical construction shows how the gain is given by $S = d/c$. Therefore a gain increase (or decrease) is represented by a shift of the focal point along the horizontal axis. This kind of graph is very useful in discussing the role of sensitivity in sensor acquisition chains. Gain plots are useful only for linear or linearized input-output relationships.

In Fig. 2.5 is shown how the increase of the gain of the sensor system is represented from the geometric point of view in the static characteristic plot and in the gain plots, respectively. In the first case, the steeper the slope of the characteristic, the higher the gain. In the second case, the increase of the gain is shown with a shift of the focal point on the left side.

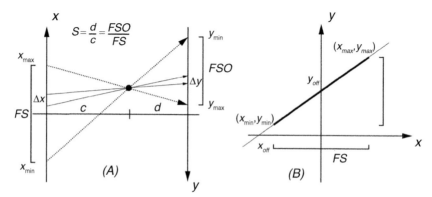

Figure 2.4   Gain plot of a transducer (A). This kind of plot is very useful to represent transduction chains, but it is valid for linear (or linearized) relationships. On the right is shown the corresponding ideal characteristic (B).

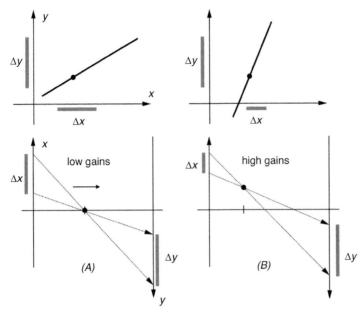

Figure 2.5 Role of the gain in signal transduction. (A) Low gains in a static characteristic plot (top) and gain plot (bottom). (B) The same situation as before but with increased gains. The corresponding input/output variations are represented in gray.

2.4 Signal Characterization

Signals are functions describing the variation of states in physical or technology domains.

A first classification could be drawn between *deterministic* and *random* signals.

- A *deterministic signal* is described by a mathematical function or rule that uniquely determines any past and any future state. The knowledge of a deterministic signal corresponds to the identification of the related function or rule model. An example is illustrated in Fig. 2.6A with a sinusoidal function.
- A *random signal* is a signal in which future values are known only under the concept of probability. These signals can be described using the mathematical tool of *random variables*. The knowledge of stochastic signals corresponds to the identification of the random variable model characteristics, such as (among many others) its probability density function (PDF). An example is shown in Fig. 2.6C.

For a correct framework, it is better to focus on two distinct viewpoints in sensor design.

- *Characterization mode.* We can characterize the system from a theoretical point of view, based on known physical models used in the design. Furthermore, from the experimental point of view, a *known* signal could be fed into the input, either deterministic (analytical function) or stochastic (by means of a known random

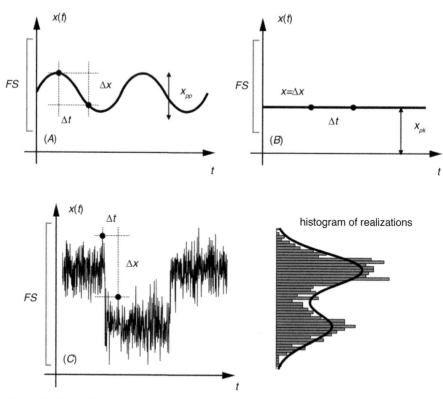

Figure 2.6   Class of signals in the time domain. (A) Deterministic AC signal. (B) DC signal. (C) Random or stochastic signal.

variable), and the output is recorded. The experimental test could validate the theoretical model, if present, or be used to characterize an experimental model in the absence of a theoretical one.

- *Operating mode.* The sensor, once characterized, monitors the environment to extract the information. In this case, the input is fed by an *unknown* signal whose knowledge will be determined on the basis of the sensor models.

Therefore, the following conclusion could be summarized. First, we should not confuse the terms "deterministic" and "stochastic" with the terms "known" and "unknown." We know the signal if we know the model (either deterministic or stochastic). Second, the deterministic signals are mostly used in the characterization process; in all other cases, signals are stochastic by nature.

We have already referred to signals as variations of a quantity versus time. Therefore, it is important to take into consideration the dependence on the time evolution of the signal. In Fig. 2.6A is shown a deterministic periodic time-varying signal mathematically defined (e.g., by an analytical function) for any given time in the past and the future. Usually, signals changing versus time are also referred to as "AC signals." Because the signal is a variation, for a given amount of time Δt we can calculate the amount of signal Δx on the describing function. The maximum swing of the amplitude of the signal is referred to as *peak-to-peak amplitude* x_{pp}.

An extreme case is a constant signal whose value is not changing with respect to time, as shown in Fig. 2.6B. These kinds of signals are referred to as "DC signals." To join in a unique framework time-varying and stationary signals, we can consider the DC signal as an asymptotic case where its value could be considered as a variation that occurred over a long time (zero-bandwidth condition) with respect to a reference that is usually chosen as the zero-input value, so that $\Delta x = x$.

An example of a *stochastic signal* is illustrated in Fig. 2.6C by the sum of a binary signal and noise. These signals cannot be characterized by using analytical functions but by *stochastic variables* to model *random processes*. Therefore, such signals could be described or characterized only from the statistical point of view. An example of experimental characterization is shown on the right of Fig. 2.6C, where sampled data of the stochastic signal are collected in a histogram that could be useful for either estimating the stochastic model or evaluating the correctness of a theoretical one.

It is useful to link the input and output evolution of signals with the quasistatic characteristic. If the signal variation is very slow (this is the reason it is referred to as a quasistatic characteristic), we could represent time variations of signals as illustrated in Fig. 2.7. Therefore

$$x(t) = x_0 + \Delta x(t)$$
$$y(t) = y(x(t)) = y_0 + \Delta y(t). \tag{2.9}$$

Figure 2.7A shows how the static characteristic could be used to map an input signal into an output one over time. Note how the variation at a time t_1 $\Delta x(t_1)$ is mapped into $\Delta y(t_1)$, which occurs at any point of the time evolution. The same signals could be represented in a time domain as an evolution of $x(t)$ and $y(t)$, as illustrated in Fig. 2.7B. The relationship mentioned above is also valid for stochastic signals.

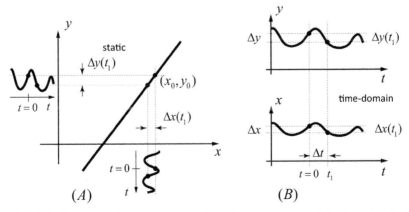

Figure 2.7 Representations of the input/output signals: (A) in the static characteristic; (B) in the time domain.

2.4.1 Limits of the Quasistatic Characteristic and Frequency Domain Representation

The input–output relationship determined by the static characteristic, as shown in Fig. 2.7, should be taken with great caution since it is valid when signal time variations are much lower than any time constant of the sensor system. If time constants of the sensor come into play, the input–output cross-point no longer follows the quasistatic characteristic due to the role of *gain* and *phase shift* operated by the system

If the system behaves as linear time-invariant (LTI), the gain and phase relationships between input and output are described by a complex function of the frequency $H(f)$ referred to as the *transfer function*. From now on, we will primarily refer to low-pass transfer functions so that the gain S, identified by the quasistatic characteristic, is equal to $H(0)$.

As shown in Fig. 2.8A, if we excite a first-order low-pass system with a sinusoidal signal having frequency components greater than the reciprocal of its characteristic time constant, we have that the static characteristic could not determine the phase shift and the amplitude of the output signal components. The input–output relationship might be described by a closed *trajectory* or *limit cycle* or *orbit* in the input–output space whose shape is determined by the system time-response description (e.g., poles and zeros) transfer function for small signals) in the bias point. In the case of a linear system, single time-constant and low-pass behavior with respect to a small sinusoidal excitation, as shown in Fig. 2.8B, the phase and amplitude scale could be described by an ellipse trajectory.[2] For very low frequency, the ellipse is squeezed on the ideal static curve; then, for increasing frequency, the phase lag increases the ellipse's minor axis. Finally, the amplitude scale is the determinant effect for very high frequency by

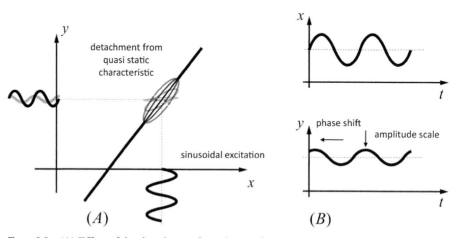

(A)   (B)

Figure 2.8 (A) Effect of the detachment from the quasistatic characteristic of the input–output description orbit in the case of (B) sinusoidal excitation with a frequency higher than the characteristic time constant of the system. A single time-constant system is assumed.

---

[2] It is a particular case of a Lissajous curve.

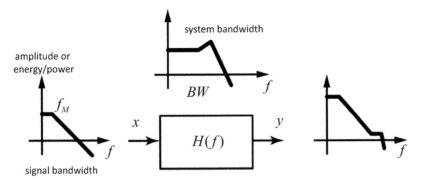

Figure 2.9 Representation of signals in the frequency domain. $H(f)$ is the linear-time-invariant (LTI) interface transfer function.

squeezing the ellipse on a horizontal line. The previous observation on the static characteristic's role points out that a frequency-domain representation is necessary when time variations come into play.

Figure 2.9 shows a graphical representation of the input and output signals in the frequency domain where only amplitude (and not phases) is represented for simplicity. It is important to note that the input frequency representation is showing the sinusoidal components' contribution to the input spectrum. Therefore, the maximum not-negligible frequency contribution sets the *signal bandwidth*. On the other hand, the system is characterized by a transfer function characteristic whose shape determines the *system bandwidth*, where – at first-order approximation – its cut-off corner is related to the dominant time constant of the system.

Note that signal and system bandwidths are *different entities*. The first is linked to the signal's characteristics and how it is decomposed in the frequency domain. The second is related to the sensor/interface and how it reacts with respect to input signals. The signal bandwidth is a typical characteristic of the required application, and the shaping of system bandwidth on signal bandwidths is a crucial point of sensor design.

Example A humidity sensor is based on the variations of the reflected microwave impedance in the spectrum from 1.5 GHz to 1.7 GHz determined by the amount of water contained in an object. A microwave generator gets a spectrum every second, and its shape is used to calculate the humidity percentage, which is expected to vary every 5 hours. The signal bandwidth is $1/(5 \cdot 3600) = 55.5 \mu Hz$ while the interface bandwidth is 1 Hz. The spectrum from 1.5 GHz to 1.7 GHz is only related to the technique we use to extract information from the environment and has nothing to do with signal bandwidth.

The time response or bandwidth of the system could be analyzed/estimated by applying a perturbation to the input and evaluating the mean time the output requires to follow/react to this change. This amount of time is referred to as the *characteristic time constant τ* of the system. For multiple time constants, the *dominant time constant approximation* considers the slowest one as the only effective on the system as if it were of the first order. From the mathematical point of view, for an LTI system, we can evaluate the time constant by applying a Dirac's impulse to the input and evaluating the average time that the system requires to relax back to the original state. The output behavior is called an *impulse response* in a time domain, and this evaluation is equivalent to the spectral analysis performed in the frequency domain. Note that the impulse input has frequency components greater than system bandwidth, which is indeed for testing the interface's frequency capabilities. An equivalent approach could be performed by applying a step excitation to the input and evaluating the output evolution, alias the system's step response, as shown in Fig. 2.10A and B. In the figures, the plots are normalized to the inputs–outputs $(x_0 \equiv y_0)$ to be readily compared by superposition of the graphs. Again, the step input excitation has frequency components much greater than system bandwidth, and the step response is a way to understand the time response of the system with respect to the perturbation.

If an input step equal to x_0 is applied to a transducer characterized by a low-pass first-order LTI system, it reacts with a typical *saturating exponential* (or *asymptotical exponential*) step response, as seen in Fig. 2.10A. This means that the output of the interface $y(t)$ would take an infinite time to reach the asymptotical value $y_0 = y(x_0)$ that lies on the static characteristic. The saturating exponential reaches 63% of the asymptotic value after one time constant and 99% after about five time constants. Therefore, it is common practice to fix an error band around the asymptotic value to

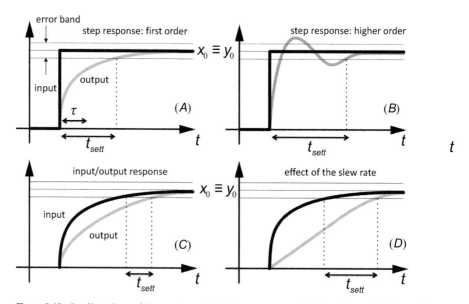

Figure 2.10 Settling time of the system. (Top) Step response in first order and higher order systems. (Bottom) Reaction of the system defined by its time constant and slew rate effect.

define the time required for the output to be within that error, referred to as *settling time* t_{sett} as shown in Fig. 2.10.

For higher-order LTI systems, the output behavior is more complex, and the definition of the dominant time constant itself becomes weaker. As shown in Fig. 2.10B, the output could have a faster response, but the settling time could be more significant due to the overshoot. In any case, the settling time calculation and compliance are fundamental to avoid systematic *errors* in the acquisition chain.

If instead of a step input, we apply an input excitation having a spectrum approaching that of the system, as shown in Fig. 2.10C, the output signal follows the behavior of the input more closely than that of the step response. The closer the signal and the system bandwidths, the smaller the settling time. In the case of Fig. 2.10C, the system bandwidth appears to be slightly smaller than the signal bandwidth.

Therefore, if we need the *transducer to follow time changes of the signal*, we have to set the bandwidth, *BW*, of the interface higher than that of the signal

$$\tau = \frac{1}{2\pi \cdot BW} < \frac{1}{2\pi \cdot f_M}, \tag{2.10}$$

where f_M is the maximum significant frequency of the input signal and τ is the characteristic time constant of the interface.

So far, we have dealt with the assumption of linear systems. Unfortunately, most systems do not have these properties, especially in dynamic behavior. For example, the settling time could depend on the amplitude of the input signal due to the *slew rate*. As shown in Fig. 2.10D, the slew rate imposes a dynamic behavior characterized by an initial constant slope, followed by a saturating exponential. Therefore, the settling time due to the slew rate *could be higher than that of linear systems*. This means that the effect is likely to have a fictitious reduction of the BW that the system offers in linear, small-signal regimes for a larger signal.

In this case, it is better to refer the comparison to the settling time (at worst case) instead of the bandwidth defined in the linear systems

$$t_{sett} \ll \frac{1}{2\pi \cdot f_M}, \tag{2.11}$$

where t_{sett} is increasing *depending on the amount of change of the input*.

However, we must be careful to enlarge too much interface bandwidth with respect to signal bandwidth due to a possible rising problem of noise folding in the case of analog-to-digital (A/D) sampling. The issue will be discussed in Chapter 3.

2.4.2 Energetic Properties of Signals

In the simplest case of measurement, we have seen that we expect from the transducer a linear response with respect to the signal "strength" intended as its amplitude. However, the transducing process requires the propagation of changes of states between the physical and the technological environments, requiring the use of energy.

This energy should be taken from the signal itself; thus, the energy of signals is one of the most significant features in the early process of sensing.

We can use examples from both simple mechanical and electronic systems to understand this intuitively. A weight scale is based on the movement of a needle pointing at a displacement resulting from the equilibrium between elastic (spring) and weight forces. This means that the weight force should move the plate scale of a certain amount for each measure tick. In other terms, it is required that the measuring system takes from the observed environment the equivalent energy to do the necessary work for displacing the weight force by one tick (level). Therefore, the energy required from the signal is proportional to the number of ticks covered by the readout displacement. What if we do not perceive any displacement? This could be seen as the "signal" is not holding the necessary energy to give a perceivable displacement; that is, its energy is too small. If we insist on seeing something and look with a lens at the displacement, we have two cases. In the first case, we see some displacement, and thus we realize that we should use the lens (i.e., increase the gain) to perceive a signal holding smaller energy. In the second case, we cannot distinguish any real displacement caused by the weight force because of "errors" caused by friction, vibration, or any other problem related to mechanical transduction despite the lens. In the latter case, we can say that the signal's energy is too weak with respect to something related to errors (and that should be measured in terms of "energy") to be physically perceived. In other words, signal energy should be larger than errors "energy."

In another example, we refer to the pixel response of a digital camera. Each pixel measures the light intensity in a time frame, and for "intensity," we refer to what is equivalently perceived from our eyes, that is, the number of photons hitting each cell of the retina over time. Since each photon has a fixed amount of energy, the "intensity" of the perceived light should be referred to as the physical concept of energy per time frame, alias power. Therefore, again, since the camera counts the photons in each pixel during the frame time, the output levels are proportional to the physical signal's power.

These two examples show how the concept of the measure, that is, the assignment of an output level to the input "strength," is related to the energy/power properties of the signal, which is in turn related to the *square* of the signal amplitudes.

We can identify the energy/power characteristics of signals in both the time domain and the frequency domain (spectrum). A first example is a signal with a limited duration in time, such as a pulse or a burst, as illustrated in Fig. 2.11A. The signal provides a limited amount of energy to a load, for example, a voltage applied to a resistor. We can calculate the *normalized signal energy* delivered to a normalized load by summing up the energy contribution of the signal $x(t)$ over time,

$$E = \int_{-\infty}^{\infty} x^2(t) \, dt. \tag{2.12}$$

For example, if the signal is measured in either volts or amperes, the integral of the square of the values represents the energy delivered to a 1 Ohm load. The *Rayleigh energy theorem* sets the energy relationship between the time and frequency domains

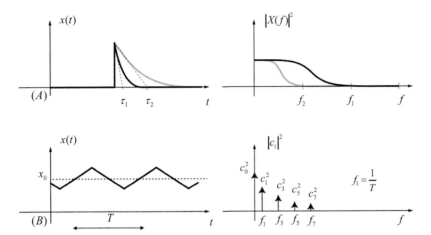

Figure 2.11 (A) Example of energy signals represented by a decaying pulse characterized by a relaxation constant in a time-domain representation (left) and frequency domain or spectrum representation (right). The signal energy spread could be identified in time and frequency. (B) Example of a periodic power signal with power distribution in time and frequency.[3]

$$E = \int_{-\infty}^{\infty} x^2(t)\, dt = \int_{-\infty}^{\infty} \left| X(f) \right|^2 df = \int_{-\infty}^{\infty} E(f)\, df, \tag{2.13}$$

where $X(f)$ is the Fourier's transform of the signal $x(t)$ and $E(f)$ is the *energy spectral density*.

Figure 2.11A shows how the energy is spread in both the time and frequency domains, and the larger the time pulse, the narrower the bandwidth.

However, if the signal lasts for an indefinite amount of time, the energy associated with it is no longer limited, and we have to think in terms of *average power*. A typical example is a periodic signal, having unlimited energy, for which we can define the signal average power or *energy per period* as

$$P = x_{rms}^2 = \langle x^2 \rangle = \frac{1}{T} \int_{-T/2}^{T/2} x^2(t)\, dt, \tag{2.14}$$

where x_{rms} is referred to as the *root mean square* (rms) of the power signal and T is its period. For nonperiodic signals (either deterministic or stochastic), we have to enlarge as much as possible the averaging timeframe for calculating the average power

$$P = \langle x^2 \rangle = \lim_{T \to \infty} \frac{1}{T} \int_{-T/2}^{T/2} x^2(t)\, dt \tag{2.15}$$

---

[3] In the present and the following figures, we have used a single-sided representation of the transforms. This is because we assumed that the original signal is real so that the representation of the transform is symmetric.

For periodic signals, their properties in the spectrum are given by the Fourier's series, and Parseval's power theorem gives the relationship of power between time and frequency

$$P = \langle x^2 \rangle = \frac{1}{T}\int_{-T/2}^{T/2} x^2(t)dt = \sum_{i=-\infty}^{\infty} |c_i^2|, \qquad (2.16)$$

where c_i are the series coefficients (harmonics) of the signal. Figure 2.11B shows an example of such signals using a triangular waveform.

Power signals, similarly to energy ones, can be characterized in the frequency domain by the *power spectral density* (PSD) defined so that

$$P = \int_{-\infty}^{\infty} S(f)df = \int_{0}^{\infty} \hat{S}(f)df, \qquad (2.17)$$

where $S(f)$ and $\hat{S}(f)$ are the double-sided and single-sided power spectral density, for which we will see that $\hat{S}(f) = 2S(f)$. In other words, PSD takes into account how the power of the signal is distributed in the spectrum. PSD could be easily understood from the experimental point of view, as shown in Fig. 2.12. The input signal is passed through an ideal bandpass filter of bandwidth Δf, and its output is squared and averaged over a time frame T.

This technique could be applied to any kind of power signal: periodic (where T is the period) or even stochastic (where T should be large enough to consider any spectral components of the signal). For a stochastic signal, the smaller the Δf, the larger T. This process provides the contribution $\hat{S}(f)\Delta f$ to the PSD. If we sweep the filter along the entire spectrum, we get the PSD shape. However, if we repeat the process using an insufficient period of time for stochastic signals, we could get a different PSD for any sweep. This is why it could be necessary to average multiple PSD estimations to better approximate the real one or increase T. Dirac's comb

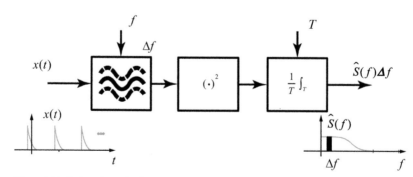

Figure 2.12 Estimation of (single-sided) power spectral density from an experimental point of view. The signal is passed throughout an ideal bandpass filter of bandwidth Δf, and the output is squared and averaged over a time frame T. This component represents the contribution of $S(f)\Delta f$ to the PSD. By sweeping the frequency throughout the spectrum, we can reconstruct the PSD figure.

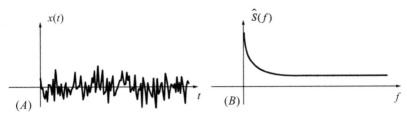

Figure 2.13 (A) Example of a stochastic (random) signal and (B) its power spectral density. In this figure, we have shown only the spectra' positive side for simplicity.

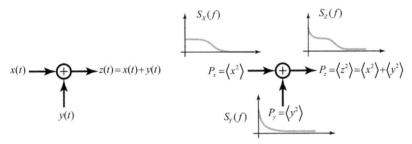

Figure 2.14 Property of power superposition in the case of an uncorrelated signal sum.

spectrum of Fig. 2.11B could be considered a characteristic power spectrum for periodic signals. In contrast, a typical estimation of a single-sided PSD of a stationary stochastic signal is shown in Fig. 2.13.

A typical case of signal processing in sensor systems is where two (or more) signals are summed together $z = x + y$ so that, using (2.14), the power of the output is

$$\langle z^2 \rangle = \langle x^2 \rangle + \langle y^2 \rangle + 2\langle xy \rangle, \tag{2.18}$$

which is the sum of the powers of the input signals plus the last term that is called *time cross-correlation* or *cross-product*

$$\langle xy \rangle = \lim_{T \to \infty} \frac{1}{T} \int_{-T/2}^{T/2} x(t)y(t)\, dt. \tag{2.19}$$

If the cross-correlation of the two signals is very small, they are said to be *uncorrelated*, and therefore the power of the output is the sum of the powers of input. This is referred to as a *property of superposition of powers* that applies to uncorrelated signals. Therefore, the property could also be applied to the spectrum figure of the power, as illustrated in Fig. 2.14. Note how the shape of the output PSD is derived by simply summing the inputs' original PSD spectra.

Referring back to Fig. 2.11B, we can notice that the power indicated by Eq. (2.16) is given by the sum of all components of the Dirac comb. However, as already discussed, AC *signals* generally carry information in the *displacements* or *perturbations* with respect to some reference. Therefore, the power related to the signal itself should be taken without the offset (DC value) power contribution. In other terms, the power of the signal that is carrying the information content should be independent of the power of the bias value, that is, the mean level around which the signal is evolving versus time

$$x_0 = \langle x \rangle = \frac{1}{T} \int_{-T/2}^{T/2} x(t)dt, \tag{2.20}$$

referred to as the *time average value* of $x(t)$.

For the reason mentioned above, a better index is the variation of the signal around its average, called the *time variance*[4] defined on a timeframe T as

$$\sigma_x^2 = \frac{1}{T} \int_{-T/2}^{T/2} \left(x(t) - \langle x \rangle \right)^2 dt = \langle x^2 \rangle - \langle x \rangle^2, \tag{2.21}$$

where the latter equivalence is a well-known property stating that the variance is given by the *mean square* (total power) value minus the *square of the mean*.

Thus, we define the *signal power* or *AC power* as

$$P_{AC} = x_{\text{rms}AC}^2 = \sigma_x^2 = \frac{1}{T} \int_{-T/2}^{T/2} x^2(t)dt - \langle x \rangle^2 \tag{2.22}$$
$$= \text{AC power} = \text{total power} - \text{DC power},$$

where $x_{\text{rms}AC}$ is the root mean square value of the AC part of the signal, which is, in the simplest case of an AC signal, the part of the power carrying the information content without taking into account the contribution of the DC offset. Referring again to Fig. 2.11B, we can identify P_{AC} with the sum of all the harmonic contributions except c_0^2 that corresponds to P_{DC}.

We can also write for AC signals

$$P = x_{\text{rms}}^2 = \underbrace{P_{AC}}_{\text{information}} + P_{DC} = x_{\text{rms}AC}^2 + x_0^2, \tag{2.23}$$

where x_{rms} is the root mean square value of the whole signal.[5]

Another essential characteristic of a periodic power signal is the ratio of its amplitude with respect to its power, called the *power factor*

$$q^2 = \frac{(\text{maximum excursion of signal})^2}{\text{power of signal}}. \tag{2.24}$$

Therefore, for AC signals, we can define q as (see Fig. 2.6)

---

[4] See also the appendix at the end of this chapter for definitions of averages and variances.
[5] In the following we will usually refer to $x_{\text{rms}AC}$ when using x_{rms}, assuming a zero mean value.

$$q = \frac{x_{pp}}{\sigma_x} = \frac{x_{pp}}{x_{rmsAC}}, \tag{2.25}$$

where $x_{pp} = 2x_{pk}$ is the maximum excursion (given by the characteristic of the function) of the signal Δx, also called *peak-to-peak (pp) value*, while its half value (for signals symmetric with respect to a reference) is called *peak value* x_{pk}. Therefore, the *power factor* is the ratio of the peak-to-peak value with the AC rms value.

Example For the signal $x(t) = A_0 + A_1 \cos(\omega t)$, the peak value is $x_{pk} = A_1$, and the peak-to-peak value is $x_{pp} = 2A_1$ while $\langle x \rangle = A_0$ and $P = \langle x^2 \rangle = A_0^2 + A_1^2/2$. Therefore $x_{rmsAC} = A_1/\sqrt{2}$ and the *power factor* is $q = 2\sqrt{2}$.

In DC signals, we do not have any P_{AC}, and the information content is in the value itself. In fact, from (2.21)

$$P = \langle x^2 \rangle = \langle x \rangle^2 = \underbrace{P_{DC}}_{\text{information}}. \tag{2.26}$$

Therefore, in the specific context of DC signals, we can use the same definition of q

$$q^2 = \frac{(\text{excursion of signal})^2}{\text{power of signal}} = \frac{(\Delta x)^2}{x^2} = \frac{x^2}{x^2} = 1. \tag{2.27}$$

We will see that several definitions in the following sections could be applied to both AC and DC signals by simply changing the value of q. In the case that the signal peak-to-peak value is covering the whole span, we have

$$FS = \Delta x_{max} = x_{pp} = q \cdot x_{rmsAC} = q \cdot \sigma_{x(max)}. \tag{2.28}$$

The preceding definition is not correct for a DC signal, since we do not have variance; however, we could interpret it under a power view with $q = 1$ using Eq. (2.27).

Finally, it is necessary to anticipate a result that will be discussed later in this book in greater detail, showing that the power components of the signal are propagated from input to the output of an LTI system by the square of the module of the transfer function

$$S_y(f) = \left| H(f) \right|^2 \cdot S_x(f). \tag{2.29}$$

2.5   Time and Amplitude Quantization

Since complex computation is mainly performed by digital processing, we need two kinds of discretization of analog signals: *time discretization* and *amplitude discretization.* Time discretization is the result of a sampling process, while amplitude discretization is the result of a quantizer. Therefore, after the elaboration of the analog signal made by a *transducer T,* the signal is processed by a *sampler* and a *quantizer* Q, as shown in Fig. 2.15.

The sampler connects on fixed *sampling periods* T_S, the output of the transducer to an analog memory (e.g., a capacitor) for a limited amount of time τ_S, as illustrated in Fig. 2.16A. When the capacitor is disconnected, it preserves the value for the entire

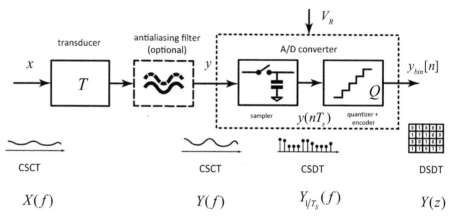

Figure 2.15 Time and amplitude discretization processes of the sensor interface. The time discretization is performed by a sampler and the amplitude discretization by a quantizer.

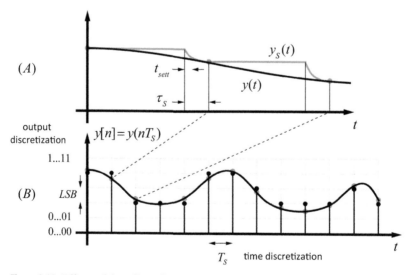

Figure 2.16 Effects of time discretization and amplitude discretization. (A) Zoom of the sampling process. During τ_S the switch is closed to a capacitor used as an analog memory to store the value when the switch is open. (B) Time and amplitude discretization with respect to the original analog waveform.

sampling period to assess the quantization. This technique is referred to as *sample-and-hold* (S&H). Of course, to avoid any systematic error, the following should hold:

$$t_{sett} \ll \tau_S \ll T_S, \tag{2.30}$$

where t_{sett} is the settling time of the sampler.

The quantization process could be represented in a graphical view, as illustrated in Fig. 2.16B. The output of the sampler is discretized in time, but an analog value still represents it in the amplitude (gray circles). The quantizer assigns the input analog value to a fixed number of levels (black circles) according to a quantization function, thus performing a rounding error. The finer the discretization, the closer the estimation of the input.

Signals are also classified into domains referred to as continuous/discrete in both signal and time. Therefore, analog signals are referred to as *continuous signal continuous-time* (CSCT), while after the sampler, they belong to the *continuous signal discrete-time* (CSDT), and after the quantizer, we have a *discrete signal discrete-time* (DSDT) status. Usually, a sensor acquisition chain's discrete-time output sample is referred to as *readout*.

We will refer to *uniform sampling* for simplicity. This means that the time-varying analog waveform is repeatedly sampled at a constant period of time:

$$T_S = \frac{1}{f_S}, \tag{2.31}$$

where f_S is called *sampling frequency*. Therefore, the sampling process encodes a real function into a *sequence of numbers* $y[n] = y(nT_S)$. If τ_S is very short compared to the sampling period, the process is close to an *ideal sampling*. Uniform ideal sampling time discretization preserves any information content of the original analog signal if $f_S > 2 \cdot f_M$ according to the *Nyquist–Shannon sampling theorem* where f_M is the maximum frequency of the signal. As known, if the requirement is not matched, the subsampling implies the folding (aliasing) of the spectral components, potentially determining an error in the estimation of the input. The aliasing problem should be carefully analyzed, and we will treat this problem in Chapter 3.

A quantizer performs amplitude discretization, and it consists of an approximation process performed by a rounding function. The union of the sampler with the quantizer implements the A/D converter that encodes the sampled information into a binary code with a fixed number of digits. A possible, but not unique, rounding to nearest integer function (in this case referred to as "mid-tread") for A/D converters characterized by N nominal bits of resolution is

$$y_{(bin)}[n] = \text{Round}\left(\frac{x}{V_R} \cdot 2^N\right) \equiv [\text{LSB}], \tag{2.32}$$

where $x \in [0, V_R]$, with V_R being a reference voltage value and where the "(bin)" subscript means that the output of the function is encoded in binary representation, quantified in *least-significant bit* units (LSB). Note in Fig. 2.15 also the typical representation in the frequency domain as Fourier's transform ($X(f)$ and $Y(f)$) for the analog domain, discrete-time Fourier's transform (DTFT $Y_{1/T_s}(f)$) for the sampled signal, and Z transform ($Y(z)$) for the digital domain.

The minimum variation of the analog input value that induces the change of the least significant bit in the output is

$$V_{\mathrm{LSB}} = \frac{V_R}{2^N} \equiv [V]. \tag{2.33}$$

However, it could be necessary to grab also negative input values, that is, $x \in [-V_R, V_R]$. In this case, we should add a *sign bit*. An example of implementing the function above (mid-tread) for $(N = 2) + sign$ bits is shown in Fig. 2.17, where the output is encoded in *two's complement* binary notation. In general, the binary encoding of N bits *plus sign* implies a maximum number of discrete levels equal to 2^{N+1}. Note that the function achieves the minimum and maximum for input values at $-V_R + 3/2V_{\mathrm{LSB}}$ and $V_R - 3/2V_{\mathrm{LSB}}$, respectively. In the case of positive-only analog inputs, we can get rid of the most significant bit and use only the function's first quadrant. Note that the maximum rounding error is $V_{\mathrm{LSB}}/2$. Therefore, from the A/D point of view, we can say that $FS = V_R$ for positive input (single input) and $FS = \pm V_R$ for signed input (e.g., a differential input).

As far as the output is concerned, we could use a derived unit for data discretization referred to as a *least significant bit*

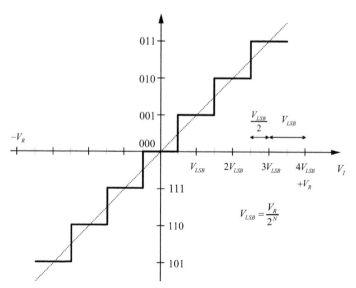

Figure 2.17 Quantization function of a ($N = 2$) + signed bits A/D converter using *two's complement* notation.

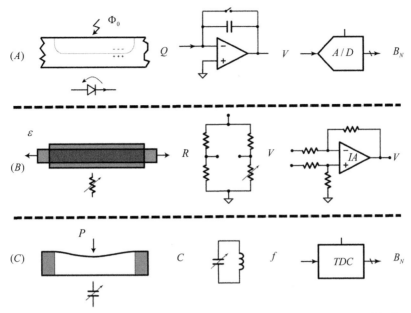

Figure 2.18 Examples of sensor acquisition chains. Light intensity sensors with direct A/D conversion (A), strain gauges interfaced with an instrumentation amplifier (B), oscillator-based pressure sensors followed by a time-to-digital converter (TDC) for frequency (period) detection (C).

$$\text{LSB} = \frac{1}{2^N} \equiv [\cdot], \tag{2.34}$$

namely, LSB units are dimensionless. In the following sections, if the output of the system is an A/D converter, we will use the LSB as the unit of measure.

2.6 Sensor Acquisition Chains and Sensor Taxonomy

Sensor taxonomy is complex. We have to deal with resistance or capacitive sensors reading out force, humidity, temperature, and so forth. Simultaneously, we have force, humidity, or temperature sensors implemented with resistance, capacitance, and other transduction processes. Therefore, sensor taxonomy is often useless: the amount of possible inputs, conversion phenomena, and sensing principles is so large that it is impossible to cover all the present and future architectures in a rationale. For this reason, we will avoid treating sensor design as a collection of single and specific cases. Instead, we will use a systematic approach in which sensor design will follow a unified scheme independent of particular implementation cases.

A tentative example of a taxonomy of sensors is shown in Table 2.1, where each column shows a possible function of the *sensor acquisition chain*: stimulus, conversion phenomena, electrical sensing, and conversion architecture. Of course, it is not

Table 2.1 Example of sensor taxonomy.

Physical input	Transduction phenomena Electrical	Sensing scheme	Signal conversion architecture
Temperature/humidity	Photoelectric	Charge sensing	A/D conversion
Pressure/stress/strain	Thermoelectric	Current/voltage sensing	Time-to-digital and digital counter conversion
Distance/proximity	Piezoelectric	Capacitive/resistance sensing	Modulation-based (lock-in, chopper, etc.) conversion
Acceleration/torque	Piezoresistance	Impedance sensing	Feedback-based (delta-sigma, etc.) conversion
Flux/viscosity	Magnetoelectric	Frequency/phase sensing	
Electric/magnetic fields	Ion/electron conversion	Time-to-event sensing	
Light/color	Optical interferometry		
Chemical concentrations/ biomolecules			
Biopotential/biocurrents			
Radiation			

feasible to cover in a list all the approaches, and it is always possible to come up with a new architecture in the future. Therefore, a specific sensor could be classified according to the collection of a single cell from each column.

The main characteristics that are used in this approach are

- *Transduction phenomena.* It is related to the process (direct or indirect) that converts physical stimuli into electric signals, namely, voltage, current, or electrical charge. For instance, an input force could displace charges by means of piezoelectricity or change material resistance by the piezoresistance effect.
- *Electrical sensing scheme.* It refers to the technique that we use to sense the transduced signal at the earliest stages of the acquisition chain. For example, a capacitive sensor could be measured by using a charge integrator or by evaluating its discharge time with a reference resistance.
- *Signal conversion architecture.* It is related to a higher level of complexity of the electronic treatment of the information. For instance, a capacitive sensor could be placed in the feedback loop of a delta-sigma modulator or be part of a resonant circuit followed by a finite state machine.

The characteristics listed so far could be used for a generic taxonomy of the sensor acquisition chain. However, the procedure is, of course, not exhaustive.

Some examples of sensor acquisition chains included in the above taxonomy are shown in Fig. 2.18, where three cases are listed. The first case (A) is related to a light intensity sensor where photon flux is first converted into a charge by means of a photodiode and then converted into a voltage before being converted by an A/D converter. The second case (B) describes a strain gauge where the induced strain of a piezoresistive element is first converted into a variable resistance that is read out by means of a Wheatstone bridge interfaced with a differential instrumentation amplifier. Finally (C), a pressure sensor is shown where the distance between the two plates of a capacitor is detected by means of its capacitance that changes the natural frequency of an oscillator measured by a time-to-digital converter. Each column represents from left to right the transduction phenomena (photoelectric, piezoresistive, capacitive), the conversion scheme (charge sensing, resistive bridge, frequency sensing), and the detection architecture (A/D conversion, analog conversion, finite state machine), respectively.

In the first column, we used a physical representation (together with its electrical device symbol) of the transducer; in the second one, an electrical representation of the sensing conversion. In the third one a functional representation (black box) of the signal conversion architecture. A black box representation of the same examples is represented in Fig. 2.19, which shows the variables at each chain stage. Before going into the details of the characterization of the sensors, we will point out the role of the units of measures since it is advantageous to follow the relationships in sensor acquisition chains.

Taking the example of Fig. 2.19A, we can draw the sensor chain's gain as illustrated in Fig. 2.20. Each section of the chain is characterized by a block gain, whose unit of measure is given by the output change ratio with respect to the input one. In this case, a change of the luminous flux per area (illuminance) induces a change of the photocharge in the photodiode by means of the gain S_1, which is measured in coulombs per lux. The change of charge induces a change in the output voltage of a charge integrator by means of S_2, whose units are volts per coulomb. Finally, a digital A/D converter provides a binary output word. The latter two variations ratio is S_3, which is measured in LSB per volt. Summarizing

$$S_1 = \frac{\Delta Q}{\Delta \Phi} \equiv \left[\frac{C}{\text{lux}}\right]; \; S_2 = \frac{\Delta V_O}{\Delta Q} \equiv \left[\frac{V}{C}\right]; \; S_3 = \frac{\Delta B_N}{\Delta V_O} \equiv \left[\frac{\text{LSB}}{V}\right]. \qquad (2.35)$$

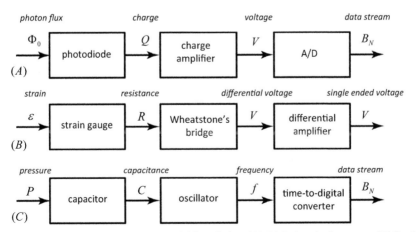

Figure 2.19 Examples of sensor acquisition chains. (A) Digital optical sensors. (B) Strain gauge interfaced with an instrumentation amplifier. (C) Oscillator-based pressure sensors.

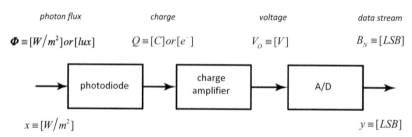

Figure 2.20 Units of measure and sensitivities of the acquisition chain.

Therefore, the total gain S_{TOT} is given by the product of the gains and is measured in LSB per lux

$$S_{TOT} = S_1 \cdot S_2 \cdot S_3 = \frac{FSO}{FS} = \frac{\Delta B_N}{\Delta \Phi} \equiv \left[\frac{LSB}{lux} \right] \equiv \left[\frac{\eta}{\xi} \right], \qquad (2.36)$$

which is the *total sensitivity of the chain*.

Example For a 22b noiseless digital pressure ($FSO = 2^{22}LSB \simeq 4M$ LSB) sensor, the input full scale is $FS = 1024$ hPa[6]. Therefore, using (2.8) its sensitivity (gain) is $S = 2^{22}/1024 \times 10^2 = 40.96$ LSB/Pa; this means that it is sufficient to put an input signal variation of $1/40.96$ Pa $= 24.4$ mPa to change the least significant bit.

Example A 11b + sign digital accelerometer has a sensitivity of 500LSB/g. Thus, it is assumed that the input full-scale range is $FS = FSO/S = 2^{12}/500 \simeq 8g = \pm 4g$.[7]

Figure 2.21 shows the mapping of the variables throughout the acquisition chain using a *gain chain plot* in the ideal case where all the input FS is mapped into the FSO of the A/D converter.

2.7   Deviations from Ideality: The Real Characteristic and Saturation

The ideal quasistationary characteristic is an abstraction. In real devices or interfaces, input and output are related by nonlinear relationships referred to as *real characteristics*, as illustrated in Fig. 2.22.

A critical source of nonlinearity is the effect of *saturation*. Any real interface is limited either by the physical or the operative point of view regarding the input and the output range. For example, power supply rails bound the output operating range of

---

[6] hPa = 100 Pa.   [7] g = 9.81 [m/s^2].

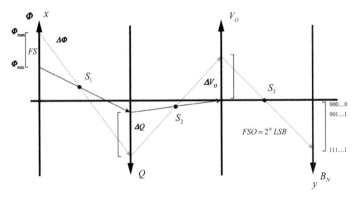

Figure 2.21 Variables mapping in a sensor acquisition chain example.

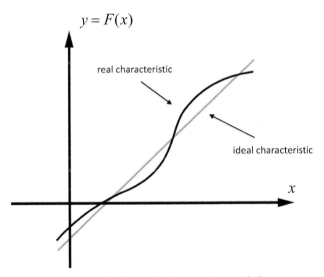

Figure 2.22 Ideal and real quasistationary characteristics.

a full-swing amplifier, and we have no signals outside those limits. In other cases, the output is limited by the saturation ranges of the amplifier beyond which the system does not convey information from input to output.

Figure 2.23A shows a simplified nonlinear S-shaped characteristic of a real amplification/transduction device where the flat ends of the plot are limited by saturation levels of the output. In this case, we have approximated the nonlinear effect of saturation using a piecewise-linear characteristic, and the input operating range is restricted to the central part of the characteristic.

Now, assume that the *required* operating range should be larger than the input operating range offered by the system, and portions of the required input values (in gray) are not falling inside and thus are not mapped into the output. This implies a loss

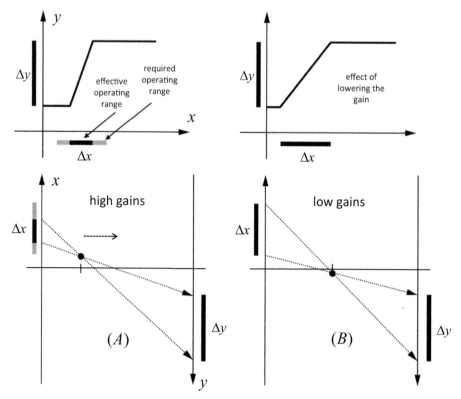

Figure 2.23 Effects of the saturation. (A) The saturation of the output prevents the coverage of the required input operating range. This means a loss of information. (B) The requirement is achieved by reducing the gain. (C, D) The same situation is seen in the gain plot representation.

of information from input to output. A solution could be found by decreasing the gain to cover the required input values, as shown in Fig. 2.23B.

An alternative view of the same situation is illustrated in Fig. 2.23C and D using the gain plot showing the same effect. A symmetric problem arises when the required operating range is smaller than the operating range offered by the system. We will see that this implies a loss of information, and an increase of sensitivity could be beneficial.

Example In a 24b noiseless digital pressure sensor, the gain S could be set to 1024, 2048, and 4096 LSB/hPa. The application requires an input operating range from 300 hPa to 16,300 hPa. To use the full output operating range, $FS = FSO/S = 2^{24}/S$ the input full-scale spans would be16,384hPa, 8192hPa, 4096hPa, respectively. Therefore, the first gain would be the most appropriate to cover the application requirement without information loss.

If only a part of the *FS* is outside the linear range, we can also *add/subtract an offset* to the input to center the input operating range.

2.8   Deviations from Ideality: Errors

The deterministic model is advantageous when dealing with transduction chains, gains, and ranges. However, for a more significant degree of knowledge, this is not sufficient: real sensors (as anything of the real world) are affected by unpredictable errors/fluctuations arising from the physical nature of the environment and the interface.

We could roughly classify errors in two main categories:

- *Random errors*: Unpredictable deviations of the output due to stochastic temporal variations. For stationary input and constant conditions of the sensing system, repeated readouts provide different results at any sample. They are realizations of random physical processes and are also referred to as *noise*.[8]
- *Systematic errors*: Constant deviations of the sensor response from the ideal characteristic. For stationary input and constant conditions of the sensing system, repeated readouts provide the same error. This is given, for example, by nonlinear deviations of the real characteristic from the ideal one or by the output variation induced by influence parameters. Even if the error is fixed for repeated conditions, it should be pointed out that it has a degree of unpredictability so that stochastic models should describe it.

An example of the difference between the random and systematic errors concepts is shown in Fig. 2.24. In this example, we will assume that an ideal sensor interface gives a null response with respect to an input that is set to zero (grounded input), as in Fig. 2.24A. The presence of random errors induces a population of different values at any readout, as shown in Fig. 2.24B. On the other hand, the systematic error is constant across different readouts under the same conditions, as shown in Fig. 2.24C. However, different conditions of the influence parameter of the error (i.e., temperature) may give rise to a population of different systematic errors on the same device, as shown in Fig. 2.24C–E. The same concept of systematic error could be applied to similar devices of the same production lot subject to fabrication process variations (e.g., operational amplifier offsets that vary across devices of the same lot), showing a population of different errors of the kind of Fig. 2.24C–E. Finally, being the errors originated by independent processes, we could represent the effect of both random and systematic errors as their sum, as shown in Fig. 2.24F.

The stochastic properties of errors could be seen from two opposite viewpoints, as shown in Fig. 2.25. If we apply a known signal Δx_S to the input (*characterization*

---

[8] The classification of systematic and random errors is frequently changing across different contexts. This book will refer to random errors as those related to repeated readouts in time.

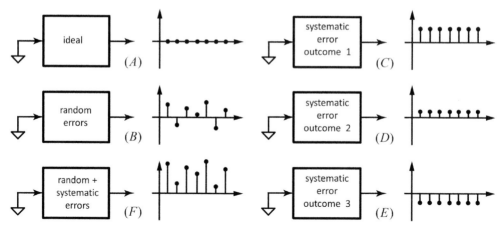

Figure 2.24   Random and systematic errors. (A) An ideal case where the output is zero for a null input. (B) Random errors. (C–E) Identically designed systems with different systematic errors. (F) Composition of random and systematic error of cases (B)+(C).

mode) in a real sensing system, it will be mapped into the output with some error added: $\Delta y_S + \Delta y_E$, as shown in Fig. 2.25A. In this case, the error could be *precisely known* and, by acquiring a large amount of data, we can characterize the *error model* using, for example, the associated *probability distributions*.

Conversely, if we know the error model and only observe the readout value (operating or prediction mode), we do not know the single error (outcome) entity even if we have characterized the error model. Therefore, since the errors have e stochastic behavior, the input could no longer be *determined* but *estimated* or *predicted*, as shown in Fig. 2.25B, under some degree of *uncertainty*.

2.8.1   The Input–Output Duality of a Single Error

Following the previous discussion, the output variation Δy of the readout could be dependent on both a signal Δy_S or error Δy_E (either systematic or random) of the system

$$\Delta y = \Delta y_S + \Delta y_E. \tag{2.37}$$

Owing to the linear behavior of the system, we can write

$$\Delta y = S \cdot \Delta x = S \cdot (\Delta x_S + \Delta x_E) \tag{2.38}$$

so that

$$\Delta x = \Delta x_S + \Delta x_E = \frac{\Delta y_S}{S} + \frac{\Delta y_E}{S}, \tag{2.39}$$

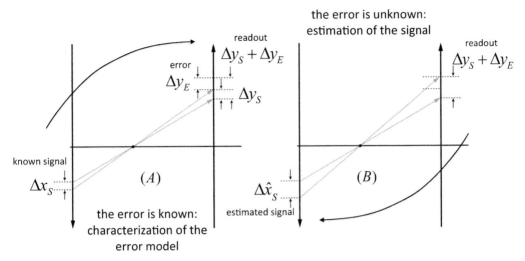

Figure 2.25   The error characterization in sensor *characterization mode* (A) and uncertainty in estimation in sensor *operating mode* (B).

where Δx_E is referred to as the *input-referred error*. Therefore, the error could be modeled as something that is summed up to the input signal to give the same deviation of the output. We will see that if the error is ascribed to noise, it is referred to as *input-referred noise* (IRN) or *equivalent input noise* (EIN) or *referred to input* (RTI) *noise*.

We can graphically represent this relationship as in Fig. 2.26A, where we can see that the error Δy_E (either systematic or random) of a real system is summed to the output signal Δy_S. Now, we can *model* the error as a contribution added to the output of an *ideal system*, as shown in Fig. 2.26B, which is also called *output-referred error*. Like the signal, we can model the same error by an additional fictitious input-referred contribution Δx_E summed to the input of an ideal system in order to get the same output result. In other terms, we can map the output error into an input-referred error by dividing it by the gain of the system. Similarly, a source of errors placed at the input could be mapped to the output by multiplying it to the gain the system.

2.8.2   Merge of Deterministic and Stochastic Models

The double viewpoint of the sensor interface in *characterization* and *operating modes* could be further illustrated using the quasistatic characteristic with respect to random errors, as illustrated in Fig. 2.27. Modeling is an essential task of the sensing design since, after its *characterization*, the model could be used to *predict* the behavior in an operating mode. The characterization could be done from either the empirical and/or theoretical perspectives. We can sweep the input values with a known reference to collect the output values in

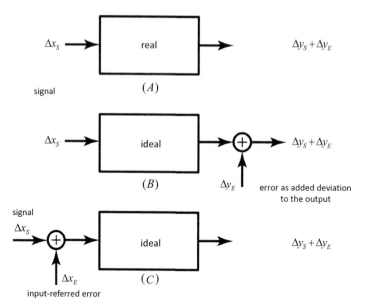

Figure 2.26 Mapping of errors from output to input and vice versa. (A) A real system with an error. (B) Model of a real system with an ideal one with the error summed to the output. (C) The real system model with an ideal one with an error summed to the input.

empirical characterization. Owing to errors, the collection of the points is not regular and follows a two-dimensional experimental distribution, as shown in Fig. 2.27A, composed of a cloud of points. This means that we can collect a certain amount of points from the "noisy" output for each known point of the input x_0, as shown in Fig. 2.27B. This experimental data collection can determine an interpolating curve following several mathematical techniques. The determination of the interpolating function $y = \hat{F}(x)$ is not unique and depends on the approach. The interpolating function of the characterization mode can then be used in operating mode to *estimate* or *predict* the input $\hat{x}_0$ based on a given output $\tilde{y}_0$, as shown in Fig. 2.27C. However, the prediction is affected by some degree of *uncertainty*. This means that possible different values of the input could give the same output due to errors.

Therefore, the overall model, including stochastic variations, could be described with a block diagram shown in Fig. 2.28A. The real transduction process is always associated with errors and stochastic fluctuations arising from *random or stochastic processes* of physical nature, residing in both the environment and in the sensing interface. In the first case, the process could only be investigated and characterized, since it resides in the nature of the observed environment. Conversely, in the second case, the designer could act on the interface in order to reduce the effects of the randomness and optimize the signal detection.

Therefore, the physical description of the process could be conveniently represented by a *mathematical model*, as shown in Fig. 2.28B. So far, we have dealt with

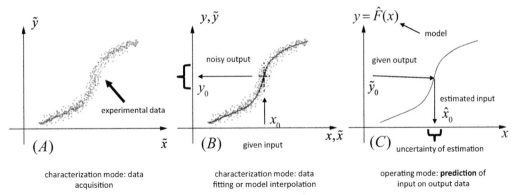

Figure 2.27 Process of modeling. (A) Collection of experimental data in characterization mode. (B) Characterization of the interpolating curve. (C) Estimation or prediction of the input by means of the model.

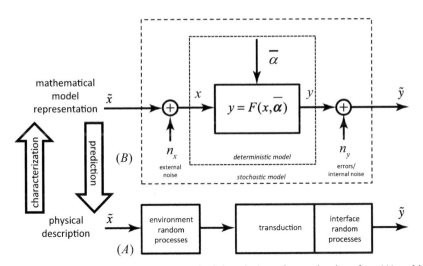

Figure 2.28 Representation of the physical description of a sensing interface (A) and its mathematical model (B).

deterministic relationships between input and output well represented by analytical functions. On the other side, random processes could be modeled by stochastic/random variables. It is said that *random variable outcomes could model realizations (i.e., values arising from experiments) of a random process.*

Thus, a more exhaustive model of the sensor is illustrated in Fig. 2.28B, where the stochastic behavior surrounds a deterministic backbone relationship. The errors are summed up to the input (if determined by environment random processes) or to the output (if determined by interface random processes). Note that both these errors could be mapped to either the input or the output according to previous transformations involving the gain of the system. As shown in Fig. 2.28, the input $\tilde{x}$ and output $\tilde{y}$ of the model are now stochastic variables since the input is unknown or known only by statistical characterization, and so it is the output due to the addition of internal and external noise. Note that also x and y variables are stochastic due to the addition of noise, even if $F(x)$ remains a deterministic function.

In the presence of errors, it is said that the input needs to be *estimated* or evaluated based on the available output data. Filtering is one of the simplest examples of estimation, where averaging the output data reduces the effects of noise. An analog filter is an example of a simple estimator implemented in the sensor interface. However, if the information to be extracted is very complex, the estimator should be implemented not in the interface/transducer but in a separate processing block working on raw data coming out of the interface.

The process of estimation could be mathematically described as a function

$$\hat{x} = \hat{x}\left(\tilde{y}(t)\right); \ t \in [0, T], \tag{2.40}$$

where $\tilde{y}$ is the observed output affected by errors, and $\hat{x}$ is referred to as the *estimated or predicted value* of the input, and T is the period of time during which the output is observed to estimate the input. In the very common case where discrete sampled data characterize the output over time, we could write

$$\hat{x}_i = \hat{x}_i(\tilde{y}_k); \ k = 1, 2, \ldots, N, \tag{2.41}$$

where $\hat{x}_i = \hat{x}[i]$ and $\tilde{y}_k = \tilde{y}[k]$ means a set of sampled output data.

2.8.3   Estimation and Effects of Averaging

Sensing is essentially a measurement process, and it is worthwhile to sketch some basics of measurement theory that are extremely important for future discussion, whose framework and formal definition are, however, beyond the scope of this text.

Assume that to make a generic measurement of a quantity that we assume fixed and stable in time. Every time we take a measurement, the result given by this procedure is different. This means that some "error" is added to the true value as a realization of a random process. A typical *model* of the random process could be, for example, a *probability distribution function* of the representative random variable. In what follows, we will also assume that the random process characteristics are stationary in time. The model could be known in several ways. On the one hand, it could be acquired with a very large (in theory infinite) number of trials in a *characterization* procedure. On the other hand, it could be the result of theoretical analysis or numerical simulation; in this case, the quality of theoretical or simulation models could be verified by experimentation.

If we assume that the error is added to the actual value and that the expected value[9] of its distribution is zero,[10] we can say that the expected value of the distribution is the actual value of the measurand.

During a measurement procedure, if we do not know anything about errors, we only have to rely on a trial sample. An example is shown in Fig. 2.29A, where we collect experimentally length measures of an object (i.e., using a calibrating gauge block). Thus, the difference of the measure with respect to the actual value of the object is referred to as the "error" of the measurement (Fig. 2.29A, left) that *we do not know* its statistical characteristics (shown as the curve on the right of each plot), except that its expected value is zero.

The task of the measurement procedure is to *estimate*, using a limited number of samples, the actual value in the presence of errors, whereas we do not know both. Thus, the problem is how this estimation can be made and how accurate it is. A fundamental theorem on which we rely is the *law of large numbers (LLN)*.

Law of Large Numbers Regardless of the error distribution function, the *sample mean*[11] of a large number of measurements converges in probability to the *expected value* as more significant numbers of samples are averaged.

This is equivalent to saying that the sample mean is an effective *estimator* of the actual value. In Fig. 2.29A, we average 10 measurements affected by error to estimate the true value (0.574 cm), resulting in 0.517 cm. Note on the right side of the figure the histogram of the acquired samples together with the true error distribution (unknown). However, if we take 10 other measurements affected by the same error process in a second trial, we obtain a different value as shown in Fig. 2.29B, where this time we estimate the true value as 0.478 cm. It signifies that the estimation process gives variable results, implicating that the estimation accuracy is low. However, following LLN, we can increase the accuracy by increasing the number of samples as shown in Fig. 2.29C, where 500 samples are averaged, giving a more accurate estimate in 0.577 cm. Note also how the histogram on the right of the figure fits better the error distribution model. If we collected infinite numbers of samples, we would determine the error distribution model as shown in the line curve.

Therefore, averaging many samples allows us to estimate the expected value better. What is the relationship between the sample size and the closeness of its estimation to

---

[9] See the appendix of this chapter for the definition of *expected value* and *variance*.
[10] This assumption is also supported by physical arguments based on energetic considerations.
[11] See the appendix of this chapter for definitions of a *sample (or empirical or population) mean*.

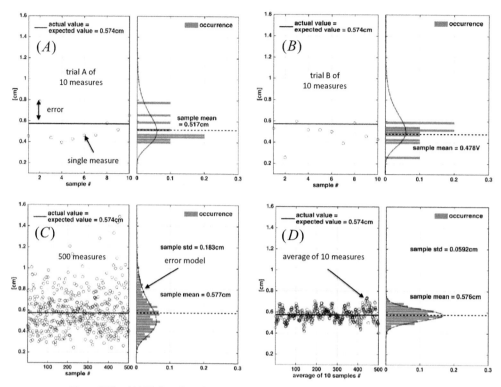

Figure 2.29 (A) Estimating the true value by averaging a trial of 10 samples of a measurement affected by a random error process. On the left, the true value and the single outcome errors. On the right, the true distribution (not a Gaussian) and the histogram of the samples. (B) Histogram and estimation of the second trial of 10 other samples of the same process. (C) Histogram and estimation by averaging of 500 samples. Note how the averaged sample is closer to the true value. (D) Histogram of 500 averages of 10 measures. Note how the distribution of averages is closer to the actual value, and their distribution tends to a Gaussian, according to CLT.

the true value? In other terms, what is the dispersion of the estimation around the expected value? Using the already used example, we can see that the estimation using 10 samples mean varies from 0.517 cm (Fig. 2.29A) to 0.478 cm (Fig. 2.29B). If we take many other averages of 10 samples, what is the standard deviation of the results? The answer to this question is given by the *central limit theorem (CLT)*.

> **Central Limit Theorem** Regardless of the error distribution function, the average of a number N of independent samples derived from the same process converges in distribution to a normal distribution whose standard deviation is $\sigma/\sqrt{N}$, where σ is the standard deviation of the original error distribution.

This law is powerful because it states how much closer we get the estimation spread to the actual value as we increase the number of averaged samples, independently from the kind of original distribution function. The effect of CLT could be seen in Fig. 2.29D, where 500 averages of 10 measures each are put into a histogram. The histogram is constructed by collecting 500 trials similar to those of Fig. 2.29A and B. As we can see, the sampled distribution is similar to a Gaussian one and has a standard deviation of 0.0592 cm, which is very close to $\sigma/\sqrt{10} = 0.0557$ cm, where σ is the standard deviation of the original (non-Gaussian) distribution.

A final note about the term "converges in probability" of LNN. This means that even if we make a considerable number of averages, we are not precisely sure to converge to the real value, but the probability of being far from it is very low.

2.8.4 Systematic Errors Due to Nonlinearity (Distortion)

Real systems are usually not following ideal linear characteristics, and we would like to understand the errors related to the deviation from ideality. We will start the discussion by referring to generic variations of input/output Δx, Δy and we will associate them with the error concept later on. Looking at Fig. 2.30, we can see that the real characteristic $y = F(x)$ differs from the ideal one (i.e., the response that we expect in the measurement process), and the difference between these two characteristics is referred to as error due to nonlinearity.

Starting from a reference point (x_0, y_0) on a real characteristic, an input variation Δx induces an output variation Δy according to the relationship

$$y_0 + \Delta y = F(x_0 + \Delta x), \tag{2.42}$$

where the input and output variations are dependent on the shape of the characteristic and the reference point. The relationship in Eq. (2.42) is valid for both linear and nonlinear characteristics, and the relationship between Δx and Δy is dependent on the reference point. In the following, we will associate the two variations with the nonlinearity errors with respect to the two variables, and finally, we will discuss the relationship between them.

In the case of Fig. 2.30, we can see that Δx and Δy are the distances between the ideal and the real characteristics. In other words, for a given and "true" input $x_T = x_0$ we get y_0 instead of the true value y_T, thus committing an error Δy. On the other hand, when we read out y_T we expect a "true" value x_T; however, the sensor system indicates the value x_0, thus committing an error Δx. These are the offset errors between the characteristics, and they are dependent on the bias point.

From the differential calculus point of view, we can express the characteristic into a power series expansion

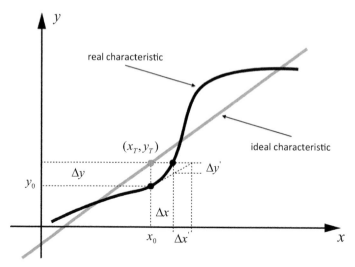

Figure 2.30 Systematic error due to the nonlinear characteristic of the sensor, also known as offset error.

$$y = F(x_0 + \Delta x) = F(x_0) + \left.\frac{dF}{dx}\right|_0 \Delta x + \frac{1}{2}\left.\frac{d^2 F}{dx^2}\right|_0 \Delta x^2 + \dots$$

$$= F(x_0) + \left.\frac{dF}{dx}\right|_0 \Delta x + \Delta y',$$

(2.43)

where Δx is a generic input variation *referring to either signal or error*, and the "0" subscript is related to the reference point. The quantity $\Delta y'$ is the higher-order nonlinearity error contribution, and it is dependent on the amount of the displacement, as shown in Fig. 2.30.

Rearranging, we get

$$F(x_0 + \Delta x) = y_0 + \Delta y = y_0 + S_0 \cdot \Delta x + \Delta y'$$
$$= y_0 + S_0 \cdot (\Delta x + \Delta x'),$$

(2.44)

where S_0 is the sensitivity of the system in the reference point

$$S_0 = \left.\frac{dF}{dx}\right|_0$$

(2.45)

and $\Delta x' = \Delta y'/S_0$ is the higher-order nonlinearity error referred to the input. If we neglect higher-order error, we have a *first-order approximation* of the relationship:

$$F(x_0 + \Delta x) = y_0 + \Delta y \approx y_0 + S_0 \cdot \Delta x.$$

(2.46)

If we associate the discrepancy of the real characteristic with the ideal one to an *error*, we will use the following notation:

$$\Delta x_{E(D)} \leftarrow \Delta x; \ \Delta y_{E(D)} \leftarrow \Delta y, \tag{2.47}$$

where subscript D stands for "distortion," since we will see that it is the effect of the nonlinearity in the AC domain.

Therefore, using first-order approximation, the *input–output relationship* referring to systematic errors induced by nonlinearity (distortion) is given by the *slope (gain)* of the real characteristic in the reference point.

In conclusion:

- The nonlinearity errors between ideal and real characteristics are functions of the reference point.
- The higher-order nonlinearity errors are functions of both the reference point and the amount of input variation.
- The input and output nonlinearity errors are related through the slope of the function (i.e., the gain in the reference point).

There are other examples of fixed systematic errors, as shown in Fig. 2.31. Figure 2.31A illustrates the uncertainty induced in the output due to a hysteresis circle. In Fig. 2.31B is shown the input-referred error induced by the quantization of an A/D converter.

2.8.5 Characterization of Random and Systematic Errors by Distributions

From the experimental point of view, we can collect a *population of errors* and characterize them using the concept of the *probability distribution*. In this approach, we can, for example, evaluate the spread of the stochastic variation with the *variance* of the variable associated with such processes and evaluate the related uncertainty of the measure.

This concept of distribution is relatively easy to understand for random errors, where the "experiment" is the output sampling, and the "outcome" is the

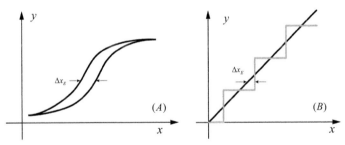

Figure 2.31 Other examples of systematic errors. (A) Hysteresis. (B) Quantization errors of A/D converters.

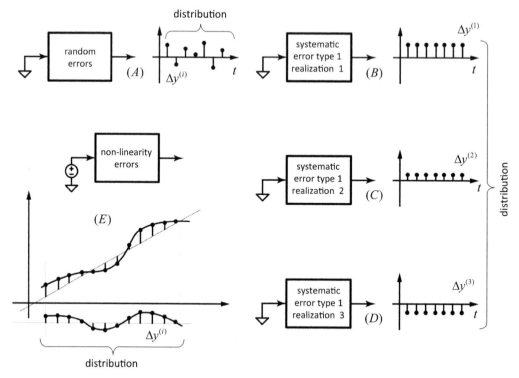

Figure 2.32   Distribution of errors. (A) Random errors, (B–D) Errors due to the variability of systematic errors. (E) Errors due to nonlinearity.

corresponding numerical value that is affected by the perturbation induced by the random process. On the other hand, the concept of distributions is subtler in systematic errors. For this reason, we have summarized the concepts in the infographic of Fig. 2.32. The main question is where the randomness of the process is arising in systematic errors since deterministic and known functions describe them. The answer is that the functions governing the systematic errors are either dependent on random parameters or have as input domain stochastic variables. We will be more specific on this with examples later on.

We will list below the characteristics of some distributions of errors as a random process outcome.

1. *Random errors.* Repeated readouts give the sample space with respect to a fixed input (Fig. 2.32A). The randomness of outcomes is due to underlying time-dependent stochastic processes such as the composition of the noise of interface devices.
2. *Systematic errors due to variability in influence parameters.* The sample space is given by the outputs of the same device due to the change of influence parameters (e.g., temperature), as shown in Fig. 2.32B–D. The randomness of the process is related to the stochastic variation of such influence parameters

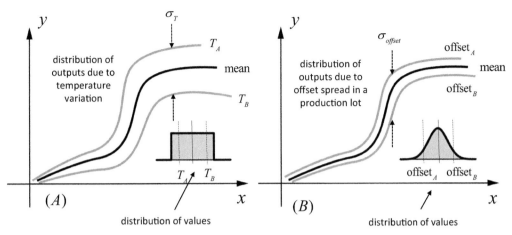

Figure 2.33   Examples of systematic errors due to variation of influence parameters such as the temperature (A) and to process variation in production such as offsets (B).

(e.g., we do not know the system's temperature in the future even if we know the relation between the temperature and the output). We can model the process on the expectations and spread of the values of such parameters.

3. *Systematic errors due to variability of production parameters.* The sample space is given by the outputs of similar devices taken from the same production lot. The distribution randomness is due to the variability of process production parameters among similar devices (e.g., offsets). The situation is similar to the previous one, as shown in Fig. 2.32B–D, but the "experiment" is to sample the output from different devices of the same production lot. The randomness is related to the fact that we do not know the parameter value for a given device. We can model the process with a distribution of values for a production lot (e.g., the distribution of offsets for integrated operational amplifiers).

4. *Systematic errors due to nonlinearity or distortion.* The sample space is given by nonlinearity errors associated with specific input. The distribution randomness is given by the distribution of the input values (Fig. 2.32E), which is unknown in operating mode. Very often, especially in DC measurement, the input distribution is assumed uniform over the input span; in operating mode, any input value is assigned the same occurrence within the full scale.

While in case 1, the sample space is defined solely by a set of time samples, in cases 2 and 3, the statistical distribution is strictly related to other variables. For example, in case 2, the distribution is related to the temperature interval variability we assume for a specific application. If we assume that the sensor will be used in an environment with temperature variations between 10°C and 30°C, its related error distribution will be different from that where the temperature variation ranges from −50°C to 125°C.

Particular attention should be taken in case 4, where *we have to assume signal distribution*. In the extreme case where input variation is highly restricted, the distribution tends to be a Dirac function in the sample space, and the error should be modeled as a *pure offset*. We will see this situation in the dithering effect of Section 2.14.1.

A graphical interpretation of systematic errors due to influence parameters and production variability is shown in Fig. 2.33. In Fig. 2.33A is shown an example of a variation of the characteristic due to a change of temperature. The temperature could be modeled by a distribution function (in this case, the temperature is uniformly distributed between two extreme values) that could be a priori assumption or result from statistical experience. The stochastic variation of the temperature will induce a variation of the characteristic with an associated stochastic error. Similarly, in Fig. 2.33B are shown characteristics given by different samples of a production. For example, in the same production lot, the output of each device is different from the others due to internal device offset variations. The distribution of offsets could be statistically evaluated at the production level and provided to the customer. Therefore, at the design level, we have to take into account the stochastic variation of the offset inducing a systematic error.

Whether we are dealing with random or systematic errors, we could characterize their distributions using statistical tools. In other words, we consider the error Δx as a realization of a random variable ΔX whose distribution should be estimated to be modeled. Note how the lower case of the variable indicates the realization, that is, the experimentally observed value, while the upper case is assigned to the random variable describing the stochastic process. For simplicity, we might use Δx in both cases in the following sections.

> **Notation** We will refer to $\Delta y_{E(\text{kind})}^{(i)}$ as the ith outcome of random variable modeling the error of a certain kind affecting the output readout.

The discrepancy between the true value given by ideal characteristics and the single readout is given by the sum of a series of errors, and in characterization mode,

$$y_T = y_0 + \Delta y_{E(1)}^{(i)} + \Delta y_{E(2)}^{(i)} + \ldots + \Delta y_{E(N)}^{(i)}, \tag{2.48}$$

where

$$y_0 = F\left(x_0(t), \overline{\alpha_0}\right); \text{ or briefly } y_0 = F(x_0) \tag{2.49}$$

is the reference or bias point and $\Delta y_{E(1)}^{(i)}, \Delta y_{E(2)}^{(i)}, \ldots \Delta y_{E(N)}^{(i)}$ are the ith realizations of N independent errors and $\overline{\alpha_0}$ is the set of reference influence parameters that have been used to determine the standard static characteristic (e.g., reference temperature = 300 K, etc.).

Referring to *random errors*, we have that for fixed input and influence parameters, the readouts $\widetilde{y}_0(t_i)$ differ from the expected value y_0 given by the characteristic for stochastic variation due to random processes

$$\widetilde{y}_0(t_i) = y_0 + \Delta y^{(i)}_{E(\text{ran})}, \tag{2.50}$$

where t_i is the acquisition time of the readout. Therefore, we can define the ith outcome of a random error as

$$\Delta y^{(i)}_{E(\text{ran})} = \widetilde{y}_0(t_i) - y_0. \tag{2.51}$$

An essential assumption about random errors, as already discussed, is that their expected value is zero; therefore

$$\lim_{N \to \infty} \frac{1}{N} \sum_{i=1}^{N} \widetilde{y}_0(t_i) \to y_0. \tag{2.52}$$

As far as the *systematic error variations* are concerned, we have that the error outcome due to the jth influence parameter is given by

$$\Delta y^{(i)}_{E(\alpha_j)} = y\left(x_0, \alpha_1, \alpha_2, \ldots, \alpha_j + \Delta \alpha^{(i)}_j, \ldots, \alpha_K\right) - y_0, \tag{2.53}$$

where we have assumed that while the jth parameter is changing, the others are kept in the standard value. Again, the influence parameter could be a technology variation dependence.

Finally, referring to nonlinearity (or distortion) errors, we have

$$\Delta y^{(i)}_{E(D)} = y_T(x_i) - y_0(x_i). \tag{2.54}$$

Therefore, referring to the output variable, we have, following LLN

$$\overline{\Delta y_E} = \frac{1}{N} \sum_i \Delta y^{(i)}_E \to E[\Delta Y_E]$$
$$s_E^2(y) = \frac{1}{N-1} \sum_i \left(\Delta y^{(i)}_E - \overline{\Delta y_E}\right)^2 \to \sigma_E^2(y) = E[(\Delta Y_E - E[\Delta Y_E])^2], \tag{2.55}$$

where ΔY is the random variable of the model and Δy is its outcome. Note that the subscript E could refer to either random errors or systematic errors. The relationships mentioned above state that the *sample* (or *empirical*) *mean* and *variance* of a limited number of sampled errors are estimators of the *expected value* and *variance* of the model distribution. The arrow means that the larger the number of observations, the closer the result (in probability). It could also be shown using the maximum likelihood principle that the mean and the sample variance for a Gaussian distribution are the *best estimators* of the expected value and the variance, respectively.

Since *noise* is considered the physical process behind random errors, the statistical dispersion of random errors is also referred to as noise standard deviation σ_N. We will use the term *noise variance* (or noise standard deviation) as a synonym for *random error variance* (or random error standard deviation).

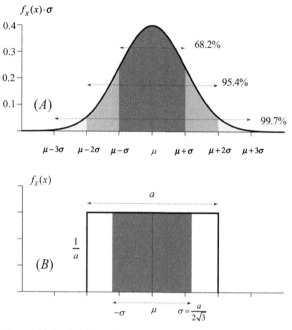

Figure 2.34 Probability of error in the error estimation. (A) Gaussian distribution. (B) uniform distribution.

We will refer to Gaussian and uniform probability distributions in the next sections since they cover most of the interest cases in sensor interfaces. The uniform distribution is usually associated with the case where a measure is made with a reference of ticks (e.g., using a yardstick): we assume that the error is uniformly distributed between two adjacent ticks as in the case of quantization noise in A/D converters. For the above distributions, it is sufficient the first and second-order moment characterization (mean and variance) for a complete description of the stochastic process model.

In Fig. 2.34A is illustrated the PDF of a Gaussian distribution whose mean and standard deviations are μ and σ, respectively. As shown, we have about 68% of probability that the outcome is outside one standard deviation around the mean μ and about 95.4% and 99.7% of probability to stay inside 2 and 3 standard deviations, respectively.[12]

Figure 2.34B illustrates the uniform distribution case. This means that the probability to get a measurement value is equal to any other values within the interval of size a. The uncertainty due to 1 standard deviation is equal to the size of the interval divided by $2\sqrt{3}$. Why this value and not the whole interval? This is because the standard deviation is derived from the average of the square values (power) of displacements (errors) with respect to the mean. Note that, differently from Gaussian, the distribution is bounded.

---

[12] The probability to be inside p standard deviations in a normal Gaussian distribution is $erf(p/\sqrt{2})$.

Table 2.2 Errors and relative distributions (see Fig. 2.32 for a graphical representation).

Type of error	ith readout outcome	Variance notation	Sample space
Random errors (noise)	$\Delta y_{E(\text{ran})}^{(i)}$	σ_N^2	Readout errors under repeated samples with constant input
Systematic errors due to variations of influence parameters	$\Delta y_{E(a_j)}^{(i)}$	$\sigma_{a_j}^2$	Readout errors of the same device due to parameters a_j variations
Systematic errors due to technology variations	$\Delta y_{E(a_j)}^{(i)}$	$\sigma_{a_j}^2$	Readout errors among all devices of a production lot due to parameters a_j variations
Systematic errors due to nonlinearity (distortion)	$\Delta y_{E(D)}^{(i)}$	σ_D^2	Readout error under different input values in the operating range

To summarize, the distributions and sample spaces of errors discussed so far are shown in Table 2.2 with the corresponding notations that will be used later on.

A way to quantify a Gaussian random process outcome or noise dispersion is to use its standard deviation σ_N as a reference for its amplitude. σ_N is also referred to as the *root mean square (rms) of the noise process* following (2.14) and (2.22). In general, as already discussed in Fig. 2.34A, we can define regions around the mean value that are multiples of the standard deviation. The larger the region, the higher the probability of getting an outcome inside such a region. A particular case is to set such an interval equal to $6.6 \cdot \sigma_N$, referred to as *peak-to-peak noise value*, to have only a 0.1% probability of getting outcomes outside that region. Of course, the term "peak-to-peak" is not strictly correct since not all outcomes are collected in such a region, but only 99.9% of them, and we have to consider that 0.01% of samples will lie outside of probability.

Example A voltage across a photodiode (Chapter 9) is read out through a long resistive path with resistance $R = 1k\Omega$ by an analog interface having a bandwidth $B = 100$ Hz and negligible noise. The entire system is affected by a thermal noise random process whose theory (Chapter 5) states that random errors due to R follows a Gaussian distribution with zero expected value and standard deviation equal to $\sigma_N = \sqrt{4kTRB}$ $= 40$nV. We will neglect any other random process. The equivalent circuit is shown in Fig. 2.35, where the interface does not have access to the source y_0, but it can only sense the output $\widetilde{y}_0(t_i)$ throughout the connection resistance.

To check the noise model, we acquire 200 readouts and collect them in a histogram shown in Fig. 2.35. Since we are in operating mode, we do not know the value of a single error $\Delta y_{E(\text{ran})}^{(i)}$ because the actual value is unknown, but we can estimate it (the expected value) and its spread of values (standard deviation) using the collected data output $\widetilde{y}_0(t_i)$. The empirical mean estimates the true value, $\overline{\widetilde{y}_0(t_i)} = \hat{y}_0 = 49.99$ mV which is very close to the actual value $y_0 = 50$mV, and the empirical standard deviation is $s_E(\tilde{y}_0(t_i)) = 39.09$ nV which is close to the theoretical model. If we increase the number of samples, we can verify that these quantities converge in probability $s_E \rightarrow \sigma_N$ and $\hat{y}_0 \rightarrow y_0$.

The noise model can also say that the rms noise is $\sigma_N = 40$ nV and its peak-to-peak value is $6.6 \cdot \sigma_N = 264$ nV.

Figure 2.35 also shows how the data collected from the random process of the previous example are distributed in the $\pm 1\sigma_N$, $\pm 2\sigma_N$, $\pm 3\sigma_N$ regions. Each bin size of the histogram identified by a straight horizontal line corresponds to the model's standard deviation.

2.8.6   Energy Properties of Random Signals

What is the power related to a random signal, such as those related to errors induced by noise? The problem is more easily treated in the discrete-time domain; we will start from this case and extend it to the continuous one in the following sections.

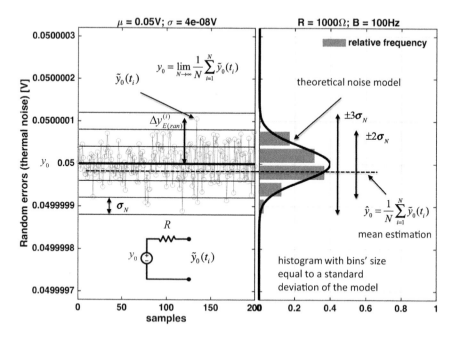

Figure 2.35 Example of random error noise generated by a photodiode voltage of 50 mV and readout through a resistance of 1 kOhm with an instrument having 100 Hz bandwidth. Two hundred samples are collected into a histogram where each bin corresponds to a standard deviation size. [The continuous graph is the PDF multiplied by the bin base to interpolate the histogram.]

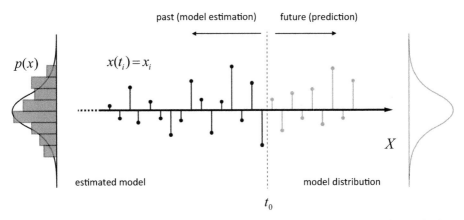

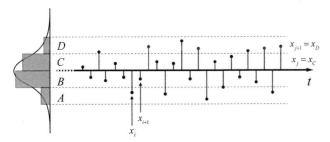

Figure 2.36 Characterization of a stationary random process: the outcome data collected in the past could be used to characterize the process distribution and predict future behavior.

Figure 2.37 Calculation of the power in a sampled signal using a statistical process.

Fig. 2.36 shows a collection of samples of a zero-mean stationary random process collected into a histogram. For what will happen in the future, we can assume that the process will behave with the same statistical characteristics of the past, and for the prediction, we can use a model estimated from the empirical one. In principle, we can determine the model distribution by collecting an infinite number of samples from the past. Now, we can calculate the power of the sampled signal as before, using the mean square of the acquired samples. However, as shown in a simple example in Fig. 2.37, we can see that the sampled data could be grouped in four bins: A, B, C, D. Therefore, the power of samples becomes

$$P = \langle x^2 \rangle = \frac{1}{N} \sum_{i=1}^{N} x^2(t_i) = \frac{N_A x_A^2 + N_B x_B^2 + N_C x_C^2 + N_D x_D^2}{N}$$

$$= \sum_{j=A}^{D} \frac{N_j}{N} x_j^2 = \overline{x^2},$$

(2.56)

where N_A, N_B, N_C, and N_D are the number of counts in the four bins;[13]

---

[13] See the appendix at the end of this chapter for definitions related to average and variance.

In the case of a nonzero average, we can calculate the power of the acquired samples using (2.21) for discrete values

$$P = \langle x^2 \rangle = s_x^2 + \langle x \rangle^2 \text{ where}$$

$$\langle x \rangle = \frac{1}{N} \sum_{i=1}^{N} x(t_i); \; s_x^2 = \frac{1}{N} \sum_{i=1}^{N} \left(x(t_i) - \langle x \rangle \right)^2, \tag{2.57}$$

where s_x^2 is the (experimental) sample variance of acquired data. Here, again, the power P is the sum of the AC power and DC power, which is the square of the *mean value*.

Now, using the same approach as before, we could easily find

$$P = \langle x^2 \rangle = \overline{x^2} = s_X^2 + \overline{x}^2 \text{ where}$$

$$\overline{x} = \frac{1}{N} \sum_{j=A}^{D} \frac{N_j}{N} x_j; \; s_X^2 = \overline{(x - \overline{x})^2}, \tag{2.58}$$

where $\overline{x}$ is the sample mean and s_X^2 is the sample variance of the discrete distribution. If we increase the number of observations, we have $N_j/N \simeq p_j$ is the frequency occupancy (i.e., probability) of the jth level. Therefore

$$\overline{x} \rightarrow E[X]; \; s_X^2 \rightarrow \sigma_X^2 = E[(E[X] - X)^2], \tag{2.59}$$

where X is the random variable of the process model. So far, we used the "bar" on top of the variable as the notation of sample means. However, in the following of the book, for simplicity, we will extend the "bar" notation also to refer to the expected value (i.e., as if a huge number of samples estimated it).

In conclusion

$$P = \overline{x^2} = \sigma_X^2 + x_0^2, \tag{2.60}$$

where σ_X^2 is a statistical variance related to the random variable. This means again that the power of the past sampled values could be calculated according to the statistical properties of the distribution.

In this view, (2.60) shows that the *power of the stationary random signal (described by the random variable X) could be calculated from the distribution's statistical properties.*

The extension to random signals of the calculation of power indicates the relationship between the entities of power and variance

$$\text{AC power} = P_{AC} \leftrightarrow \text{variance of random variable} = \sigma_X^2. \tag{2.61}$$

The previous discussion assumed that the statistical characterization of the past is a good estimator of the parent distribution that characterizes the entire process of *this specific device*. In other words, we created a *model* of the random process of one device. However, what we achieved is only the first step since the model, to be helpful, should be valid for any other device with the same physical characteristics. If the

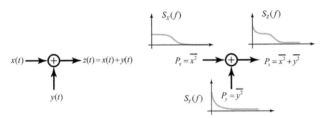

Figure 2.38 Property of power superposition for uncorrelated random signals.

model is valid for any other device governed by the same process, it is referred to as *ergodic*. In this case, the model characterized on statistical time samples on one device is valid for any other systems governed by the same process. Therefore, the statistical averages will be applied to the sample space given by different devices governed by the same process. In this case, the statistical operators will be referred to as *ensemble averages*. In the case the operators are applied to time samples, they are referred to as *time averages*. We will return on this in Chapter 4.

What is the meaning of power for a random process from the physical perspective? As far as deterministic signals are concerned, it is clear. For instance, we can drive a resistance with such a signal and measure that power by evaluating the generated heat. We will see that random signals such as noise are also generated by purely passive devices such as a resistor. If we devise an experiment to shortcut the terminals of a resistor to calculate the power of the random signal generated by its own noise, we discover that – obviously – it does not heat it at all. So, where is the power? The answer is that noise power is an expression of the mean kinetic energy of the particles composing the matter in thermal equilibrium. However, we cannot extract power unless we change the thermal equilibrium of the system according to the second principle of thermodynamics.

As a summary, we have found that in the case of random signals, we can use statistical distribution averages operators in place of time averages operators to characterize signal powers based on the statistical properties of the processes. Furthermore, we will see that we could use the statistical concept of covariance to determine the correlation between them. In the case the two signals are statistically uncorrelated, we can apply the property of superposition of powers as shown Fig. 2.38, where we use as notation the symbol $\overline{x^2}$ (statistical averages) instead of $\langle x^2 \rangle$ (time averages).

2.9 Input–Output Relationships for Random Error Distributions

Random errors could be modeled as stochastic outcomes added to the deterministic model at the input and output as shown in Fig. 2.39, where the stochastic model incorporates the deterministic one as in Fig. 2.28. We might wonder what the relationship between the two aspects is. In real cases, since errors and noise are always present

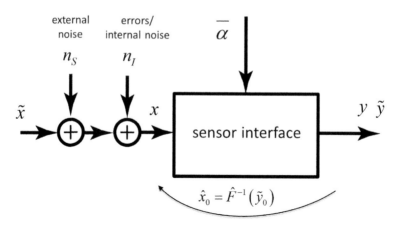

Figure 2.39 Hybrid model of the sensing system. Input and interface noises are grouped together at the input with respect to Fig. 2.28.

in physical systems, we cannot measure the characteristic directly, and thus we have to infer it from the input and output knowledge. Usually, this is accomplished by using a suitable amount of observed data to get the error model and infer the real characteristic. The characteristic could also be estimated theoretically, but usually, this has to be confirmed experimentally to consider the theoretical framework consistent. The input and interface noise introduction requires that random variables $\tilde{x}$ and $\tilde{y}$ describe the input and output. It should be pointed out that due to the addition of stochastic outcomes, x and y are random variables as well, even if their relationship, given by $y = F(x)$, is deterministic.

As far as *random errors* (or noise) are concerned, let us assume in *characterization mode* to take multiple consecutive readouts $\tilde{y}_0(t_i)$ for a fixed and known input value $\tilde{x} = x_0$. We can now imagine repeating it for different input values and collecting them in a two-dimensional space, as shown in Fig. 2.40. Since errors are random, we expect that they are sparsely distributed in two dimensions to be considered samples of an unknown joint probability distribution $p(\tilde{x}, \tilde{y})$. Moreover, since random errors are deviations from a real characteristic, we expect they are somehow accumulated around it. Note that the deviations of the experimental data from the characteristic are given by *both input (external) and interface noises* n_S, n_I.

Conversely, in the *operating mode*, we have to estimate the *unknown* value x_0 that is observed by the input $\tilde{x} = x_0$ based on the readouts $\tilde{y}_0(t_i)$.

In what follows, we will take the following simplifying assumptions:

- Random noise is a realization of a physical process modeled with uniform PSD and Gaussian noise distributions with zero expected value.[14]
- The errors are small perturbations with respect to a static characteristic.
- There is an absence of systematic error offsets in the measurements.

---

[14] This is based on thermodynamic considerations and according to the scheme of Fig. 2.39.

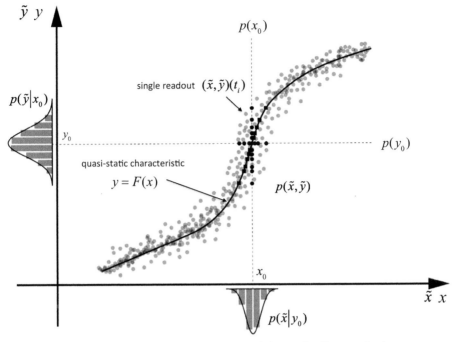

Figure 2.40 Quasistatic characteristic and experimental dataset dot diagram. On the axes are represented histograms of the dataset for a specific input value and conditional distributions of the noise model.

Notation

n_S, n_I: Noise associated with input stimulus and the interface, respectively. The noise sources summing points could be moved from input to output and vice versa by means of the function of the deterministic model.

$\tilde{x}$, $\tilde{y}$: Observed input and output variables.

x, y: Input and output variables of the deterministic model; due to summation of random variables, they are random variables as well, therefore $y = F(x)$ is a function of random variables.

x_0: Known (characterization mode) or unknown (operating mode) input value.

$\tilde{x}_0(t_i)$, $\tilde{y}_0(t_i)$: Samples of the input/output taken at $t = t_i$ for a given input $\tilde{x} = x_0$ or for a given output $\tilde{y} = y_0$, respectively (characterization mode). They could be visualized in vertical and horizontal dark points of Fig. 2.40.

$\hat{x}_0(t_i)$: Estimation of the input taken $t = t_i$ by way of the characteristic function based on the output values (operating mode).

$p(\tilde{x}, \tilde{y})$: Joint probability distribution of observed variables, estimated by acquired empirical distribution.

$p(\tilde{y}|x_0)$, $p(\tilde{x}|y_0)$: Conditional distributions for given input and output, respectively.

The limited dataset of observed points prevents an exact determination of the characteristic that *should be estimated* following several mathematical tools aimed at reducing the error between the estimated $y = \hat{F}(x)$ and the true characteristics, where the larger the number, the better the estimation. This process could be seen in Fig. 2.39, where the sampled data are taken from the system's boundaries, and therefore we do not know the exact form of the quasistatic characteristic that should be *interpolated*. In measurement science, the best fitting (interpolated) characteristic is also referred to as a *calibration curve* $y = \hat{F}(x)$.

Working on the scheme of Fig. 2.39, we can tie the input of the system to a known/unknown (depending on the characterization or operating mode) true value x_0 so that $\tilde{x} = x_0$. Then, we move any output noise source at the input (after a transformation $\hat{F}^{-1}(x)$) to be added to source noise before the input of the deterministic block. Therefore, the output random variable $\tilde{y}$ is a function $\hat{F}(x)$ of a random variable that summarizes all the random processes. Thus, for the LNN, it is

$$
\begin{aligned}
x_0 &= \mu_x = E[x], \quad \text{and}: \\
\sigma(x_0) &= E[(x - \mu_x)^2] = E[(x - x_0)^2]; \qquad \forall \tilde{x} = x_0 \in FS
\end{aligned}
\tag{2.62}
$$

Furthermore, following (2.55), the acquired dataset could be used as an estimator of the mean and the variance (spread) of the distributions at the output

$$
\begin{aligned}
y_0 &= \mu_y = E[\tilde{y}], \quad \text{and}: \\
\sigma(y_0) &= E[(\tilde{y} - \mu_y)^2] = E[(\tilde{y} - y_0)^2]; \qquad \forall \tilde{x} = x_0 \in FS; .
\end{aligned}
\tag{2.63}
$$

Once we have acquired the experimental points dataset, the duality between input and output is better illustrated in Fig. 2.41, an unfolded version of Fig. 2.39. If we ideally set the input to the value, x_0 Fig. 2.41A shows that many experimental points are taken at different times, having different outputs. The empirical distribution is an estimator of $p(\tilde{y}|x_0)$. Conversely, as shown in Fig. 2.41B, a certain number of experimental points are taken at different times, that for a given output correspond to different input values due to errors. If we had no errors, the model would estimate the true (unknown) input value x_0 based on the output y_0, but due to errors, different inputs would give the same observed output y_0. This empirical distribution is an estimator of the distribution $p(\tilde{x}|y_0)$. The key objective is now to understand the relationship between the expected values and variances of the two distributions. This is very important since this will allow us to understand (1) how good the estimation is and (2) how we can improve it.

Once we have moved all random sources to the input, we can use the characteristic $\tilde{y} = \hat{F}(x)$ to predict the input $\hat{x}_0(t_1)$ from a single sample of the output

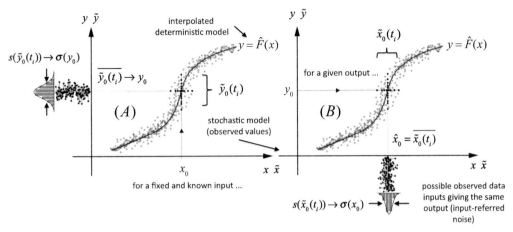

Figure 2.41 Statistical distributions of the random errors with respect to the input and the output. (A) The error could be characterized at the output. (B) Based on the characterized model, we could estimate (predict) input values with a degree of confidence.

$$\hat{x}_0(t_1) = \hat{F}^{-1}(\tilde{y}_0(t_1)).$$ (2.64)

However, we need to understand how the estimated values $\hat{x}_0(t_1)$ are related to the true value $\tilde{x} = x_0$. For the properties of functions of a random variable (x), we have at first order

$$\tilde{y} = \hat{F}\Big|_{x=x_0}(x) \simeq \hat{F}(\mu_x) + \frac{d\hat{F}}{dx}\Big|_{x=\mu_x}(x-\mu_x) = a_0 + a_1(x-\mu_x),$$ (2.65)

from which

$$y_0 = \mu_y = E[\tilde{y}] = E[a_0] + a_1 E[x] - a_1 E[\mu_x] = E[a_0] = \hat{F}(\mu_x) = \hat{F}(x_0); \quad \text{and,}$$

$$\sigma^2(y_0) = E[(\tilde{y}-\mu_y)^2] = a_1^2 E[(x-\mu_x)^2] = \left(\frac{d\hat{F}}{dx}\Big|_{x=\mu_x}\right)^2 \sigma^2(x_0).$$ (2.66)

The above relationships show that (1) $x_0 = \mu_x = \hat{F}^{-1}(\mu_y)$ estimated by $\hat{x}_0 = \hat{F}^{-1}\left(\overline{\tilde{y}_0(t_1)}\right)$ which tells us how to improve (2.64) by means of LNN and (2) the spreads of the two input and output variables are related according to

$$\sigma^2(x_0) = \left|S_0^{-1}\right|^2 \cdot \sigma^2(y_0),$$ (2.67)

which is the simplest expression of the *law of propagation of errors* where $S_0 = (d\hat{F}/dx)_{x=\mu_x}$ is the sensitivity of characteristic in the reference point.

Since we can use the variable x as an estimator of the unknown input by way of the characteristic function, $\sigma(x_0)$ is the spread of the input estimations based on the output data $\hat{x}_0(t_1) = \hat{F}^{-1}(\tilde{y}_0(t_1))$ given for the value $\tilde{x} = x_0$. Therefore, the distribution described by $p(\tilde{x}|y_0)$ is the distribution of the input values giving the same output value y_0 and the distribution of the single readout estimators $\hat{x}_0$ of the true value x_0. However, since (2.66) is a first-order approximation, the spread equivalence is done on variances only, while in general, it should comprise higher orders. Therefore, $p(\tilde{x}|y_0)$ represents the distribution of the *input-referred noise*.

To summarize:

- In the characterization mode, the statistical properties of the errors are characterized together with the characteristic static function.
- In the operating mode, using the characterized model, we can estimate the input on the basis of output data: $\hat{x}_0 = \hat{F}^{-1}(\tilde{y}_0(t_i))$ as well as its spread $\sigma(x_0) = |S_0^{-1}|\sigma(y_0)$ (input-referred noise), by means of the slope (sensitivity) of the characteristic.

We can visualize the input–output distribution relationship in the gain plot, as shown in Fig. 2.42. If we fix the input, we can acquire a certain amount of data to characterize the characteristic and the noise model (characterization mode). Conversely, based on the model and readout data, we can estimate the input under some degree of uncertainty (operating mode).

2.9.1   Concept of Input-Referred Resolution Due to Random Errors

The term "resolution" commonly refers to the *smallest input variation that a sensing system can detect*. On the basis of previous discussions, the term "smallest" should be defined in the probabilistic context. In the characterization mode, we can set two inputs $\tilde{x} = x_1$ and $\tilde{x} = x_2$ that are fairly close together. Due to random errors, the outputs are distributed following the probability density functions (PDFs) $p(\tilde{y}|x_1)$ and $p(\tilde{y}|x_2)$ whose expected values are $y_1 = F(x_1)$ and $y_2 = F(x_2)$, respectively, shown in Fig. 2.43. The noise model of the system gives the spread of those probability density functions. We now associate the two output values with two adjacent *classification regions* $\mathcal{R}_1$ and $\mathcal{R}_2$ (of equal length) to partially cover the distribution of output values using a given number of standard deviations of the noise model: $\mathcal{R}_i = [y_i \pm p \cdot \sigma(y)]$, where p is referred to as the *coverage factor*. Therefore, the coverage factor sets the size of the classification regions.

Then, the inference rule is

$$\text{IF } \tilde{y}(t_i) \in \mathcal{R}_1 \rightarrow \tilde{x} = x_1; \quad \text{IF } \tilde{y}(t_i) \in \mathcal{R}_2 \rightarrow \tilde{x} = x_2, \tag{2.68}$$

which means that any output sample falling in the region $\mathcal{R}_1$ is associated with an input value equal to x_1 and similarly for x_2.

Unfortunately, when the distributions are unbounded (e.g., Gaussian distributions), a tailing part of the distribution lies outside of the region, causing a possible

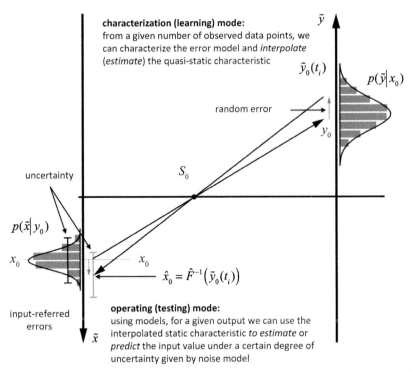

Figure 2.42 Relationship between input and output error distributions. For an input equal to x_0 the output is different from y_0. Conversely, we can estimate the input as $\hat{x}_0(t_i) = F^{-1}(\tilde{y}_0(t_i))$ different from the actual value x_0 by using the readouts $\tilde{y}_0(t_i)$. $p(\tilde{x}|y_0)$ is the distribution of estimated values referred to x_0.

misclassification error due to the random error. Assuming the equal probability of occurrence between the two values, we can set a decision threshold $y_{12} = (y_1 + y_2)/2$ between the two output values $y_1 = F(x_1)$ and $y_2 = F(x_2)$ where the two distributions are equal. Using this approach, the misclassification error is

$$\text{misclassification error} = Pr(\tilde{x} = x_1, \tilde{y}(t_i) \in \mathcal{R}_2) + Pr(\tilde{x} = x_2, \tilde{y}(t_i) \in \mathcal{R}_1)$$

$$= erfc\left(\frac{p}{\sqrt{2}}\right) \quad \text{[for Gaussian distribution],} \qquad (2.69)$$

where $Pr(\tilde{x} = x_1, \tilde{y}(t_i) \in \mathcal{R}_2)$ is the probability of assigning to $\tilde{x} = x_1$ for an output belonging to $\mathcal{R}_2$ and $Pr(x = x_2, \tilde{y}(t_i) \in \mathcal{R}_1)$ is the probability to assign to $\tilde{x} = x_2$ for an output belonging to $\mathcal{R}_1$. The misclassification error is evidenced in Fig. 2.43A, in the gray area, and depends on the coverage factor.

We have now to refer the resolution to the input as shown in Fig. 2.43 using the gain

$$\text{RTI or input-referred resolution} = \Delta x_{min} = \frac{2p \cdot \sigma(y)}{S} = 2p \cdot \sigma(x) \equiv [\xi], \qquad (2.70)$$

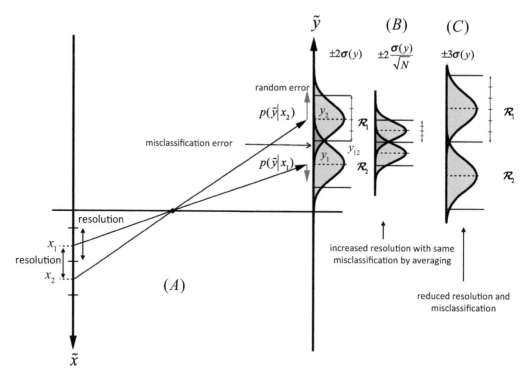

Figure 2.43 Determination of the input-referred resolution using the output probability distributions. The discrimination between two input values is determined according to the misclassification error between the two adjacent classification regions $\mathcal{R}_1$ and $\mathcal{R}_2$ of output values (A). The resolution could be increased by averaging N outputs to reduce standard deviation according to CLT (B). On the other side, we can decrease misclassification by increasing the number of standard deviations delimiting the regions at the expense of a lower resolution (C).

where p is again the coverage factor, and RTI stands for "referred to input."

> **Hint** In common usage, a resolution is *higher* than another if we can discriminate input values closer together, that is, when $\Delta x_{\min}$ is smaller.

Therefore, the resolution concept could not be disjointed from that of the misclassification error, which is related to the coverage factor p. From (2.69) and (2.70), it is clear that we can increase the resolution at the misclassification error expenses. Usually, it is better to set the value of p to clarify the misclassification error before defining the resolution – for example, if $p = 1/2 \rightarrow error = 62\%$ (effective resolution) $p = 2 \rightarrow error = 4.5\%$ and $p = 3.3 \rightarrow error = 0.1\%$ (noise-free resolution).

From the previous discussion, *resolution and misclassification errors* are strictly related, and they should be traded one for the other. How can we increase the resolution (i.e., increase the capability to distinguish smaller input values)? A possible approach (Fig. 2.43B) is to reduce the standard deviation by averaging on N readouts so that the spread is reduced by the factor $\sqrt{N}$. In this case, for an equal misclassification error (i.e., an equal number of sigmas), we can reduce the regions' size, thus increasing the resolution, as can be seen in (2.70).

Conversely, for the same noise model, characterized by a standard deviation σ, we can set the length of the regions to a higher number of standard deviations by increasing p. However, as shown in Fig. 2.43C, this implies decreasing the resolution, according to (2.70). In conclusion, the resolution is dependent on p, and thus on the misclassification rate, and on the amount of averaging.

Example An pressure sensor has an input operating range of ± 25kPa and an output operating range from 0.2 V to 4.7 V with a response time in a bandwidth of 1 Hz. Therefore, its gain is $S = (4.7 - 0.2)/50 = 90$mV/Pa. The static characteristic is given by $y = 90 \times 10^{-3} \cdot x + 2.45$. The rms output noise in the bandwidth is $\sigma(y) = 1$mV with Gaussian distribution. The input resolution with a misclassification error of 4.5% ($p = 2$) is $4 \cdot 1 \times 10^{-3}/90 \times 10^{-3} = 44$Pa. This means that we can discriminate two measures with a difference of 44 Pa with a probability to fail in 4.5% of cases. Therefore, if we have two input values at a resolution distance: $x_1 = 10.000$ kPa and $x_2 = 10.044$ kPa, we can identify two output values $y_1 = 3.350$ V and $y_2 = 3.354$ V. In other words, any output value from 3.348 V to 3.352 V ($\pm 2\sigma$) is associated with 10.000 kPa, while any value from 3.352 V to 3.356 V is associated with 10.044 kPa. However, the association has a probability error of 4.5%. We can decrease the misclassification error to 0.27% by choosing a classification region of $\pm 3\sigma (p = 3)$. However, in this case, the resolution is lowered to $6 \cdot 1 \times 10^{-3}/90 \times 10^{-3} = 67$Pa. On the other hand, we can increase the resolution by averaging 10 readouts that for $\pm 3\sigma$ becomes $67/\sqrt{10} = 21.1$Pa.

2.9.2   Concept of Uncertainty and Its Relationship with Resolution

We can use the same arguments about the resolution to introduce the concept of *uncertainty* as shown in Fig. 2.44:

- If we had no errors, the characteristic would be known, and the output y_0 would uniquely identify the input $x_0 = F^{-1}(y_0)$.
- Due to errors, we get different values $\tilde{y}_0(t_i)$ for any ith sample for a given input value $\tilde{x} = x_0$. For each of them, we could say "the input is estimated by $\hat{x}_0(t_1) = \hat{F}^{-1}\left(\tilde{y}_0(t_1)\right)$." As discussed, the estimated values $\hat{x}_0$ are distributed around x_0 as $p(\tilde{x}|y_0)$ whose spread could be measured as $\sigma(x_0)$. To have a reference, we can define a *standard/expanded uncertainty* (here defined for random errors only) as

$$
\begin{aligned}
\text{standard uncertainty} &= u(x_0) = \sigma(x_0) \text{ and :} \\
\text{expanded uncertainty} &= U(x_0) = p \cdot \sigma(x_0) \text{ where :} \\
\text{coverage interval} &= 2U = 2p \cdot \sigma(x),
\end{aligned}
\tag{2.71}
$$

where p is again the *coverage factor. The coverage interval is twice the expanded uncertainty.*

Therefore, in the operating mode, we can say that the estimated value $\hat{x}_0$ could encompass the true value in a *coverage interval* of uncertainty so that

$$x_0 \in \hat{x}_0 \pm U(x_0) = \hat{x}_0 \pm p \cdot \sigma(x) \text{ meaning}$$
$$\hat{x}_0 - p \cdot \sigma(x) \leq x_0 \leq \hat{x}_0 + p \cdot \sigma(x). \tag{2.72}$$

Therefore, the uncertainty *characterizes a region* where the true input value is expected to lie with a certain degree of confidence. Again, for Gaussian distributions, for an estimated value, $\hat{x}_0(t_1) = \hat{F}^{-1}(\tilde{y}_0(t_1))$ we have 99.9% confidence that the true value x_0 is within the interval $\hat{x}_0 \pm 3.3 \cdot \sigma(x)$.

It should be pointed out that the coverage interval *is not* centered in the "true value." Instead, as shown in Fig. 2.44A, the coverage interval is centered on the estimated value to expect to comprehend the true value. For example, the figure indicates six estimated inputs $\hat{x}_0(t_1), \ldots, \hat{x}_0(t_6)$ (the actual value x_0 is fixed). Five of them encompass the true value, but one $\hat{x}_0(t_2)$ does not. *If we take a large number of estimations, the ratio of successes over the total number converges in probability to the percent mentioned above of confidence.* The fact that the coverage interval is centered on the estimated

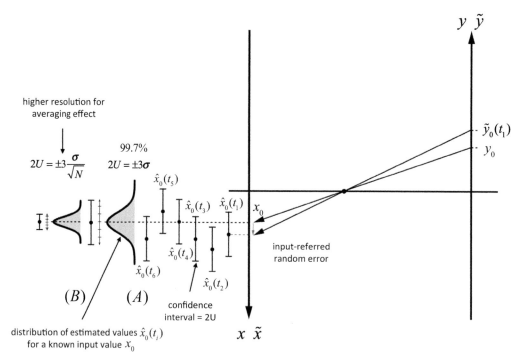

Figure 2.44 Concept of input-referred uncertainty and relationship with resolution. (A) Distribution of estimated values for a fixed input x_0 and coverage interval set to $\pm 3\sigma$ with confidence level of 99.7%. (B) Increased resolution by reducing the noise standard deviation by averaging.

value and not the true value is essential because, once we have one readout value and a corresponding estimation, *we do not know a priori where the true value is placed within the coverage interval.*

> **Hint** So far, we referred to the *uncertainty* solely due to the random errors or noise. We will see that the concept could also embrace systematic errors using the *generalized uncertainty.*

Now, referring back to the resolution concept introduced in (2.70), we can relate it to the uncertainty

$$\text{RTI resolution} = \text{coverage interval}$$
$$= 2U = 2p \cdot \sigma(x) \equiv [\xi]. \tag{2.73}$$

Note how the *input-referred resolution [or referred to input (RTI) resolution]* is *equivalent to the coverage interval.* Depending on the probability error, it is common to find two definitions of input-referred resolutions. In the first case, the smallest detectable signal is identified as the amplitude of the signal corresponding to the rms value of noise, alias *minimum detectable signal.* In other terms, the minimum detectable signal is the one that has the same power of noise. Therefore

$$\text{RTI effective resolution} = \text{minimum detectable signal (MDS)} = \sigma(x) \equiv [\xi].$$
$$\tag{2.74}$$

Thus, it is an input-referred coverage interval with $p = 1/2$. Now, we have eventually found a definition of minimum detectable signal introduced in Section 2.2 for an analog interface.

A second definition is related to a coverage interval given by 6.6 sigmas. This is referred to as a *noise-free resolution*

$$\text{RTI noise-free resolution} = 6.6 \cdot \sigma(x) \equiv [\xi]. \tag{2.75}$$

According to the CLT theorem, if we want to get a smaller uncertainty (higher resolution), we may perform an average of the output samples to estimate the input $\hat{x}_0 = F^{-1}(\overline{y}_0(t_1))$. In this case, the CLT theorem ensures that the uncertainty is reduced as the number of samples increases, as shown in Fig. 2.44B. For example, we can decrease the probability estimation error by averaging N outputs. This is useful for both the absolute determination, by virtue of LNN, and for the corresponding spread, which is reduced to $\sigma/\sqrt{N}$ for the CLT.

> **Hint** Resolution is the smallest input variation a sensing system can detect. The *RTI effective resolution* is given by the minimum detectable signal ($p = 1/2$), while the *RTI noise-free resolution* raises the confidence level to 99.9% ($p = 3.3$). Note that RTI resolutions are always referred to the input stimulus.

Example An amplifier characterized by a gain $G = 1000$ shows an output standard deviation noise of $10 mV$. Its *RTI effective resolution* is $10 \times 10^{-3}/1000 = 10\mu V$ while its *RTI noise-free resolution* is $10\mu V \cdot 6.6 = 60.6\mu V$. This means that we can detect an input variation of $66.0\mu V$ with a 99.9% of confidence level.

> **Hint** When we are in the operating mode and have a specific estimation of one input, the noise model tells us that the true value will be within the coverage interval around the estimation with some degree of *confidence*. The true value could be in any position of the interval. *That is why all the values of the defined interval are associated with a reference point disregarding their position.* Furthermore, it *does not mean* that the following estimations for a given estimation $\hat{x}_0$ will be in the coverage interval with the probability equal to the confidence one (see Fig. 2.44). Instead, we should interpret that as we have the probability equal to the confidence level in a large number of estimations to cover the true value over the total samples. That is why the term "confidence" is preferred to the term "probability."

Example An analog temperature sensor gives an output in the range $1.1 \div 1.6$ V for an input in the range $-40 \div 85°C$ and has an output noise of $8~\mu V$ rms in its system bandwidth. Therefore, its sensitivity is $(1.6 - 1.1)/(85 + 40) = 4$ mV/°C, and the offset is $y_{off} = 1.1 + 40 \cdot 4 \times 10^{-3} = 1.26$ V. For an output readout $y_0 = 1.5$ V we estimate the input $\hat{x}_0 = (1.5 - 1.26)/4 \times 10^{-3} = 60°C$. However, the input-referred effective resolution is $8 \times 10^{-6}/4 \times 10^{-3} = 2 \times 10^{-3}$ °C (2 mK) so the correct estimation should include the coverage interval $(\pm 3\sigma)$: $\hat{x}_0 = 60 \pm 0.006°C$. This means that the true value is within the interval with 99% of confidence, and it is not necessarily 60°C. Furthermore, we do not know a priori where the true value sits within the interval. We can have an idea of that only by acquiring a large number of values to estimate the true value with LLN.

2.9.3 Discretization of the Measurement Using Resolution Levels in the Analog Domain

In this section, we will make a link with Chapter 1 related to the concept of the resolution levels. The approach is as follows:

- We can partition the *FS* and the *FSO* into subdomains $C_1, C_2, \ldots C_N$ and $R_1, R_2, \ldots R_N$ whose size is related to a number of input/output referred noise standard deviations.

- As discussed before, we do not know a priori where the true value is in the coverage interval; therefore, all the values within an uncertainty subdomain are associated with a reference point, typically the midpoint one.
- Therefore, the noise uncertainty discretizes the analog domain into a finite number of values. The size of the subdomains depends on the misclassification error. The larger the size, the lower the misclassification.
- The discretization domains could be shifted according to suitable input/output references.
- The association of analog points of a domain to a single point is performed by the quantization process (e.g., A/D converters) and will be discussed in the next sections.

The confidence intervals or resolutions defined so far could be used to divide the full scales into regions or classes for both the input and the output, as shown in Fig. 2.45. *The aim is to understand the maximum number of distinguishable levels that the system could detect.* Thus, it is a measure of the information conveyed in the analog domain.

We can refer resolutions to the full scale and call this the *number of effective resolution levels*, NL, or *effective resolution in a percent or ppm*

$$NL = \sqrt{\frac{\sigma_{S(max)}^2 + \sigma_N^2}{\sigma_N^2}} \approx \frac{\sigma_{S(max)}}{\sigma_N} \equiv [\cdot];$$

and for deterministic signals

$$NL = \frac{\sigma_{S(max)}}{\sigma_N} = \frac{FS/q}{RTI\ effective\ resolution} = \frac{FS}{q \cdot \sigma_N} \equiv [\cdot]; \qquad (2.76)$$

while in percent or ppm is

$$effective\ resolution\ in percent = \frac{\sigma_N}{FS} \cdot 100 \equiv [\%];$$

$$effective\ resolution\ in\ ppm = \frac{\sigma_N}{FS} \cdot 10^6 \equiv [ppm]; ,$$

where we used (2.28). The introduction of q for deterministic signals is dictated by the fact that we have to consider the power of AC deterministic signals in characterization mode, as shown later on. The form of the first expression of (2.76) resides in information theory and is discussed in Chapter 3.

> **Hint** *RTI effective resolution* and *number of resolution levels* refer to the same concept, but they are expressed in different units: in the first case, the unit of resolution is the input variable, while in the second, it is the ratio between the *FS* and the previous quantity.

As indicated in the previous section, we can define output NLs, dividing the *FSO* into several regions $\mathcal{R}_i$ whose size is one standard deviation of output noise. We can do the same with the input, dividing the *FS* into many regions $\mathbf{C}_i$. If the characteristic is linear, the number of resolution levels is the same, and the ratio of the sizes of the two regions is equal to the gain.

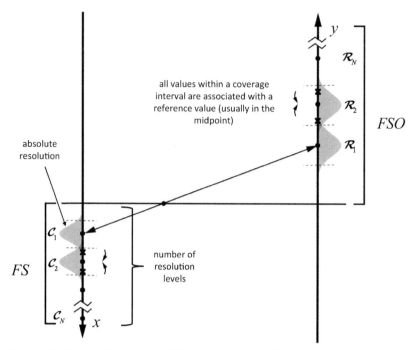

Figure 2.45 Relationship between RTI resolution (referred to input units) and the number of resolution levels (NL). The association of analog values to discrete levels due to noise uncertainty was anticipated in Chapter 1.

2.10 Systematic Errors Due to Nonlinearity: The DC Approach

Section 2.8.4 analyzes several methods for characterizing systematic errors due to nonlinearity using the quasistatic characteristic properties, alias DC methods. In the DC mode, the error is calculated using very slow input variations on the quasistatic characteristic. Therefore, no time-dependence characterization is involved. More specifically, we have referred to errors related to a single input value of the characteristic. However, it is useful to characterize the systematic error considering the variation (distribution) of the signal in the whole domain or full scale in order to have an average value and a more compact reference of this kind of error.

There are two approaches for defining the systematic error due to nonlinearity based on the quasistatic characteristic (DC mode):

1. Using the *ideal curve as a reference*. In this case, the error is calculated as a bare discrepancy between the two curves.
2. Using the *regression line as a reference*. The error is calculated as the difference between the real curve and its regression. In most cases, the regression line is *not* identical to the ideal one, and then we should take into account the *gain* and *offset errors* of the interpolating line with respect to the ideal one.

What we measure in the characterization mode is the output; therefore, it is easier to calculate the error on the output $\Delta y_{E(D)} = y_T - y_0$, where y_T is the output given by the reference line and y_0 is the readout value given by the sensor. This is also easy from the analytical point of view, where the difference between the reference and real characteristics could be numerically computed. However, the evaluation should eventually be referred to the input to understand the limits of detection and combine it with other uncertainties. An easy way is to refer the error back using the gain of the reference line, but this should be done carefully, as will be discussed later.

We can devise very simple ways to be warned about the nonlinearity errors; for example, we can characterize "safety limits" defined by an uncertainty given by the maximum displacement of the real characteristic with respect to the ideal one

$$\Delta y_{E(D)\text{max}} = \max_{x \in FS}\{\Delta y_{E(D)}\}, \tag{2.77}$$

For example, a temperature sensor is characterized by an error of $\pm 0.5°C$ in the full-scale range, implying that the maximum nonlinearity error is $0.5°C$. Safety limits could be graphically represented by parallel lines to the reference line, as shown in Fig. 2.46A.

Alternatively, the error value is characterized by a value that is proportional to the reading value, as shown in Fig. 2.46B and described by the following relationship

$$\Delta y_{E(D)}(x) = \alpha + \beta \cdot x, \tag{2.78}$$

where α and β are coefficients defining a minimum value and a constant of proportionality. Sometimes, the second constant is also provided in percent. For example, a temperature sensor is characterized by an offset error given in $\pm 0.5°C + 0.1\%$ *of reading*. This evaluation is also provided in the worst-case referring to the full scale, such as $\pm 0.3°C + 0.1\%$ *of the full scale*. The latter kind of characterization is motivated by the fact that the real characteristic is quite linear around the origin, and the deflection from ideality increases for larger input values in a large number of cases. It should be pointed out that safety limits are improperly referred to as *uncertainties* because they are not related to statistical characterization.

Another simplified way to express the deformation of the real characteristic with respect to the regression line (case 2 in the preceding) is to use the *integral nonlinearity* (INL) defined as

$$INL = \frac{\max_{x \in FS}\{\Delta y_{E(D)}\}}{FSO} \cdot 100 \equiv [\%], \tag{2.79}$$

where $\Delta y_{E(D)}$ is the offset *between the regression line and the real characteristic*.

Even if the preceding characterizations are widely used, most proper approaches should consider the statistical properties of the nonlinearity using the concept of *average*. More specifically, instead of using the maximum error as a measure of a safety limit, it should be better to characterize the error on

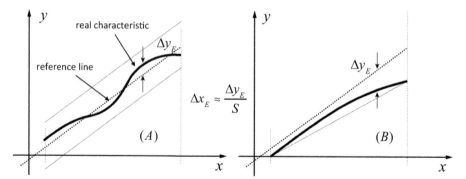

Figure 2.46 Evaluation of fixed systematic errors by boundary safety limits (A) and input dependence (B).

average over the full scale. We will start making a nonlinear characterization characteristic with respect to a generic reference line. Then, we will distinguish between two cases: (1) the reference line is the ideal characteristic, or (2) the reference line is the real characteristic's regression line. In the second case, if the two lines are not identical, we have to consider the required *offset* and *gain* corrections (errors) of the regression line with respect to the ideal one. The regression line is used instead of the ideal one (if they are different) because from the implementation point of view, offset and gain corrections can be easily adjusted (both in analog and digital domains). Furthermore, we will see that the AC characterization of nonlinearity (based on the concept of distortion) is invariant with respect to gain and a shift correction.

If we compare a nonlinear characteristic versus a reference line, as shown in Fig. 2.47, we can see the evolution of the error with respect to the output. If we take a single point, this kind of error has no statistical attribution since it is just a displacement evaluation. However, if we assume that the signal *has uniform probability distribution within* the input full-scale, we can infer these statistical properties.

Thus, we first evaluate the average error in the *FS* (Fig. 2.47)

$$\overline{\Delta y_{E(D)}} = \frac{1}{FS} \int_{FS} \Delta y_{E(D)}(x) dx. \tag{2.80}$$

If the reference line is the ideal one, the resulting value of (2.80) could be different from zero. If the reference line is the *regression* one, the value is zero by definition because it is calculated so that the average of residuals (the difference between the characteristic line and the reference) is zero.

Then, we evaluate its variance error

$$\sigma_D^2(y) = \frac{1}{FS} \int_{FS} \left(\Delta y_{E(D)}(x) - \overline{\Delta y_{E(D)}} \right)^2 dx. \tag{2.81}$$

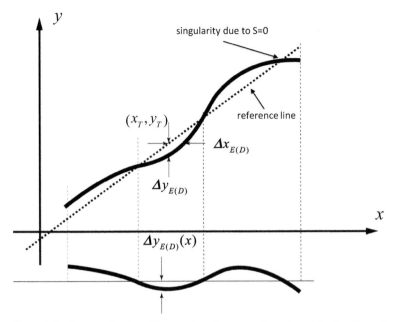

Figure 2.47 Characterization of uncertainty due to nonlinearity. Note that if we directly map the x-axis error, we might encounter problems due to the function's noninvertible zone.

In these expressions, we have used the letter D to relate this error to the concept of *distortion* since we will see that fixed systematic errors are referred to this concept.

> **Hint** The aforementioned approach stems from the assumption that the input is characterized by a uniform distribution of values over the full-scale span (this is a common assumption for measuring instruments). In other terms, *the error distribution is dependent on the input distribution.*

Single errors could also be calculated with respect to the x variable using the relationship we already derived

$$\Delta x_{E(D)}(x) = \frac{\Delta y_{E(D)}(x)}{S(x)}, \tag{2.82}$$

where $S(x)$ is the gain of the characteristic. However, this approach may encounter problems in singularities where $S = 0$ (see Fig. 2.47). This is a frequent case, especially when there is a saturation of the characteristic. Therefore, it is better to calculate the mean values of the error on the output using (2.80) and (2.81), and then refer the whole result to the input by means of the gain of the reference line:

$$\sigma_D(x) \approx \frac{\sigma_D(y)}{S}, \tag{2.83}$$

where S is the gain (sensitivity) of the reference line. Of course, this expression is the result of a first-order approximation.

As an example of the approach, we will calculate the offset error of the A/D conversion technique (which is a quantization) since it is a systematic error. In this case, we will use a variable of integration the x one, since it refers to analog values. This is also referred to as *quantization error*. Note that in this case, the ideal line is also the regression line. As shown in Fig. 2.48A, the offset error is given by the difference between the ideal and the real curves, plotted in Fig. 2.48B. Since the error is periodic of the period T

$$\Delta x_{E(D)} = V_{\text{LSB}}(-x/T); \ t \in [-T/2; T/2;], \tag{2.84}$$

where V_{LSB} is the least significant bit voltage. We can calculate the variance in just one period

$$\overline{\Delta x_{E(D)}} = 0$$

$$\sigma_D^2(x) = \frac{1}{T} \int\limits_{-T/2}^{T/2} V_{\text{LSB}}^2(-x/T) \, dx = \frac{V_{\text{LSB}}^2}{12} = \sigma_Q^2, \tag{2.85}$$

from which we get the typical uncertainty of the A/D conversion. Here, the letter Q is used to indicate that this is also called the *quantization error variance* of an A/D converter.

Note that we could have reached the same result assuming a uniform distribution of error in the period (since we assume a uniform distribution of input values) and calculate the variance as

$$\sigma_D^2 = \int\limits_{-\infty}^{\infty} p(\Delta x_E) x^2 dx = \frac{1}{V_{\text{LSB}}} \int\limits_{-V_{\text{LSB}}/2}^{V_{\text{LSB}}/2} x^2 dx = \frac{V_{\text{LSB}}^2}{12} = \sigma_Q^2$$

$$\sigma_Q = \frac{V_{\text{LSB}}}{2\sqrt{3}}. \tag{2.86}$$

It should be pointed out that the result is identical to the generic standard deviation of uniform distribution, as illustrated in Fig. 2.34B.

As far as the reference line is the regression one, the characterization of the fixed systematic errors approach is split into two phases (Fig. 2.49):

1. A first error is calculated by estimating the "deformation" of the real characteristic with respect to its *regression line* using the distortion concept. We can use both INL or variance estimation of error following (2.81).
2. A second error is calculated by the composition of the gain and offset discrepancy between the regression line and the ideal one.

The graphical representation of the concept is shown in Fig. 2.49A.

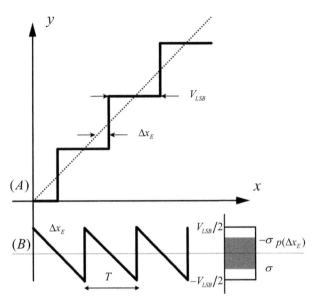

Figure 2.48 Calculation of the offset error in A/D conversion. This is also referred to as *quantization error or noise.*

Then, regarding point (2), we can evaluate the errors separately regarding both the gain and the offset by comparing the interpolating line with the ideal one as shown in Fig. 2.49B and C and combining them. It is common practice to take the worst case as

$$\Delta y_{E(OG)} = \max\{\Delta y_{E(\text{off})} + \Delta y_{E(\text{gain})}\}. \tag{2.87}$$

2.11   Generalized Uncertainty and the Law of Propagation of Errors

So far, we have referred to the concept of *uncertainty* to random errors only, and we linked it to the *resolution* concept. However, we can expand the uncertainty concept to other sources of errors, such as those related to nonlinearities and parameter variations.

Therefore, we can similarly associate the spread of nonlinearity errors with a specific uncertainty at the input: $u_D^2(x) = \sigma_D^2(x)$. We can also associate an input-referred uncertainty with errors due to influence parameters $u_{\alpha_j}^2(x) = \sigma_{\alpha_j}^2(x)$ where j is related to the jth influence parameter.

Assuming that the errors are statistically uncorrelated, the total *uncertainty* or *combined uncertainty or generalized uncertainty* (in power) is given by the sum of uncertainties (in power), or, equivalently, the combined uncertainty could be expressed as the combinations of single ith uncorrelated uncertainties

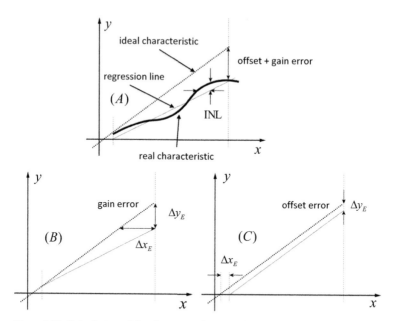

Figure 2.49 Calculation of fixed systematic errors by using integral nonlinearity (INL) and gain and offset errors. (A) Concept of INL. (B) Gain error (C) Offset error.

$$u_C^2(x) = u_N^2(x) + u_D^2(x) + \sum_{j=1}^{K} u_{a_j}^2(x); \text{ or}$$

$$u_C(x) = \sqrt{\sum_{i=1}^{N} u_i^2(x)}.$$

(2.88)

The last expression of (2.88) is also referred to as the *root-sum-of-squares* (RSS) of single uncertainties. If necessary, we can also use the expanded uncertainty $U = p \cdot u$ by means of the coverage factor, p.

The uncertainty is essentially referred to as the input, so it is important to find a method for mapping output-referred error spread into an input uncertainty $u_i^2(x)$. We have found that nonlinearity errors, such as random errors, are mapped between input and output by means of the gain of the characteristic. The problem now concerns errors induced by variations of the influence parameters where we have to refer to the derivative of the characteristic with respect to that parameter.

The *law of propagation of errors* is stating that the variance of the *independent* scattered data is propagated from input to output by means of the square of the derivative of the characteristic, with respect to the referred variable

$$y = F(x, \underbrace{\alpha_1 \dots \alpha_K}_{\text{parameters}})$$

$$u^2(y_0) = \sum_{i=1}^{N} \left(\frac{dF}{dx} \Big|_0 \right)^2 u_i^2(x_0) + \sum_{j=1}^{K} \left(\frac{dF}{d\alpha_j} \Big|_0 \right)^2 u_{\alpha_j}^2(x_0),$$

(2.89)

where u_i^2 are the uncertainties related to the input variable, and $u_{\alpha_j}^2$ are the uncertainties due to the jth influence parameters. This law was derived in the simplest form in Section 2.9. Therefore, using (2.89), we can refer to the input any kind of error calculated at the output and combine them all using (2.88).

2.12 Comparing Signals with Errors in Power

2.12.1 The Signal-to-Noise Ratio

One of the most important aspects of the early stages of transduction is understanding the strength of the noise generated by the sensor interface with respect to the signal because this greatly helps to evaluate the amount of information conveyed by the system. Therefore, we need to compare the power of signal with the power of noise using a ratio on an equal basis.

Assume for the moment a noiseless signal to be measured by a linear sensor interface as shown in Fig. 2.50. Figure 2.50A illustrates how the interface introduces noise that is added to the amplified signal to the output. The output signal AC power is represented by its variance σ_S^2, and the noise power is indicated with σ_N^2. That source of uncertainty does not come from the input but from random processes generated inside the system.

We can model the situation mentioned earlier with an equivalent system where a noiseless bock interface with noise $\sigma_N^2(y)$ is added in power to the output, as shown in Fig. 2.50B. Using the law of propagation of errors, we can further model the original situation by adding an input-referred noise $\sigma_N^2(x)$ that is directly added to the input signal, as shown in Fig. 2.50C. Note that we can do the opposite; a source of error/noise on the input could be mapped to the output by multiplying it by the gain of the system.

The comparison between signal and noise is evaluated by using a ratio of their powers, called *signal-to-noise ratio* (*SNR*)

$$\text{SNR} = \frac{\text{signal power}}{\text{noise power}} = \frac{\sigma_S^2}{\sigma_N^2}. \tag{2.90}$$

The same quantity could be expressed in terms of decibels as

$$\text{SNR}_{dB} = 10\log\frac{\text{signal power}}{\text{noise power}} = 10\log\frac{P_S}{P_N} = 10\log\frac{\sigma_S^2}{\sigma_N^2} = P_{S(dB)} - P_{N(dB)}. \tag{2.91}$$

In the foregoing interpretations, the strength of the signal whose power is equal to that of noise, that is $\text{SNR}_{dB} = 0$, is referred to as *minimum detectable signal* (MDS), and we have encountered this also in the definition of the effective resolution.

In terms of signal amplitudes, we have

$$\text{SNR}_{dB} = 20\log\frac{\sigma_S}{\sigma_N} = 20\log\frac{\text{signal ampitude}}{\text{noise stan dard deviation}}. \tag{2.92}$$

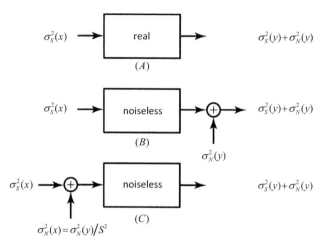

Figure 2.50 Concept of input-referred noise with a noiseless input signal. (A) A real system. (B) The same system is modeled with a noiseless system and an output-referred noise added to the output. (C) The same system is modeled by an input-referred noise added to the input.

We must be very careful in the equivalence between input/output noises/errors because this is true for a given gain S. However, if we modify the gain, the effects are very different whether the original source is at the output or at the input. In other words, noise/errors observed at the output are dependent on S. More specifically, it is apparent that if the original noise is physically added to the output, the SNR could be improved by increasing S. Conversely, if the original noise is due to sources placed at the input, the SNR is not affected by S since we amplify both signal and noise of an equal amount. However, we will see in Chapter 7 that in complex systems, the most effective noise sources are placed at the input according to the Friis formula. Therefore, the model with an original noise source at the input is the most appropriate, and the SNR is weakly dependent or independent on S as in the example of Chapter 1.

The SNR is one of the most used figures of merit of sensing systems even if it merely compares the power of the desired quantity (signal) with that of an undesired quantity (noise). This ratio is primarily used to evaluate the performance of the early analog interface of the sensing systems. However, it should be pointed out that it is not necessarily related to the whole sensor system's ability to detect the information, as discussed in Chapter 1.

The decibel is related to a measure of power in watts using a reference 1 watt. However, the reference could also be expressed in amplitude units

$$P_{x(\mathrm{dB[ref]})} = 10\log\frac{\text{power } x}{\text{power reference}} = 20\log\frac{\text{amplitude } x}{\text{amplitude reference}}. \tag{2.93}$$

Other references are used, for example, if the reference is 1 mW, the unit is called dBm

$$P_{x(\text{dBm})} = 10\log\frac{\text{power } x}{1\text{mW}}, \tag{2.94}$$

or if the reference is the full scale, the unit is called dB*FS*

$$P_{x(\text{dB}_{FS})} = 20\log\frac{\text{amplitude } x}{FS} \tag{2.95}$$

or as a reference for acoustic sensing (e.g., microphones), the *sound pressure level* (SPL)

$$P_{x(\text{dB}_{\text{SPL}})} = 20\log\frac{\text{sound pressure } x@1\text{ kHz}}{\text{sound pressure} = 20\mu\text{Pa}@1\text{ kHz}}, \tag{2.96}$$

where the sound pressure of $20\mu\text{Pa}@1$ kHz is the human hearing threshold. This means that the sound pressure power $P_x = 0$ dB$_{\text{SPL}}$ is at the limit of human perception.

The decibel notation is very useful, especially for graphical representation of ratios, since the logarithmic law allows visualizing a ratio as a geometrical distance, which is valid irrespective of the reference.

An example of graphical representation could be seen in the example shown in Fig. 2.51 regarding the SNR of a sensor interface. In this example, the maximum amplitude of the input signal is 3.3 V that coincides with the full-scale, and the input-referred noise of the interface is 113 mV, which is equivalent to about −89.3 dB*FS*. In this case, the noise is independent of the signal strength, and it constitutes the *noise floor* of the system. On the other hand, the signal power is increasing by +20 dB per decade, as for (2.95). Therefore, the SNR expressed in dB is the geometric distance between the noise power and the signal power curves and increases with the signal strength. Note how the MDS is identified by the intersection of the two curves, where SNR$_{\text{dB}} = 0$, and that the maximum SNR is identified by the maximum allowed input signal.

Another important aspect of the SNR is that *for noiseless input signals* and LTI interfaces, it could be calculated either to the output or to the input

$$\text{SNR}(y) = 10\log\frac{\sigma_S^2(y)}{\sigma_N^2(y)} = 10\log\frac{S^2\,\sigma_S^2(x)}{S^2\,\sigma_N^2(x)} = 10\log\frac{\sigma_S^2(x)}{\sigma_N^2(x)} = \text{SNR}_{EQ}(x), \tag{2.97}$$

where $SNR(y)$ is the signal-to-noise ratio calculated at the output and SNR$_{EQ}(x)$ is the equivalent signal-to-noise ratio calculated at the input. Of course, in the case of noisy signals, in the presence of noise not generated by the interface, the equivalence is no longer valid, as we will see later.

We can notice that if the interface only generates the noise, we can measure noise at the output, but we cannot measure any noise at the input since the input-referred noise results from a model. For this reason, we have used the name *equivalent* SNR, as it is the result of a model and not a ratio of signals that could be measured.

As shown in Fig. 2.52, the situation is different for noisy signals. A real sensor interface amplifies both signal power σ_S^2 and input noise power σ_{SN}^2. Note

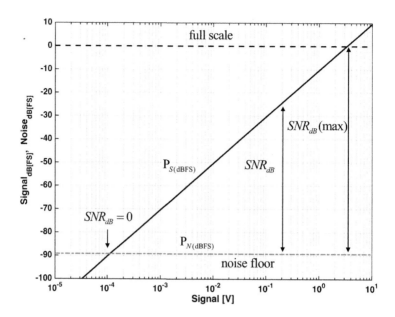

Figure 2.51 Log-log plot graphical representation of the signal-to-noise ratio of a voltage sensor interface (amplifier) with a maximum input signal of 3.3 V and a noise floor of 113 mV rms equivalent to a resistance of 1 MOhm in a 200 kHz bandwidth.

how the input noise power σ_{SN}^2 is generated by external sources from the interface and has nothing to do with interface noise σ_N^2. For this reason, we can now measure a real noise at the input and a real noise at the output, and thus define real input signal-to-noise ratio

$$\text{SNR}(x) = 10\log\frac{\sigma_S^2(x)}{\sigma_{SN}^2(x)} \neq \text{SNR}(y) = 10\log\frac{\sigma_S^2(y)}{\sigma_N^2(y)}$$

$$= 10\log\frac{\sigma_S^2(x)}{\sigma_N^2(x)} = \text{SNR}_{EQ}(x). \tag{2.98}$$

We will see that the difference between input and output SNR is taken into account by the *noise figure*.

2.12.2   The Concept of Dynamic Range

Nothing could be referred to as "large" or "small" without a reference. Any system dealing with input/output intensities should refer to a reference for a *scale* of values. Since the concept of sensing is derived from biological perception, we will make an example referring back to two human senses: *hearing* and *sight*.

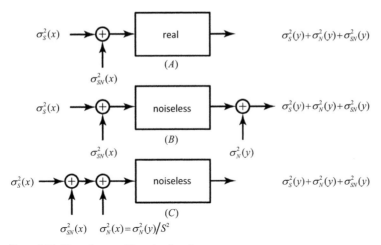

Figure 2.52 Transducer with noisy input.

The hearing is based on the perception of signals represented by pressure waves and, more specifically, by the difference between its maxima and minima. Again, since the signal should create some deformation in the sensing system to be recognized (e.g., a membrane deformation), it should perform work; therefore, it is more important to think about *signal power* (or intensity) instead of signal amplitude. In Table 2.3 are shown the levels of sound that human ears could detect.

The minimum pressure level of the signal that we can sense is as low as 20 μPa, and the threshold of discomfort is as high as 20 Pa. Therefore, there is a huge variation of signals that could be detected; it is about 13 decades in power. Thus, we can express the variation of sound referring to the hearing threshold as a reference, as shown in the last column of the table where decibels are used.

As far as the perception of sight is concerned, as shown in Table 2.4, we can see a similar capability to sense very large input variations, spanning more than 9 decades of light power. It is also interesting to show the power in terms of the number of photons per second per square micrometer, whose value is about the area of the pixels of electronic cameras. Note how for low-intensity levels, the signal's discretization is determinant, and we will see the implications in the next sections.

Therefore, we can use the minimum perceivable signal as the reference of variation, as shown in the last column. To quantify the overall span of the signal, we can define the *dynamic range* (DR) of the sensing system as

$$DR_{dB} = \text{dynamic range} = 20\log\frac{\text{max detectable signal amplitude}}{\text{min detectable signal amplitude}}$$

$$= 10\log\frac{\text{max detectable intensity (power)}}{\text{min detectable intensity (power)}}.$$

(2.99)

Table 2.3 Scales of the human hearing.

Source	Sound pressure [Pa]	Sound intensity [W/m^2]	Intensity [dB]
Threshold of pain	63	10	130
Threshold of discomfort	20	1	120
Chainsaw, 1 m distance	6.3	1E-1	110
Disco, 1 m from the speaker	2	1E-2	100
Diesel truck, 10 m away	0.63	1E-3	90
Curbside of busy road, 5 m	0.2	1E-4	80
Vacuum cleaner, distance 1 m	0.063	1E-5	70
Conversational speech, 1 m	0.02	1E-6	60
Average home	0.0063	1E-7	50
Quiet library	0.002	1E-8	40
Quiet bedroom at night	0.00063	1E-9	30
Background in TV studio	0.0002	1E-10	20
Rustling leaves in the distance	0.000063	1E-11	10
Hearing threshold	0.00002	1E-12	0

From www.sengpielaudio.com.

Note how *DR* is always a ratio of power/energy.[15] Therefore, according to the previous definition, we can state that human hearing should cope with 130 dB of dynamic range and human sight covers about 90 dB of DR.

As far as electronic interfaces are concerned, we can better define DR considering as reference the power of interface noise: what we have already referred to as minimum detectable signal in the definition of SNR. In other words, we can set the threshold of the sensing system as that defined by a signal having a power equal to the power of the interface noise (SNR = 1 or SNR$_{(dB)}$ = 0). On the other hand, we can set the upper bound as the maximum undistorted signal power.

In the case in which the *minimum detectable signal* is given by *random noise*, the dynamic range of an electronic sensing system is

$$DR_{dB} = 10\log\frac{\text{max undistorted signal power}}{\text{system noise power @ signal} = 0} = 10\log\frac{\sigma^2_{S(max)}}{\sigma^2_N}$$
$$= 20\log\frac{\sigma_{S(max)}}{\sigma_N} = 20\log\frac{\sigma_{S(max)}}{\text{RTI effective resolution}},$$
(2.100)

where $\sigma^2_N = \sigma^2_N(x = 0)$ is the *input-referred noise power* at zero signal, which we also have linked to the *RTI effective resolution*. For undistorted signal, it is referred to the signal that is transformed by the system in a linear region of the characteristic. It is due to the processes internal to the interface and not to any other external sources. Similarly to the SNR, as shown in the previous section in the relationship (2.97), we can define

---

[15] In some specific cases such as audio DR it could also be considered as the ratio of amplitudes.

Table 2.4 Scales of human sight.

Source	Illuminance [lx]	[Photons/μm^2/s1]	Light power [dB]
Direct sunlight	100,000	1.2E9	90
Full daylight	10,000	1.2E8	80
Overcast day	1000	1.2E7	70
Very dark day/artificial lamp	100	1.2E6	60
Twilight	10	1.2E5	50
Deep twilight	1	1.2E4	40
Full moon	0.1	1.2E3	30
Quarter moon	0.01	1.2E2	20
Moonless starlight	0.001	12	10
Moonless overcast night sky	0.0001	1.2	0

DR on either the output or the input (in the preceding case, it is an input-referred equivalent *DR* that is the most used form).

Depending on the characteristic of the signal, we have a relationship between the maximum variance of the signal and the *FS* value by means of the power factor as given by (2.28). Thus, for deterministic signals, we have $\sigma_{S(\text{max})} = FS/q$ and then

$$DR_{\text{dB}} = 20\log\frac{FS}{q \cdot \sigma_N}, \tag{2.101}$$

where q is the *power factor* (e.g., $q = 2\sqrt{2}$ for sinusoidal waveforms and $q = 1$ for DC signals).

Hint Do not confuse dynamic range with the full-scale span, even if they are both related to input ranges. The *DR* is referred to the minimum detectable signal, and it expresses the number of resolution levels of information. Conversely, the *FS* is defined by the input operating range of the sensor.

Example A pressure sensor operates between −25 kPa and 25 kPa with an RTI resolution of 1.25 Pa. The dynamic range is $DR_{\text{dB}} = 20\log(50 \times 10^3/1.25) = 92.04$ dB where we used $q = 1$ since the sensing is de facto without constraints on the bandwidth.

Example A temperature sensor operates between −15°C and 85°C; thus, the $FS = 100$°C. The RTI resolution is estimated in 0.05°C. The $DR_{\text{dB}} = 20\log(100/0.05) = 66.02$ dB, where again, we assumed a DC sensing.

Example An analog sensor interface operates with input signals having a swing of ±100 mV. The minimum detectable signal is 100 nV. Therefore, the $DR_{dB} = 20\log$ $\left(200 \times 10^{-3}/(2\sqrt{2} \cdot 100 \times 10^{-9})\right) = 117$ dB. In this case, we referred the DR to a sinusoidal regime since it is an AC sensing.

Since the standard deviation of system noise appears in both the definitions of SNR and the number of resolution levels, there is a relationship between them

$$\text{resolution}_b = NL_b = \log_2(NL) = \log_2\left(\frac{\sqrt{\sigma^2_{S(\max)} + \sigma^2_N}}{\sigma_N}\right); \ \text{SNR}_{\max} = \frac{\sigma^2_{S(\max)}}{\sigma^2_N} \rightarrow$$

$$NL_b = \frac{1}{2}\log_2(1 + \text{SNR}_{\max}) = \frac{1}{2}\log_2(1 + DR).$$

$$(2.102)$$

This relationship applies when the maximum SNR equals the DR; however, a better discussion of this issue will be provided in the next section.

The last relationship states that the effective resolution levels NL and the SNR in terms of bits are closely related. Therefore, any discussion related to the SNR or DR will also affect the resolution levels in terms of bits. Note that for *deterministic signals*, we have $\sigma_{S(\max)} = FS/q$ and therefore, we can approximate (2.102) with

$$\text{effective resolution}_b = \log_2(NL) \simeq \log_2\left(\frac{FS}{q \cdot \sigma_N}\right); \ \text{SNR}_{\max} = \frac{FS^2}{q^2 \cdot \sigma^2_N}. \quad (2.103)$$

The resemblance of (2.102) with the information theory *channel capacity* will be discussed in Chapter 3.

It is interesting to point out that

$$NL = 2^{NL_b} = \sqrt{1 + DR} \simeq \sqrt{DR}. \quad (2.104)$$

In DC ($q = 1$), this is calculated by counting the number of levels given by dividing FS by the minimum detectable signal value.

2.12.3   Is Dynamic Range the Maximum Signal-to-Noise Ratio?

From the definition of the dynamic range of (2.100), one could argue if DR is simply the maximum SNR of a system. To answer this question, let us take the same example used in Fig. 2.51 and rearranged in Fig. 2.53. Using the dB notation, as clearly shown, DR is the distance between the full scale (where we assume that the signal becomes distorted) and the noise floor (which is the power

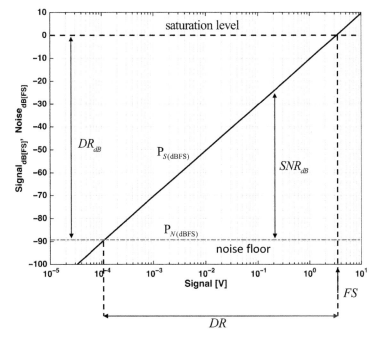

Figure 2.53 Comparison between *DR* and SNR in the case where the noise of the system is constant with respect to the input value. Note also the graphical expression of the input and output *DR*s.

of the noise surface). Since the noise power is constant, the SNR is incremented as long as the signal increases, until its maximum, where it is equal to the *DR*. Note how the power of the signal is proportional to the square of the signal; thus, the slope of the signal line is +20 dB/decade.

The case mentioned earlier is where the signal power is proportional to the square of the signal (as usual), and the noise interface power is constant:

$$P_S \propto x^2; \ P_N = k. \tag{2.105}$$

Of course, we assumed that the signal is noiseless, and the only source of noise is the interface itself.

However, other cases in which signal is intrinsically carrying noise power to be added with noise interface. Therefore, the noise power is in part proportional to the signal itself:

$$P_S \propto x^2; \ P_N = k_1 + k_2 x, \tag{2.106}$$

where k_1 and k_2 are constants. The first sets the noise floor of the system, while the second set the proportionality with the signal. The latter case is very common for noise affecting current signals, following Poisson's distribution processes. This is referred to as *current noise*. In Fig. 2.54, we present an example of this situation where a photosite (typically a photodiode) is creating a current photogenerated by the incoming light. The current is very weak, in the order of fAs or pAs, and due to the intrinsic quantization of the electric charge, the signal holds noise that is proportional to the signal itself, as we will analyze in Chapter 10. The maximum level of current generated by the photosite is that one considered to saturate the device.

Therefore, the output *DR* range allowed by this device is about 92 dB, which is the distance between the two dotted lines corresponding to the saturation level and the noise floor. Note that the noise floor is the noise power corresponding to the noise arising from the interface, that is, to very low signal levels, as in the *DR* definition. Note how the signal power is rising with +20 dB/decade as usual and the noise power with +10 dB/decade following (2.106).

However, the SNR is the distance between the noise power and the signal power, represented by solid lines in Fig. 2.54. Therefore, SNR depends on the signal strength, and the maximum value, corresponding to about 53 dB, is given by the maximum current for achieving saturation (~1 pA). In conclusion, *DR* is not, in this case, the maximum SNR of the system.

Another example of the difference between *DR* and SNR is in acoustic sensing and in microphone design, and it is due to a difference between references and not on the nature of noise. As already stated, following the relationship (2.96), the pressure of sound at the limit of hearing is $P_x = 20\ \mu\text{Pa} \rightarrow P_{x(\text{dB}_{\text{SPL}})} = 0\ \text{dB}_{\text{SPL}}$. Another reference in acoustic techniques is where the sound pressure is 1 Pascal, that is,

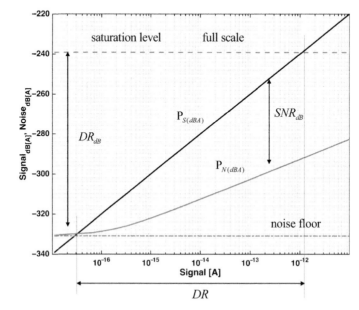

Figure 2.54 Graphic representation of the SNR and DR in a logarithmic plot in which the noise depends on the signal strength.

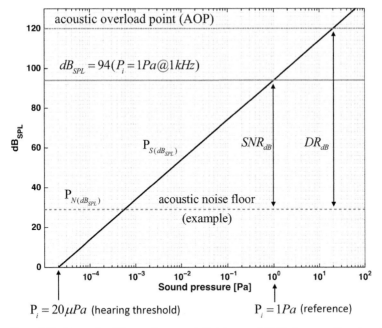

Figure 2.55 Example of dynamic range definitions and SNR for acoustic microphones.

$P_x = 1$ Pa$\rightarrow P_{x(\text{dB}_{\text{SPL}})} = 94$ dB$_{\text{SPL}}$. A scenario regarding the characteristics of a microphone is illustrated in Fig. 2.55. Using the above notation, the microphone is characterized as having a noise floor $P_{N(\text{dB}_{\text{SPL}})} = 29$ dB$_{\text{SPL}}$, which means that the signal power at the output of the microphone is below the power of noise below that level.

On the other hand, the microphone has physical limitations for high sound levels, and the maximum pressure before distortion (referred to as *acoustic overload point* (AOP)) in this case is given by $P_{S(\text{dB}_{\text{SPL}})} = 120$ dB$_{\text{SPL}}$; therefore, its dynamic range is $DR = 120 - 29 = 91$ dB$_{\text{SPL}}$. As long as the SNR is concerned, since the AOP is changing among devices, it is customary to use as a reference the sound pressure of 1 Pascal, that is, 94 dB$_{\text{SPL}}$. In conclusion, the SNR is SNR $= 94 - 29 = 65$ dB$_{\text{SPL}}$.

2.12.4 Relationship Between Signal-to-Noise Ratio and Dynamic Range for Defined Operating Ranges

We should point out some interesting characteristics for the *DR* requirements in DC measurements. If, for example, we have to make a DC measurement of an impedance between two values indicated by the application, $Z_{\text{min}}, Z_{\text{max}}$ and we want to get measurements with a required SNR (i.e., a required resolution), what should be the required DR of the interface?

The best way to analyze this is to look at the log-log plot of Fig. 2.56. The two extreme points of the full scale are indicated in the abscissa as x_{min} to x_{max}. Therefore, their distance in log space represents the ratio between those values: $x_{\text{max}}/x_{\text{min}}$. Now, if

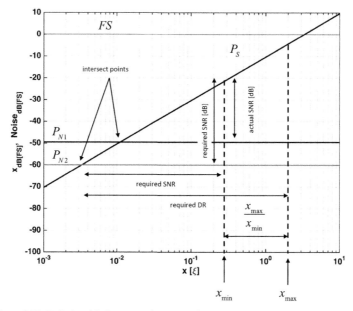

Figure 2.56 Relationship between the DR and the SNR of an interface sensing DC measurement.

we want to measure all values with a required SNR, our interface should detect the smaller value times SNR. Therefore, if the status of the noise floor of the interface is P_{N1}, the vertical distance between the noise floor and the signal evaluated at x_{min} is the worst case of SNR. Since the signal line slope is unitary in the [20 dB/dec] units, we can also see the SNR in the horizontal axis, between the intersect point and x_{min}. Note that the distance between the intersecting point and x_{max} is the DR of the system.

Now, assume that the actual SNR is smaller than the required one. Thus, we have to lower the noise of the interface from P_{N1} to P_{N2} increase the worst SNR and increase the required DR, as shown in Fig. 2.56. In mathematical expressions

$$DR_{dB} = 20\log\left(SNR \cdot \frac{x_{max}}{x_{min}}\right) = SNR_{dB} + 20\log\frac{x_{max}}{x_{min}}. \tag{2.107}$$

We will see that this relationship is very useful to achieve precision requirements of the interface, and, substituting SNR with SNDR, also accuracy requirements.

See Section 2.15.1 for the application of (2.107) in DC sensing with required system precision and accuracy.

2.13   Systematic Errors Due to Nonlinearity: The AC Approach

The DC characterization of errors is based on evaluating the difference between a reference characteristic and the real one using very slow variations of the input.

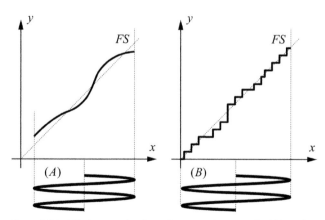

Figure 2.57 AC characterization of systematic errors in (A) analog transducers and (B) A/D converters.

Instead, the AC characterization of systematic errors due to nonlinearity is based on the concept of evaluating the "deformation" (or nonlinearity) of the characteristic measuring the output response in the frequency domain with respect to a sinusoidal input signal. That response is not dependent on the orientation or shift of the curve but only on its nonlinear characteristics—the more "deformation," the more distortion. AC characterization has the advantage of giving more insights into the system with respect to the DC one by also evaluating the distortion with respect to the input frequency and amplitude.

In AC characterization approaches, a deterministic sinusoidal input signal is applied to the systems, and its output is analyzed using fast Fourier transforms (FFTs), as shown in Fig. 2.57. The approach could be applied to both analog transducers as illustrated in Fig. 2.57A and to A/D converters, as shown in Fig. 2.57B. In the case of an A/D converter, the FFT is performed directly on the digital output. For the transducer, the output should be sampled and converted by an "ideal" (i.e., having a resolution much higher than nonidealities of the characteristic) A/D converter. The input signal could be variable in both amplitude (up to the *FS*) and frequency (up to the system's bandwidth). The idea is to analyze the power of the harmonics in order to characterize its distortion associated with systematic errors.

The FFT can be seen as sending a time signal through a bank of M filters, where the size of each frequency bin is $\Delta f = f_S/M$ (FFT resolution), where f_S is the sampling frequency. Since the signal is real, half of the spectrum could be discarded for symmetry, and the bandwidth covered by the plot is from DC to $f_S/2$. A typical example of an FFT plot is shown in Fig. 2.58: the levels in dB of the signal, the harmonics, and the noise floor provide a valuable and fast approach to evaluate SNR and distortion. However, the actual noise floor is not related to the noise floor shown in FFT plots, and it should be calculated on the number of bins M. In fact if we increase the number of frequency bins (i.e., the number of filters), the power of each of them is

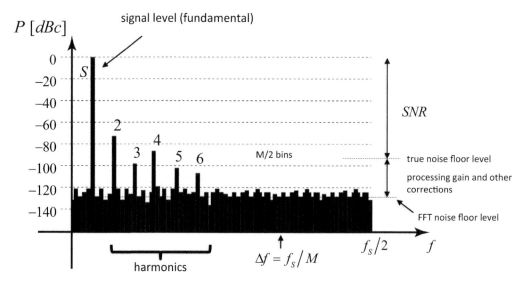

Figure 2.58   Power spectrum of the sensor output. The power of harmonics is measured in decibels relative to the carrier (dBc), meaning that the fundamental harmonic (carrier) power is used as a reference.

decreased due to noise averaging. This noise floor shift is called the *processing gain* of FFT, which is equal to $10\log(M/2)$.

Furthermore, other corrections are necessary for the windowing filter necessary for noncoherent sampling. Therefore, we should add several corrections (among which processing gain is the most important) to get the true noise floor, as shown in Fig. 2.58. The effect of noise gain is beneficial since we can lower the displayed noise floor to identify each harmonic's power contribution better.

In principle, if the real characteristic were linear, only the fundamental power response would be present. However, other power harmonic contributions are present due to systematic errors in the nonlinearities. One of the main figures of merit is called *total harmonic distortion* (THD), defined as

$$\text{THD}_{\text{dB}} = -10\log\frac{\text{signal power}}{\text{distortion power}} = -10\log\frac{P_S}{P_D} = -10\log\frac{P_S}{\sum_{i>1} P_i}, \qquad (2.108)$$

where $P_S = P_1$ is the power of the signal (fundamental) and $P_{i>1}$ are the powers of the harmonics.[16] It should be pointed out that the power of distortion calculated according to this AC method *depends on both the amplitude and frequency of the excitation signal*.

For slow excitation, we can see a convergence between the DC and the AC methods of characterization of systematic errors so that neglecting offset and gain errors,

---

[16] THD is also frequently expressed in percent: $\text{THD}_\% = P_D/P_S \cdot 100 \equiv [\%]$.

$$P_D \approx \sigma_D^2 \qquad (2.109)$$

Hint The FFT characterization of systematic errors *does not take into account offset and gain errors* because the harmonic power distribution is invariant with respect to gain and shift of the static characteristic; thus, we should add those in the total error budget calculation.

The FFT plot is also helpful in merging the systematic errors implied in distortion and the random errors causing the noise. For this purpose, it is defined the *signal-to-noise-and-distortion ratio* (*SNDR or SINAD*)

$$\text{SNDR}_{dB} = 10\log\frac{\text{signal power}}{\text{noise power} + \text{distortion power}} = 10\log\frac{P_S}{P_N + P_D}. \qquad (2.110)$$

The combination of random noise power and distortion power is possible because the related errors are considered uncorrelated, and thus they could be summed-up in power. It is easy to show that

$$10^{-\text{SNDR}_{dB}/10} = 10^{-\text{SNR}_{dB}/10} + 10^{\text{THD}_{dB}/10} \qquad (2.111)$$

Note THD and SNDR could be characterized versus either input amplitude or frequency. Usually, distortion is higher for larger input values and frequency, so these functions are usually monotonically decreasing functions.

Example An FFT power plot of a transducer performed over $M = 8192$ samples shows a noise floor of -102 dBc. The power of the second, third, and sixth harmonics are H2 $= -65.95$ dBc, H3 $= -78.38$ dBc, H6 $= -90.94$ dBc, respectively. Calculate the SNR and the SNDR.

The true noise floor could be calculated with a processing gain correction of $10\log(M/2) = 36$ dB. We do consider other corrections due to windowing negligible. Therefore, the actual noise floor is given by $\text{SNR}_{dBc} = -102$ dBc $+ 36$ dBc $= -66$ dBc. The total power of harmonics is $P_{D(dBc)} = 10\log(10^{(H1)/10} + 10^{(H3)/10} + 10^{(H6)/10}) = -65.7$ dBc $= \text{THD}_{dBc}$ while for the signal to noise and distortion ratios, it is $\text{SNDR}_{dBc} = 10\log(10^{(\text{SNR}_{dBc})/10} + 10^{(\text{THD}_{dBc})/10}) = -62.8$ dB.

[Note: In this example, we used as reference the signal for using dBc; thus, we should take into account a minus in front of each expression.]

The SNDR definition could be further expanded considering all the possible powers of random and systematic errors

$$\text{SNDRT}_{\text{dB}} = 10\log\frac{P_S}{P_N + P_D + P_{sys}}, \tag{2.112}$$

where P_{sys} is the sum of the powers of other systematic errors

$$P_{sys} = \sum_{j=1}^{J}\sigma_{\alpha_j}^2 + \Delta x_{E(OG)}^2 + \dots, \tag{2.113}$$

where $\sum_{j=1}^{J}\sigma_{\alpha_j}^2$ is the sum of variances of all influence parameters (i.e., change of temperature, parameter variations, etc.) and $\Delta x_{E(OG)}^2$ is the square of the worst case of offset and gain error. Note that the denominator of (2.112) is the power of the *combined uncertainty* at the input. To summarize, we could use a mixed approach, where the nonlinearities are characterized by means of distortion with AC testing, while temperature and device production distributions are characterized by DC procedures or other means.

2.14   The Quantization Process

The quantization process is essential in an electronic sensing system because the data elaboration (either simple or complex) is performed in the digital environment.

In the early stages of sensing systems, the quantization process is usually performed by A/D converters transforming an analog value (a real number) into a discrete value (natural number) by using a *rounding function*. The quantization process introduces a systematic error called *quantization error* or *quantization noise*.

Both analog and quantization noises are present at the interface level, as shown in Fig. 2.59, where the analog noise is represented by a bell-shaped *probability density function* (PDF) model. For the moment, we will assume that the analog noise is due only to the interface and the A/D converter is noiseless and therefore affected only by quantization noise. Therefore, the measure is affected by both the analog output noise and the A/D converter quantization noise at the transduction chain's output. How quantization and random noises are related together to define the measurement's combined uncertainty? We will make the discussion starting from a DC measurement.

To understand this, we have to recall the estimation concept following Fig. 2.60 that shows an example of the relationship between the analog noise model, represented by a Gaussian PDF, at the input of the quantizer and the discrete level assignment among three adjacent binary levels, in this example "001," "010," "011." The noise model has been characterized from either the experimental or the theoretical perspective. We plan to store a set of samples from the A/D converter in a trial and collect them into a histogram. If we repeat other trials and collect the datasets, the histograms will be different every time due to the random process at the quantizer input. Therefore, if we want to estimate the expected value using the mean of the samples, we are subject to an error that is accounted for in the coverage interval of the uncertainty (related to the standard deviation of the model) as already discussed.

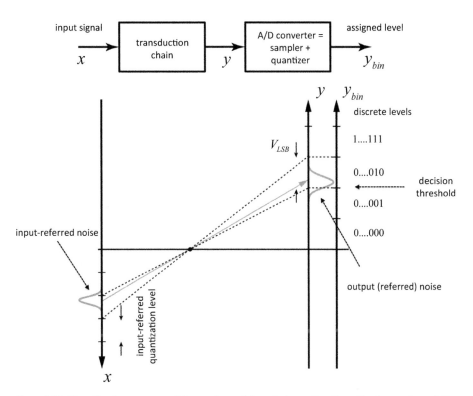

Figure 2.59 Quantization process. The main problem is how the discretization noise of the quantizer should be combined with the random noise of the transducer for the general characterization of the uncertainty.

We can determine the exact value without any error only in two cases. The first is where the value of the histogram's bin equals the area covered by the PDF function for each level; however, this occurs with a very small probability among all the trials. The second is if we collect an infinite number of samples so that the empirical characteristics of the distribution tend toward the PDF of the model.

Now we take two cases where the amount of noise is quite different, as shown in Fig. 2.61. In the first case, the analog noise spreads on several quantizing levels, while in the second case, it is almost confined to a single discretizing level. In the example of Fig. 2.61A, we have most of the cases to get 010 and fair cases to get also 001 and 011 values. Instead, for the noise model of Fig. 2.61B, almost all samples fall in the 010 division.

We want to evaluate the resolution of the system collecting more samples and change the A/D input through a signal variation $\Delta y = y_2 - y_1$ smaller than the quantization level. In the case of Fig. 2.61A, the noise spread is higher than the quantization level, and we take two sets of samples corresponding to the two input values. The two histograms of Fig. 2.61A will allow an estimation of the two actual values with some degree of confidence since all three bins will change with respect to the input variation.

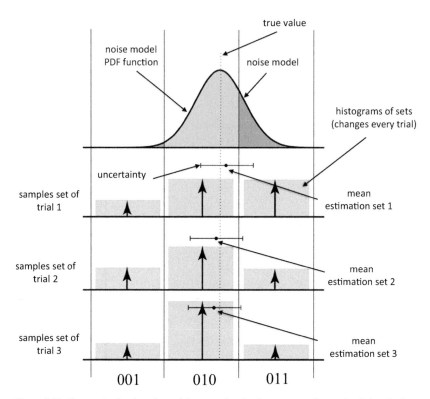

Figure 2.60 Concept of estimation of the actual value by means of quantized data in the presence of random analog noise. We can collect a certain number of samples from a trial to construct an *output code histogram.* Every time we repeat the trial, the histogram would be different.

Of course, the ability to sense that difference correctly depends on the amount of data collected. On the other hand, if the noise spread is smaller than the quantization level, as shown in Fig. 2.61B, even if we have the same A/D input variation, the histogram is the same, made of just one bin. In this case, differently from the previous one, we can collect as much data as we want, but the inability to sense the variation (smaller than a quantization level) still holds. Conversely, in the case with higher noise (Fig. 2.61A), if we collect a sufficient amount of data, the resolution is better with respect to the case with lower noise (Fig. 2.61B).

To better study the relationships between errors, we introduce a factor k relating the two errors

$$V_{\mathrm{LSB}} = k \cdot \sigma_N, \tag{2.114}$$

where the standard deviation is referred to the analog noise at the input of the quantizer, that is, $\sigma_N = \sigma_N(y)$.

Two cases with different values of k are illustrated in the *output code histograms* of Fig. 2.62, where an analog noise Gaussian time evolution (continuous line) is sampled

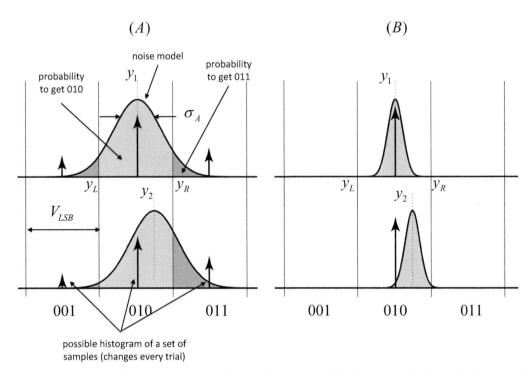

(A) (B)

possible histogram of a set of
samples (changes every trial)

Figure 2.61 Assignment of levels in a noisy system. (A) A Gaussian distribution noise is centered in two analog values y_1 and y_2 within the discrete level coded by the 010 binary value. (B) The same analog values but with one-half of the standard deviation in the noise model than before.

by a quantizer (circle dots). Each circle's position is placed in the level corresponding to the quantization law illustrated in Fig. 2.17. On the right side of the plots are illustrated the relative frequency of samples into a histogram.

Note that in Fig. 2.62 (top), we have a 16b A/D converter that is encoding on $65536 = 2^{16}$ levels with a reference voltage $V_R = 1.2$ V resulting in a $V_{LSB} = 1.2/2^{16} = 18.2\mu V$. The analog test DC input is $y = 0.633$ V and is affected by noise with $\sigma = 56.5\mu V$. Therefore, the amount of noise is more than three times the size of the discrete level and $k = 0.32$. Without noise, the right level should be assigned to the value $Round(0.633 \cdot 2^{16}/1.2) = Round(34570.24) = 34570$. However, due to noise, the assignment is spread around adjacent levels. The noise is scattering around about 16 levels, equivalent to about 4 bits. This means that we should expect that some of the least significant bits are always changing their conditions. These bits are referred to as *noisy (or toggling) bits*.

Conversely, in Fig. 2.62 (bottom), we have the same A/D converter, but the noise level is $\sigma = 5.65\mu V$, that is, 10 times smaller than before. In this case, $k = 3.2$ and the assigned levels are spread on about three levels. Note how the 34,570th level is more populated than the 34,571st level because the first one is closer than the second one to the input level. This situation is similar to the case described in Fig. 2.61.

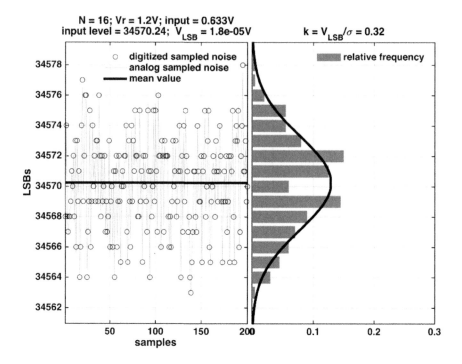

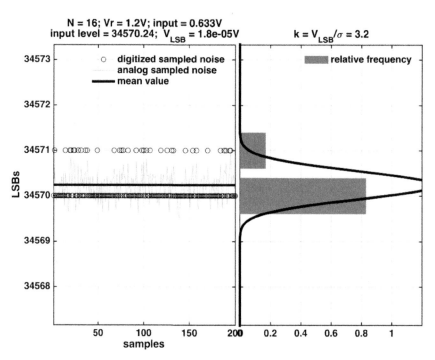

Figure 2.62 Output code histograms. Quantization process where V_{LSB} is one half of the standard deviation of the noise of the analog value (top) and three times the standard deviation (bottom).

2.14.1   Composition of Random and Quantization Noise and Dithering

The main task of this section is to understand the best approach to evaluate uncertainty in the presence of both random and quantization noise. To do so, we have to take a step back and look at the mathematical description of quantization's statistical implications on a generic random signal, as shown in Fig. 2.63.

In Fig. 2.63A, a generic random signal is described by the PDF $f(x)$ at the input of a quantizer. As we have seen, the output PDF $f(y)$ is a Dirac's comb where each arrow represents the probability for the output to occupy a quantization value (in the figure, we have assumed a very large amount of acquisition data).

Now, the idea is to use the tools already employed for combining noise sources. Therefore, we analyze the scheme of Fig. 2.63B, where the input PDF is added to an *uncorrelated* uniform PDF of a quantization level. This second scheme is called the *pseudo-quantization noise* (PQN) model. The output PDF is $f(x+q)$, and it is a continuous density function given by the *convolution* of the input components; therefore, the two functions $f(y)$ and $f(x+q)$ are discrete and continuous PDFs, respectively. However, if the relationship between the two PDFs satisfies the *first and second Widrow quantizing theorems*, all the moments of the two PDFs are equal. Therefore we can use the PQN model to infer the statistical characteristics of the input by using the sum of contributions as already used in noise analysis. In other words, if the input is a signal satisfying such theorems, we can fully reconstruct the input PDF from the quantized output and use *averaging to reduce uncertainty*

When we apply the PQN model to random noise, we consider $f(x)$ as the PDF of an input random noise. This means that, as far as the theorems hold, the two models of Fig. 2.63 are statistically equivalent, and *we can combine quantization noise with random noise quadratically as independent contributions*. When are quantization theorems satisfied? In our specific case, assuming uniform PSD, Gaussian noise input, it can be shown that if $k < 2$ (i.e., random noise standard deviation is higher

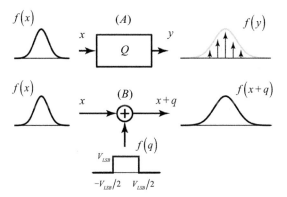

Figure 2.63 The pseudo quantization noise (PQN) model. If $k < 2$, the effect of the quantizer could be modeled by a uniform distribution between $\pm V_{LSB}/2$ that is added in power to the input.

than quantization level), we can reasonably satisfy quantization theorems, and therefore we can use the PQN model.

Conversely, if quantization theorems are not satisfied, the overall behavior is that one given by an ideal A/D converter whose uncertainty is $u_Q = V_{LSB}/\sqrt{12}$. Therefore, if the quantizing theorems do not hold, from the uncertainty point of view, we can say that when the noise is small compared to quantization, the uncertainty is related to that of an ideal A/D converter.

To summarize, the *total* (or *generalized*) *uncertainty* for a single sample at the input of the interface/quantizer could be calculated by the relationship

$$u(y) = \sqrt{u_N^2 + u_Q^2}, \tag{2.115}$$

where $u_N = \sigma_N(y)$ is the uncertainty of the analog noise and $u_Q = V_{LSB}/\sqrt{12}$ is the uncertainty due to quantization. If we use the average of K repetitive samples, we can reduce the uncertainty as

$$u(y) = \frac{\sqrt{u_N^2 + u_Q^2}}{\sqrt{K}}; \quad \text{for } k < 2, \tag{2.116}$$

but this is valid *only if the PQN model holds*, that is, $k < 2$.

The spread of noise around adjacent levels implies the increase of "noisy bits" with the number of the total bits of the A/D converter: the smaller the k, the larger number of noisy bits. How can we handle noisy bits? We have two possible approaches:

- Truncate noisy bits. This also means using a quantizer with a lower number of bits. This approach allows saving power consumption but has a limited resolution.
- Reduce noisy bits by averaging multiple samples. This approach needs more power consumption and computational resources but allows greater resolutions.

The two approaches are illustrated in Fig. 2.64. In Fig. 2.64A, we have the case of low noise whose distribution is contained in a quantizing level. Almost all the single samples fall in the discrete level, and averaging the output of the A/D converter is useless. The quantization error limits the uncertainty according to (2.115). In Fig. 2.64B is shown the case where the noise is larger than the quantization level. The PQN model holds that uncertainty is more significant than in the first case. However, we can use averaging to have a better resolution and lower uncertainty using (2.116). Therefore, using averages, we could achieve better resolutions than the first case.

On the other hand, we can choose to truncate 1 bit as in Fig. 2.64C. In this case, the size of the discretization level doubles, and the quantization error becomes the same size as the spread of analog noise. Therefore, the PQN model is not applicable, and we cannot average output values. If we further truncate 1 other bit as in Fig. 2.64D, we have a situation similar to the one of Fig. 2.64A but with a quantization error that is four times higher.

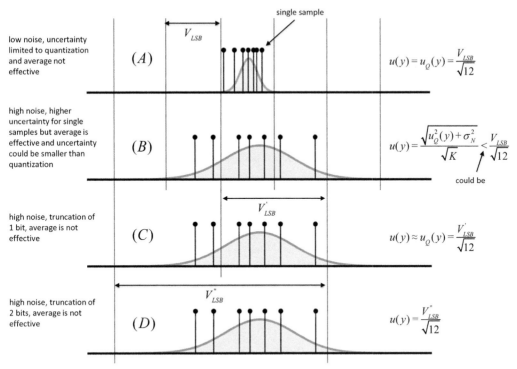

single sample

low noise, uncertainty limited to quantization and average not effective

(A)

$$u(y)=u_Q(y)=\frac{V_{LSB}}{\sqrt{12}}$$

high noise, higher uncertainty for single samples but average is effective and uncertainty could be smaller than quantization

(B)

$$u(y)=\frac{\sqrt{u_Q^2(y)+\sigma_N^2}}{\sqrt{K}}\underset{\text{could be}}{<}\frac{V_{LSB}}{\sqrt{12}}$$

high noise, truncation of 1 bit, average is not effective

(C)

$$u(y)\approx u_Q(y)=\frac{V'_{LSB}}{\sqrt{12}}$$

high noise, truncation of 2 bits, average is not effective

(D)

$$u(y)=\frac{V''_{LSB}}{\sqrt{12}}$$

Figure 2.64 Effect of truncation of least significant bits. (A) Low-noise, single samples are contained in a quantization level. The average is not effective, and uncertainty could not be smaller than quantization. (B) High noise; the single sample is spread around, but averaging could decrease uncertainty below quantization. (C) Truncation of 1 bit; the uncertainty returns to being limited by quantization, and averaging is not effective. (D) Truncation of 2 bits.

The effect of quantization on a noisy input with variable quantizing levels is also illustrated in Fig. 2.65A–C, where 500 samples of an input characterized by the same noise level are digitized with a variable quantizer resolution. As we can see, the bins' spread is almost constant since it is due to the same level of noise; however, the distribution is changing according to the number of bits of the A/D converter. This implies a different number of "noisy bits" of the code, which are frequently and randomly changing due to noise. In Fig. 2.65A, the digitized output of a 12b+ sign bits A/D converter where $V_R = 1.2$ V is shown. Therefore, $V_{LSB} = 1.2/2^{12} = 0.29$ mV. A DC signal of 0.932 V is applied to the input, but it is corrupted with a Gaussian noise characterized by a standard deviation of 1.26 mV so that $k = 0.23$. There are about 4.8 noisy bits due to the frequent transitions between levels, as clearly shown. If we take a single sample, we could have an error as large as 28 levels due to analog noise. The estimation of the input could be increased by averaging 500 samples. In doing so, we expect to reduce the error by a factor $\sim \sqrt{500} = 22.3$. However, since averaging is expensive in computational resources, the idea is to save power by truncating the last 3 bits as shown in Fig. 2.65B so that we have $k = 1.9$. If we decide to

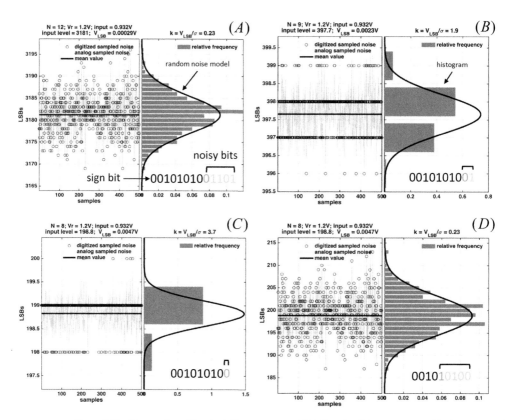

Figure 2.65 Truncation effect in A/D converters. (A) 12b +sign bits converter with about 4.5 noisy bits. (B) Truncation of the 2 least significant bits from the previous example in a 9b+sign converter. (C) Truncation of 1 more bit, so it behaves as an 8b+sign converter. (D) Dithering effect of the previous example by increasing the analog noise level.

cut off another bit, as shown in Fig. 2.65C, we can achieve $k = 3.7$. However, we should be careful because if we reduce the number of bits, the quantization error becomes larger than the standard deviation of analog noise, and the resolution approaches that one of an A/D converter of 8 bits. The main drawback of this situation is that averaging is no longer effective in estimating the true value, and we cannot eliminate a systematic quantization error.

If noise is small enough with respect to quantization noise (as in the truncation process), it makes averaging useless. This effect is better illustrated in Fig. 2.66A, where we collected several trials of the average of 500 sampled outputs with variable quantizer resolution. Then, the average is compared to the true value to get the estimation error. We can see that the estimation error is much reduced with respect to a single sample (by a factor equal to the square root of the number of samples), and the result is almost the same using resolutions of either 14b or 12b or 9b. It is consistent with the relationship (2.115), where the analog noise gives the uncertainty and not by the quantization one. As

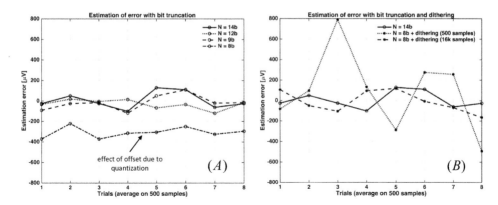

Figure 2.66 Repetitive estimation errors for (A) progressive bit truncation from 14b to 8b and (B) with the addition of 16x dithering noise and increasing of the average samples to 16,000 points. The noise data are the same as those of Fig. 2.65, where $V_{LSB} = V_R/2^N$ and $\sigma = 1.26$ mV. The reader could check that (2.116) holds, considering the reduction of $\sqrt{K}$ where K is the number of averages.

a first conclusion, we can say that it could be redundant (and power-hungry) to increase the number of bits of the A/D converter for a given average and required accuracy.

On the other hand, if we keep decreasing up to 8b, we have that $k > 2$, and averaging does not reduce the uncertainty, as clearly visible in Fig. 2.66A. This amount is strictly dependent on the input, and in the worst case, it is equal to the least significant bit voltage.

A strategy to avoid this is to artificially increase the noise, keeping a reduced number of bits of the A/D converter as shown in Fig. 2.65D, where the noise standard deviation is incremented 16 times to 20.2 mV on an 8b quantizer so that $k = 0.23$ again as in the first example. The result given by averaging 500 samples is shown in Fig. 2.66B showing an increase of the error among single trials. This fact suggests increasing the averaging: if we further increase the number of samples to average, up to 16,000, we get an error, which is similar to that of using 14b. This approach is referred to as *dithering*.[17] Of course, dithering requires reducing the signal bandwidth or an oversampling.

To be more explicit about the dithering effect for DC signals, we have represented the approach in Fig. 2.67, using two cases. In the first one of Fig. 2.67A, the analog noise is lower than that quantization error ($k > 2$). The result of the A/D conversion is that we have a constant offset, and applying average to the output does not increase the estimation since it is always the same.

Conversely, as shown in Fig. 2.67B, when the analog noise is greater than that of the quantization step ($k < 2$), the output sample distribution might better estimate the true

[17] As already pointed out, averaging consumes power and reduces bandwidth. Among dithering approaches, there are techniques such as oversampling signal and/or introducing noise with spectral components outside the signal bandwidth.

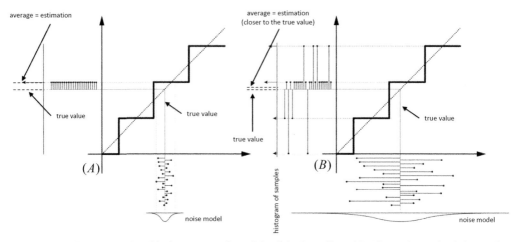

Figure 2.67 Graphical representation of the dithering effect. (A) The analog noise is lower than that of the quantization step, and the output PMF is a pure offset. (B) The analog noise is greater than that of the quantization step. In this case, the output PMF could better estimate the true value due to the higher spread.

value due to a broader spread of values. Note that this representation is another way to look at the situation shown in Fig. 2.61.

To summarize:

- If the standard deviation of analog noise is sufficiently greater than that of quantization, uncertainty could be combined with quantization as an uncorrelated noise. The uncertainty could be reduced by averaging.
- We can reduce the number of bits N as much as PQN holds to get rid of redundant least significant bits. Of course, we should check the uncertainty by means of (2.115).
- If analog noise is smaller than quantization, PQN does not hold, and we have a fixed bias of the estimator (depending on the input) that *could not be reduced by averaging*. Therefore, to avoid this, we can increase the noise artificially by the *dithering* technique so as to apply averaging. Dithering plus averaging could achieve results similar to that at higher A/D resolutions but at the cost of bandwidth reduction and overall power consumption.

To conclude, we can show the modeling of analog and quantization noise in behavioral blocks, as shown in Fig. 2.68. The quantizer could be modeled as a summing node of quantization noise, according to the PQN model. Then the two noises could be summed together and referred to the input by means of the sensitivity.

Example The output of an amplifier with $G = 1000$ has a noise of $\sigma_N(y) = 30$ μV rms, and it is interfaced with a noiseless A/D 16b converter with $V_R = 3.3$ V. The input FS is $3.3/1000 = 3.3$ mV. Therefore, $V_{LSB} = 3.3/2^{16} = 50.4$ μV and $k = 1.68 < 2$. Therefore, we can apply the PQN model. The total input noise at the input of the A/D converter is $\sigma(y) = \sqrt{\sigma_N^2 + V_{LSB}^2/12} = 33.3$ μV, and the input-referred noise at the

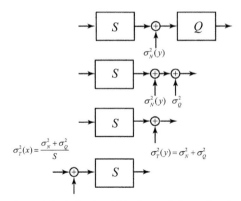

Figure 2.68 Behavioral blocks modeling of analog and quantization noise.

input of the amplifier is $\sigma(x) = 33.3 \times 10^{-6}/1000 = 33$ nV. Now, if we average the output by a low-pass filter in 50 samples, we have an input-referred resolution of $33/\sqrt{50} = 4.7$ nV. However, we should take into account higher power consumption and reduction of the bandwidth required by averaging.

2.14.2   DC Resolution in A/D Converters

A/D converters are very complex systems that might have intrinsic internal random noise that can be referred to the input and affect the quantization process. In this section, we will refer to A/D alone; therefore, *the analog noise is referred to the internal one of the converter*. In general, we will assume that noise standard deviation is always greater than quantization, that is $\sigma_N > V_{\mathrm{LSB}}$, ensuring the PQN model.

Therefore, the following RTI resolutions $u(y)$ at the input y of the A/D converter are defined:

- *RTI nominal resolution* is the maximum error of the quantizer without taking into account analog random noise.
- *RTI effective resolution* is the resolution defined by the rms value of the noise; in this case, it is greater than quantizing levels.
- *RTI noise-free resolution* is the resolution defined by the peak-to-peak value of noise, given by $p = 3.3$ (confidence interval equal to 6.6 noise standard deviations) so that we get 99.9% of values within the uncertainty boundaries.

These definitions could be summarized by the following for $\sigma_N > V_{\mathrm{LSB}}$:

$$
\begin{aligned}
\text{RTI nominal resolution} &\quad\leftarrow\quad u(y) = V_{\mathrm{LSB}} \\
\text{RTI effective resolution} &\quad\leftarrow\quad u(y) = \sigma_N; \quad (\sigma_N > V_{\mathrm{LSB}}) \\
\text{RTI noise-free resolution} &\quad\leftarrow\quad u(y) = 6.6\sigma_N; \quad (\sigma_N > V_{\mathrm{LSB}}).
\end{aligned}
\tag{2.117}
$$

Note for $u(y)$ it is *common usage* not using (2.115) even if this is not strictly correct, especially for the effective resolution.

Example A 16b A/D converter with $FS = V_R = 3.3$ V is affected by an input-referred noise of 88 μV rms. Its *RTI nominal resolution* is $3.3/2^{16} = 50.4$ μV, the *RTI effective resolution* is 88 μV, and the *RTI noise-free resolution* is 581 μV. Note that if we had half random noise of 44 μV the definition of the effective resolution would not apply since it is lower than the nominal one ($\sigma_N < V_{LSB}$).

Again, another way to express the resolution is to compare it with the full scale, that is, with the reference voltage, and then expressed it in bits

$$\text{resolution(s)}_b = \log_2\left(\frac{FS}{u(y)}\right) = \log_2\left(\frac{FS}{q \cdot 2p \cdot \sigma_N}\right)$$

$$\text{where}: p = \begin{cases} 1/2 & \text{for effective resolution} \\ 3.3 & \text{for noise-free resolution} \end{cases}. \tag{2.118}$$

Example For the same A/D converter of above, its *nominal resolution* is $\log_2(2^{16}) = 16$b, the *effective resolution* is $\log_2(3.3/88 \times 10^{-6}) = 15.2$b, and the *noise-free resolution* is $\log_2(3.3/(6.6 \cdot 88 \times 10^{-6})) = 12.5$b.

Note that if we use a coverage factor equal to 3.3, the *noise-free resolution* levels are about 2.7b less than *effective resolution* levels since

$$\log_2(1/6.6) = 2.72\text{b}. \tag{2.119}$$

2.14.3   AC Characterization of A/D Converters by Effective Number of Bits

Similar to analog systems, the DR of ideal A/D converters can be calculated on the maximum signal power (limited by FS) and the quantization noise *as the only source of noise*

$$\begin{aligned} DR_{dB} &= 20\log\frac{2^N V_{LSB}}{q \cdot V_{LSB}/\sqrt{12}} \\ &= N \cdot 20\log(2) + 20\log(\sqrt{12}) - 20\log(q) \\ &= N \cdot 6.02 + 10.79 - 20\log(q), \end{aligned} \tag{2.120}$$

where q is the power factor, useful to quantify the maximum signal power based on the kind of signal. In the case we use as reference a *sinusoidal waveform* ($q = 2\sqrt{2}$), we have

$$DR_{dB} = N \cdot 6.02 + 1.76. \tag{2.121}$$

We can reverse this relationship to define the *effective number of bits* (ENOB) as

$$N = \text{ENOB} = \frac{DR_{\text{dB}} - 1.76}{6.02}. \tag{2.122}$$

This definition could be better explained as follows: ENOB is *the number of bits of a virtual and ideal A/D converter (thus limited by only quantization noise) matching a given DR*. It should be added that this definition is based on calculating *DR* using AC sinusoidal waveforms. In the case we use different signals, we should take this into account using (2.120). To better understand the above concept, we will provide an example.

Example We have already shown that human hearing is characterized by 130 dB of dynamic range. Therefore, the effective number of bits is ENOB = (130 − 1.76)/6.02 = 21.3b. This means that we can match the *DR* of hearing with an ideal (i.e., characterized by only quantization noise) of 22b. Thus, we are induced to conclude that if we want to use an A/D converter for audio purposes, we have to use *at least* an *ideal* A/D converter with 22b. We will see that the term "at least" is important for two reasons. First, the A/D converter is limited not only by quantization noise but also by random noise. Second, we have to consider the *resolution rule of acquisition chains*, as will be discussed in Chapter 3.

Unfortunately, there are no ideal A/D converters. Considering an A/D converter also affected by random noise, we can use in place of the *DR* of (2.122), limited by only quantization noise, the experimentally characterized maximum SNR of the real A/D

$$\text{ENOB}_{\text{SNR}} = \frac{\text{SNR}_{\text{dB(max)}} - 1.76}{6.02}. \tag{2.123}$$

This relationship could be better explained by looking at Fig. 2.69. Assume that we have a noisy A/D converter of N bits and that noise exceeds the quantization level V'_{LSB}. Now the idea is to equal this situation in terms of SNR to a noiseless (i.e., limited by only quantization noise) A/D converter with coarser quantization levels, as shown in Fig. 2.69. Therefore, the original 2^N levels are reduced into 2^{ENOB} discrete levels.

Comparing the two SNRs, we have

$$\text{SNR}_{\text{dB(max)}} \text{of a noisy A/D converter} = \text{SNR}_{\text{dB(max)}} \text{of a ideal A/D converter}$$

$$= 20\log\left(\frac{2^N V'_{\text{LSB}}}{q \cdot \sigma_N}\right) = 20\log\left(\frac{2^{\text{ENOB}} V_{\text{LSB}}}{q \cdot V_{\text{LSB}}/\sqrt{12}}\right)$$

$$\rightarrow 2^N = 2^{\text{ENOB}} \frac{\sqrt{12}\sigma_N}{V'_{\text{LSB}}}, \tag{2.124}$$

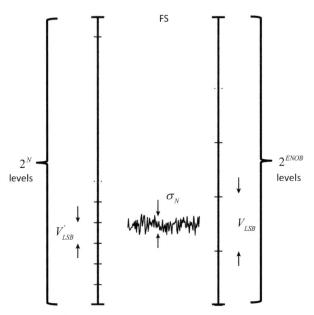

Figure 2.69 Concept of ENOB for A/D converters.

where V'_{LSB} is the least significant input voltage of the nominal N bit converter, while V_{LSB} is the one of an ideal converter of ENOB bits. Of course, the relationship is valid if $\sigma_N \geq V'_{\text{LSB}}/\sqrt{12}$ and we have used ENOB as a notation of ENOB_{SNR} for simplicity.

Therefore, the ENOB of a real A/D converter is *the number of bits of an ideal A/D converter (limited only by quantization noise) whose SNR equals that of the real one*, and it is given by making the substitution

$$N \leftrightarrow \text{ENOB}; \quad \sigma_N \leftrightarrow \frac{V_{\text{LSB}}}{\sqrt{12}}. \tag{2.125}$$

It is important to underline that this definition is based on the assumption that the noise is greater than quantization; otherwise, obviously, for a noiseless converter, $\text{ENOB} = N$.

It is easy to show from (2.124) that the relationship between the nominal bits N and ENOB in a noisy A/D converter is

$$N = \text{ENOB} + \log_2\left(\frac{\sigma_N \cdot \sqrt{12}}{V_{\text{LSB}}}\right). \tag{2.126}$$

Therefore, since usually, random noise is more significant than quantization noise, ENOB is smaller than N. Following the previous example, if we want to cope with audio applications with noisy A/D converters, for example, having $N = 24\text{b}$, we have to ensure that at least it offers $\text{ENOB} = 22\text{b}$.

Furthermore, the same relationship could be used in an even more generic way, also considering the distortion of the A/D converter. Therefore, we can calculate its signal-to-noise-and-distortion ratio (SNDR or SINAD) according to (2.110) using AC tools such as FFT spectra and substitute in (2.122) in place of SNR the SNDR

$$\text{ENOB}_{\text{SNDR}} = \frac{\text{SNDR}_{\text{dB(max)}} - 1.76}{6.02}, \tag{2.127}$$

where in the calculation of the SNDR, we used the generalized uncertainty $u(y)$, including distortion errors. We also used $\text{ENOB}_{\text{SNDR}}$ to distinguish those from ENOB_{SNR}, which is calculated based on SNR only. The preceding relationship is the most used approach to characterize the real A/D converters' characteristics. In other words, SNDR encodes both random and systematic errors that are translated into a certain number of least significant either *noisy bits* (random errors) or *distortion bits* (systematic errors) that should be distinguished from the number of bits that are used to encode the information.

It should be pointed out that SNDR is changing with respect to both the amplitude and the frequency of the input. Thus, ENOB is changing (usually decreasing) with respect to incremental values of these parameters as well.

Example A 16b A/D converter with $V_R = 5$ V is tested using an input DC value. The output code histogram shows a standard deviation due to noise of 0.78b that is equal to $2^{(-0.78)} V_{\text{LSB}} = 0.58\ V_{\text{LSB}}$. This corresponds to an input-referred thermal noise of $0.58 \cdot 5/2^{16} = \sigma_N = 44.5$ μVrms. The ratio k of $V_{\text{LSB}} = 5/2^{16} = 76.4$ μV to the noise is below 2; thus, we can combine the two noises in power: $\sqrt{(44.5 \times 10^{-6})^2}$ $+(74.6 \times 10^{-6})^2/12 = 49.6$ μV. For a sinusoidal input, this corresponds to SNR $= 20\log(5/((49.6 \times 10^{-6}) \cdot 2\sqrt{2})) = 91.0$ dB. This value is about the same value of 91.9 dB given by the experimental spectrum of low-frequency FFTs. The same spectrum shows a THD $= -106.1$ dB calculated using (2.108). Thus, the SNDR could be calculated using (2.111): $\text{SNDR}_{\text{dB}} = -10\log(10^{-91.9/10} + 10^{-106.1/10}) = 91.7$ dB. Thus, the effective number of bits is ENOB $= (91.7 - 1.76)/6.02 = 14.94$b. Note that the use of SNDR in place of the SNR has a weak effect on the ENOB, meaning that distortion has a little (but not negligible) effect with respect to ENOB compared to noise.

In conclusion, we have more than one noisy/distortion bit out of 16. This situation changes as the testing frequency increases and could be as low as ENOB $= 13.5$b at the limit of the A/D bandwidth. Note also that (2.126) holds. [example extrapolated from ADAQ7980 datasheet by Analog Devices].

The maximum SNR calculation could also be expressed in terms of the number of discretized levels $K = V_R/V_{\text{LSB}}$. In fact, the calculation of the dynamic range is for a sinusoidal input signal ($q = 2\sqrt{2}$)

$$\mathrm{SNR}_{(\mathrm{max})} = \frac{V_R^2}{q(V_R^2/12K^2)} = \frac{4}{3}K^2 \ \text{and}$$

$$\mathrm{SNR}_{\mathrm{dB(max)}} = 20\log(K) + 1.25. \tag{2.128}$$

2.14.4   Relationship Between Resolution and Effective Number of Bits

The idea is to use the concept of ENOB not only for characterizing A/D converters but also in the analog domain to quantify its resolution levels. Thus, we equate the generic expression of the resolution with that one of an ideal A/D of ENOB bits

$$\mathrm{resolution(s)}_b = \frac{1}{2}\log_2\left(\frac{FS}{q \cdot 2p \cdot \sigma_N}\right) = \frac{1}{2}\log_2(DR) = \frac{1}{2}\log_2(DR_{A/D}) =$$

$$= \frac{1}{2}\log_2\left(\frac{2^{\mathrm{ENOB}}V_{\mathrm{LSB}}}{2\sqrt{2} \cdot 2p \cdot V_{\mathrm{LSB}}/\sqrt{12}}\right) = \mathrm{ENOB} + \log_2\left(\sqrt{\frac{3}{2}}\frac{1}{2p}\right), \tag{2.129}$$

where q is the power factor used in the definition of resolution (i.e., $q = 1$ in the case the resolution is calculated in DC). As used in the definition of ENOB, we have used ENOB as a notation of $\mathrm{ENOB}_{\mathrm{SNR}}$ for simplicity.

Therefore, using (2.129), with $p = 3.3$, we have

$$\mathrm{noise-free\,resolution}_b = \mathrm{ENOB}_{\mathrm{SNR}} + 2.43 \tag{2.130}$$

while for $p = 0.5$

$$\mathrm{effective\,resolution}_b = NL_b = \mathrm{ENOB}_{\mathrm{SNR}} + 0.29 \tag{2.131}$$

or equivalently

$$2^{NL_b} = 2^{\mathrm{ENOB}} \cdot \frac{\sqrt{12}}{2\sqrt{2}} = 2^{\mathrm{ENOB}} \cdot \sqrt{\frac{3}{2}}. \tag{2.132}$$

Note that (2.132) and (2.131) are consistent with (2.119).

Hint For any DR derived using ENOB according to (2.123) and similar, assume that DR is calculated in an AC sinusoidal case with a power factor $2\sqrt{2}$. However, when this "A/D like" dynamic range is equated to the DR used for defining the resolution in the analog domain as in (2.129), in the latter one, we should use the q appropriate to the definition of resolution ($q = 1$ for DC and $q = 2\sqrt{2}$ for sinusoidal AC). In other terms, if we need to derive σ_N from the dynamic range used for the definition of the resolution (and where $DR = DR_{A/D}$ according to (2.129)), we have to use the appropriate q factor $\mathrm{resolution}(s) \leftrightarrow^q DR = DR_{A/D} \overset{q=2\sqrt{2}}{\leftrightarrow} \mathrm{ENOB}$.

Example A weight scale needs to have a noise-free resolution (i.e., stable digits within a resolution level) of 10 g in a 100-kg of full-scale weight. Therefore, the *noise-free resolution* in bits is $\log_2(100/0.01) = 13.2b$. Hence, the effective number of bits of the analog interface is ENOB = noise-free resolution$_b$ + 2.43 = 15.6b.

To summarize, we have two ways to calculate the number of resolution levels in analog domains. The first is related to information theory where

$$DR = \frac{FS^2}{q^2\sigma_N^2} \rightarrow NL = 2^{NL_b} = \sqrt{1 + DR}$$

$$NL_b = \text{effective resolution}_b.$$

(2.133)

Another approach is to use ENOB

$$DR_{dB} = 10\log\frac{FS^2}{q^2\sigma_N^2} \rightarrow ENOB = \frac{DR_{dB} - 1.76}{6.02} \rightarrow NL_{ENOB} = 2^{ENOB}$$

(2.134)

and following (2.132), we have

$$NL = 1.22 \cdot NL_{ENOB} \quad \text{or}:$$
$$NL_b = ENOB + 0.29.$$

(2.135)

In the first case, we have NL resolution levels. In the second one, we have a smaller number since we have to adapt the same noise to the quantization noise as shown in Fig. 2.69, where $NL_b = N$.

The relationship between ENOB and resolution is illustrated in Fig. 2.70, where, using common parameters, *ENOB is about 1b more than noise-free resolution*. It should be pointed out that all the above expressions relating to resolutions and ENOB do *not* consider distortion.

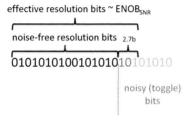

Figure 2.70 Resolution and ENOB in the case in which distortion is negligible with respect to random noise.

> **Hint** Equation (2.132) tells that we can use as a metric of resolution levels of analog interfaces both NL_b and ENOB, but what is the difference? In the first case NL_b is strictly related to information theory concepts where the resolution is indexed to the noise standard deviation. In the second case ENOB is a resolution indexed to the quantization noise of an ideal A/D converter. We can use both in an acquisition chain design, but the metric should be uniform. Since an A/D converter usually ends the acquisition chain, it is preferable to use the second metric. For this reason, in the next section, we will use ENOB to characterize precision and accuracy.

It should be emphasized that "noisy bits" or "toggle bits" could be misleading and should be taken within a probabilistic mean: for example, the change of a V_{LSB} implies the change of the least significant bit in all cases, but also the changes of most significant bits with the rarest occurrence.

Example A 14b A/D converter is characterized SNR = 71.9 dB. The ENOB given by SNR is given by $ENOB_{SNR} = (71.9 - 1.76)/6.02 = 11.6b$. Therefore, we have about 3 noisy bits, the effective resolution is about 12b, and the noise-free resolution is about 9b.

2.15   Precision, Accuracy, and Trueness

The uncertainty given by the spread of errors could be used to characterize the sensor effectiveness in sensing the correct value using the concepts of *precision* and *accuracy*. Unfortunately, these definitions belong to one of the most confusing and misused aspects of sensor design. In the most rigorous references of measurement science, it is stated that precision and accuracy should be considered as a qualitative concept of the characterization process. However, in most sensor and sensor interfaces datasheets, precision and/or accuracy are often quantitatively characterized.

The concept of uncertainty determines the following characteristics of sensors (see Fig. 2.71):

- *Precision* indicates the spread between the values obtained by repeated readouts with the same system conditions and input value. Precision depends only on the distribution of random errors and does not relate to the closeness of the estimated value to the actual value.
- *Trueness* is the closeness of agreement between the estimated value obtained from a series of readouts and the actual value. It is related to systematic and not to random errors.

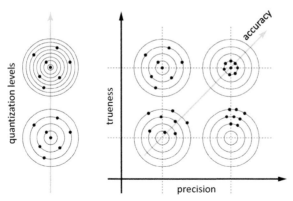

Figure 2.71 Qualitative characterization of sensing: precision, trueness, accuracy, and resolution. The true value is represented by the bullseye and each readout by black points.

- *Accuracy* is the closeness of agreement between the estimated value and the actual value. It is a combination of trueness and precision, thus combining random components and systematic errors.

  In simple terms, accuracy can be defined as the correctness of the estimated result, while precision is a measure of readouts variability.

  A graphical representation of the concepts is illustrated in Fig. 2.71. The center of the target is the "actual value," while the dots represent a single sample of the measurement or readout from the sensor. Since we assume that the readout is digitized, each concentric grid circle represents a quantized output. Precision is shown with respect to the x-axis, and it is a degree of spread around a barycenter of a population (mean) of readouts, disregarding where it is located. Trueness is the degree of displacement of the barycenter with respect to the true value, and it is shown with respect to the y-axis. From this graphical representation, we can see that accuracy is the combination of both trueness and precision.

  We could characterize the aforementioned concepts using the uncertainty

$$
\begin{aligned}
\text{precision} \;&\leftarrow\; & u_p(x) &= 2p \cdot \sigma_N \\
\text{trueness} \;&\leftarrow\; & u_t(x) &= 2p \cdot \sqrt{\sigma_\alpha^2 + \sigma_D^2 + \Delta x_{E(OG)}^2} \\
\text{accuray} \;&\leftarrow\; & u_a(x) &= 2p \cdot \sqrt{\sigma_N^2 + \sigma_\alpha^2 + \sigma_D^2 + \Delta x_{E(OG)}^2}.
\end{aligned} \tag{2.136}
$$

Therefore, precision, trueness, and accuracy are related to the random, systematic, and combined uncertainties, respectively.

Hint When precision and accuracy are quantitatively defined, it is necessary to indicate the number of sigmas to which they are referred (i.e., the factor p). For example, the accuracy at 6σ or $\pm 3\sigma$ means that all the uncertainties are first summed up in power and then weighted using $p = 3$ in (2.136). However, other rules (e.g., using the factor p only for random errors or different p values between precision and accuracy) could be used.

A way to transform the definitions mentioned above into a quantitative measure is to use the "effective number of bits" concept familiar to the A/D systems. The main advantage of this approach is that in doing so, we can easily compare the characteristics of the analog transducer with those of the final A/D converter.

Therefore, we could associate the concept of precision with a number of "virtual bits" or " information bits" using this expression

$$\text{ENOB}_p = \frac{\text{SNR}_p - 1.76}{6.02}; \quad \text{where} \quad \text{SNR}_p = 20\log\left(\frac{FS}{q \cdot u_p(x)}\right). \qquad (2.137)$$

In other words, the precision is characterized as if its uncertainty were equivalent to that of an ideal A/D converter having ENOB_p bits of nominal resolution using (2.125).

Following this approach, we could extend it to the concept of accuracy. Therefore, we have

$$\text{ENOB}_a = \frac{\text{SNR}_a - 1.76}{6.02}; \quad \text{SNR}_a = 20\log\left(\frac{FS}{q \cdot u_a(x)}\right). \qquad (2.138)$$

Note that SNR_a is substantially the $\text{SNRDT}_{\text{dB(max)}}$, since it uses the combined uncertainties of all the possible source of errors

$$\text{SNR}_a = \text{SNDRT}_{\text{dB(max)}} = 10\log\frac{P_S}{P_N + P_D + P_{sys}} = 20\log\left(\frac{FS}{q \cdot u_c(x)}\right) \qquad (2.139)$$

where $u_c(x)$ is the combined uncertainty. It should be pointed out that we can omit offset and gain errors in the calculation of uncertainty so as to correct them later in a digital environment.

A graphical representation of the preceding approach is illustrated in Fig. 2.72. As shown, the systematic errors might be more significant than random errors, thus defining a range of "systematic error bits" that are more significant than noisy ones. If we can reduce noisy bits by averaging (thus decreasing the bandwidth), we can reduce systematic errors bits only by using calibration. Thus, precision bits are the minimum number of bits that are not affected by flickering (in

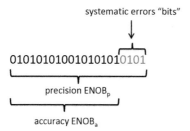

Figure 2.72 Quantitative representation of precision and accuracy of an analog interface in terms of "virtual bits."

a probabilistic fashion). Therefore, the systematic error bits are the difference between the precision and the accuracy bits encoding wrong information from systematic errors.

Example A Wheatstone bridge offers an output full-scale $FSO = 100$ mV and a noise of $\sigma_N = 198$ nV. The static characteristic shows a gain and offset error $\Delta y_{E(OG)} = 444$ µV and a nonlinearity error of $\sigma_D = 856$ µV. The equivalent precision (with $p = 3.3$, so at "noise-free" conditions) could be calculated using first $SNR_p = (100 \times 10^{-3}/(2\sqrt{2} \cdot 6.6 \cdot 198 \times 10^{-9})) = 88.6$ dB from which $ENOB_p = (88.6 - 1.76)/6.02 = 14.4$b. For the accuracy, we have first to calculate the total combined uncertainty as

$$u_a(x) = \sqrt{(198 \times 10^{-6})^2 + (444 \times 10^{-3})^2 + (856 \times 10^{-3})^2} = 964 \text{ µV so that}$$

$SNR_a = \left(100 \times 10^{-3}/(2\sqrt{2} \cdot 964 \times 10^{-6})\right) = 31.2.6$ dB and $ENOB_a = (31.2 - 1.76)/6.02 = 4.9$b. We used for accuracy $p = 1/2$. In conclusion, we can see that systematic errors are dominant over random errors, thus reducing the accuracy of about 9 equivalent bits with respect to precision.

A summary picture of the definition of precision and accuracy is shown in Fig. 2.73, where different levels of errors are taken into account. The first level is given by the noise standard deviation that is identified by the P_N line, which fixes both SNR and DR. The noise-free resolution level is then determined by 6.6 sigmas that fix the "precision" and thus the noise-free resolution levels. Finally, the sum of all noise sources fixes the bottom line of the "accuracy." On the right scale of the plot is indicated the "equivalent number of bits" measure of previous quantities using (2.102) or (2.138).

A graphical representation of the "virtual or information bit" concept is shown in Fig. 2.74. A noisy characteristic is compared with that of a virtual A/D converter. Suppose we want to represent the analog characteristic with that of an A/D converter; we have to contain the errors in the $V_{LSB}/\sqrt{12}$ zone so that the total errors are smaller than the quantization one. In that case, random and systematic errors could, on average induce at most an error of the least significant bit. In Fig. 2.74A, we can see that even if random errors are contained in a quantization step (thus, the number of bits represented by the characteristic is enough to represent the system's precision), systematic errors induce a relevant deviation from the A/D characteristic. Therefore, more than one bit of the A/D converter could be affected by systematic errors so that a certain number of least significant bits of the virtual converter could change. If we want to contain the errors in the least significant bit, we have to reduce the number of nominal bits, as shown in Fig. 2.74B.

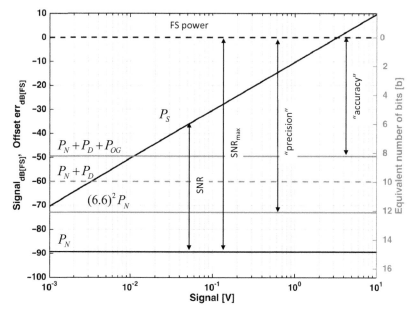

Figure 2.73 Visual description of the precision and accuracy concepts in terms of the equivalent number of bits with respect to the DR.

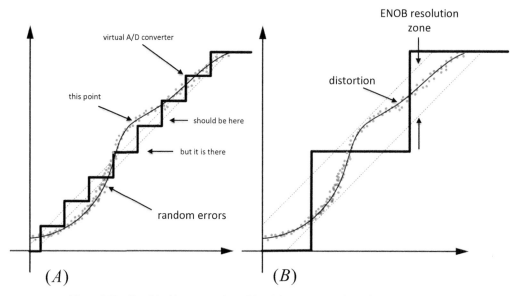

Figure 2.74 Graphical interpretation of the "bit" representation of an analog interface's precision and accuracy in the case of nonlinearity and random errors. (A) The number of bits of the "virtual converter" represented in the plot is enough for precision but not for accuracy. (B) The case where a reduced ENOB could be enough to cover nonlinearities and random errors.

Hint The fact that the ENOB of accuracy is lower than that of precision seems contradictory. Actually, what is represented in this model is the number of bits that are significant from the information point of view. Therefore, the number of bits required for accuracy is lower because we disregard the bits of systematic errors as non-carrying information. For example, we can take the case of a digital weight scale where we are interested in a smaller number of stable and significant digits related to the accuracy only. We can increase the number of digits, but in that case, they would be noisy and meaningless for our application.

2.15.1 Relationship Between Precision, Accuracy, and Dynamic Range for Defined Operating Range

Following Section 2.12.4 on the relationship between SNR and *DR* in DC measurements, we would like to link these issues to the concept of precision and accuracy.

As we have seen, the *precision* concept is related to SNR. On the other side, *accuracy* to the concept of SNDR (or SNRDT in the case of offsets). Therefore, assume that we want to measure DC values in the range $x_{min} \div x_{max}$ with some degree of precision/accuracy, we need to be able to read those values with a percentage of the minimum value given by SNR or SNDR

$$DR_{dB} = 20\log\left(SN(D)R \cdot \frac{x_{max}}{x_{min}}\right) = SN(D)R_{dB} + 20\log\frac{x_{max}}{x_{min}}, \qquad (2.140)$$

where the term SN(D)R means that we can use SNR for precision requirements and SNDR for accuracy requirements.

The *DR*'s knowledge is crucial because it sets the ENOBs necessary for the A/D converter (in the case we use just the A/D converter as interface).

Example We need an impedance interface reading impedances with no change of setup from 10 kOhm to 100 MOhm, thus ranging for 4 decades. Furthermore, we need 1% of accuracy; therefore, we need an SNDR of 40 dB. Thus, using (2.140), we have that the

necessary DR should be $DR_{dB} = 20\log\left(\frac{1}{0.01} \cdot \frac{100E6}{10E3}\right) = 120$ dB. This means that if

we need to use an A/D converter, we have to use one with *at least* ENOB =

$\frac{120 - 1.76}{6.02} = 19.6$b.

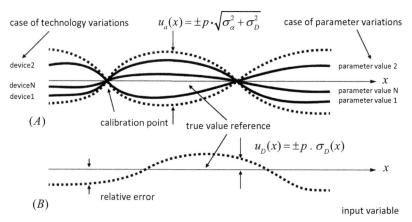

Figure 2.75 Inaccuracy plot examples. (A) Inaccuracy plot for the case of technology and/or parameter variations with two-point calibrated devices. The plot also considers nonlinearity errors. (B) Inaccuracy plot just for nonlinear characteristic errors.

2.15.2   Inaccuracy Plots

A useful experimental estimation of the accuracy (sometimes referred to as "inaccuracy") is given by evaluating the error, referred to the input, with respect to different calibrated values of the input stimulus and for the distribution of influence parameters. This experimental technique considers also distortion errors. The starting point refers to errors induced by influence parameters, disregarding distortion (but they could be considered a second time) and random errors. We consider variations in the parameter or the populations of devices illustrated before in Fig. 2.33. Then we plot the error versus the input stimulus as shown in Fig. 2.75A considering the boundaries of a coverage interval (e.g., $\pm 3\sigma$). The maximum values of those boundaries in the operating range identify the accuracy (or "inaccuracy") uncertainty of the interface.

Very often, devices under test are also previously calibrated. Therefore, in the calibration points, the relative error is by definition zero, as shown in Fig. 2.75A. If we are interested only in nonlinearity errors, we can estimate the inaccuracy referred to the distortion by starting from Fig. 2.47. The output error is then referred to the input and plotted as in Fig. 2.75B with the option of considering expanded uncertainty by means of the coverage factor p.

Example Temperature sensors (i.e., input stimulus is temperature) of the same lot are individually calibrated. However, despite calibration, a residual error gives a 3σ inaccuracy of $0.55°C$ in the $-40 \div 85°C$ operating range. This means that a collection of calibrated temperature sensors is characterized for different temperatures in the input operating range. The distribution of errors is evaluated for each temperature by means of the variance $\sigma_\alpha^2(x)$. Therefore, the inaccuracy of the system is given by the maximum value $\max\left(\pm 3\sigma_\alpha(x)\right) = \pm 0.55°C \quad \forall x \in [-40, 85]$.

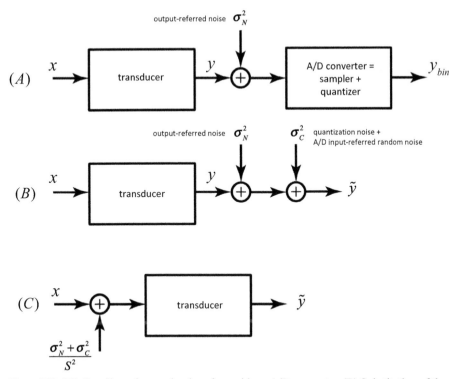

Figure 2.76 (A) Coupling of an analog interface with an A/D converter. (B) Substitution of the A/D converter with quantization noise adder. (C) Input-referred overall noise.

2.15.3 Analysis of Interface and A/D Converter Chains

As shown in Fig. 2.76, an analog transducer is followed by a quantizer. The analog noise is represented by a point that summarizes both the output noise of the transducer with the input-referred random noise of the A/D converter (Fig. 2.76A). The quantization noise could be represented by another summing point in the chain, as shown in Fig. 2.76B. If we set the quantization noise of the A/D converter lower than that of the random process such that $(k < 2)$, we can use the PQN model and combine them quadratically and refer them to the input (Fig. 2.76C).

Example An amplifier of gain $G = 100$ is characterized by an output noise of 120 µV rms in a range of $FSO = 5$ V. The amplifier is interfaced with a 16b A/D having 45 µV rms of input-referred noise. The quantization noise is $5/(2^{16} \cdot \sqrt{12}) = 22$ µV r.m.s. Since $k < 2$, we can apply the PQN model and combine them in RSS and refer to the input-referred effective resolution:

$u(x) = \left(\sqrt{(120 \cdot 10^{-6})^2 + (45 \cdot 10^{-6})^2 + (22 \cdot 10^{-6})^2} \right)/100 = 1.3$µV. Since the

input FS is $5/100 = 50$ mV we have that $SNR_p = 20\log\left(\dfrac{FS}{q \cdot u(x)}\right) = 82.7$ dB.

Therefore, the *precision* (of the whole system) is $ENOB_p = (82.7 - 1.76)/6.02 = 13.4$b.

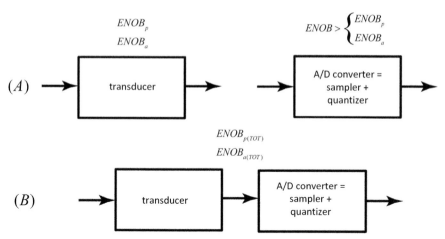

Figure 2.77 Coupling an analog interface with an A/D converter. (A) The two systems are characterized alone. (B): the two systems are coupled together as a unique system, characterized by a global effective number of bits ENOB$_{(TOT)}$.

2.15.4   Design of the Interface with an A/D System

When we couple an analog interface characterized by either a given precision and accuracy with an A/D converter, we may be led to use an A/D converter with the same characteristics. However, this is not correct because we will see in the next chapter that two pipelined stages with equivalent resolution will give a total resolution reduced with respect to a single stage.

Therefore, the question is: *For a given interface characterized by certain precision/ accuracy, which is the A/D converter ENOB for preserving the resolution characteristics (see Fig. 2.77)?*

The *resolution rule of acquisition chains* (RRC) states that the overall precision/ accuracy of a two-stage pipeline (transducer + A/D converter) is lower with respect to the transducer's characteristic, whichever is the characteristic of the A/D converter. Thus, to reduce this drawback, we have to choose an A/D converter with ENOB greater than those characterizing the transducer. A good compromise is to use one with at least 1b more than that of the transducer. We will check this in the following example.

Example Using the same data as the previous example, the effective input resolution of the transducer is $u(x) = (120 \cdot 10^{-6})/100 = 1.2$ µV; therefore, its signal-to-noise ratio is SNR$_p$ = $20\log(5 \times 10^{-2}/(1.2 \times 10^{-6} \cdot 2\sqrt{2})) = 83.3$ dB, and thus its precision is ENOB$_p$ = 13.6b. Therefore, to preserve its precision quality, we should use an A/D converter with *more than that number of bits*. It is interesting to note that if we use the A/D converter of the previous example with an input-referred noise-free uncertainty, $u(y) = \sqrt{(45 \cdot 10^{-6})^2 + (22 \cdot 10^{-6})^2} = 50.1$ µV and ENOB = 14.8, which satisfies the constraint mentioned above. This is confirmed by the fact that by using this A/D converter, the overall input noise-free resolution is degraded only to 1.3 µV (see the previous example) from the transducer one of 1.2 µV.

A final summary about this section: as a rule of thumb, we will choose an A/D converter with greater resolution than the transducer. However, this process should be quantified because it could be expensive in cost and power consumption. In other terms, *how much should we increase the resolution of the A/D converter with respect to that of the transducer?* Furthermore, *what is the contribution of resolutions of single blocks in the overall chain?*

The answers to these questions will be discussed in Chapter 3.

2.16   Appendix: Mean and Variance in Different Contexts

Here are summarized some notations used in this chapter and in the following. For more details, see Chapter 4. An important aspect concerns using the same variable x in two different contexts: if it is considered a sampled value $x_i = x(t_i)$ we can have several identical values. However, if it is related to a numbered outcome value $x_j (j = 1, 2, \ldots, K)$ of a random variable, it is necessarily $x_1 < \cdots < x_j < x_{j+1} < \cdots < x_k$.

Name	Expression
Continuous-time variable	$x(t)$
Sampled-time variable	$x_i = x(t_i); \quad i = 1, \ldots, N$
Random variable	X
Random variable outcome	$x = X(s_i); \quad s_i = \text{i-th experiment outcome}$
Random variable outcome value	$X(s_i) = x \in \{x_1, x_2, \ldots, x_j, \ldots x_K\};$ where : $x_1 < \ldots < x_j < x_{j+1} < \ldots < x_K$
Probability of discrete variable	$p_j = p_X(x_j)$

The following are some definitions of mean and variance in different contexts.

	Mean/Expected value	Variance
Sample, population, empirical $x_i = 1, \ldots, N$ (equally likely outcomes)	$\bar{x} = \dfrac{1}{N} \sum_{i=1}^{N} x_i$	$s_x^2 = \dfrac{1}{(N-1)} \sum_{i=1}^{N} (x_i - \bar{x})^2$
Time periodic variable	$\langle x \rangle = \dfrac{1}{T} \int_{-T/2}^{T/2} x(t)dt$	$\sigma_x^2 = \dfrac{1}{T} \int_{-T/2}^{T/2} \left(x(t) - \langle x \rangle\right)^2 dt$
Discrete random variable	$E[X] = \sum_{j=1}^{K} p_j x_j$ $E[X] = \text{expected value}$	$s_X^2 = \sum_{j=1}^{K} p_j (x_j - E[X])^2$
Continuous random variable	$E[X] = \int_{-\infty}^{\infty} x p_X(x)dx$	$\sigma_X^2 = E[(X - E[X]^2)]$

Further Reading

Carlson, A. B., *Communication Systems: An Introduction to Signal and Noise in Electrical Communication*. New York: McGraw-Hill, 1986.

Duda, R., Hart, P., and David, S., *Pattern Classification*. New York: John Wiley & Sons, 2001.

Gregorian, R. and Temes, G. C., *Analog MOS Integrated Circuits*. New York: John Wiley & Sons, 1986.

Johns, D., and Martin, K., *Analog Integrated Circuit Design*. New York: John Wiley & Sons, 1997.

Joint Committee for Guides in Metrology, Evaluation of measurement data – Guide to the expression of uncertainty in measurement (GUM). Working Paper, Geneva, 2008.

Kester, W., Ed., *The Data Conversion Handbook*. Philadelphia: Elsevier, 2004.

Maloberti, F., *Data Converters*. New York: Springer Science+Business Media, 2007.

Taylor, J. R., *An Introduction to Error Analysis*. Sausalito, CA: University Science Books, 1997.

Widrow, B., and Kollar, I., *Quantization Noise*. Cambridge: Cambridge University Press, 2008.

3 Sensor Design Optimization and Tradeoffs

This chapter will start by first describing techniques to reduce errors. As far as the random ones are concerned, reduction approaches oriented to increase the signal-to-noise ratio on the spectrum domain and their strict relationship with sample averaging are discussed. Next, strategies for the limitation of systematic errors are presented, primarily based on the feedback concept. However, since the error reduction techniques allow several degrees of freedom, this chapter will discuss the tradeoffs in optimizing sensing systems from the standpoint of the resolution, bandwidth, and power consumption. More specifically, the resolution optimization of the sensing process will be treated under the information theory standpoint, and the approach will be extended to acquisition chains to understand the role of single building blocks.

3.1 Reduction of Random Errors by Averaging

The signal-to-noise ratio (SNR) is a usual reference in comparing signals and noise power. In this section, we will analyze SNR from the spectrum viewpoint. Figure 3.1 shows the example of the sum of a signal made of *random exponentially decaying pulses* with uncorrelated noise in time and frequency domains. The signal holds the characteristic that we can easily approximate it as a *bandwidth-limited signal* with maximum frequency f_M (see Chapter 4).

In this example, noise is characterized by uniform single-side power spectral density (PSD) $S_N(f)$ up to the frequency f_N. The system bandwidth (BW) of Fig. 3.1A is related to the low-pass filtering effect introduced by the interface. A possible approach for increasing SNR is to reduce the noise power in regions of the spectrum not covered by signal components by restricting the system bandwidth as shown in Fig. 3.1B, where we assume for simplicity that an ideal low-pass filter gives the filtering effect.

Now, if we compare the SNR of Fig. 3.1A with SNR$'$ as in Fig. 3.1B, we have

$$\text{SNR} \simeq \frac{P_S}{S_N(f) \cdot BW}; \quad \text{SNR}' \simeq \frac{P_S}{S_N(f) \cdot BW'}$$

$$\rightarrow \text{SNR}' \simeq \text{SNR} \cdot \frac{BW}{BW'}; \quad BW \geq f_M. \tag{3.1}$$

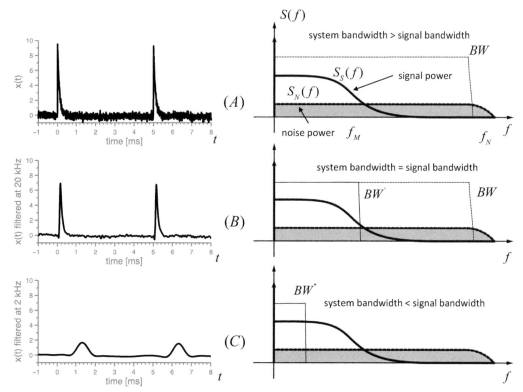

Figure 3.1 Effect of low-pass filtering on the composition of signal and noise. (A) Signal and noise as determined by the system bandwidth. (B) Reduction of the system bandwidth to the signal bandwidth to reduce noise. (C) Reduction of the system bandwidth below the signal bandwidth influencing signal integrity.

Therefore, reducing system bandwidth (down to the maximum frequency of the signal) corresponds to a proportional increase of the SNR.

We have already seen that the SNRs could also be expressed in terms of noise variance. Thus, we can sample the signal and the noise at a frequency twice the maximum frequency of the noise $f_S = 2 \cdot f_N$ without any signal and noise loss. This is important because it avoids *noise folding*, as will be discussed in Section 3.5. Then, we use a *moving average* filter in which samples are averaged in a window of the K consecutive values. We have to choose a value of K not to expect a variation of signal in the average window. This means that averaging does not change the signal characteristics (and thus the power) under the sampling theorem. On the other hand, noise has a higher bandwidth than signal, so the averaging process smoothens faster variation of the noise, as shown in the spectrum of Fig. 3.1.

Therefore, following the central limit theorem on averaging, we have

$$\text{SNR} = \frac{P_S}{\sigma_N^2}; \quad \text{SNR}' = \frac{P_S}{\sigma_N'^2}; \quad \sigma_N'^2 = \sigma_N^2/K \tag{3.2}$$
$$\rightarrow \text{SNR}' = \text{SNR} \cdot K.$$

This result shows that the moving average has the same effect of low-pass filtering in increasing the SNR under the assumptions mentioned above. We must be careful about this conclusion: it only says that the effect is the same. The effect of a moving average filter is not like an ideal one, and we could find an infinite number of low-pass filters having a similar effect on the variance in the reduction of noise.

> **Averaging and Filtering** Averaging sampled signals in the time domain has the effect of a low-pass filter in the frequency domain.

However, this approach to increase the SNR by reducing the bandwidth has a limit. If we further decrease the bandwidth of the filter to BW'', as shown in Fig. 3.1C, we cut off not only noise power but also signal power components. Therefore, we have a degradation of the signal integrity.

Figure 3.2 shows the effect of a moving average filter on a uniform distribution using windows (kernels) of 5, 10, and 50 samples showing how the standard deviation

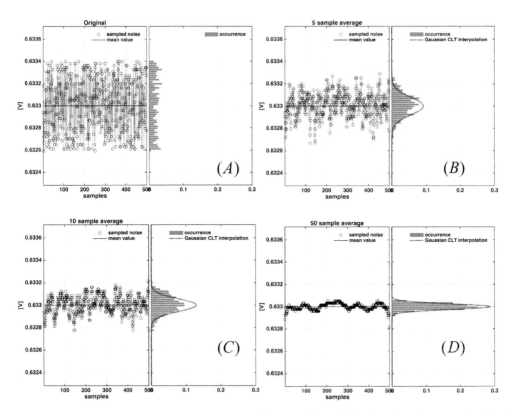

Figure 3.2 Effects of a moving average filter on uniformly distributed random errors (noise). The original uniform distribution (characterized by $\mu = 0.633$ V and $\sigma = (0.6334 - 0.6326)/(2\sqrt{3}) = 230.9\,\mu$V) shown in (A) is treated by a moving average filter of 5 samples (B), 10 samples (C), and 50 samples (D). The standard deviation is reduced by a factor $\sigma/\sqrt{K}$, where K is the number of average samples. Note how the histograms fit with Gaussian models.

is reduced by a factor of $\sigma/\sqrt{K}$, where K is the size of the kernel. Note also how, following the central limit theorem, the distribution of 500 filtered samples tends to the Gaussian irrespective of the original distribution.

As already discussed, SNR is the primary reference parameter for information detection at the *early stages of transduction*. The simple low-pass filtering technique could be easily implemented in analog processing at the physical/circuital levels of the initial stages.

However, there are much more sophisticated techniques that can be used to increase the extracted information at the algorithmic level. For example, we can shape the filter (thus, not only a bare low-pass filter) on the characteristic of signal and noise such as *adaptive filtering* (e.g., inverse, deconvolution, optimal, matching, and Wiener filtering). These techniques should be employed at the deepest stages of the acquisition chain on raw data using *signal processing*. Therefore, the detection performances employing these techniques could be much higher than those obtained with a simple energy spectrum–based approach, even if at the expense of a higher computational cost.

Hint The importance of the SNR should be regarded only at the earliest stages of signal acquisition chains and should not be considered related to the ultimate detection limit of the sensing process. Indeed, the information extraction can be effectively performed using computational techniques at deeper acquisition chain levels despite low SNR levels.

3.2   Reduction of Systematic Errors

Systematic errors are difficult to reduce since we cannot use averaging techniques due to their not time-dependent population distribution. We will see three approaches for reducing systematic errors: *feedback sensing*, *differential sensing*, and *calibration*.

3.2.1   Feedback Sensing

The sensing systems seen so far are based on a direct flow of information from input to output, also called *direct sensing* (or *open-loop sensing*). However, we could use an alternative approach in which a part of the output is fed back and subtracted from the input. The idea is to "neutralize" the input quantity by counteracting an action that we better control or know to measure the stimulus. Thus, we measure the stimulus indirectly by the amount of the neutralizing quantity. The latter approach is called *indirect or feedback sensing*. The trick dates back to simple mechanical measuring systems such as weight scales, as shown in Fig. 3.3. Figure 3.3A shows the easiest way to build a direct sensing scale by using a spring. The equilibrium between the gravity force and the elastic force gives rise to a displacement of the plate with respect to the original reference by an

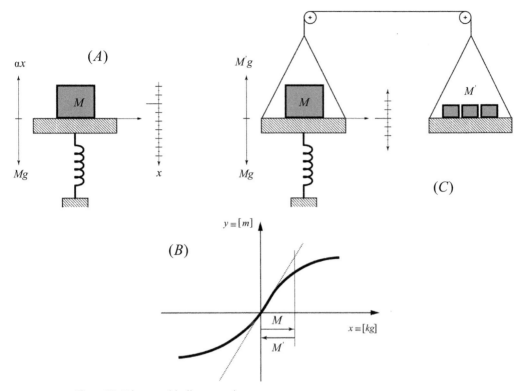

Figure 3.3 Direct and indirect sensing.

amount proportional to the weight of the mass to be measured. Figure 3.3B shows the real characteristic of the spring scale where the input quantity is the mass, and the output one is the displacement. Unfortunately, due to the nonlinear effects of the spring, the characteristic could be deviated by the ideal one, resulting in a systematic error due to distortion.

One of the oldest ways to avoid the deviation mentioned above from ideality is by implementing indirect sensing in the "two-pan balance scale" shown in Fig. 3.3C. The weight M is sensed using an equal force given by a reference weight M' to be balanced in a mechanical system. Once we ensure the balance of the two forces by a zero-offset reference, we can say that the "the weight of M is given by the value of M'," thus performing indirect sensing. The effect on the real characteristic is shown in Fig. 3.3B, where, thanks to the effect of the measuring weights, the spring is forced to act on small displacements around its steady-state point, where it offers the maximum linearity behavior. It is apparent that the indirect sensing resolution is translated into the resolution we can achieve to detect the zero-offset point.

An indirect sensing evolution is given by the approach shown in Fig. 3.4A, where a system is used to adjust the neutralizing effect. Referring again to the case of the scale, we can think about an automatic system that is generating a force that has an opposite effect to the weight one. The counteracting force is proportional to the plate's displacement error with respect to the zero-offset reference. The weight is measured

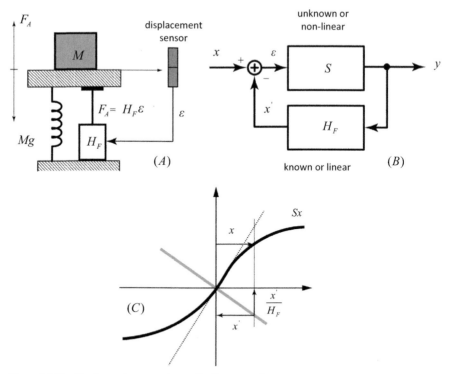

Figure 3.4 Feedback sensing as a class of indirect sensing.

indirectly by quantifying the amount of force that is necessary to compensate for the input quantity. Thus, the system feeds back an amount of counteracting action "as much as it is needed" to consider the error negligible.

A feedback sensing scheme model is shown in Fig. 3.4B, where a nonlinear (thus affected by systematic errors) sensor behavior is fed back by a linear system block in a loop scheme. The input/output and error/input relationship could be calculated as

$$y = S(x - x') = S\varepsilon = Sx - S \cdot H_F y \Rightarrow$$
$$\frac{y}{x} = \frac{S}{1 + S \cdot H_F} \xrightarrow{S \cdot H_F \gg 1} \frac{1}{H_F}$$
$$\frac{\varepsilon}{x} = \frac{1}{1 + S \cdot H_F} \xrightarrow{S \cdot H_F \gg 1} 0,$$

$$(3.3)$$

where H_F is the feedback block transfer function, and the product SH_F is called *open-loop gain*. If the open-loop gain is high, the input/output characteristic is given by the reciprocal value of the feedback gain. The geometrical representation of what has been explained so far is illustrated in Fig. 3.4C, where the sensor's nonlinear characteristic is evidenced concerning the linear behavior of the feedback block.

An extension of this approach could be applied even when the sensor characteristic is unknown (but monotonic) through the *known* (linear or monotonic) feedback

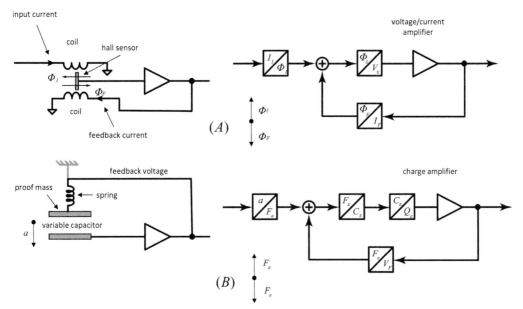

Figure 3.5 Examples of feedback sensing. (A) Closed-loop current sensing by Hall's sensor. (B) Electrostatic force feedback accelerometer.

characteristics. Therefore, if the feedback characteristic is linear, we get a linear input/output relationship disregarding the nonlinearities of the forward block.

Two application examples of feedback sensing are shown in Fig. 3.5. In Fig. 3.5A is shown a closed-loop current sensing. A current I_I that we would like to sense is flowing in a solenoid that is generating a magnetic flux Φ_I. However, a feedback I_F current is generating an opposite magnetic flux Φ_F by means of another coil. The total difference of magnetic fluxes is sensed by a Hall's sensor that transduces it in an error voltage that is amplified by a voltage-to-current amplifier and fed back. Therefore, we can see the system functionality as follows: a feedback magnetic feedback field is generated to neglect as much as possible the input one. The input current is sensed by measuring the output current necessary to balance the two magnetic fluxes.

In Fig. 3.5B is shown an electrostatic force feedback accelerometer. A capacitor plate is acting as a proof mass that is subject to a force due to acceleration (vertical direction) that we want to measure. The plate is anchored to a mechanical reference through an elastic force. Therefore, when the elastic and the acceleration forces are in equilibrium, we have a plate displacement with respect to the steady-state position. Instead of allowing the free displacement, we counteract the acceleration force F_a using an electrostatic force F_e that acts in the opposite direction of the acceleration one by applying a feedback voltage V_F to the plate. We have to amplify the error between the two forces by amplifying the capacitance variation and sensing it using a charge amplifier. Therefore, we measure the acceleration indirectly through the voltage to apply to the capacitor to counteract the acceleration force. The higher the acceleration, the higher the feedback voltage. The feedback force's effect is to make the movable plate more rigid with respect to the resting place.

In both examples, the feedback difference is applied to the physical environment thanks to the forces principle's superposition. The above examples' architectures are intentionally simple to illustrate the concepts, while the real integration is much more complex.

3.2.2   Dummy Differential Sensing

A widespread problem occurs when a sensor is dependent on influence parameters (such as temperature), and we could not distinguish if the output changes with respect to the input stimulus or with respect to such parameters. In Fig. 3.6A are shown two characteristics of the same sensor where the influence parameter has changed from α_1 to α_2. As clearly shown, an error due to a change of the parameter α could be added to the signal output variation Δy_S corresponding to the stimulus Δx_S.

For instance, a resistance sensor that has been designed to be implemented as a strain sensor is also sensitive to the temperature: in case the change of temperature is not known, the output is affected by an error that could not be distinguished by the signal.

A technique to compensate for that error is called *dummy differential sensing*. A second sensor, called the *reference or dummy* sensor, is exposed to the same parameter α but not to the input, as shown in Fig. 3.6B. We have at the first order of approximation

$$y_R(x_0 + \Delta x_S, \alpha_0) \cong y_0 + \frac{dy_R}{dx}\bigg|_0 \Delta x_S + \frac{dy_R}{d\alpha}\bigg|_0 \Delta\alpha$$

$$y_D(x_0, \alpha_0) \cong y_0 + \frac{dy_D}{d\alpha}\bigg|_0 \Delta\alpha,$$

(3.4)

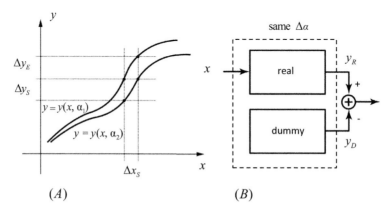

(A)   (B)

Figure 3.6 Differential sensing approach. (A) Errors due to influence parameters could be indistinguishable with respect to the input signal. (B) The differential sensing uses a dummy sensor (identical to the real one) exposed to the same influence parameter variation. Note that in (A), the real and dummy characteristics are not shown, but only the influence parameter's effect on the real characteristic.

where y_R is the output of the real sensor and y_D is that of the dummy sensor. Thus assuming (first-order approximation) that $\dfrac{dy_R}{da}\bigg|_0 \approx \dfrac{dy_D}{da}\bigg|_0$ and subtracting both outputs, we have

$$\Delta y_S = y_R - y_D \cong \frac{dy_R}{dx}\bigg|_0 \Delta x_S, \tag{3.5}$$

showing that the output variation is independent of the influence parameter variations. If the assumption about first-order variations is not negligible, the final equivalence should be verified.

3.2.3 Electronic Calibration

When the goal of a design is accuracy, the reduction of nonlinearities becomes a major task. In modern architectures, we can use a mixed-signal technique referred to as *electronic calibration*.

In Fig. 3.7 is illustrated a possible simplified architecture of an electronic digital calibration. The idea is that the fixed systematic error is dependent on the output (or input) value, as shown in Fig. 3.7A so that we can correct it runtime according to the input variation. Therefore, after an A/D converter has digitized the readout, that value is used to address a lookup table (LUT) where the errors have been previously stored in

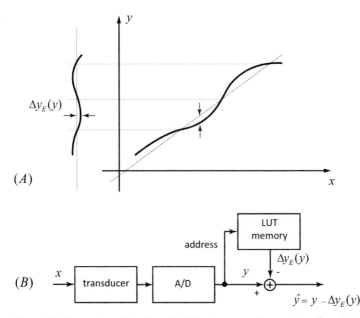

Figure 3.7 Electronic calibration. (A) Fixed systematic error (nonlinearity) as a function of the output. (B) Simplified architecture of electronic calibration.

characterization mode. Then, in operating mode, the offset values are subtracted from the raw output in order to estimate the true value $\hat{x}$.

There are several other architectures in which, for example, the fixed systematic errors are treated in terms of offset and gain value corrections; however, the principle is very similar to what has been shown.

3.3   The Role of Information in Sensor Acquisition Chains

We will shortly review some theoretical background of information theory in order to understand its role in sensor characterization from the resolution point of view and how this framework could be used in sensor design. More specifically, we will show that the sensing process could be optimized by maximizing the amount of "information" conveyed by the acquisition chain. To make "information" a measurable quantity, we will briefly introduce the concept of Shannon's *information entropy*. The concept will be discussed in the base-2 numeral system (even if other bases are possible) since electronic computation works on binary numbers.

Starting from an informal approach, using the graphs of Fig. 3.8, suppose that a completely uninformed traveler has to reach one of four destinations (d, e, f, g) from a starting point (a) using only binary decisions (binary decision tree). Assume also that the traveler chooses the way as in a "random walk" (i.e., equal chance to take a right or left direction). Thus the structure of the graph determines the mass probability density of the final destinations that can also be estimated after a large number of trials (for example, in the left figure, we have $Pr(d) = 1/4$, $Pr(e) = 1/8$, etc.).

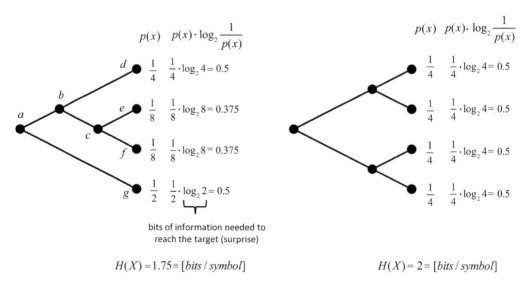

$$H(X) = 1.75 \equiv [bits\,/\,symbol]\qquad\qquad H(X) = 2 \equiv [bits\,/\,symbol]$$

Figure 3.8 Graphical interpretation of information entropy. We need to arrive at four points using only binary decisions from a starting point. The Shannon information is also referred to as "surprise" of the outcome: the less probable, the more surprise, and the longer path and higher information should be provided.

Now, assume that we achieve the same result (probability distribution) under the supervision of a "wise" instructor, telling each traveler the right directions to reach the final destination. For example, the traveler that has to go from a to e is instructed with 3 bits of information (left, right, left). On the other hand, a traveler that has to go from a to g needs only 1 bit of information (right). We will refer to the information associated with a particular outcome (destination point) as *Shannon information*. Note that the number of bits of information (number of crossroads to a final destination) is related to the probability of that destination: the larger the information needed for the destination, the lower the probability. For example, if we need $3 = \log_2(8)$ bits (i.e., three crossroads instructions) for the travel a–f, the final destination also has the probability of being reached in a random walk equal to $p(x) = 1/8$. So, for a given graph [i.e., a probability mass function (PMF)], what is the *average* amount of information that the instructor has to communicate to a large population of travelers to reach their destinations? In other words, we want to quantify the average information that is needed to "create" the given distribution from a large number of trials. From the "nature" viewpoint, this average information is also the average amount of changes induced on many trials of random walks; thus, it is also an *average uncertainty*. This double viewpoint is essential since the decrease in average uncertainty of an observed distribution implies the gain in the information of an observing system.

Following the previous discussion, such average of *uncertainty* (or *surprise*) is referred to as *information entropy* and could be evaluated as

$$E\left[\log_2 \frac{1}{p(x)}\right] = \sum_{i=1}^{N} p(x_i)\log_2 \frac{1}{p(x_i)} = H(X) \equiv [\text{bits/symbol}], \qquad (3.6)$$

where "bits" is a binary measure of choices, while "symbols" are, in our example, the destinations. It is interesting to note that in the graph (probability mass function) on the left of Fig. 3.8, we have $H = 1.75$, while for the right one, where the PMF is uniform, we have $H = 2$. It could be shown that a uniform distribution gives maximum entropy. It means that, in this latter case, we encounter the maximum average number of choices per destination for many random walk trials. Of course, the preceding discussion is focused on giving an intuitive approach for understanding information entropy, while the most formal approach should be based on an *axiomatic* approach, where (3.6) is the definition of information entropy. Furthermore, in the following discussion, we will associate the input and output signals with discrete levels for simplicity and because we will see that the resolution model due to noise will reduce the information measurement to discrete levels.

Now, since we are usually referring to relationships between an input and an output of a system, we can generalize the definition mentioned above with the *entropy of the joint distribution* between two (discrete) variables, x and y,

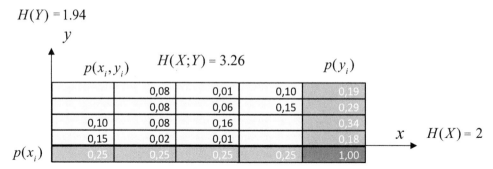

Figure 3.9 Example of entropy of a joint distribution.

$$E\left[\log_2 \frac{1}{p(x,y)}\right]$$

$$= \sum_{i=1}^{N}\sum_{j=1}^{M} p(x_i, y_j)\log_2 \frac{1}{p(x_i, y_j)} = H(X;Y) \equiv [\text{bits/pair}].$$

(3.7)

Note that for independent variables, $p(x_i, y_j) = p(x_i)p(y_j)$ the joint entropy is $H(X;Y) = H(X) + H(Y)$. An example of the application of this function is shown in Fig. 3.9, where we have associated, as usual, x and y with the input and output of a system. Note that, since the x variable is uniformly distributed, its entropy $H(X)$ is maximum.

We would like now to understand how much information is conveyed from an input to the output. In other words, how much entropy of the input is mapped to the output. The extreme case is where there is no relationship at all between input and output so that $H(X;Y) = H(X) + H(Y)$. Therefore, we define *mutual information* $I(X,Y)$ so that

$$I(X;Y) = H(X) + H(Y) - H(X;Y).$$

(3.8)

It can be easily found that the expression is

$$I(X;Y) = E\left[\log_2 \frac{p(x,y)}{p(x)p(y)}\right] = E\left[\log_2 \frac{p(y|x)}{p(y)}\right] \equiv [\text{bits/pair}].$$

(3.9)

The second expression suggests another relationship that can be easily demonstrated

$$I(X;Y) = H(Y) - H(Y|X) = H(X) - H(X|Y) \quad \text{where}$$

$$H(X|Y) = E\left[\log_2 \frac{1}{p(x|y)}\right]; \quad H(Y|X) = E\left[\log_2 \frac{1}{p(y|x)}\right],$$

(3.10)

where $H(Y|X)$ and $H(X|Y)$ are referred to as *conditional entropies*. This quantity could be explained as follows. If we do not know anything about the input X, the entropy (average uncertainty) about the output is $H(Y)$. But if we know the distribution

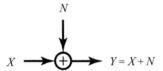

Figure 3.10 Effect of noise in reducing the information conveyed.

of X, the output's entropy is reduced to $H(Y|X)$, which is the average uncertainty of Y after X is observed. Therefore, Eq. (3.10) tells us that the information conveyed to the output is the entropy of the output minus the conditional entropy of the same output for a given input.

Let us take an extreme case where we have no measurement error. If we are before any choice (i.e., before taking any measure), we have an average uncertainty given by $H(X)$. However, if we take a measure without uncertainty is, $H(X|Y) = 0$. This could also be seen by means of the last expression of (3.10); if *for any* given x we know exactly y, then $p(y|x) = 1$, then $H(X|Y) = 0$. Thus, according to Eq. (3.10), we have that the mutual information gained by the system is equal to H(X). For example, the input of a system offers 32 equal probable choices. It is like having a meter with 32 ticks where we assume that the future measurement will fall with equal probability in one of the 32 sections of the meter. We have an input entropy (uncertainty) of 5b, and it is like to make a measurement with an ideal a 5b A/D converter. Once we make a sample (a measure) with no uncertainty, we have H $(X|Y) = 0$, and the information gained by the system sample is 5b – 0b = 5b. It is equivalent to say that our system has 5b of resolution capability. However, if we take the output with uncertainty $H(X|Y)$, then the gained information is $H(X)$ reduced by that amount.

The effect of reducing gained information could be better understood by looking at the case of noise summed to a signal as illustrated in Fig. 3.10, which is the typical situation of noise degradation of signal in sensing process.

Therefore, following the notation of the figure

$$I(X;Y) = H(Y) - H((X+N)|X) = H(Y) - H(N|X) = H(Y) - H(N), \quad (3.11)$$

where $H((X+N)|X) = H(N|X)$ because the average uncertainty of knowing X after X has been observed is zero. Therefore, comparing with Eq. (3.10), we have

$$H(N) = H(Y|X), \quad (3.12)$$

which means that the conditional entropy is the entropy of noise.

Relationship (3.11) is important since it shows that

- The conveyed information is a *variation* of entropy: it is given by the entropy of the output reduced by the noise's entropy.
- Noise is one of the first causes of reducing the conveyed information in sensing systems.

To better understand the concepts, we will provide an example.

Example We expect a source to be uniformly distributed in 256 levels of the full scale. Therefore, we have the same probability of occupancy, and the entropy of the source is 8 bits of information. We attempt to read the source with a noisy 8b A/D converter. The noise is making the last two bits flicker. Thus we still have an average uncertainty of 4 levels (2 bits) after the measure is done $H(X|Y) = 2$. Therefore, the information conveyed by the system is 8b − 2b = 6b.

We now make some examples of the information conveyed by different 2b discrete input–output relationships, as shown in Fig. 3.11. We have at the moment chosen uniformly distributed inputs so that $H(X)$ is maximum in all cases.

In the "deterministic" case (no noise), we have the maximum information transfer $H(X) = H(Y) = I(X;Y) = 2$. If now we introduce some noise, we have a fairly decrease of the conveyed information, and the "noisy" case gives $I(X;Y) = H(Y) - H(N) = 1.94$. However, we also decrease the information if we have a "saturated" characteristic. For "saturated," we mean that the characteristic is not linear and does not map the entire full-scale (*FS*) into the output, as shown in the table. This is because the joint entropy is the same but $H(Y)$ is reduced since not all the values are reached. Finally, the worst case is given where we have a completely unrelated relationship between input and output. In this case, we have $H(X) = H(Y) = 2$ but $I(X;Y) = 0$.

The expression of (3.10) could be expanded for $H(X|Y)$ as well, and the result is shown in a Venn diagram as in Fig. 3.12A, where we can see that for a given joint entropy, the mutual information is constrained between the two conditional entropies. We can better visualize the situation in a corresponding diagram, Fig. 3.12B, where *the entropy of y given x*, $H(Y|X)$, could be seen as a number of levels $2^{H(Y|X)}$ in which a given input is spread into the output. This is what happens typically (but not only) with random noise. Symmetrically, $H(X|Y)$ is seen as the number $2^{H(X|Y)}$ of input levels that converge into a unique output level. When does it happen? The most typical case is where we have saturation; however, even coarse quantization increases such a conditional entropy. In the case of random noise only, the two conditional entropies are related by means of the interpolating function because they are expressing the spread of the input and the output-referred noise, as already shown in Chapter 2.

Hint The conditional entropy $H(Y|X)$ is the average "uncertainty" of the output given the input value, while the conditional entropy $H(X|Y)$ is the average "uncertainty" of the input given the output.

It is useful to refer back to Fig. 3.11 to understand the role of the conditional entropies. Note that in the "deterministic" case, the conditional entropies are zero since the function is a one-to-one correspondence and there is no spread between the

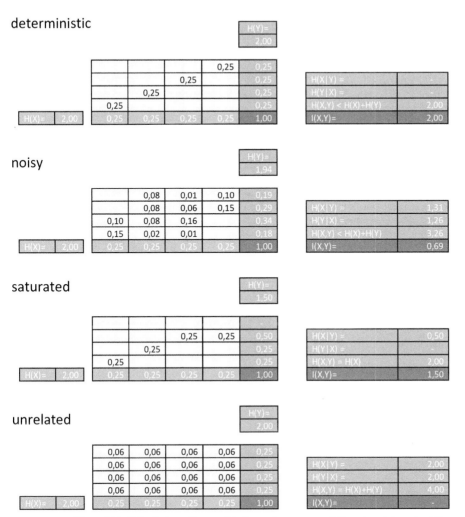

Figure 3.11 Calculation of mutual information conveyed from input to output for several characteristics.

two variables. The situation changes in the *saturated* case, where two input values are mapping a unique output value so that $H(X|Y) \neq 0$. We have both conditional entropies different than zero in the noisy case, as expected. Finally, we have maximum conditional entropies and zero mutual information in the case of completely uncorrelated transformation.

From what we have seen so far, the amount of information conveyed from input to output is limited due to various causes such as random errors and saturation; therefore, the main task of sensor design is to maximize mutual information from input to output.

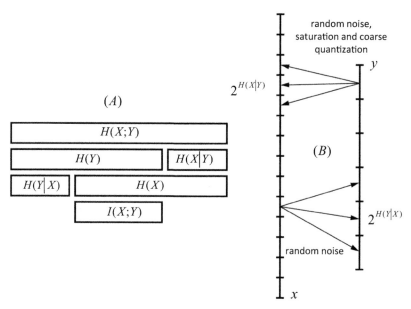

Figure 3.12 (A) Relationship between joint, conditional entropies and mutual information. (B) Visual interpretation of conditional entropies with a correspondence diagram between input and output levels.

Hint As far as the resolution of the system is concerned, and for given constraints (noise, errors, etc.), the sensor design optimization from the resolution point of view is equal to *maximize the mutual information from input to output.*

We have dealt with uniform input probability mass function because this is a typical measurement case. However, what happens in the case of the nonuniform input mass probability function? We can see this in Fig. 3.13, where clearly the input entropy is no longer maximum $H(X) < 2$.

Referring to Eq. (3.11), we can see that, for given noise, maximizing $I(X; Y)$ is equal to maximize $H(Y)$, that is, *making things so that the output probability distribution is uniform.* Looking at the sensor acquisition chain case, the above sentence is equal to tell that any level of the final A/D converter is equally probable with respect to the input probability distribution.

So far, we have dealt with discrete variables. However, it could be shown that most of the concepts are still valid even for continuous variables with the definition of *differential entropy.* Here, we will omit the discussion for simplicity because the essential concepts illustrated in Fig. 3.12 remain unchanged for continuous variables. Furthermore, measurement errors and noise from the source restrict the entropy of a continuous variable (in theory infinite as modeled by an infinite number of infinitesimal intervals) to the one defined by a finite number of discrete intervals.

noisy and non-uniform input

H(Y)= 1,96

	0,12	0,02	0,10	0,23
	0,16	0,08	0,03	0,27
0,09	0,08	0,16		0,33
0,03	0,02	0,12		0,16
0,12	0,38	0,38	0,13	1,00

H(X)= 1,81

H(X|Y) =	1,34
H(Y|X) =	1,48
H(X,Y) < H(X)+H(Y)	3,29
I(X,Y)=	0,47

Figure 3.13 Calculation of mutual entropies in the case of the nonuniform input distribution.

Therefore, if the input's probability distribution function is not uniform, is there an input/output function that maximizes the mutual information, that is, makes the output probability distribution uniform?

It can be shown that the optimal characteristic is the integral of the probability distribution function of the input

$$y = F(x) = \int_{-\infty}^{x} p_X(t)\, dt, \tag{3.13}$$

where t is a dummy variable. In other terms, the best characteristic function is the *cumulative distribution function* of the input distribution.

We can see the effects of this optimization in Fig. 3.14, where the input probability density function (PDF) $p_X(x)$ is a Gaussian distribution. Therefore, to get a uniform output density function, $p_Y(y)$ we have to use Eq. (3.13), which is the cumulative distribution function of the Gaussian (closely related to the $erf(.)$ function). In this case, a linear characteristic is no longer the best function for acquiring signals that have a nonuniform distribution. It has been shown that biological sensory systems follow this kind of optimization using nonlinear characteristics.

To summarize from the information theory point of view:

- The linear characteristic is the best function for a sensing system whenever the input variable is uniformly distributed. This is the typical case of DC measurement instruments, where no assumptions on the input variable could be made.
- For nonuniform input distributions, the optimal characteristic could be found to get maximum mutual information by means of a uniform PDF at the output.

Hint In the case of a uniform input probability distribution, the best sensor characteristic to maximize the information conveyed by the system is the linear one.

Now we can revisit some definitions of previous chapters with information theory concepts.

Recalling the relationship between NL and the SNR, from Eq. (2.100)[1], we have

---

[1] Actually, according to (2.100) we should use SNR_{max}, which is the condition where the signal assumes the maximum value for a given full scale (*FS*). However, to avoid confusion with the maximization of SNR in the following discussion, we have taken out the subscript.

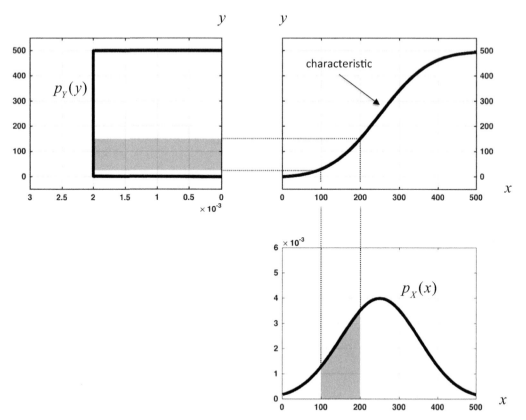

Figure 3.14 Maximization of the mutual information conveyed by a sensing system based on the input probability density function. The maximization is obtained by ensuring a uniform probability density on the output value. In this example, we used a Gaussian PDF at the input. [Note: Arbitrary units used for input and outputs.]

$$NL = \frac{\sqrt{\sigma_S^2 + \sigma_N^2}}{\sigma_N}; \ \text{SNR} = \frac{\sigma_S^2}{\sigma_N^2} \rightarrow$$

$$NL_b = \frac{1}{2}\log_2(1 + \text{SNR}),$$

(3.14)

where the calculation of the number of levels σ_S (given by the application) assumes the maximum value with respect to the full scale (*FS*).

Now, the information theory in telecommunication defines the *channel capacity* C for a *given* noisy channel

$$C = \frac{1}{2}\log_2(1 + \text{SNR}) = \max_{p_X(x)} I(X;Y) \equiv [\text{bit/sample}].$$

(3.15)

It should be pointed that here we have a fixed amount of noise coming from the input and from the channel itself. However, the task of sensing is different from that of telecommunications. In telecommunication, we have to characterize a channel taking

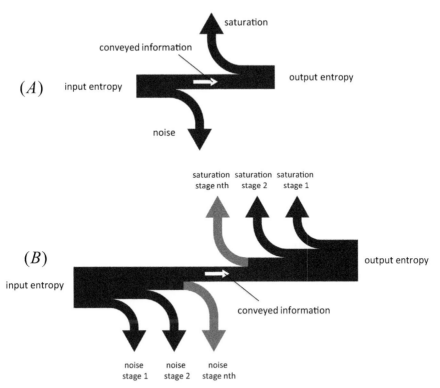

Figure 3.15 Schematic diagram of the information degradation. (A) Single-stage. (B) Multiple stages.

over all possible choices of input signal probability distribution $p_X(x)$. Once we have found it, that is the best possible way to transmit information. Therefore, C is a characteristic of the channel fixed by SNR.

On the other hand, in sensing systems, the function $p_X(x)$ is given by the nature of the signal, and we cannot change it. Conversely, we have to optimize the noise (SNR) of the system (channel) to convey as much mutual information as possible to the output. Therefore, we can read Eq. (3.15) in another way; since $p_X(x)$ is fixed, we have

$$C = \frac{1}{2}\log_2(1 + \text{SNR}) = \text{NL}_b. \tag{3.16}$$

We can maximize the capacity channel (which is no longer fixed) by maximizing SNR. It means maximizing the sensor acquisition chain from the resolution point of view.

The expression could be interpreted as follows. Assume that the analog domain input is approximated by a huge number of equally probable discrete values, and we have no noise. In this case, its entropy is enormous, and every value of the input could convey very large information, according to Eq. (3.11). Unfortunately, the resolution of the measurement is limited by noise (both from the environment and the interface), limiting the capacity of the system to convey information. Measurement noise divides the ranges of probabilities of an analog value into a finite number of discrete intervals, alias *resolution levels NL*. The number of levels increases as the noise decreases. This implies that the number of equiprobable input that can be discriminated using a single output value is

$$\frac{\sqrt{\sigma_S^2 + \sigma_N^2}}{\sigma_N} = NL = 2^C. \tag{3.17}$$

> **Hint** The noise of the measurement effectively divides the number of possibilities of a continuous variable (with infinite entropy) into a number of discrete levels with a finite entropy.

Therefore, we can get the optimization of the channel as

$$C = \max_{\text{SNR}} I(X; Y) = \frac{1}{2}\log_2\left(1 + \max(\text{SNR})\right). \tag{3.18}$$

We can make a relationship with deterministic signals (in characterization mode) by considering $\sigma_S = FS/q$; thus, the number of resolution levels and SNR is related to the FS by the power factor q:

$$NL \approx \frac{FS}{q\sigma_N}; \quad \text{SNR} = \left(\frac{FS}{q\sigma_N}\right)^2, \tag{3.19}$$

where, for example, $q = 2\sqrt{2}$ for sinusoidal signals, and $q = 1$ for DC signals. Therefore, the maximization of the SNR could be given by (1) reduction of noise and (2) maximization of the full-scale swing. Therefore, *the maximization of the number of the resolution levels encoded to the output corresponds to the conveyed information's maximization.*

> **Hint** For a given noise, maximizing the mutual information in a sensing system corresponds to maximize the number of resolution levels encoded to the output.

So far, we have referred to one sample (this is why we have used a DC reference). If we refer to sampling at 2BW, we get from Eq. (3.18)

$$C = BW \log_2(1 + \text{SNR}) \equiv [\text{bit/s}], \tag{3.20}$$

where *BW* is the bandwidth of the system. It shows that the system capacity is linearly related to the bandwidth and logarithmically related to the SNR.

We can show the role of information in a sensor acquisition chain as illustrated in the graphical representation of Fig. 3.15. In real environments, the signal is affected by noise, and we have seen that this fixes the input entropy of the sensing system. Then,

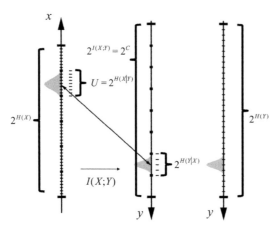

Figure 3.16 Resolution and uncertainty under information theory viewpoint.

the sensor introduces its noise, which could be seen as a corruption of the information, as shown in Fig. 3.15A. Again, the same information could be limited by the saturation of the output or coarse quantization. In the case of multiple stages, each of them contributes to decreasing the information, as shown in Fig. 3.15B.

Previously discussed concepts of uncertainty could be viewed under the information theory viewpoint as illustrated in Fig. 3.16. Following the approach of Fig. 3.12 and the definitions of resolution and uncertainty treated in Section 2.9 of Chapter 2, we can look at the relationships of Fig. 3.16, where an analog input x is mapped into a digitized output y. Assume we would like to calculate a system resolution where we know internal noise and we have to deal with an A/D converter with $N = H(Y)$ bits. Due to noise, for a given x (characterization mode), we have more levels of the converter covering in particular $2^{H(Y|X)}$ levels. Similarly, a given output y corresponds to more input levels of an input having entropy $H(X)$. This is referred to as uncertainty in the operation mode and corresponds to $2^{H(X|Y)}$ levels. Therefore, the information conveyed from the input to the output is $I(X; Y) = H(Y) - H(Y|X)$ bits in this case. In the specific case of Fig. 3.16, we have that the original A/D converter of 5 bits (32 levels) could only read 3 bits since the noise level covers 4 levels (2 bits): $I(X; Y) = 5b - 2b = 3b$. We can also view the argument from a different perspective: it is useless to use a 5b converter since the information is just 3 bits. In this case, refering to the input we have that if we increase the number of levels (in order to approximate an analog continuous variable), we increase the input entropy $H(X)$, but also the number of levels covered by the conditional uncertainty $H(X|Y)$, so that the information, using Eq. (3.10) is the same of before: $I(X; Y) = H(X) - H(X|Y)$. In the latter example, the system resolution is limited by internal noise. If we assume to reduce to zero internal noise and increase the bits of the converter, the overall resolution (i.e., conveyed information) is limited by source noise; therefore, it is easy to show that input entropy $H(X)$ should be limited by source noise and could not be infinite as in an ideal analog input.

3.4   Resolution in Acquisition Chains

3.4.1   Gain and Resolution

Let us take a generic system where the internal noise is represented by an input-referred power σ_N^2 and where the input signal x is affected by an external noise whose variance is σ_{SN}^2, as illustrated in Fig. 3.17A. We will assume in this section to deal with thermal noise where there is no dependence of noise on signal amplitude (as in shot or current noises). We have chosen to model system noise to the input instead of the output (even if we can exchange the notation for a fixed gain) because this is the most common case in real systems internally composed of a chain of stages of various gains and where the dominant noise contribution is given by the first one, according to Friis noise formula (see Section 7.4). This is congruent with the discussion of Section 2.12.1, showing that if the noise is modeled at the input, the SNR is in general insensitive to gain variations. However, the preceding statements do not consider the effects of bounded operating ranges at the input/output. We may wonder what the relationship of the resolution versus the gain of the system is in general. We will first tackle the problem by changing FS in an unbounded system for constant noise and gain and then consider the constraints of the real architectures requiring a change of gain.

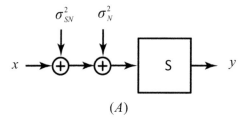

(A)

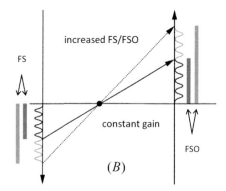

(B)

Figure 3.17 Increase in the number of resolution levels in *unbounded systems* acting on the FS. (A) Approximation of the system for high-gain stages, where most of the input-referred noise is given by the front-end stage. (B) Increase of the NL by increase of FS or FSO under equal gain. The higher the FS, the higher the information encoded because it increases the output information entropy. We assumed here physical noise sources placed at the input.

For unbounded systems, we refer to a system where hypothetically there is no limitation in the full scales of the input and of the output.

Using the gain plots, as shown in Fig. 3.17B, the resolution levels are depicted as bumps, as discussed in Chapter 2. This is just a qualitative representation that serves to illustrate the concept better. To recollect some issues:

1. The *RTI resolution* is a measure given in input units of the input-referred noise amount, and it is useful to compare the amount of noise with the amount of the signal (e.g., in the SNR, where it is also referred to as minimum detectable signal). Therefore, it is a property of the system noise and not of the *FS*. We can geometrically and qualitatively associate it with the size of the "bump."
2. The *number of resolution levels* (*NL*) is given by a ratio of the *FS* with the RTI resolution, and it is geometrically associated with the number of "bumps" in the input/output *FS* range. In this case, the *FS* plays a fundamental role. The *NL* is related to the number of analog states to be read; thus, it is an index of the information encoded, as discussed in Section 3.3.

A strategy to increase the number of resolution levels for a given noise is based on *increasing the FS* in order to cover as many resolution levels as possible, as shown in Fig. 3.17B. The increase of *FS* also implies an increase in the SNR. From the information theory point of view, we have that the higher the *NL*, the higher the information conveyed since the higher the output information entropy (see Fig. 3.12A). The number of resolution levels is linearly proportional to the *FS* change for constant gain from the figures.

The preceding arguments assume no limits in the input and output full scales. However, in real systems design, we have two important constraints:

1. The input range swing is determined by the *application*; therefore, the full-scale input (*FS*) is a primary and fixed constraint of the design.
2. The full-scale output (*FSO*) is limited by physical and/or technological constraints.

Examples of limitations related to the last statement are voltage supply rails, output linearity range, and input maximum allowed swing of the next stage (e.g., the input range of an A/D converter). As noticed in Fig. 3.17, the higher the full scales, the higher the information encoded. In the case of unbounded ranges and constant gain (Fig. 3.17B), the resolution levels linearly increase with full scales. Therefore, for a maximum information transfer and with a given noise budget, we have to maximize *FSO*. Assume that the output of the chain is digitally encoded, and the last block of an acquisition chain is a 16b A/D converter with $V_R = 3.3$ V. If we want to maximize the information transfer, we need to span all the 2^{16} levels on the output full-scale *FSO* = 3.3 V. If the output is analog, the maximum information content is achieved with the maximum output SNR, that is, using the maximum output swing before saturation (or distortion).

The design constraints set the first important parameter of the design, the *required total gain of the acquisition chain*

$$S_T = \text{required gain of the acquisition chain} = \frac{FSO_{max}}{FS_{APP}}, \qquad (3.21)$$

where FSO_{max} is the *FSO* that achieves the maximum information transfer content and FS_{APP} is the input signal swing required by the application.

Example A digital pressure sensor covers a 100 Pa range with a 16b digital encoding. The total gain is $2^{16}/100 = 655.3$ LSB/Pa.

In the presence of the constraints mentioned earlier, the gain should be carefully adjusted, since if it is too large or too small, we have a loss of information, as will be apparent in a while.

We illustrate the tradeoffs related to the gain with an example where an analog input is mapped into a digital output by means of an A/D converter. As shown in Fig. 3.18, a transducer should be designed onto a defined range at the input; therefore, we will assume that the input *FS* is fixed at the moment. The transducer is characterized by some internal noise that is represented by the resolution bumps.

Note how in Fig. 3.18A, the *FSO* is not fully mapped into the optimal output swing, for example, the input range of an A/D converter. In this case, the application range is mapped only in the first levels of the converter, resulting in a certain number of unused levels, with a loss of information transfer. This is due to the fact that reducing the number of A/D levels implies a reduction of coded bits and thus the number of possible combinations, that is, the output information entropy.

If we decrease the gain as shown in Fig. 3.18B for equal *FS* (the application range is fixed), we have a decrease of *FSO*, and we keep worsening the situation because the number of mapped A/D levels is further reduced.

Conversely, if we increase the gain as shown in Fig. 3.18C, we can cover the optimal output space at the same number of resolution levels of the input. This is an optimization of the chain acquisition gain by achieving the *required gain* of the sensor system.

However, if we keep increasing the gain, the saturation point of the output plays its role. As illustrated in Fig. 3.19B, a further increase in the gain reduces the number of resolution levels at the input, which is reflected at the output. This means a decrease of mutual information as in the previous case because we reduce the input entropy (see Fig. 3.12A). The effect is similar to having a thinner optical focus on the input resolution, where the surrounding levels are lost.

To summarize, the gain acts differently on the number of resolution levels. For a given overall input *FS* that is imposed by the application:

- If the gain is not sufficiently high, the resolution levels are mapped into the output on a reduced range (Fig. 3.18A). In this case, the *FSO* is mapped into a subdomain of the

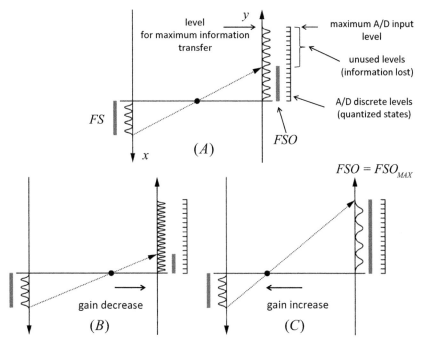

Figure 3.18 Effect of the gain on the resolution in the presence of constraints. (A) The original system, where the input *FS* is mapped into *FSO* which is, however, smaller than the optimal value for the maximum information transfer. The gap implies unused coded levels. (B) Case in which the gain is decreased. The number of unused levels increases[2]. (C) The case where the gain increases. The gain is optimized to make *FSO* achieve the saturation level.

A/D quantization levels. Thus, there is a loss of information due to reducing the output information entropy.

- We can increase the gain up to an optimal value S_T so that *FSO* reaches the maximum level to be detected (Fig. 3.18C). The overall gain is referred to as *required gain*.
- If we further increase the gain, we have a decrease of the input *FS* (against the application constraints), and the number of resolution levels starts to decrease (Fig. 3.19B). We have a loss of information due to the decrease of the input information entropy.

In general, what is the relationship between *NL* and full scales? Moreover, for maximizing *NL* (that is, the mutual information), is it better to look at the gain or the full scales? This is important since noise could be a function of the gain $\sigma_N^2 = \sigma_N^2(S)$. However, in general, we will see that an increase of *FS* due to a change of the gain induces a (monotonically) increase of the *NL* and vice versa. Therefore *NL(FS)*, in

---

[2] The two axes' units are different, so a thinner bump at the output does not imply that the output effective resolution is greater than the input one since they are different things: this is a geometrical artifact.

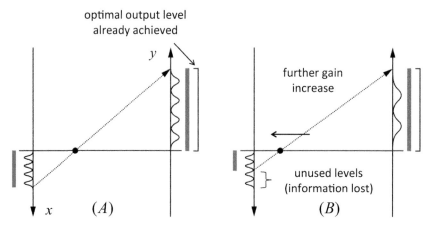

Figure 3.19 Effect of the gain increase when the output saturation is achieved. (A) The starting point, where the output saturation has been achieved. (B) A further increase.

most cases, is a monotonic increasing function. We see this in two cases by changing S with a variation of either *FS* or *FSO*:

- We are dealing with amplifiers with constant bandwidth. If the physical noise source is set at the input of the amplifier (this is also an approximation of the Friis formula for most of the practical cases), and we change the gain by keeping constant *FS*, the ratio (*NL*) between *FSO* and the output noise standard deviation is constant at the output. If we change the gain by keeping constant *FSO*, the ratio (*NL*) between *FS* and the input noise standard deviation is proportional to *FS*. Similar symmetric conclusions are deducted if the physical noise is placed at the amplifier's end stages. In conclusion, *NL(FS, FSO)* is a monotonic increasing function.
- We are dealing with amplifiers with a constant gain-bandwidth product. The discussion is similar to the aforementioned point; however, noise depends on the gain. In any case, the change of noise standard deviation is related to the square root of the change of *FS* or *FSO*. Therefore, *NL(FS, FSO)* is again a monotonic increasing function.

Therefore, the increase of resolution levels by the increase of full scales is ensured in most practical cases, and in conclusion, to maximize mutual information, we have to look at the full scales of the single stages and determine the gains accordingly, as will be apparent in the following section.

3.4.2   The Resolution Rule in Acquisition Chains

The previous example was based on a single block case. What happens for complex chains composed of multiple stages, illustrated in the plot diagram of Fig. 3.20? For a system composed of *K* stages, the total required gain chain is given by

$$S_T = S_1 \cdot S_2 \cdots S_K = \frac{FSO_{max}}{FS_{APP}}. \tag{3.22}$$

In Eq. (3.22), we can see that the main constraints are related to the first and the last stages; however, we have several degrees of freedom related to each internal stage. In practice, several combinations of gains producing the same total gain.

Consider a sensor acquisition chain composed of K known and characterized linear blocks as illustrated in Fig. 3.21A. On each of them is modeled the internal noise that input-referred noises have represented at the input of each block. The input is mixed with an external noise σ_N^2 while the input-referred noise of each block is indicated with σ_{iN}^2; $i = 1, \ldots, K$. Following the arguments of Section 3.3, we can transpose the noise generators from the output to the input by dividing their noise power contribution for the square of the gain (sensitivity) of each block. Therefore, by applying this rule to each block, we can obtain the representation of Fig. 3.21B, where $\sigma_{N(tot)}^2(x)$ is the input-referred noise of the whole system.

Therefore,

$$\sigma_{N(tot)}^2 = \sigma_N^2 + \sigma_{1N}^2 + \frac{\sigma_{2N}^2}{S_1^2} + \frac{\sigma_{3N}^2}{S_1^2 S_2^2} + \cdots + \frac{\sigma_{KN}^2}{S_1^2 S_2^2 \ldots S_{K-1}^2}. \tag{3.23}$$

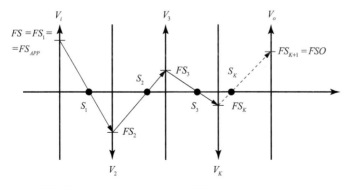

Figure 3.20 Gain path of a sensor acquisition chain.

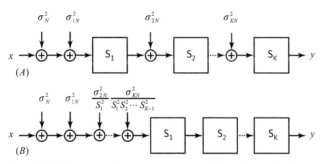

Figure 3.21 Representation of a sensor acquisition chain with input noise and input-referred noise blocks (A). Equivalent representation (B).

Noting that

$$FS_j = FS \cdot S_1 \cdot S_2 \cdots S_{j-1} \tag{3.24}$$

and assuming $NL_k = FS/(q \cdot \sigma_{kN}) = \sqrt{1 + DR_k}$ as the number of resolution levels of the kth section, we can calculate the following (using $q = 1$ for the moment for simplicity of notation),

$$\frac{1}{NL_T^2} = \frac{\sigma_{N(tot)}^2}{FS^2} = \frac{\sigma_N^2 + \sigma_{1N}^2 + \dfrac{\sigma_{2N}^2}{S_1^2} + \dfrac{\sigma_{3N}^2}{S_1^2 S_2^2} + \cdots + \dfrac{\sigma_{KN}^2}{S_1^2 S_2^2 \cdots S_{K-1}^2}}{FS^2}$$

$$= \frac{\sigma_N^2}{FS^2} + \frac{\sigma_{1N}^2}{FS^2} + \frac{\sigma_{2N}^2}{FS^2 \cdot S_1^2} + \frac{\sigma_{3N}^2}{FS^2 \cdot S_1^2 S_2^2} + \cdots + \frac{\sigma_{KN}^2}{FS^2 \cdot S_1^2 S_2^2 \cdots S_{K-1}^2} \tag{3.25}$$

$$= \frac{\sigma_N^2}{FS^2} + \frac{\sigma_{1N}^2}{FS^2} + \frac{\sigma_{2N}^2}{FS_2^2} + \frac{\sigma_{3N}^2}{FS_3^2} + \cdots + \frac{\sigma_{KN}^2}{FS_K^2}$$

$$= \frac{1}{NL_0^2} + \frac{1}{NL_1^2} + \frac{1}{NL_2^2} + \cdots + \frac{1}{NL_K^2},$$

where $NL_0 = FS_{APP}/(q \cdot \sigma_N)$ is the natural resolution of the source, that is, the information levels seen by the sensor for a given source noise.

The relationship (3.25) is the *resolution rule in acquisition chains (RRC)*.

Therefore, the reciprocal of the total resolution levels' square is given by the sum of reciprocals of each stage's square resolution levels. This means that the *lowest resolution stage limits the total resolution*,[3] and therefore *the increase of the overall resolution could be achieved only by increasing the resolution of any single block*.

Hint The stage of lowest resolution limits the total resolution of the system.

It should be noted that the first term of Eq. (3.25) collects both the input-referred noise of the first block and the input noise associated with the signal. Note that the resolution levels contribute to the relationship as their squared value because the noise should be accounted for in power. For instance, two cascade-connected blocks with identical resolution give a lower resolution system by a square-of-two factor. If the last block is an A/D converter, we should consider the resolution given by the effective number of bits (ENOB) as indicated in Chapter 2. From the information point of view, to convey the maximum information from the input to output, we need to avoid any bottleneck that reduces the number of levels during single-stage transductions.

It should be recalled that if the last block is an A/D converter, it could be better to use the number of resolution levels expressed using ENOB for any block of the chain:

---

[3] This comes from the fact that noises are composed as the root of the sum of squares (Pythagorean's theorem), and the result (hypotenuse) is always greater than the largest component (cathetus).

$$NL_{\mathrm{ENOB}_k} = 2^{\mathrm{ENOB}_k} \quad \text{where,}$$

$$\mathrm{ENOB}_k = \frac{DR_{dB(k)} - 1.76}{6.02} \quad \text{and} \tag{3.26}$$

$$DR_{dB(k)} = 20\log\frac{FS_k}{q \cdot \sigma_{kN}}$$

where for the last block, we use the ENOB of the A/D converter directly. It should also be reminded that NL_{ENOB} is about 20% lower than NL_k, as explained in Section 2.14.4.

The resolution rule for acquisition chains could also be seen in terms of dynamic ranges

$$\frac{1}{DR_T} = \frac{1}{DR_0} + \frac{1}{DR_1} + \frac{1}{DR_2} + \cdots + \frac{1}{DR_K}. \tag{3.27}$$

The approach for calculating the resolution rule could be better illustrated as in Fig. 3.22. Each block is represented by a noiseless stage with the associated input-referred noise. The input full-scale defines the number of resolution levels. Therefore, we can define the DR and the NL for each block to which it is strictly related.

3.4.3   Approach and Example of Application of the Resolution Rule in Acquisition Chains

The partitioning of the acquisition chain evidenced in Fig. 3.22 suggests using the resolution rule (3.27) for easily defining the resolution/precision/accuracy[4] characteristics of one or more blocks on the basis of the requirements. To put the approach in points:

1. The resolution design of the entire chain should cope with the application requirements summarized in the overall dynamic range DR_T, which is related to the overall resolution or total ENOB_T.

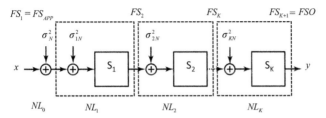

Figure 3.22 Sensor acquisition path where each stage is characterized in its own resolution.

---

[4] At the moment, these considerations are related only to resolution requirements and do not take into account the bandwidth constraints that will be addressed in the next sections.

Table 3.1 Resolution levels definition in different contexts.

$$DR = \frac{\sigma_{S(\text{max})}^2}{\sigma_N^2} = \frac{FS^2}{q^2\sigma_N^2}$$

Information/ theory notation	**A/D converter notation**
$NL = \sqrt{1+DR} \approx \sqrt{DR}$	$NL_{\text{ENOB}} = 2^{\text{ENOB}}$
effec. resolution $= NL_b = \dfrac{1}{2}\log_2(1+DR)$	$\text{ENOB} = \dfrac{DR - 1.76}{6.02}$

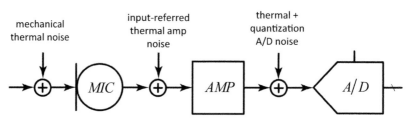

Figure 3.23 Block diagram of the exercise. The microphone is characterized by a $DR = 91$ dB given by thermal-mechanical noise and other sources. The goal is to find the amplifier and A/D converter characteristics to get a 12b noise-free resolution in 20-kHz bandwidth.

2. We partition the design by first treating the analog interface with its total dynamic range DR_{AT}; then, we will consider the choice of the A/D converter characterized by its dynamic range $DR_{A/D}$. Of course, we will start using $DR_{AT} > DR_T$ due to RRC.
3. For the analog part, define the DR_k for each block of the analog chain. For some of them (e.g., a third-party sensor), the DR is fixed, while for others, the DR is an expression of input-referred noise.
4. Based on DR_{AT}, we find the noise requirements of the blocks to be designed, using RRC for the analog part
5. Finally, we choose the A/D converter to cope with DR_T, using RRC for the whole system

In every step, we can control the total or the section resolution characteristics using RRC by means of NL_b or NL_{ENOB} as summarized in Table 3.1.

We will explain this better with an example.

Example An analog microphone is characterized by $DR_{M(\text{dB})} = 91$ dB and output at an acoustic overload point $FS_M = 25$ mV. The building blocks are illustrated in Fig. 3.23. We want to design an acquisition chain to achieve 11b noise-free resolution in bandwidth $BW = 20$ kHz and, in particular, to calculate the required input-referred noise of the amplifier. The first question is whether we can achieve this. The answer is positive because the ENOB of the total chain should be (Eq. 2.130) 11+2.43=13.43b, corresponding to 82.6 dB of total dynamic range, which is lower than that of the mic.

Following RRC, we start by setting the ENOB of the analog part 1b more than the total chain, that is $ENOB_{AT} = 13.43 + 1 = 14.43b$, so that the corresponding dynamic range is $DR_{AT(dB)} = 14.43 \cdot 6.02 + 1.76 = 88.6$ dB. We find the dynamic range of the amplifier using (3.27): $DR_{AMP} = 1/(1/DR_{AT} - 1/DR_M) = 1.7 \times 10^9$ from which we determine the input-referred noise density: $\sigma_N(f) = \sqrt{FS_M^2/(q^2 \cdot DR_{AMP} \cdot BW)} = 1.5 nV/\sqrt{Hz}$, the required characteristic of the amplifier and the main result of the design. Note that we used $q = 2\sqrt{2}$ since we are working in AC mode.

Now we must choose the A/D converter. We can start with an A/D resolution and check it with RRC to know if the total dynamic range DR_T is as close as possible to $DR_{T(dB)} = 82.6$ dB or $ENOB_T = 13.43b$. If we start using an A/D converter with ENOB = 14b, we get $DR_T = 1/(1/DR_M + 1/DR_{AMP} + 1/DR_{A/D}) \rightarrow DR_{T(dB)} = 84.0$ dB$\rightarrow ENOB_T = 13.6b$ which is right with the specs. If we lower the constraint to ENOB = 13b, we get d $DR_{T(dB)} = 79.4$ dB$\rightarrow ENOB_T = 12.9b$ that does not comply with specifications.

3.4.4 Optimization of the Acquisition Chain from the Resolution Point of View

Following the final discussion of Section 3.4.1, note that resolution levels are monotonic functions of the input and output full scales (i.e., gain)

$$NL_K = NL_K(FS_{K+1}, FS_K). \tag{3.28}$$

We have referred to the fact that, in most circuital cases, an *FS* increase (either at the input and/or output) corresponds to a monotonical increase of the resolution levels. Therefore

$$\max\{NL_K\} = NL_K(\max\{FS_{K+1}\}, \max\{FS_K\}). \tag{3.29}$$

This turns out to be related to the fact that by increasing the full-scale swings, we increase the input and output information entropy with increased mutual information.

Another critical issue is that changing each block's gain could involve changing the *bandwidth* of each block with a corresponding change of the input-referred noise. We will be more explicit about this in the following.

From the design standpoint, it could be convenient to follow these steps:

- Fix $FS = FS_{APP}$ so that the input full scale corresponds to the input swing required by the application. This is the primary goal of the design. For example, we have to ensure that the saturation points of the first block of the chain shall cope with the required range. If not, we have to decrease its gain.
- For the maximum information transfer, we have to reach the maximum optimal output swing; therefore FSO = FSO$_{max}$. For example, if an analog transduction

chain should be interfaced with an A/D converter with $V_R = 3.3$V, the optimal point is given by $FSO = 3.3$ V.
- The above points fix the required total gain. Given that, we have many combinations of stage gains resulting in the same total gain $FSO/FS = S_T = S_1 \cdot S_2 \cdots S_K$.

Now the question is: which is the best combination of single gains to achieve the best resolution number of levels? Eqs. (3.25) and (3.29) show that the best result could be achieved *by maximizing the FSs at each stage junction*. Of course, this is an optimization that considers only the resolution point of view, and we will see in the next sections how to relate this with bandwidth and power consumption.

> **Hint** In a sensor acquisition chain, the optimal design in terms of resolution is achieved by maximizing the signal swings (*FSs*) at each stage within the given constraints. This is related to the fact that this rule increases the output information entropy at each stage and increases the mutual information.

It is interesting to note that in the overall optimization of the resolution, given by the *resolution rule* in the acquisition chain (3.25), the sensitivity should be increased for some stages, while in other cases, the gain should be decreased. We will be more specific using a graphical approach of optimization as follows. In Fig. 3.24 are graphically represented the main limitations of the gain variations.

The main constraints related to each block are

- High gain constraints. We are not allowed to get a higher gain than a certain value. This could be imposed for several reasons. For example, our technology could not achieve values higher than a certain level, or the gain is related to the bandwidth, and we cannot increase the gain unless restricting the bandwidth that is required by the

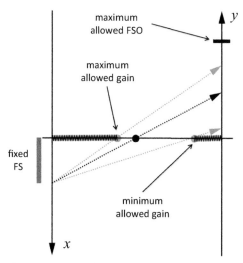

Figure 3.24 Gain plot representation of the system constraints.

application. The forbidden range for a high gain is shown as a thick line on the left of Fig. 3.24. That boundary is due because the increase of the gain implies a shift on the left point.

- Low gain constraints. This is less frequent; however, there are cases where, for example, to avoid stability problems of the stage, we cannot decrease the gain below a certain level. The forbidden range for a high gain is shown as a thick line on the right of Fig. 3.24. Such a boundary is due because the decrease of the gain implies a shift on the right of the point.
- Full-scale constraints. In addition to the maximum/minimum constraints of the gain, we could encounter the case where there is a maximum allowed output swing. This could also be related to a constraint of the input swing of the subsequent stage. For example, we cannot surpass rail-to-rail power supply boundaries, which is a hard limit of real electronic devices. This constraint is shown as a tick of the right axis of Fig. 3.24.

In an acquisition chain, we can draw the constraints as in Fig. 3.25A, where both the gain and the *FS* constraints are represented. Note that even if the total gain is achieved, it is not optimized from the resolution point of view. Therefore, the goal is to increase the *FS*s at the intersections of each stage as much as possible, respecting the

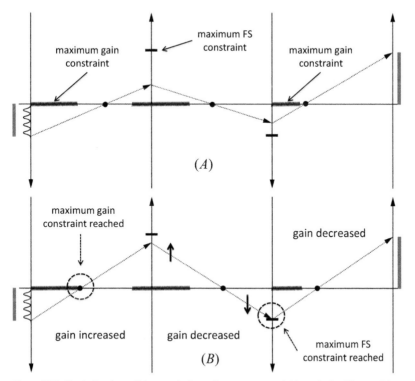

Figure 3.25 Optimization of the resolution of a sensor acquisition chain. The goal is to increase the *FS*s at each stage intersections. (A) Unoptimized chain. (B) The optimized chain where the boundary constraints have been achieved.

constraints. In the optimized version of the chain of Fig. 3.25B, we have to increase as much as possible the gain of the first amplifier; however, we cannot reach the *FS* constraint since we achieve the gain constraint before. Also, note that we decrease the gain of the second and third stages.

Example A strain gauge is interfaced with a programmable gain amplifier (PGA) to implement a weight scale, as illustrated in Fig. 3.26. The force exerted by the weight strains a rod on which is applied a strain gauge whose change of resistance is transduced into a voltage by a Wheatstone bridge that a PGA finally amplifies. The application requires $FS_{APP} = 100$ kg full-scale, and the optimal analog output should be FSO $_{max} = 3.3$ V because it should be interfaced to an A/D converter with $V_R = 3.3$ V; therefore, the required gain should be $S_T = 3.3/100 = 0.033$ V/kg. We will disregard at the moment the resolution of the A/D converter. The gain of the transducer and the bridge together is $S_B = 1.45 \times 10^{-5}$ V/kg while the default gain of the PGA is $S_E = 1000$ V/V. The input-referred noise of the PGA is $\sigma_E = 12$ nV at 1 Hz of bandwidth and does not depend on the gain. The noise at the bridge could be calculated for bridge resistances of 120 Ω in a bandwidth of 1 Hz as $\sigma = \sqrt{4kRT} = 1.4 \times 10^{-9}$ V that once referred to the input by mechanical gain becomes $\sigma_B = 1.4 \times 10^{-9}/1.45 \times 10^{-5} = 9.6 \times 10^{-5}$ kg and it is quite lower with respect to that of the PGA. In fact, $NL_B = 100/9.6 \times 10^{-5} = 1.035\,k \rightarrow NL_{B(b)} = 19.9$b and $NL_E = 1.4$ mV/12 nV $= 121\,k \rightarrow NL_{E(b)} = 16.8$b. Applying (3.25) we have $NL_T = 120\,k \rightarrow NL_{T(b)} = 16.8$b; therefore, the resolution is essentially determined by the PGA and minimally by the bridge.

If we apply a full-scale input of 100 kg, we thus get about 1.4 mV at the input of the PGA, thus just 1.4 V $< FSO_{max}$ at the output. Therefore, we need to increase the total required gain. We might increase the gain of the amplifier to 2.270 to get 3.3 V. However, we do not have any advantage of the overall resolution, as shown in Fig. 3.26B.

Following the previous discussion, a better strategy is to increase the *FS* of the input of the PGA by increasing the gain of the bridge. How can we do that? We can operate at the *mechanical level*, as shown in Fig. 3.27, by reducing the size of the bar. It could be shown that the gain increases linearly with the reduction of the cross area of the bar. Therefore, by keeping the gain of the PGA equal to 1000, we have to decrease the size of the bar of a factor $3.3/1.4 = 2.27$. In doing so, we have an increase of the overall resolution because now $NL_B = 100/9.6 \times 10^{-5} \cdot 2.27 = 2.350\,k \rightarrow NL_{B(b)} = 21.1$b and $NL_E = 3.3$ mV/12 nV $= 275\,k \rightarrow NL_{E(b)} = 18.0$b. Applying (3.25) we have $NL_T = 273\,k \rightarrow NL_{T(b)} = 18.0$b, with an increment of more than 1 virtual bit.

We must be careful not to increase the gain of the PGA. For example, if we double the gain to 2000, we lose half of the range of the data.

On the other hand, we might think to increase further the gain of the bridge to get higher resolution levels. However, this is not possible without a limit. More specifically, if we decrease too much the size of the bar, the strain is so high that the

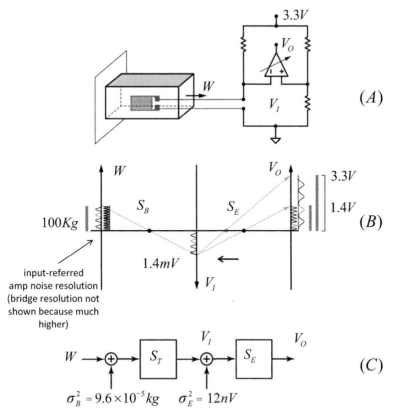

Figure 3.26 Strain gauge example. The optimal gain is set in the electronic part of the system.

material could encounter nonreversible damages. Therefore, we must limit the gain to a threshold, as shown in Fig. 3.27.

In any case, we can *further increase* the gain, up to physical limits, of the first stage and *decrease* the gain of the second one (light arrows Fig. 3.26) to get even more resolution.

The lesson learned in the previous example is that we must act at increasing the full scales at the intersections of all stages for optimized resolution. The gains of the respective stages should adapt (by increase or decrease) accordingly.

3.4.5 Optimal Choice of the A/D Converter

If we take the case where a real A/D converter should terminate a transduction chain, as shown in Fig. 3.28, we may wonder what the right choice of the A/D converter to cope with the required precision/or accuracy given by the transducer is.

In Chapter 2, we discussed how an analog interface could be characterized in its precision and accuracy using "virtual bits." (in the sense that they are not binary digits

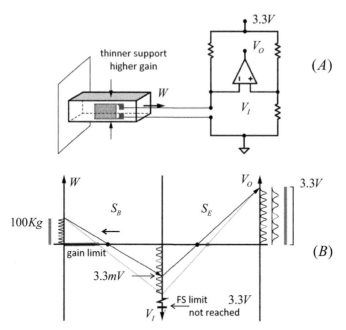

Figure 3.27 Strain gauge example. The optimal gain is set in the physical part of the system. However, we cannot further increase the mechanical gain due to irreversible damage to the material.

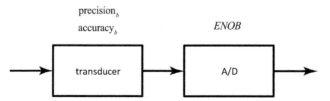

Figure 3.28 Interfacing a characterized transducer with an A/D converter.

but related to a unit of information). Therefore, at first guess, we could be induced to use a converter with an ENOB equal to the same virtual bits of the transducer. This would lead to a design of the transduction chain that is not optimized due to the *resolution rule in acquisition chains* (*RRC*). This is because we have to look at the overall chain's resolution and not only of the single blocks. For example, if we make a system by a cascade of two stages of equal resolution, the overall resolution of the system will be that of single stages divided by the square of two under Eq. (3.25).

For the above reason, it is better to set the ENOB of the converter at least one order of magnitude (in binary notation, i.e., one bit) greater than that of the resolution of the transducer to get an overall system having a total resolution as close as possible to that of the transducer. As far as precision is concerned, we choose an A/D converter with an ENOB greater than the ENOB of the transducer (in terms of "virtual bits" precision)

$$\text{ENOB}_{\text{SNR}} > \text{precision}_b, \tag{3.30}$$

where ENOB_{SNR} is the effective number of bits calculated using SNR alone.

A similar conclusion could be achieved as far as the accuracy is concerned:

$$\text{ENOB}_{\text{SNDR}} > \text{accuracy}_b, \tag{3.31}$$

where $\text{ENOB}_{\text{SNDR}}$ is the effective number of bits calculated using the SNDRT. Of course, if we assume that offset and gain errors have been corrected before the quantizer and do not consider systematic error variations, we can substitute for accuracy SNDRT with SNDR.

The number of bits should be checked by means of the *resolution rule of the acquisition chain*, and for the case of precision, it is

$$
\begin{aligned}
DR_{AT(\text{dB})} &= \text{precision}_b \cdot 6.02 + 1.76; \\
DR_{A/D(\text{dB})} &= ENOB \cdot 6.02 + 1.76; \\
\frac{1}{DR_T} &= \frac{1}{DR_{AT}} + \frac{1}{DR_{A/D}}
\end{aligned}
\tag{3.32}
$$

and we have to check if the ENOB of the A/D converter is consistent with the DR_T requirements. The same reasoning could be applied for accuracy. In general, 1–3 bits are sufficient to achieve this requirement, depending on the case.

A final note should be pointed out. If we intend to correct by calibration the offset and systematic errors using digital techniques, it is useless to adhere to Eq. (3.31): it is enough to use Eq. (3.30) and correct both the transducer and A/D converter nonlinearities at the end of the chain as we have seen in the Section 3.2.3.

Example The AC characterization of an interface sensor shows an $\text{SNR}_{\text{dB}} = 85$ dB and a THD of -42 dB. This is consistent with the measured INL error, that is 63 ppm $= 10^{-42/10} \cdot 1 \times 10^6$. The SNDR could be found by using $\text{SNDR}_{\text{dB}} = -10\log(10^{-\text{SNR}_{\text{dB}}/10} + 10^{\text{THD}_{\text{dB}}/10}) = 42$ dB. This value should not surprise since THD is much worse than DR, so SNDR is de facto $-$THD. This means that the precision of the analog interface is $\text{precision}_b = (\text{SNR}_{\text{dB}} - 1.76)/6.02 = 13.8b$. To understand what A/D to couple with, we should use Eq. (3.32), where $DR_{AT(\text{dB})} = \text{SNR}_{\text{dB}}$ and $DR_{A/D(\text{dB})} = 6.02 \cdot ENOB + 1.76$. Thus, if we use an A/D converter with ENOB $=17b$, we get $DR_T = 1/(1/DR_{AT} + 1/DR_{A/D}) \rightarrow DR_{T(\text{dB})} = 84.95$ dB, which is close to the 85 dB of the analog interface. Therefore, we need about 3 bits more in this case. If we had used the same number of equivalent bits of the analog part, the total DR would decrease by -3 dB, as already explained. If we used 16b and 15b converters, we would have achieved 84.79 dB and 84.22 dB, respectively.

As far as the accuracy is concerned, we have $\text{accuracy}_b = (\text{SNDR}_{\text{dB}} - 1.76)/6.02 = 6.7b$. Following the same reasoning as before, using an A/D converter with ENOB $= 9b$, we get $\text{SNRT}_{T(\text{dB})} = 41.8$ dB which is close to the 42 dB of the analog interface. Even in this case, we use about 3 more bits of the converter with respect to the calculated accuracy of the analog interface.

> **Hint** For a characterized transducer in terms of noise and systematic errors, the final A/D converter should have an ENOB greater than the equivalent number of bits in precision and accuracy of the analog interface.

3.5   Sampling, Undersampling, Oversampling, and Aliasing Filters

3.5.1   Oversampling and Quantization

The Nyquist–Shannon sampling theorem states that we can preserve all the characteristics of a signal with limited bandwidth f_M by sampling it at a frequency $f_S \geq 2f_M$ referred to as Nyquist frequency. If f_S is sufficient to catch all the information of the signal, why sample the signal at a frequency higher than Nyquist's one? We will show that this approach could be beneficial in the case of quantization noise.

Referring to Fig. 3.29, we have seen in Chapter 2 that the quantization noise PSD could be assumed as uniform in the baseband if we assume an equal distribution of input values in the FS,[5] where the total noise power is $P_N = V_{\mathrm{LSB}}^2/12$. Now, if we oversample to OSR $\cdot f_S$ by the factor *oversampling ratio OSR*, the same quantization power P_N is uniformly distributed between $-\mathrm{OSR} \cdot f_M$ and OSR $\cdot f_M$. However, by filtering the result back to the baseband, the quantizing noise is reduced to $V_{\mathrm{LSB}}^2/(12 \cdot \mathrm{OSR})$.

Therefore, under the cited assumption, oversampling is beneficial to reduce quantization noise.

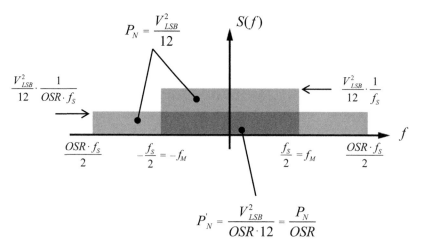

Figure 3.29 Concept of oversampling.

---

[5] Actually, this is an abrupt assumption. For a clear understanding of this assumption's limits, refer to Widrow and Kollar's book.

3.5.2   Oversampling and Undersampling of White Noise

Let assume that a source of white noise is bandwidth-limited to f_N (e.g., an amplifier that has a cutoff frequency equal to f_N). In Fig. 3.30A is shown a band-limited noise oversampled at a frequency $f_S > 2f_N$. Even if we are in the power domain, thanks to the fact that we can apply superposition of power for uncorrelated sources (Chapter 4), we can visualize the effect of sampling as in the amplitude modulation (AM). Therefore, a copy of the bandlimited white noise bandwidth is rotated around the origin and shifted to f_S as a sideband replica. The effect is shown in Fig. 3.30A, right. If we now filter at baseband, we have no advantage in the process of oversampling, as in the quantization noise. The approach could also be explained in this way. If we are oversampling a band-limited white noise, we get fairly correlated values because the effect of the band limitation of white noise spreads its autocorrelation function for a time constant (Chapter 4). Therefore, when we filter to the baseband, we have no advantage in reducing noise power.

The problems arise if we downsample, as we can see in Fig. 3.30C–E. In that case, the samples are uncorrelated, and then they sum them up, creating what is referred to as *noise folding or noise aliasing*. Therefore when we filter back to the baseband, we have an effect of multiplicating the noise. The amount of noise folding is given by the $(2f_N)/f_S$ factor. We will discuss this effect in detail in Chapter 6.

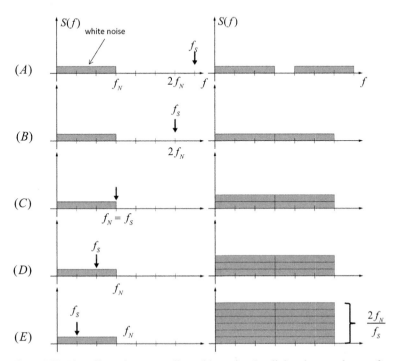

Figure 3.30 The effect of oversampling white noise. In all the plots, we have referred only to the positive frequency for simplicity. Furthermore, only the first two replicas are shown.

3.5.3   Oversampling and Downsampling of Signals and Noise

There are frequent cases where signal and noise have different bandwidths. An example is shown in Fig. 3.31, where the bandwidth of the interface noise is greater than that of the signal because we have chosen an interface with different properties with respect to those of the signal. In this case, to avoid noise folding, we need to sample at a frequency that is at least double that of the noise bandwidth as shown in Fig. 3.32A, B. As explained in Section 3.5.2, oversampling could be beneficial for quantization noise only.

However, if we overlook the problem by looking at the signal only, we might get noise folding, as is shown in Fig. 3.32C and D, seriously degrading the SNR. This problem could be avoided using an anti-aliasing filter, as shown in Fig. 3.31. The key point is to filter the composition of signal and interface noise at the signal bandwidth *before sampling*.

3.6   Power, Resolution, and Bandwidth Tradeoffs in Sensing

This section will investigate how the basic parameters of sensing systems are related to each other. Most of these tradeoffs are based on thermal noise, which is why these constraints are frequently referred to as *thermal noise tradeoffs*.

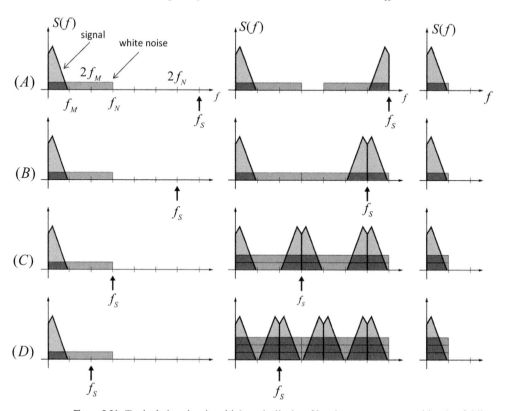

Figure 3.31 Typical situation in which anti-aliasing filter is necessary to avoid noise folding.

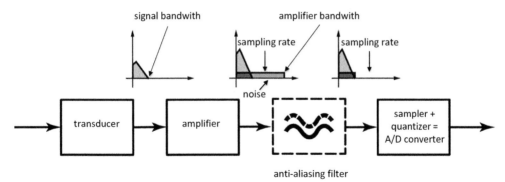

Figure 3.32 Sampling signal and noise with different sampling rates. As clearly shown, if noise is downsampled, noise folding could seriously lower the SNR at the baseband.

Usually, we will refer to *figures of merit* (FoMs) and how they are composed. Figures of merit cannot, in general, be demonstrated, but they are tentative laws based on experiences regarding a large number of examples. Of course, they have hidden basic principles; however, they are mainly focused on technology and/or architecture performance. One approach could be in considering as few parameters as possible, trying to make the relationship as general as possible. Of course, if many examples could help determine a trend, a counterexample could be exhaustive to determine that something is not true in general, indicating new possible relationships.

We will refer to a new quantity in the next discussion, which is the electric power P consumed by the system, which is an important issue in modern autonomous sensors. This is a power that is necessary to supply sensor systems, which is used for computation and is finally dissipated into heat.

To summarize, in the next sections, we will refer to the following entities.

Notation

P: Electrical power consumed by the system in the sensing process. Not to be confused with the power of signal or noise.

BW: Bandwidth of the system.

DR: Dynamic range, related to the number of resolution levels (NL).

3.6.1　The Role of Time

If we reduce the bandwidth of the same system by a factor α we have an equivalent reduction of the noise power, as discussed in Section 3.1. Therefore, we have an increase of the DR/SNR of the same factor. At the same time, in general, reduction of bandwidth does not imply a change of power consumption, as in the case of using a higher-order passive filter.

> **Bandwidth/DR (Resolution) Tradeoff** For equal power consumption, the product
> of the *bandwidth* for the *dynamic range* (resolution) is constant.

The relationships are

$$
\begin{aligned}
&P = \text{constant} \\
&BW \leftarrow BW/\alpha; \\
&\sigma_N^2 \leftarrow \sigma_N^2/\alpha; \\
&DR \leftarrow DR \cdot \alpha \\
&BW \cdot DR = \text{constant}.
\end{aligned}
\tag{3.33}
$$

3.6.2   The Role of Power

If we desire to increase the interface's bandwidth by keeping the same resolution,
a possible approach is to use sampling in a new fashion. We have available an A/D
converter with a specified resolution that consumes P but can only achieve half the
sampling frequency f_S of the input signal. We can increase the BW by sampling the
same signal using another identical ADC at interleaved intervals as illustrated in Fig.
3.33 and then collecting the data at the output. Each of them is sampling the original
signal at half the Nyquist frequency to achieve the right sampling frequency. However,
the total power consumption doubles with respect to the original approach, showing
a linear relationship versus bandwidth. The same approach could be used to increase
the frequency at the cost of increasing the power consumed by the same factor. This
approach is referred to as *interleaving sampling*.

Another example is in complementary metal–oxide–semiconductor (CMOS) technology, where it is well known that any switched capacitor circuit has a power
consumption that is given by

$$
P = k \cdot CV_{DD}^2 f,
\tag{3.34}
$$

where k is a constant, C is the load capacitance, V_{DD} is the power supply voltage, and f
is the switching frequency. Eq. (3.34) shows a linear relationship between power
consumption and the interface's bandwidth again.

> **Bandwidth/Power Tradeoff** For equal *dynamic range* (resolution), the higher the
> *bandwidth*, the higher the *power consumption*.

Therefore, using the same scaling constant α

$$
\begin{aligned}
&DR = \text{constant} \\
&BW \leftarrow BW \cdot \alpha; \\
&P \leftarrow P \cdot \alpha; \\
&P/BW = \text{constant}.
\end{aligned}
\tag{3.35}
$$

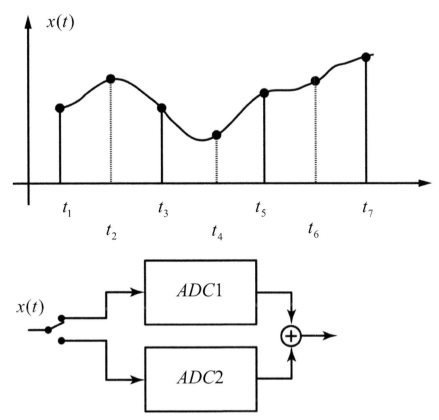

Figure 3.33 Interleaved ADC conversion. ADC1 samples the signal during t_1, t_3, t_5, ... while ADC2 during t_2, t_4, t_6, ...

As a result, we note that the ratio between the power consumed and the bandwidth is, in general, constant.

3.6.3   The Role of the Dynamic Range

The dynamic range and the power consumption give another important relationship. Assume we need to increase the dynamic range, that is, the resolution, by keeping the same bandwidth. How to cope with that? We can take an example of an "averaging" amplifiers. We already have discussed that averaging reduces the mean square of a random process such as noise. The idea is that we sample the same signal $x(t)$ using two identical amplifiers, A1 and A2, characterized by uncorrelated input-referred noise σ_{1N}^2 and σ_{2N}^2 and power consumption P. The outputs of the amplifiers are averaged, as illustrated in Fig. 3.34.

   The result is that the mean square of the noise introduced by the interface is reduced by a factor of 2 by averaging with respect to the single interface case. However, we need to consume double power to achieve this. Therefore, we could double the DR by

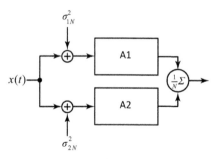

Figure 3.34 Averaged interface approach.

doubling the consumed power. The simple case above is used to explain the concept and could not be applied for an undefined number of interfaces in real systems for several constraints.

Averaging images could give another example. Assume that we need to take a picture in poor light conditions, but we cannot increase the shutter time (decrease bandwidth) because of moving objects. To decrease noise, we can think of taking the same image using several cameras with the same shutter time and then averaging the result as shown in Fig. 3.35.

As in the previous example, we expect that noise is reduced by a factor of N in power, where N is the number of cameras. However, parallel cameras increase the consumed power by a factor of N.

The same example could be used for the previous tradeoff. If we had no problems with moving objects, we could imagine increasing shutter time of a factor of N, giving the same reduction of noise but reducing the bandwidth of the same amount. In this case, the power consumption is the same since the single camera has to be N times slower (consuming N times less) for N times higher period.

> ***DR*/Power Tradeoff** For equal *bandwidth*, the higher the *dynamic range*, the higher the power *consumption*.

Therefore

$$\begin{aligned}
BW &= \text{constant} \\
DR &\leftarrow DR \cdot \alpha; \\
P &\leftarrow P \cdot \alpha; \\
P/DR &= \text{constant}.
\end{aligned} \tag{3.36}$$

3.6.4   Putting It All Together

The tradeoffs mentioned earlier, supported by the examples, could be merged in the following expression

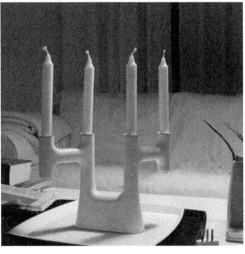

(A) (B)

Figure 3.35 Parallel camera acquisition. (A) Image acquired by a single camera in low light conditions. The same image is taken by the other six cameras using the same shutter time and then averaged in the image (B).

$$\text{FoM}_A = \frac{P}{DR \cdot BW} = \frac{P}{(NL)^2 \cdot BW} = \text{constant} \equiv [J], \tag{3.37}$$

where P is the power consumed, DR is the dynamic range, and BW is the bandwidth. Therefore, *the lower* the FoM, *the better*.

Another way to express the previous relationship is to use the part-per-notation (%, ppm, ppb, etc.) in place of the resolution levels. An example using ppm is

$$\text{FoM}_A = \frac{P}{(NL)^2 \cdot BW} = \frac{P \cdot \tau}{(NL)^2} \equiv [J] \rightarrow \text{FoM}'_A = \frac{P}{(NL)^2 \cdot BW} 1 \times 10^{12} \equiv [J \cdot \text{ppm}^2],$$
$$\tag{3.38}$$

where τ is the system response time and we used the resolution in ppm as $10^6/NL$ and $[\text{ppm}] \equiv 10^6 [\cdot]$.

If we multiply both terms of Eq. (3.37) by the square of the full scale and express DR in the number of resolution levels, we have

$$\text{FoM}_B = \frac{P \cdot FS^2}{NL^2 \cdot BW} = \frac{P \cdot \sigma_{iN}^2}{BW} = \text{energy} \cdot \text{resolution}^2 \equiv [J \cdot \xi^2], \tag{3.39}$$

where "resolution" is referred to as the effective resolution. This could be understood as the energy required to "resolve" (or read) an input power quantity equal to the input-referred noise power. Again, the lower, the better.

Example A temperature sensor and humidity sensors have the following figures of merit: 62 fJ $\cdot K^2$ and 0.83 pJ $\cdot (\%RH)^2$, respectively.

FoM_B should be used with caution because by multiplying for the full scale, it loses the concept of the amount of information (number of information levels) exchanged by the interface, which is instead considered in FoM_A. Therefore, FoM_B is useful to compare architecture within the same application (e.g., temperature and humidity sensors), where the full scale is almost the same.

We can also reverse the relationship (3.37) and put it into the log domain as

$$FoM_C = DR_{(dB)} + 10\log\left(\frac{BW}{P}\right) \equiv [dB]. \tag{3.40}$$

Again, the hypothesis underlying the constraints expressed by Eqs. (3.37) and (3.40) is that the noise characterizing the system interface is thermal noise. This is why the limits expressed by those relationships are called *thermal noise limited* FoMs.

As far as A/D converters are concerned, many FoMs regarding ADCs have been proposed so far. One of the first, and mostly used FoM, is related to the number of discrete levels $NL = 2^N$ in which the output of the ADC is segmented, where N is the number of bits, to the sampling frequency f_S, so that the energy is related to each conversion step

$$FoM_W = \frac{P}{2^N \cdot f_S} = \frac{P}{2^N \cdot 2BW} \equiv [J/\text{dicrete levels}], \tag{3.41}$$

which is also referred to as Walden's FoM, where the unit of measure is in Joules per sample. Typical values are in the order of 1 pJ/sample or less for a number of bits lower than 10.

Recently, for a significant number of bits ($N > 10$), it has been shown an FoM that better follows the technology and architecture trend. Following Eq. (3.37), we have

$$FoM_S = SNDR + 10\log\left(\frac{BW}{P}\right) \equiv [dB], \tag{3.42}$$

which is called Schreirer's FoM. The log-log form of it is nothing else than (3.40), and this is why it is also referred to as *thermal-noise limited* FoM.

The above FOMs could be converted in the energy per conversion-step unit

$$\frac{P}{f_S} = FoM_W \cdot 2^{\left(\frac{SNDR - 1.76}{6.02}\right)}; \quad \text{and} \quad \frac{P}{f_S} = \frac{1}{2}10^{-\frac{(FoM_S - SNDR)}{10}}. \tag{3.43}$$

Figure 3.36 shows the plot of the two aforementioned FOMs with respect to the A/D performances presented in the most prestigious integrated electronics conferences.

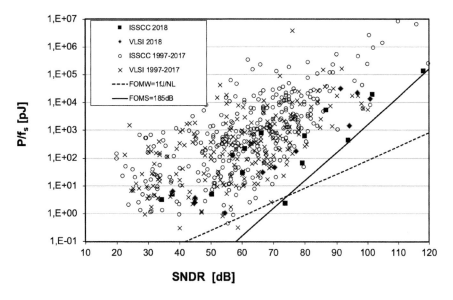

Figure 3.36 Log-log plot of a figure of merit limits for FOM$_W$ and FOM$_S$. As shown, FOM$_S$ is better fitting data for a higher number of bits. Elaborated from: Murmann, "ADC Performance Survey 1997–2018." Available at:http://web.stanford.edu/~murmann/adcsurvey.html.

That is the reason why we used ENOBs in Eq. (3.43). As clearly shown, the thermal FOM sets the boundary for a larger number of bits. As long as the technology trend is evolving, new boundaries are shifting down parallel to the lines traced in the picture, showing a better performance.

Example A 20-kHz bandwidth sensor has an output full-scale $FSO = 1$ V with a noise of $\sigma_N = 1.8\ \mu V$. The static characteristic shows a gain and offset error $\Delta y_{E(OG)} = 30\ \mu V$ and a nonlinearity error of $\sigma_D = 20\ \mu V$. The SNR (with a 6.6 coverage factor) could be calculated as $20\log(1/(2\sqrt{2} \cdot 6.6 \cdot 1.80 \times 10^{-6})) = 89$dB while for the accuracy, the signal to noise and distortion (SNDRT) (with a 6.6 coverage factor) is

$$10\log\left(1\Big/\left((2\sqrt{2} \cdot 6.6)^2 \cdot \left((1.8 \times 10^{-6})^2 + (20 \times 10^{-6})^2 + (30 \times 10^{-6})^2\right)\right)\right) = 63 \text{ dB}.$$

If we want to interface such a sensor with an A/D converter, we have to choose one with at least 6 dB more (1 bit more in ENOB) not to decrease the resolution too much; thus, we have to sample the system at 40 kS/s. Therefore, to the best to date technology (see Fig. 3.36, where P/fs@95 dB~300 pJ and P/fs@70 dB~3 pJ), we should consume at least ~ 300 pJ $\cdot 40 \times 10^3 = 12\ \mu W$ for the precision. If we are interested only in accuracy, we should consume at least ~ 3 pJ $\cdot 40 \times 10^3 = 1.2\ \mu W$. We consume less in accuracy because the distortion of the system makes the least significant bits nonsignificant; therefore, we get lower consumption if we are not interested in them. On the other hand, if we are interested in the noise-free resolution that characterizes the precision and disregards the distortion and offset issues, we should take into account more bits, consuming more.

3.6.5   Beyond Thermal Noise Limited Figures of Merit

Ultimate physical limits show constraints between time and energy. The most known constraint is derived from the Heisenberg uncertainty principle that states that in a physical system, the change of energy ΔE in a time Δt should be limited by the product

$$\Delta E \cdot \Delta t \geq \text{constant} = \hbar/2, \tag{3.44}$$

where $\hbar$ is the reduced Plank's constant. If we refer to the A/D conversion, we can put things in this way:

$$
\begin{aligned}
&\Delta P \cdot \Delta t^2 \geq \hbar/2 \\
&\rightarrow (\Delta V \cdot \Delta t)^2 \geq \hbar/2 \cdot Z \\
&\rightarrow \Delta V \cdot \Delta t \geq \sqrt{\hbar/2 \cdot Z} \\
&\rightarrow \frac{V_{FS}}{NL} \cdot \frac{1}{2BW} \geq \sqrt{\hbar/2 \cdot Z} \\
&\rightarrow NL \cdot BW \leq V_{FS}/2\sqrt{\hbar/2 \cdot Z} = \text{constant},
\end{aligned} \tag{3.45}
$$

where Z is the impedance of the sensor interface, V_{FS} is the input voltage full scale, and NL is the number of discrete levels. Note that for $V_{FS} = 1$, and $Z = 50\,\Omega$ we have $NL \cdot BW \leq 6,100 THz$. Of course, this is a limit very far from today's technologies, but it is, in any case, an ultimate resolution/bandwidth tradeoff. If we apply this to A/D conversion, we cannot achieve more than 18 bits with 10 GS/s.

3.6.6   The Role of Bandwidth in Acquisition Chains and Overall Optimization

So far, we have dealt with optimization in acquisition chains as far as resolution is concerned by means of the *resolution rule in acquisition chains*. However, we have to cope also with bandwidth, and the two things may be strictly related (for example, we may reduce noise by reducing bandwidth, if possible).

In Fig. 3.37 is shown an acquisition chain where each stage is characterized by a time constant (i.e., bandwidth) $\tau_1, \tau_2, \ldots, \tau_K$. Therefore, using a dominant pole approximation, we may define the overall time constant and bandwidth as

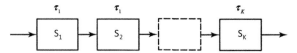

Figure 3.37 Acquisition chain composed of several stages with different time constants.

$$\tau_T \simeq \tau_1 + \tau_2 + \cdots \tau_K;$$

$$\frac{1}{BW_T} \simeq \frac{1}{BW_1} + \frac{1}{BW_2} + \cdots + \frac{1}{BW_K}. \tag{3.46}$$

Once again, the overall bandwidth is defined by the block with narrower bandwidth, similarly to the resolutions rule.

Therefore, we may recollect all the rules for acquisition chains in the following set of equations:

$$\begin{cases} \dfrac{1}{NL_T^2} = \dfrac{1}{NL_1^2} + \dfrac{1}{NL_2^2} + \cdots + \dfrac{1}{NL_K^2} \\[2mm] \dfrac{1}{BW_T} \simeq \dfrac{1}{BW_1} + \dfrac{1}{BW_2} + \cdots + \dfrac{1}{BW_K} \\[2mm] \mathrm{FoM} = \dfrac{P}{(NL_T)^2 \cdot BW_T} \end{cases} \tag{3.47}$$

The set of equations could be explained as follows. First, we may set the design's objective in either the total bandwidth or total resolution. Those are linked with the power consumed by means of the FoM, which is given by the technology employed. Then, we have to design single stages (and in particular the gains) so as to achieve the final targets in terms of both resolution and bandwidth.

3.6.7   Example: Noise Optimization in a Two-Stage Sensor Interface

This section will take a simple example of two stages whose behavior is characterized by constant gain-bandwidth products $GBW = S \cdot BW$, where S is the gain and BW the bandwidth. This is a very common case in OPAMP feedback amplifiers. We want to understand the tradeoffs in the gains S_1, S_2 of single amplifiers for a constant total gain.

In Fig. 3.38 is shown the system together with the gain plot. Since we want to keep the total gain S_T constant, the increase of a factor k of the gain of the first implies a reduction of the same factor of the second one.

Now, referring to the number of levels of each stage, if we assume uniform spectral noise density $N(f) = N$ since we assume constant FS and FSO, we have

$$NL_1^2 = \frac{FS^2}{\sigma_N^2} = \frac{FS^2}{N \cdot BW} = \frac{FS^2 \cdot S_1}{N \cdot \mathrm{GBW}};$$

$$NL_2^2 = \frac{FS_1^2}{\sigma_N^2} = \frac{FSO^2}{S_2^2 \sigma_N^2} = \frac{FSO \cdot S_2}{S_2^2 \cdot N \cdot \mathrm{GBW}} = \frac{FSO}{S_2 \cdot N \cdot \mathrm{GBW}}. \tag{3.48}$$

Now, if we set

$$S_1' = S_1 \cdot k; \quad S_2' = \frac{S_2}{k}$$

$$\rightarrow NL_1'^2 = NL_1^2 \cdot k; \quad NL_2'^2 = NL_2^2 \cdot k; \tag{3.49}$$

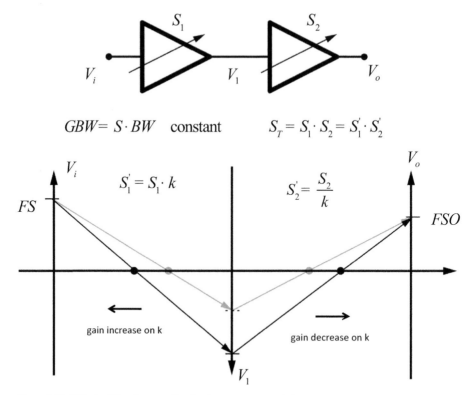

Figure 3.38 Effect of the change of gains in a two-stage chain.

using the first of (3.47) (rule of resolution in acquisition chains), we have

$$\frac{1}{NL_T^2} = \frac{1}{NL_1^2} + \frac{1}{NL_2^2};$$

$$\frac{1}{NL_T^{2'}} = \frac{1}{k \cdot NL_1^2} + \frac{1}{k \cdot NL_2^2} = \frac{1}{k \cdot NL_T^2}.$$

(3.50)

Therefore, the overall number of levels, that is, the overall resolution, increases as $\sqrt{k}$.

> **Hint** In a two-stage constant gain-bandwidth product signal acquisition chain, the resolution is optimized for a fixed total gain by incrementing as much as possible the gain of the first stage.

Now, we can try to understand what happens for the bandwidth. Using the second equation of the system (3.47), we have

$$\frac{1}{BW_T} = \frac{1}{BW_1} + \frac{1}{BW_2} = \frac{GBW}{S_1} + \frac{GBW}{S_2}.$$

(3.51)

Now, using Lagrange multipliers, it is easy to show that under the constraint, $S_T = S_1 \cdot S_2 = $ constant the maximum total bandwidth is achieved for equal gains.

> **Hint** In a two-stage constant gain-bandwidth product signal acquisition chain, the bandwidth is optimized for a fixed total gain by equally sharing the gain between the two stages.

Of course, the two rules are in contrast to each other. Thus, the final application will dictate the optimal choice. We will see an application of this in an OPAMP circuit case in Chapter 7.

3.6.8 On the Role of the Sensitivity

What about the role of gain or *sensitivity* on the FoMs? Never was mentioned in the tradeoffs such as Eq. (3.37). To understand this, let us take an example of a mechanical sensor as illustrated in Fig. 3.39 and already introduced in Chapter 1. A cantilever is a very tiny and flexible bar of material anchored on one end and free to flex (or oscillate) at the other. We can use it to sense a very small force applied at the cantilever's free end by reflecting a laser beam onto the surface and looking at its projection on a distant plane.

Due to mechanical thermal noise, the cantilever is subject to a random movement that induces uncertainty in the measurement that is sketched as a Gaussian distribution. If we choose a close projection as shown in Fig. 3.39A, given

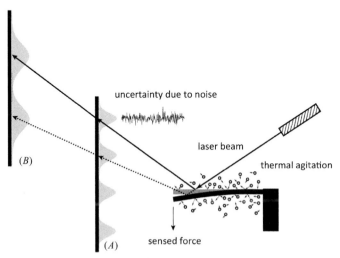

Figure 3.39 Sensitivity in a mechanical force sensor. A laser beam is reflected onto a cantilever where a force to be measured is applied. The projection of the reflected laser beam is a measure of the force applied. Due to thermomechanical noise, the measure is subject to uncertainty. (A) Short distance projection (lower sensitivity) and (B) long-distance projection (higher sensitivity).

the uncertainty induced by mechanical noise, as illustrated, we can get about four discrete levels (NL = 4) in the measurement with a good degree of confidence.

Now, assume that we want to look better into the levels (for instance, you cannot clearly see the levels) and enlarge them. A solution could be to move further away from the projection plane, as shown in Fig. 3.39B, thus increasing the *sensitivity* of the sensor. If we use the same plane size, the discrete levels are magnified as well as the output uncertainty; thus, the total number of levels is reduced (NL = 2) even if the absolute input resolution is unchanged. However, in this case, we also reduce the input full scale, likewise to what happens in Fig. 3.19. Alternatively, we can keep the same input FS and the same resolution in terms of the number of levels by increasing the size of the projection plane.

An essential aspect of this example is that increasing the sensitivity does not require any additional power supplied to the system. Therefore, sensitivity is some sort of an amplifying lens of the resolution levels.

The illustrated case is quite specific since the amplifying means (reflected laser beams on a projection plane) do not introduce significant noise with respect to the source. It is like having in an electronic system where the only noise source is at the input and we have negligible interface noise. However, this is a case where we can asymptotically tend, even in real electronic systems.

Role of Sensitivity We can reduce the sensitivity's role as much as possible not to impact either power consumption or resolutions of a sensing system.

The above statement should be better understood in this way: we can devise physical systems that are able to increase their gain and dynamic ranges with negligible power consumption. The above example indicates a case of why there is no role of sensitivity in power and resolution relationships and suggests why the sensitivity variable is absent in FoMs.

3.7   General Rules for Sensing Design

The FOM (3.37) shows an interesting relationship between the main constraints in the sensor design in the case of thermal noise limited systems. Figure 3.40 shows the interrelationships between the essential constraints of sensor design *for a given FoM*. To better explain the concept, we will provide two examples.

Example A capacitive sensing interface is characterized by the FoM: $\text{FoM}_A = 165 \times 10^{-6}$ J $\cdot$ ppm^2. Now assume that the full scale of the sensing is $C_0 = 2$ pF and that we need to resolve a capacitance variation of $\Delta C = 1$ fF in a bandwidth of 100 Hz; what is the minimum required power consumption? We have first to characterize the dynamic range, which is $DR = 20\log(2 \times 10^{-12}/1 \times$

$$FOM = \frac{P}{DR \cdot BW}$$

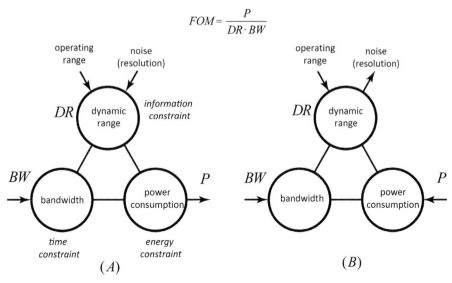

Figure 3.40 General rules for sensor design in the case of thermal noise limited systems. (A) Case in which the minimum power consumption is determined on the basis of resolution and bandwidth. (B) Case in which the resolution is determined on the basis of power consumption and bandwidth.

$10^{-15}/2\sqrt{2} = 76.9$ dB corresponding to a resolution in terms of ENOB $= (78 - 1.76)/6.09 = 12.5$b. Therefore, the minimum required power is $P = \text{FoM}_A \cdot BW \cdot NL^2 = 165 \times 10^{-6} \cdot 100/(2^{12.5}/1 \times 10^6)^2 = 0.55$ µW, where we used ppm $= 10^6/NL$. This corresponds to the diagram flow of Fig. 3.40A.

Example An electronic accelerometer has the following performances: it consumes 6.6 µW at 50 Hz of bandwidth (100 samples/s); the noise floor is 0.04 µg/$\sqrt{\text{Hz}}$ and operates in a full scale of $FS = 2$ g. Therefore the input-referred noise power is $0.04 \times 10^{-6}\sqrt{50} = 2.8$ mg, and the dynamic range is $DR = 20\log(2/2.8 \times 10^{-3}/2\sqrt{2}) = 48$ dB corresponding to ENOB of 7.7b, where we used $q = 2\sqrt{2}$. Therefore the figure of merit is $\text{FoM}_A = 6.6 \times 10^{-6}/50/(10^{48/20}) = 2.1$ pJ. Now, assuming we have the same power budget and bandwidth of $1Hz$, which is the maximum achievable resolution? The dynamic range could be calculated as $DR = 6.6 \times 10^{-6}/1/2.1 \times 10^{-12} = 3.1 \times 10^6$, considering that the input-referred effective resolution is $\sigma_N = FS/2\sqrt{2}/\sqrt{DR} = 400$ µg in this bandwidth. This corresponds to the diagram flow of Fig. 3.40B. (Data were extrapolated from Analog Devices ADXL362 datasheet.)

Further Reading

Cover, J. A., and Thomas, T. M., *Elements of Information Theory*. New York: John Wiley & Sons, 1991.

Gregorian, R., and Temes, G. C., *Analog MOS Integrated Circuits*. New York: John Wiley & Sons, 1986.

Kester, W., Ed., *The Data Conversion Handbook*. Philadelphia: Elsevier, 2004.

Schreier, R., and Temes, G. C., *Understanding Delta-Sigma Data Converters*. New York: IEEE Press, 2005.

Stone, J. V., *Information Theory: A Tutorial Introduction*. Sebtel Press, 2015.

Walden, R. H., Analog-to-digital converter survey and analysis, *IEEE J. Sel. Areas Commun.*, vol. 17, no. 4, pp. 539–550, 1999.

Widrow, B., and Kollar, I., *Quantization Noise*. Cambridge: Cambridge University Press, 2008.

Zhirnov, V., and Cavin III, R. K., *Microsystems for Bioelectronics*. Philadelphia: Elsevier, 2015.

4 Overview of Mathematical Tools

This chapter will give a short summary of mathematical instruments required to model sensor systems in the presence of both deterministic and random processes. The concepts will be organized in a compact overview for a more rapid consultation, emphasizing the convergences between different contexts.

4.1 Deterministic and Random Signals

Signals are functions conveying information between different environments. The information could be apparent or hidden in them depending on its complexity level. Signals in sensing systems might be deterministic and random (stochastic). Deterministic signals are characterized by the complete knowledge of the past and future evolutions. Deterministic signals are instrumental in characterizing sensing systems in time and frequency response and evaluating errors. On the other side, random signals involve the concept of probability. The knowledge of random signals implies the knowledge of their statistical characteristics. Of course, we can store and thus know the past evolution, but we know the future progress only under a probabilistic premise. We know the likelihood of any future value, but we do not know when it will happen. Therefore, random signals could be modeled only using the mathematical tools of random variables. They are essential for describing noise and errors and signals since, in sensor operating mode, the input is unknown by definition.

4.1.1 Characterization of Electrical Deterministic Signals

Signals could be characterized by their energetic properties.

From the qualitative point of view, we have to distinguish between two classes of electrical signals: (1) *energy signals* characterized by transient behavior, where the area covered by the signal curve has a limited size as in Fig. 4.1A; (2) *power signals* characterized by continuous waveform versus time with the infinite area covered by the signal function as in the example of Fig. 4.1B (in this case is periodic). We will characterize better the two classes from the mathematical viewpoint.

> **Hint** We will restrict the mathematical models and formal definitions to real signals $x(t) \in \mathbb{R}$ even if the following definitions could be extended to complex signals. Therefore, this assumption will be implied unless specified.

The *time average* (or *mean*) of a signal is defined as

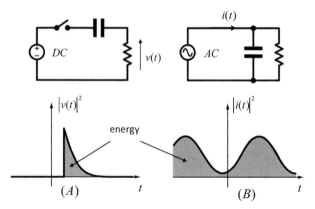

Figure 4.1 Example of energy signal (A) and periodic power signal (B).

$$\langle x\rangle \triangleq \lim_{T\to\infty} \frac{1}{T} \int_{-T/2}^{T/2} x(t)dt, \tag{4.1}$$

where T is the time on which the average is calculated, and the lim function could be dropped out in the case the signal is periodic with period T. In analog design, the time average is also referred to as the *signal's DC component*.

Since often energy is proportional to the square of physical quantities (kinetic energy with the square of velocity, elastic potential energy with the square of the displacement, electrical energy with the square of the voltage, etc.), we can define the *energy* of a signal as

$$E \triangleq \int_{-\infty}^{+\infty} x^2(t)dt. \tag{4.2}$$

If $x(t)$ is expressed in voltage or current, Eq. (4.2) is from the physical viewpoint the energy dissipated in a standard resistive load $R = 1\Omega$.

Signals having finite energy $0 < E < +\infty$ are called *energy signals*. Typically, energy signals are burst or impulsive signals that correspond to a finite balance of energy transfer, such as the transfer of a charge on a capacitor. Energy signals *could not* be periodic because their energy would be infinite.

We can also define the *power P* (energy averaged over a time slot T) of a signal $x(t)$ as

$$P \triangleq \langle x^2\rangle = \lim_{T\to\infty} \frac{1}{T} \int_T x^2(t)dt, \tag{4.3}$$

If $0 < P < +\infty$ the signal is referred to as a *power signal*. To better characterize it on the time evolution, we can extend the time slot as much as possible; however, the lim function could be dropped out in the case the signal is periodic with period T. Periodic signals characterized by the power as in Eq. (4.3) are referred to as *periodic power signals*.

Power is also a *time mean square* of the signal. As already seen, we can also characterize the power of the variation with respect to the mean as the *time-dependent (or experimental) variance*, alias AC power, as

$$\sigma_x^2 = P_{AC} = \lim_{T \to \infty} \frac{1}{T} \int_T \left(x(t) - \langle x \rangle \right)^2 dt = \langle x^2 \rangle - \langle x \rangle^2 = P - P_{DC}, \quad (4.4)$$

where P_{DC} is the power of the DC component of the signal.

A *periodic power signal* and could be expressed with the sum of harmonics by means of *Fourier's series*

$$x(t) = \sum_{n=-\infty}^{\infty} c_n e^{jn2\pi f_0 t}; \ c_n = \frac{1}{T} \int_T x(t) e^{-jn2\pi f_0 t} dt; \ f_0 = 1/T, \quad (4.5)$$

where T is the signal period and c_n are *complex* numbers. Thus, a periodic power signal is composed of a sum of multiples or *harmonics* of the ordinary fundamental frequency f_0, which is the frequency defined by periodicity. Note how

- $c_0 = \frac{1}{T} \int_T x(t) dt = \langle x \rangle$ is the DC component of the signal.

- The terms $|c_n|^2$ are the power components of the harmonics and, according to Parseval's theorem

$$P = \sum_{n=-\infty}^{+\infty} |c_n|^2. \quad (4.6)$$

Equation (4.6) reveals that the *total power* is given by the sum of each harmonic's average power, thus obeying a superposition property, since harmonics are orthogonal to each other from a mathematical point of view.

On the other side, in general, any *energy signal* has a *Fourier transform* and its *inverse*

$$X(f) = F[x(t)] = \int_{-\infty}^{+\infty} x(t) e^{-j2\pi f t} dt$$

$$x(t) = F^{-1}[X(f)] = \int_{-\infty}^{+\infty} X(f) e^{j2\pi f t} df, \quad (4.7)$$

where t is time and f is the ordinary frequency.

The Fourier transform $X(f)$ is also called the spectrum of $x(t)$, and since it is complex, it is composed of the amplitude spectrum $|X(f)|$ and of the phase spectrum $\sphericalangle X(f)$. Note how

- $X(0)$ is the total area delimited by $x(t)$ so it is the DC average value of the signal.
- If $x(t)$ is real, then $X(-f) = X^*(f)$, so $|X(-f)| = |X(f)|$ and $\sphericalangle X(-f) = -\sphericalangle X(f)$ so we have even amplitude and odd phase symmetry, respectively. This is why we will always assume that the amplitude spectrum is symmetric with respect to the origin. We can represent the spectrum of a periodic signal $x(t)$ using Dirac's function by F-transforming Eq. (4.5), getting

$$X(f) = \sum_{n=-\infty}^{\infty} c_n \delta(f - nf_0), \tag{4.8}$$

where c_n are calculated according to (4.5), showing how the spectrum of a periodic signal is a modulated Dirac's comb function of harmonics.

Rayleigh's theorem is somehow Parseval's theorem applied to energy signals, stating that[1] the total energy of the signal is

$$E = \int_{-\infty}^{+\infty} |X(f)|^2 df; \tag{4.9}$$

that is, $|X(f)|^2$ is the (double-sided) *energy spectrum* of the energy signal $x(t)$. Note how Eq. (4.9) is equivalent to (4.6) even if one is expressed in power and the other in energy units. To understand this duality, we must recall that Eq. (4.6) is actually *energy per period*. Actually, when the period becomes infinite, becoming aperiodic, the energy per period is the energy of the signal itself.

Example Figure 4.2 shows a comparison between a periodic power signal and an aperiodic energy signal. In the first case (Fig. 4.2A), a rectangular pulse of height k and width τ is repeated every period $T_0 = 1/f_0$ were in this specific case $T_0 = 5\tau$. Its average power is $P = 1/T_0 \cdot \int_{-\tau/2}^{\tau/2} k^2 dt = (k^2\tau)/T_0$, and the Fourier series is given by harmonic coefficients $c_n = (k\tau/T_0)\mathrm{sinc}(nf_0\tau)$. The *power spectral density* is a comb of Dirac's delta function modulated by[2] $(k\tau/T_0)^2 |\mathrm{sinc}(nf\tau)|^2$, and the sum of the pulses over the whole spectrum is again $P = \sum_{n=-\infty}^{+\infty} |c_n|^2 = (k^2\tau)/T_0$ for Parseval's theorem.

In the second case (Fig. 4.2B), a single rectangular pulse is characterized by the energy given by $E = \int_{-\tau/2}^{\tau/2} k^2 dt = k^2\tau$. Its module of the F-transform is given by $(k\tau)\mathrm{sinc}(f\tau)$ whose square is the *energy spectral density*. Again, the overall integral[3] in the frequency domain gives the same energy $E = \int_{-\infty}^{\infty} (k\tau)^2 |\mathrm{sinc}(f\tau)|^2 df = k^2\tau$. Note that

---

[1] From the relationship $\int_{-\infty}^{\infty} x(t)y^*(t)dt = \int_{-\infty}^{\infty} X(f)Y^*(f)df$ [2] $\mathrm{sinc}(x) = \sin(\pi x)/(\pi x)$.

[3] Recall that $\int_{-\infty}^{\infty} \mathrm{sinc}^2(\alpha x)dx = 1/\alpha$.

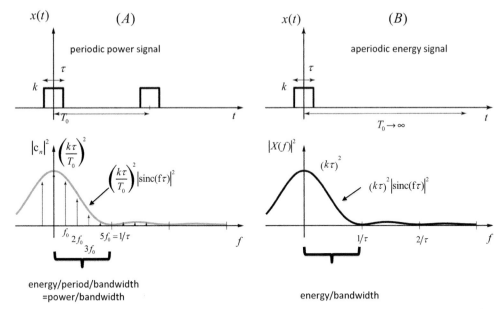

Figure 4.2 Energy signal as a limit of a periodic power signal. (A) Periodic power signal and its transform. (B) Energy signal whose shape is identical to that of the power signal period.

we can consider the energy case as the limit for $T_0 \to \infty$ of the periodic power signal case. In the limit, we can see that since $c_n \simeq X(f_n)\Delta f_0 \to |c_n|^2 \simeq |X(f_n)|^2 \Delta f_0^2$ the energy of each Dirac's pulse decreases with respect to T_0 while its density per bandwidth increases by the same amount so that the limit of the envelope goes to the *sinc* function of Fig. 4.2B expressed in energy per bandwidth units. Note also that the interpolating function of Fig. 4.2A is not a power spectral density but an envelope of Fourier's coefficients.

Hint If we use the *angular frequency* $\omega = 2\pi f$, the "nonunitary" Fourier transform

becomes $X_\omega(\omega) = \int\limits_{-\infty}^{+\infty} x(t)e^{-j\omega t}dt$ and $x(t) = \dfrac{1}{2\pi} \int\limits_{-\infty}^{+\infty} X_\omega(\omega)e^{j\omega t}d\omega$. Therefore, we

must be careful to use the $1/2\pi$ factor in the inverse transform of $X_\omega(\omega)$.

Linear time-invariant (LTI) systems are characterized by an input–output functional relationship $y(t) = f[x(t)]$ that obeys (1) the *principle of superposition* and (2) the time-invariance property that is

$$y(t) = \sum_i a_i f[x_i(t)] \text{ for } x(t) = \sum_i a_i x_i(t) \ \forall \ a_i$$
$$y(t - t_d) = f[x(t - t_d)] \ \forall \ t_d,$$

(4.10)

for which we can observe that, for well-known Fourier's transform properties

$$y(t) = (h^*x)(t) = \int_{-\infty}^{\infty} h(\tau)x(t-\tau)d\tau \text{ then}$$

$$Y(f) = H(f)X(f) \text{ where } Y(f) = F[y(t)]; \ X(f) = F[x(t)]; \ H(f) = F[h(t)],$$
(4.11)

where the $*$ operator stands for the convolution[4] and $H(f)$ is referred to as the *transfer function* of the LTI system. From the second line of Eq. (4.11), if we apply an impulse to the input $x(t) = \delta(t)$, then $X(f) = 1$; thus the output spectrum takes the same shape of $H(f)$. The inverse transform of $h(t) = F^{-1}[H(f)]$ is the impulse response transform of this system.

If $x(t)$ is an energy signal, by using Rayleigh's theorem, we get

$$|Y(f)|^2 = |H(f)|^2 |X(f)|^2 \text{ and}$$

$$E_y = \int_{-\infty}^{\infty} |H(f)|^2 |X(f)|^2 df; \quad E_x = \int_{-\infty}^{\infty} |X(f)|^2 df.$$
(4.12)

Thus, the energy spectrum of the output is given by the energy spectrum of the input multiplied by $|H(f)|^2$. Similarly, for periodic power signals, by using (4.8) and the superposition of harmonic powers, we can see that $|H(f)|^2$ maps each harmonic power $|c_n|^2$ from the input to the output.

One of the best ways to investigate the dependence of two signals characterized by N samples such as $x = \{x_1, x_2, \ldots x_N\}$ and $y = \{y_1, y_2, \ldots y_N\}$ is the averaged cross-correlation of an N-sampled discrete signal as

$$\langle xy \rangle = \frac{1}{N} \sum_{i=1}^{N} x_i y_i.$$
(4.13)

Since the relationship is an averaged inner product summation,[5] we can look at the relationship from a geometrical point of view, where each sample is a variable component of the signal. If any component of x has no dependence (projection) with respect to the related components of y, then $\langle xy \rangle = 0$ and the two signals are said to be orthogonal. This means that any change in x does not correspond to any change into y.

> **Hint** *Correlation does not imply causation.* If two events have a cause-and-effect relationship, a correlation could be detected; however, it is not true vice versa. In other terms, the result of the correlation operation does not necessarily imply a causal relationship among events.

---

[4] The convolution between two functions is defined as $(x * y)(t) = \int_{-\infty}^{\infty} x(\tau)y(t-\tau)d\tau = (h(t) * x(-\tau))(t)$.

[5] Note how the inner product brackets notation $\langle .. \rangle$ is also used for time-averaged quantities.

We can now extend the scalar product from the discrete-time (DT) case to the continuous-time (CT) one for two analog signals x and y in a time frame T

$$\langle xy \rangle = \frac{1}{T} \int_T x(t)y(t)dt. \tag{4.14}$$

To be more generic, we could compare the two signals where one is shifted of a variable amount of time τ

$$R_{xy}(\tau) = \langle x(t+\tau)y(t) \rangle \doteq \int_{-\infty}^{\infty} x(t+\tau)y(t)dt \qquad \text{for energy signals}$$

$$\doteq \lim_{T \to \infty} \frac{1}{T} \int_T x(t+\tau)y(t)dt \quad \text{for power signals,} \tag{4.15}$$

which is called *time* cross-*correlation*[6] between the two signals x and y. Again, the $\lim\limits_{T \to \infty}$ is dropped out if the signal is periodic of period T. If we compare the signal with a delayed copy of itself, using $y = x$ in (4.15), we get the *time autocorrelation*

$$R_{xx}(\tau) = \langle x(t+\tau)x(t) \rangle \doteq \int_{-\infty}^{\infty} x(t+\tau)x(t)dt \qquad \text{for energy signals}$$

$$\doteq \lim_{T \to \infty} \frac{1}{T} \int_T x(t+\tau)x(t)dt \qquad \text{for power signals.} \tag{4.16}$$

Note how for power signals for $\tau = 0$ we have $R_{xx}(0) = \langle xx \rangle = \langle x^2 \rangle = P$, which explains why we used for *power* or *time mean square* the scalar product brackets. For simplicity of notation, we will also use in place of $R_{xx}(\tau)$ the compact form $R_x(\tau)$. Therefore, the following properties for the autocorrelation could be shown:

$$\text{(i)} \quad R_{xx}(0) = \begin{cases} P & \text{for power signals} \\ E = \int_{-\infty}^{\infty} |X(f)|^2 df & \text{for energy signals} \end{cases}$$

(ii) $|R_{xx}(\tau)| \leq R_{xx}(0)$

(iii) $R_{xx}(\tau) = R_{xx}(-\tau)$ for real signals, $\tag{4.17}$

meaning that the autocorrelation is a symmetric function whose value at the origin represents either the power or the energy of the signal.

---

[6] A more complete definition of cross-correlation in the complex domain is $R_{xy} = \langle x(t)y^*(t-\tau) \rangle = \langle x(t+\tau)y^*(t) \rangle$ that for real signals $(x^*(t) = x(t))$ becomes $R_{xy} = \langle x(t+\tau)y(t) \rangle$. Furthermore, there is a strong relationship with convolution because $R_{xy}(\tau) = (x(t) * y^*(-t))(\tau)$ and for real signals: $R_{xy}(\tau) = (x(t) * y(-t))(\tau)$.

A similar conclusion could be derived for the cross-correlation, where *for real signals*, it is

$$
\begin{array}{ll}
\text{(i)} & R_{xy}(\tau) = R_{yx}(-\tau) \\
\text{(ii)} & R_{xy}(0) = R_{yx}(0) = \langle xy \rangle.
\end{array}
\tag{4.18}
$$

In the case of LTI systems, defined by $h(t)$, if we call $R_y(\tau)$ and $R_x(\tau)$ the autocorrelation functions of the output and input, respectively, it can be easily shown that for real signals

$$
R_{yy}(\tau) = \Big(h(-t) * h(t) \Big)(\tau) * R_{xx}(\tau).
\tag{4.19}
$$

Now, if we take the Fourier's transform of both terms of Eq. (4.19) and considering the convolution theorem,[7] we get

$$
F[R_{yy}(\tau)] = |H(f)|^2 F[R_{xx}(\tau)],
\tag{4.20}
$$

which is very similar to the first line of Eq. (4.12), but it is more general since it could be applied to both energy and power signals.

This suggests defining a function as

$$
S_x(f) = F[R_{xx}(\tau)] = \int_{-\infty}^{\infty} R_{xx}(\tau)e^{-j2\pi f\tau}d\tau.
\tag{4.21}
$$

If we now calculate the inverse transform of $S_x(f)$, it is

$$
R_{xx}(\tau) = F^{-1}[S_x(f)] = \int_{-\infty}^{\infty} S_x(f)e^{j2\pi f\tau}df
\tag{4.22}
$$

but since

$$
R_{xx}(0) = \int_{-\infty}^{\infty} S_x(f)df = \begin{cases} E_x & \text{for energy signals} \\ P_x & \text{for periodic power signals} \end{cases},
\tag{4.23}
$$

then $S_x(f)$ could be either *energy spectral density* or *power spectral density* depending on the kind of signal. The relationship (4.21) is referred to as the *Wiener–Khinchin theorem*. Thus, we have

$$
S_y(f) = |H(f)|^2 S_x(f),
\tag{4.24}
$$

showing how the energy or power spectra are related to each other by means of the square of the transfer function. If we consider power density functions, by integrating Eq. (4.24) over the entire spectrum, we get

$$
P_y = \int_{0}^{\infty} |H(f)|^2 S_x(f)df.
\tag{4.25}
$$

[7] $F[x \cdot y] = F[x] * F[y]$.

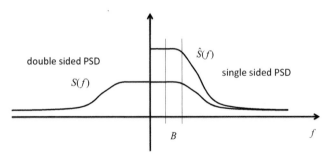

Figure 4.3 Single-sided and double-sided power spectral density (PSD) relationship.

The power (energy) spectral density (PSD), since it is related to the Fourier's transform, is defined on both positive and negative frequency. For this reason, it is called *bilateral (or double-sided) power spectral density*. However, in practical use and especially in electrical engineering, since it is a symmetric function (because related to a real signal), it is often used as defined on the positive axis only. In that case, it is called *unilateral* (or *single-sided*) *power spectral density*. The relationship, due to symmetry, is

$$S_x(f) = \frac{1}{2}\hat{S}_x(f), \qquad (4.26)$$

where $\hat{S}_x(f)$ is the *single-sided PSD*, which is graphically shown in Fig. 4.3.

Since the total area subtended by the two curves should be the same, the single-sided height should be double than the double-sided. Based on this observation, we can see that

$$R_{xx}(\tau) = F^{-1}[S_x] = \frac{1}{2}F^{-1}[\hat{S}_x] \text{ and}$$
$$\hat{S}_x = 2F[R_{xx}(\tau)]. \qquad (4.27)$$

Also, remember to pay attention to the integration variable so that if we use angular frequency, we have[8]

$$P = \int_0^\infty \hat{S}_x(f)df = \int_{-\infty}^\infty S_x(f)df = \frac{1}{2\pi}\int_0^\infty \hat{S}_x(\omega)d\omega = \frac{1}{2\pi}\int_{-\infty}^\infty S_x(\omega)d\omega. \qquad (4.28)$$

In the next chapters, especially in noise discussion, we will mostly refer to "power spectral density" as single-sided PSD, omitting the "hat" symbol.

---

[8] The change of variable could be a bit tricky. The basic concept is that because of Eq. (4.28) we must be careful on the measurement units. Take, for example, the expression $\hat{S}(f) = k^2/\left(1 + (f/f_0)^2\right)$. In this case $k^2 \equiv [W/Hz]$. We can then change the variable to ω so that $\hat{S}(\omega) = k^2/\left(1 + (\omega/\omega_0)^2\right)$ but in this case $k'^2 \equiv [W/rad/sec]$.

Since $R_{xx}(\tau)$, which is derived by a real function, is an even function as shown in Eqs. (4.17) and (4.67), the relationships (4.21) and (4.22) can be rewritten in the more compact form[9]

$$R_{xx}(\tau) = \int_0^\infty \hat{S}_x \cdot \cos(2\pi f \tau) df$$

$$\hat{S}_x = 4 \int_0^\infty R_{xx}(\tau) \cdot \cos(2\pi f \tau) d\tau,$$

(4.29)

where $\hat{S}_x$ is the *single-sided* PSD.

Example The exponential decaying energy signal $x(t) = ke^{-t/\tau}u(t)$, where $u(t)$ is the Heaviside unit step function (Fig. 4.1A), has energy given by $E = \int_0^\infty k^2 e^{-2t/\tau} dt = \dfrac{k^2\tau}{2}$.

Its *bilateral* energy spectrum can be found by means of the Fourier's transform $X(\omega) = \int_{-\infty}^\infty ke^{-t/\tau}e^{-j\omega t}d\omega = \dfrac{k\tau}{1+j\omega\tau}$. Therefore, its bilateral energy spectrum is $E(\omega) = |X(\omega)|^2 = \dfrac{k^2\tau^2}{1+\omega^2\tau^2}$ which is a Lorentzian form (see Appendix). From this, we can reconstruct the energy again by integration[10] in the spectrum

$$E = \frac{1}{2\pi} \int_{-\infty}^\infty \frac{k^2\tau^2}{1+\omega^2\tau^2} d\omega = \frac{k^2\tau}{2}.$$

Finally, it is easy to show using Eq. (4.18) that if we have a power signal composed of $z(t) = ax(t) + by(t)$ we get

$$P_z = a^2\langle x^2\rangle + b^2\langle y^2\rangle + 2ab\langle xy\rangle = a^2 P_x + b^2 P_y + 2ab\langle xy\rangle,$$

(4.30)

showing how the power of the weighted sum of two signals is equal to the sum of the power of single signals plus the weighted correlation of the two signals. The same expression in terms of (auto)correlation is

$$P_z = a^2 R_{xx}(0) + b^2 R_{yy}(0) + 2ab R_{xy}(0).$$

(4.31)

Note that this relationship is comprehensive of the DC power component.

Following the Wiener–Kinchine theorem relationship, we could now introduce the *cross-spectral density* as the Fourier transform of the cross-correlation

---

[9] Considering that for even real functions $x(t)$ we have $X(f) = 2\int_0^\infty x(t) \cdot \cos(2\pi ft)dt$.

[10] Remember that $\int_0^\infty \dfrac{1}{1+\omega^2\tau^2}d\omega = \dfrac{\pi}{2}\dfrac{1}{\tau}$.

$$S_{xy}(f) = \int\limits_{-\infty}^{\infty} R_{xy}(\tau)e^{-j2\pi ft}dt. \tag{4.32}$$

However, in comparison with the spectral density, since it is not symmetric as in Eq. (4.18), it is a *complex function even if the generating functions are real*

$$S_{xy}(f) = \mathrm{Re}\{S_{xy}(f)\} + j\mathrm{Im}\{S_{xy}(f)\} \tag{4.33}$$

and could be shown that, for (4.18) is

$$S_{xy}(f) = S_{xy}^{*}(f). \tag{4.34}$$

In Section 4.8 we will see the physical interpretation of this relationship.

4.1.2   Characterization of Random Signals

When we deal with stochastic processes, we must refer to the probability of an event referred to as an experiment outcome. The most robust and at the same time simplest definition of probability is the axiomatic one, saying that the probability of an event is a positive number, the probability of the entire space of events is 1, and the probability of mutually exclusive event sets is given by the sum of the probability of each of them. The axiomatic theory formally defines what probability is, but it does not regard how probabilities have to be interpreted, especially in the physical environment.

As the interpretation of the probability, from the experimental point of view, if we repeat the same experiment N times and N_A is the number of times that the event A occurs, we refer to N_A/N as the *relative frequency of occurrence A* or *empirical probability Pr(A)* of the event A for that set of trials. If N becomes very large, and for any possible set of trials, according to the *law of large numbers* (LLN), the ratio N_A/N converges in probability to the expected value. According to the *frequentist (or physical)* interpretation, the probability of an event corresponds to its expected value

$$Pr(A) = \frac{N_A}{N} \; N{\rightarrow}\infty \text{ with } 0 \le Pr(A) \le 1 \text{ for any event } A. \tag{4.35}$$

Equation (4.35) implies that the empirical probability is an estimation of what we consider probability: we need an infinite number of trials so as we can *expect* that the number of trials in which A occurs is $N_A \approx P_R(A) \cdot N$. Even if this interpretation does not cover several cases well, we will use this interpretation of probability for the scope of this book.

The mathematical tool that is used for describing random processes is the *random variable*. A random variable X is a *function* that maps outcomes s of the sample (experiment) space S into the ordered set of real numbers; thus $x = X(s)$ is called *realization* or *statistical sample* of random variable X. By using this mapping, a particular event A that defines a subset into S also defines a subset into the real

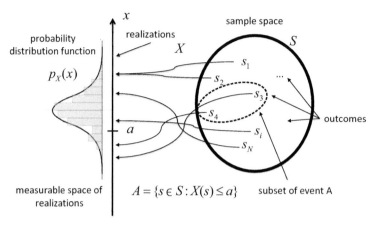

Figure 4.4 Random variable concept.

number axis. Conversely, a subset of the ordered space of the variable x defines a subset of the sample space. For example, we can define event A as the set of outcomes such that $X(s) \leq a$, where a is a numerical value of the real variable x. Note how formally X (upper case), the random variable, should not be confused with x (lowercase), which is its realization. However, in the following discussions, we might use lowercase for both of them for simplicity of notation.

The concept of a random variable is shown in Fig. 4.4, where a set of experimental results is connected with real numbers by means of a random variable called X. The key aspect of the random variable is to link a usually not-ordered set of outcomes with a *measurable space* such as real numbers.

To summarize:

- *Experiment*: a procedure performed by assuming the same initial conditions and producing a result or experimental *outcome* s_i of the ith experiment.
- *Sample space S*: the set of all the possible outcomes s_i. The sample space could contain a finite or infinite number of elements. It is not the set of the trials but the set of all possible outcomes or results of the experiment.
- *Event*: a subset of S, say A. Event A occurs if and only if the outcome $s_i \in A \subset S$.
- *Random variable*: the function associating the space of experiments with a corresponding measurable space of numbers. The *realization* of a random variable is written $x^{(i)} = X(s_i)$.

It could be shown that the function X qualifies as a random variable if the event $A_x = \{s \in S : X(s) < x\}$ is defined for every real number x.

It should be noted that in the same experiment, we could define several sample spaces. For example, on the experiment "coin tossing", we might define a sample space consisting of two elements, H and T, or another one containing all the possible coordinates and positions in which the coin rested. In the latter case, we could later define a random variable defining two events, H and T, that are subsets of the sample space.

Example The experiment consists of testing the voltage of battery production and check if it is below a critical value v_L. Therefore, the sample space could be defined as all the possible outcomes of analog values s in a given space S on which the random variable V identifies the realization v for any experiment (sample of the variable V). Mathematically, $x^{(j)} = v_j = V(s_j) = X(s_j)$ is the realization (sample) of the random variable V representing the outcome of the ith experiment. On this random variable, we can evaluate the event "the tested battery has a critical voltage" expressed formally, with $A_v = \{s \in S : V(s) = v < v_L\}$ on which we can estimate the probability on a large set of experiments in accordance with Eq. (4.35).

Thus, the mapping on real numbers of an experiment allows defining well-known functions that characterize the random variable, such as the *cumulative distribution function F_X* and the *probability density function (PDF)p_X*

$$F_X(x) \doteq \Pr(X \le x); \quad p_X(x) \doteq \frac{dF_X(x)}{dx} \tag{4.36}$$

so that

$$p_X(x)\,dx = \Pr(x < X < x + dx) \tag{4.37}$$

where Pr is the probability of the event referred to so far.

One of the most important characterizations of a discrete random variable is the *expectation value* that is *defined* for a discrete variable as

$$E[X] \doteq \sum_{k=1}^{M} x_k p_X(x_k), \tag{4.38}$$

where $Pr(X = x_k) = p_X(x_k)$ is the probability of getting the value x_k from the sample space of the random variable X. The same definition applies to continuous random variables using (4.37)

$$E[X] \doteq \int_{-\infty}^{+\infty} x p_X(x)\,dx. \tag{4.39}$$

The reason for this definition will be apparent as follows. If we have N independent observations (or N possible discrete possible values of the variable), and if the event $X_k = x_k$ occurs N_k times, the *sample average* or *mean value* $\bar{x}$ of the observations is given by

$$\bar{x} \doteq \frac{1}{N}\sum_{i=1}^{N} x_i = \frac{N_1 x_1 + N_2 x_2 + \ldots N_N x_N}{N} = \sum_{k=1}^{M} \frac{N_k}{N} x_k. \tag{4.40}$$

Now, for large values of N, according to LLN, the relative frequency $N_k/N \rightarrow P_X(x_k)$; thus

$$\bar{x} \rightarrow E[X] \text{ for } N \rightarrow \infty. \tag{4.41}$$

Another way to describe this concept is to say that the *sample mean* $\bar{x}$ is an *estimator* of the *expected value or probabilistic mean* $E[X]$. In the following, we will also use the notation $\bar{x}$ to refer to the asymptotic limit $E[X]$ for simplicity of notation.

Expected value is a linear operator, that is, $E[\alpha X + \beta Y] = \alpha E[X] + \beta E[Y]$ and for a function of random variable $Z = g(X)$ the expected value is

$$E[g(X)] = \int_{-\infty}^{+\infty} g(x) p_X(x) dx. \tag{4.42}$$

When $g(X) = X^n$, its expected value is called nth *order moment* of the random variable X, whose estimator is $\overline{x^n} \rightarrow E[X^n]$. A significant value is the second-order moment of the random variable $\overline{x^2} \rightarrow E[X^2]$, which is called the *mean square* of the random variable X.

Another important characterization of a random variable is related to the spread of the observed values around the mean value

$$\sigma_X^2 = E[(X - E[X])^2] = \int_{-\infty}^{+\infty} p_X(x)(x - E[X])^2 dx = E[X^2] - E[X]^2, \tag{4.43}$$

which is called the *variance* of the second moment, meaning that the variance is given by the *mean square* minus the square of the *mean value*.

For a discrete variable, an estimator of the variance is[11]

$$\overline{(X - \bar{x})^2} = \frac{1}{N}\sum_{i=1}^{N}(x_i - \bar{x})^2 = \frac{1}{N}\sum_{i=1}^{N}x_i^2 - \bar{x}^2 = \sum_{k=1}^{M}\frac{N_k}{N}x_k^2 - \bar{x}^2 = \overline{x^2} - \bar{x}^2. \tag{4.44}$$

Note how the expression of Eq. (4.44) is used as an estimator of Eq. (4.43) based on the LNN.

An important relationship regarding variance is the following, called *Chebyshev's inequality*

$$P(|X - E[X]| \geq k\sigma_X) \leq \frac{1}{k^2}, \tag{4.45}$$

meaning that the probability that an experiment outcome is outside its expected value by k standard deviations $\sigma_X = \sqrt{\sigma_X^2}$ is no larger than $1/k^2$.

---

[11] Actually, in statistics the sum should be divided by $(N-1)$ for better estimator properties. However, for the sake of simplicity we will assume from now on to have N large enough to use N instead on $(N-1)$.

If we observe two random variables simultaneously, we can test if they are correlated with each other, following the previous discussion on deterministic variables. To understand this, we must introduce the *joint probability* density function, which is defined as

$$p_{XY}(x,y)\, dxdy = \Pr(x < X < x + dx,\ y < Y < y + dy), \tag{4.46}$$

where, as well known, if the two variables X and Y are independent, then $p_{XY}(x,y) = p_X(x)p_Y(y)$. A useful test of the relationship between the two variables is given by the *cross-correlation* of two random variables, X and Y

$$R_{XY} = E[XY] = \iint_{\infty} xy\, p_{XY}(x,y)dxdy \tag{4.47}$$

and similarly, to the single variable case, it is defined

$$\sigma_{XY}^2 = E[(X - E[X])(Y - E[Y])] = \iint_{\infty}(x - E[X])(y - E[Y])\, p_{XY}(x,y)dxdy =$$

$$\iint_{\infty} xy\, p_{XY}(x,y)dxdy - E[X]E[Y] = E[XY] - E[X]E[Y]$$

$$= R_{XY} - E[X]E[Y], \tag{4.48}$$

which is called *covariance*[12] (also indicated as C_{XY}) between the two random variables X and Y. Note that $\sigma_{XX}^2 = \sigma_X^2$. In simple words, the test evaluates how *on average*, the displacements of the two variables around the means are correlated by taking their product. Thus, the *covariance* is given by the *cross-correlation* minus the product of the *two variables' expected* values. In the case that two variables are uncorrelated, then $\sigma_{XY} = 0$.

For a sum of random variables, similarly to Eq. (4.30), if we have $Z = aX + bY$, then

$$E[Z^2] = a^2 E[X^2] + b^2 E[Y^2] + 2abR_{XY}$$

$$= a^2 E[X^2] + b^2 E[Y^2] + 2ab \cdot \sigma_{XY}^2 + E[X]E[Y]. \tag{4.49}$$

Note that the last term is *not* the power DC component of the resulting variable since it is also distributed in the first two terms.

Again, its experimental (sample) discrete estimator is

$$\hat{\sigma}_{XY}^2 = \overline{(X - \bar{x})(Y - \bar{y})} = \sum_{i=1}^{N} \frac{(x_i - \bar{x})(y_i - \bar{y})}{N}$$

$$= \sum_{i=1}^{N} \frac{x_i y_i}{N} - \bar{x} \cdot \bar{y} = \sum_{j=1}^{L}\sum_{k=1}^{M} \frac{N_{jk}}{N} y_j x_k - \bar{x} \cdot \bar{y}, \tag{4.50}$$

---

[12] Pay attention to the fact that in different disciplines the terms "correlation" and "covariance" are sometimes used interchangeably, and the "correlation" term is frequently used in the normalized form of Pearson coefficient.

where L and M are the discrete bins of possible values of the two variables, and the expression N_{jk}/N is the relative frequency that the event $Y_j = y_j$ and $X_k = x_k$ occurs. Note how the expression (4.50) is an estimator in probability to (4.48) according to LLN.

In general, the expectation for a relationship between two random variables linked together by the relationship $Z = g(X, Y)$ is given by using the joint probability density function

$$E[g(X, Y)] = \iint_{-\infty} g(x, y) p_{XY}(x, y) dx dy. \tag{4.51}$$

An important relationship could be shown by using Eq. (4.51) in the case of the weighted sum of two random variables $Z = aX + bY$

$$\sigma^2(aX + bY) = a^2 \sigma_X^2 + b^2 \sigma_Y^2 + 2ab \cdot \sigma_{XY}^2. \tag{4.52}$$

Note how in the case of independent variables, since $\sigma_{XY} = 0$ we get $\sigma_Z^2 = \sigma_X^2 + \sigma_Y^2$.

From (4.52) follows that a weighted sum of *independent* variables X_i characterized by same distributions, $Z = \sum_{i=1}^{N} X_i/N$, we get

$$\sigma_Z^2 = \frac{\sigma_X^2}{N} \quad \text{and} \quad \sigma_Z = \frac{\sigma_X}{\sqrt{N}}, \tag{4.53}$$

where σ_X^2 is the variance of the X variable. This relationship is behind the arguments discussed in previous sections also stated by the *central limit theorem (CLT)*, stating that *if X is the sum of N independent random variables, then its probability density function (PDF) approaches a Gaussian probability density function regardless the distribution of individual components as N becomes very large*. The Gaussian probability density function is given by

$$p_X(x) = \frac{1}{\sqrt{2\pi\sigma^2}} e^{-\frac{(x - \mu)^2}{2\sigma^2}}, \tag{4.54}$$

where μ and σ are the mean and standard deviation of the distribution, respectively. From Eq. (4.54), we can conclude that if a Gaussian distribution describes a random variable, it is completely defined by its first- and second-order moment: *mean* and *variance*. CLT is fundamental in any measurement process: since each measurement (whatever its distribution) is independent of each other, the final result is normally distributed.

Equation (4.53) is in accordance with CLT stating that an *averaged sum* of N measurements has a spread around its mean (called "true value") that is reduced by the $\sqrt{N}$ factor, as previously discussed.

4.2 Random Processes

So far, we have dealt with probabilistic concepts related to a single phenomenon or device on which an experiment has been performed several times. The question is whether the probabilistic model in our experiment is valid for other physically different devices having the same properties.

Similarly, suppose we set up a model using a priori assumptions about the time evolution of stochastic signals. In that case, we need to imagine the same signal as the outcome of many other "identical" (i.e., having the same physical properties) systems performed under the same conditions. In other words, the random variable should be extended to a "population" of similar experiments performed under identical conditions on different devices having the same properties. Such a characterization is based on the paradigm of a *random process*.

The aforementioned concept could be explained better with an example. Suppose that we have to characterize the noise performance of a sensor that is going into production.[13] In that case, we not only have to forecast its behavior on a single specific sample, but we also need to be sure that the model is valid (in a probabilistic sense) for all the samples coming out of production. Thus, from the mathematical point of view, we have to extend the random variable associated with output noise to any possible outcomes generated by a population of "identical" sensors. In other words, even if statistical inference could be performed on only one sensor, the conclusion should be conceptually applied to the entire population of sensors having the same model.

A *random process* is a random variable $X(t, s)$ function of both the sample space and of time. Therefore, for each outcome (or sample point) it is defined a random variable of time $X(\cdot, s)$ referred to as a *sample function*. Furthermore, for each time, t it defines a random variable $X(t, \cdot)$. In the specific case of electronic devices, the sample space is made of the output of all possible physically distinguishable devices having the same characteristics. We will use $x^{(s)}(t) = X(\cdot, s)$ and $x(t) = X(t, \cdot)$ for the random variables defined on the sample and time spaces, respectively, for the economy of notation.

Example We have a set of physically different noisy resistors characterized by the same electrical model. The experiment of sampling the voltage across terminals in a jth resistor at a time t_i is a realization mathematically described by $v^{(j)}(t_i)$. Therefore, the *random process* could be defined as the random variable $V(t, s)$ whose realizations are $v^{(j)}(t_i) = V(t_i, s_j)$.

The collection of sampled functions is called an *ensemble*. In essence, a random process is the probabilistic model of an ensemble of time waveforms.

---

[13] This example does not take into account variations of systematic errors and related distributions.

> **Hint** We have chosen to use the bar notation ‾ as an operator for ensembles averages and the bracket operator $\langle \cdot \rangle$ for time averages since it is more congruent with electronics literature. However, in other contexts, especially in physics, it could be the opposite.

An example of ensembles of a random process is shown in Fig. 4.5. In this case, we ideally take a population of N "identical" devices (i.e., having the same physical properties even if they are different devices) by monitoring the outputs at the same time with the assumption of the same setup and input conditions. Following the notation chosen, we have that the realization of the experiment "readout of the device number j at the time t_i" is

$$x^{(j)}(t_i) = X(t_i, s_j). \tag{4.55}$$

Then, we will further simplify the realization $X_i = X(t_i)$ as

$$x_i = X(t_i). \tag{4.56}$$

It should be noted how for each waveform of the ensemble, we have a specific time average and variance given by $\langle x_j \rangle$ and $\langle x_j^2 \rangle$, shown in the histograms on the right of Fig. 4.5, should be distinguished from the ensemble averages, as will be defined in the text that follows.

Applying Eq. (4.39) to the random process with the above notation, we get

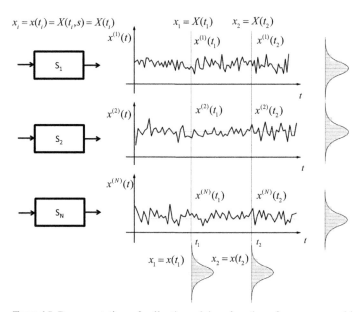

Figure 4.5 Representation of collection of time functions from an ensemble.

$$\overline{x(t)} = E[X(t)] \triangleq \int_{-\infty}^{\infty} x p_X(x, t) dx, \tag{4.57}$$

which is the expected value of $x(t)$ over the ensemble and for this reason is called the *ensemble average*. Another important average of random processes is called the *ensemble autocorrelation* and is based on Eq. (4.47)

$$R_{XX}(t_1, t_2) = \overline{x_1 x_2} \triangleq E[X(t_1)X(t_2)] = \iint_{\infty} x_1 x_2\, p_{X_1 X_2}(x_1, x_2) dx_1 dx_2, \tag{4.58}$$

where again $x_1 = x(t_1)$ and so forth, as shown in Fig. 4.5. If we set $t_1 = t$; $t_2 = t + \tau$, we get another way to write Eq. (4.58)

$$R_{XX}(t, \tau) \triangleq E[X(t)X(t + \tau)], \tag{4.59}$$

which is specular to the time autocorrelation $R_{xx}(\tau)$ of Eq. (4.16) even if, in this case, it could be in general dependent on time. For brevity of notation, also $R_x(t, \tau)$ is used in place of $R_{xx}(t, \tau)$. We can apply the definition of variance σ_x^2 given by Eq. (4.43) to estimate the spread around the average along with the ensemble. An illustrative sketch of the ensemble dispersion is shown in the histograms at the bottom of Fig. 4.5.

Another interesting ensemble averaging properties is the ensemble *cross-correlation* given by

$$R_{XY}(t_1, t_2) = \overline{x_1 y_2} \triangleq E[X(t_1)Y(t_2)] = \iint_{\infty} x_1 y_2\, p_{X_1 Y_2}(x_1, y_2) dx_1 dy_2, \tag{4.60}$$

which is similar to the time cross-correlation R_{xy} given by Eq. (4.15), as will be explained better in the text that follows. In the case we consider as a reference of the mean value, we have the *cross-covariance*

$$\begin{aligned} \sigma_{XY}^2(t_1, t_2) &= E[(X(t_1) - E[X(t_1)])(Y(t_2) - E[Y(t_2)])] \\ &= R_{XY}(t_1, t_2) - E[X(t_1)]E[Y(t_1)]. \end{aligned} \tag{4.61}$$

Instead, if we make a correlation on the same variable, that is, $Y = X$ at different times $R_{XX}(t_1, t_2)$, we have the *autocovariance* defined as

$$\begin{aligned} \sigma_{XX}^2(t_1, t_2) &= E[(X(t_1) - E[X(t_1)])(X(t_2) - E[X(t_2)])] \\ &= R(t_1, t_2) - E[X(t_1)]E[X(t_2)]. \end{aligned} \tag{4.62}$$

Another important parameter of the relationship between two random variables is the *correlation (or Pearson) coefficient*, defined as

$$\rho_{XY} = \frac{\sigma_{XY}^2}{\sigma_X \sigma_X} \quad \text{where:}\ |\rho_{XY}| \le 1. \tag{4.63}$$

Therefore for $Z = aX + bY$, Eq. (4.49) becomes

$$E[Z^2] = a^2 E[X^2] + b^2 E[Y^2] + 2ab \cdot \rho_{XY}\sigma_X\sigma_Y + E[X]E[Y]. \tag{4.64}$$

Now we define the *stationary random process* in a *"wide sense"* (WSS), the random process having the following properties, from Eqs. (4.57), and (4.58)

$$
\begin{aligned}
&\text{(i)} \quad E[X(t)] = \overline{x(t)} = \overline{x} \qquad\qquad\qquad \forall t \\
&\text{(ii)} \quad E[X(t)X(t+\tau)] = R(t,\tau) = R(\tau) \quad \forall t,
\end{aligned}
\tag{4.65}
$$

meaning that the ensemble mean and autocorrelation are independent of the time we refer to. Therefore, for a WSS process, *mean* and *variance* are time-independent.

Thus, Eq. (4.58) becomes

$$
R_{XX}(\tau) \doteq E[X(t)X(t+\tau)],
\tag{4.66}
$$

that is specular to $R_{xx}(\tau)$ of Eq. (4.16). We will discuss better the nature of this similarity in the next sections. From Eq. (4.66) is easy to show that

$$
\begin{aligned}
&\text{(i)} \quad R_{XX}(0) = E[X^2(t)] = \overline{x^2} \\
&\text{(ii)} \quad |R_{XX}(\tau)| \leq R_{XX}(0) \\
&\text{(iii)} \quad R_{XX}(\tau) = R_{XX}(-\tau),
\end{aligned}
\tag{4.67}
$$

as in Eq. (4.17).

Note that for WSS processes, the cross-correlation $R_{XY}(t_1, t_2)$ for $\tau = t_2 - t_1$ and $X = Y$ gives the autocorrelation $R_{XX}(t_1, t_2) = R_{XX}(\tau)$ and the auto-covariance $\sigma^2_{XX}(t_1, t_2) = \sigma^2_{XX}(\tau)$.

According to the definition mentioned earlier, we can state that a *stationary random process in a wide sense is a power signal*. In fact, if it were an energy signal, the signal itself should sometimes vanish to zero, that is, $|x_i(t)| \to 0$ as $t \to \infty$, but in that case, its ensemble averages should change with respect to time, in contradiction with the stationary statement.

Since we refer to a random process and since the *concept of power is related to time averages*, we would like to extend this concept to the entire ensemble. To do this, we extend the definition of power, estimating it on a time period T not only to a single waveform of a WSS process

$$
P(s_i) = P(x^{(i)}) = P_i = \frac{1}{T} \int_T \left(x^{(i)} \right)^2 (t) dt
\tag{4.68}
$$

but also to the entire ensemble by taking the average over the entire ensemble on a time period as large as possible

$$
\overline{P} = \lim_{T \to \infty} E[P_T(x)] = \lim_{T \to \infty} \frac{1}{T} \overline{\int_T x^2(t) dt}.
\tag{4.69}
$$

However, since the average could be performed after the square value, we can write

$$\bar{P} = \lim_{T \to \infty} \frac{1}{T} \int_T \overline{x^2(t)} dt = \lim_{T \to \infty} \frac{1}{T} \int_T \overline{x^2(t)} dt = \langle \overline{x^2(t)} \rangle, \tag{4.70}$$

and since the process is stationary, the square value does not depend on time, so

$$\bar{P} = \lim_{T \to \infty} \frac{1}{T} \int_T \overline{x^2(t)} dt = \langle \overline{x^2(t)} \rangle = \lim_{T \to \infty} \frac{1}{T} \int_T \overline{x^2} dt = \overline{x^2}. \tag{4.71}$$

This is a significant result since we can state that a WSS process is a *power signal* and therefore

$$\underbrace{\bar{P}}_{\text{average power}} = R_{XX}(0) = \overline{x^2} = \underbrace{\sigma_X^2}_{\text{AC power}} + \underbrace{\bar{x}^2}_{\text{DC power}}, \tag{4.72}$$

showing that the average power is given by the random process variance plus the square of its mean. Eq. (4.72) is a very important result since it states that the *energetic properties of a WSS random signal could be estimated by the statistical characterization of the random process.*

4.3   Concept of Ergodicity

We can now consider signals where all the ensemble averages equal the corresponding time averages. More specifically, we will refer to random processes where mean and autocorrelation either in the ensemble and in the time are equal for any waveform of the ensemble

(i) $\langle x^{(j)}(t) \rangle = \lim_{T \to \infty} \frac{1}{T} \int_T x^{(j)}(t) dt = \bar{x}$

$$\tag{4.73}$$

(ii) $\langle x^{(j)}(t+\tau)x^{(j)}(t) \rangle = R_{xx}(\tau) = \lim_{T \to \infty} \frac{1}{T} \int_T x^{(j)}(t+\tau)x^{(j)}(t) dt = R_{XX}(\tau).$

In this case, the process is called *ergodic in the mean and autocorrelation*. In the next sections, we will mostly refer to this case by simply calling it ergodic for simplicity.

For an ergodic signal, relationship (4.72) becomes

$$\underbrace{P = \langle x^{(j)} \rangle^2 = \langle x \rangle^2}_{\text{time average concept}} = \underbrace{R_X(0) = \overline{x^2} = \sigma_X^2 + \bar{x}^2}_{\text{ensemble average concept}}; \tag{4.74}$$

that is,

$$\begin{aligned} \text{Power of arandom signal} &= \overline{x^2} = \sigma_X^2 + \overline{x}^2 \\ \text{DC power of arandom signal} &= \overline{x}^2 \\ \text{AC power of arandom signal} &= \sigma_X^2, \end{aligned} \qquad (4.75)$$

With respect to Eq. (4.72), Eq. (4.74) tells us that the power experimentally determined on every single device is characteristic of the entire population and vice versa. Thus, random processes ergodicity allows us to *get a stochastic characterization of the entire process from time data on a single waveform.* An ergodic process implies stationarity; however, not all stationary processes are ergodic. To explain this better, we will use two examples.

Example A collection of batteries where each battery is providing different voltages. The set's average power is given by the square of the average DC values plus a fluctuation given by the diversity among the set that could be accounted for by the variance. However, the average power (DC value) of each battery is not equal to the average values of the collection; therefore, Eq. (4.72) applies while (4.74) does not. Thus, *it is not ergodic even if it is stationary.*

Maybe the best way to understand the concept of an ergodic process is to make an example of a very common nonergodic process. Assume that an ergodic random process with zero-mean Gaussian noise is fed into a set of electronic integrators, as shown in Fig. 4.6. We integrate 500 input samples of such noise with all of them. Alternatively, we can repeat several times the same experiment with the same integrator with the same number of samples. As shown in Fig. 4.6, the output of the integrators appears to have different random behavior in the three illustrated experiments. It could be shown that the output diverges from the origin and the amount of the divergence tends (in a probabilistic sense) to be higher the longer is the integration time. Therefore, the process is not stationary, and thus it is not ergodic.

The nonstationarity could be perceived by the output trend that could take either direction with an increasing divergence in behavior. This is a *random walk* process shown in Brownian motion (see Appendix A of this section). We can better show this by superimposing the same three outputs as in Fig. 4.6. As we can see, the divergence between the final points increases on average (it can be shown as the square root of time). This behavior should be taken in a probabilistic sense, meaning that it might happen in a large number of trials to get a very small spread with respect to the reference, but this has a very low probability of occurring.

Therefore, the mean and the mean square of the three curves are different: the time-variant characterization of a single curve is not indicative of the ensemble statistics. From the mathematical viewpoint, we can say that

$$\langle x^{(i)} \rangle \neq \overline{x}; \ \langle x^{(i)2} \rangle \neq \overline{x^2}. \qquad (4.76)$$

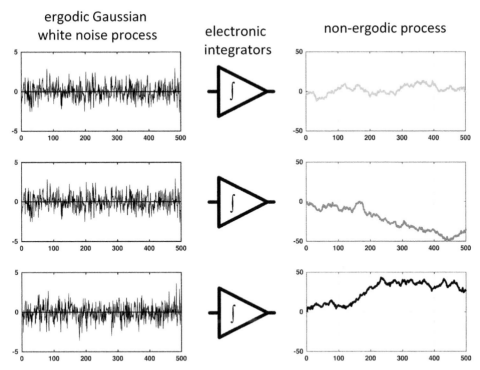

Figure. 4.6 Example of the relationship between an ergodic and nonergodic random process. Different integrators integrate a zero-mean Gaussian noise for the same integrating time.

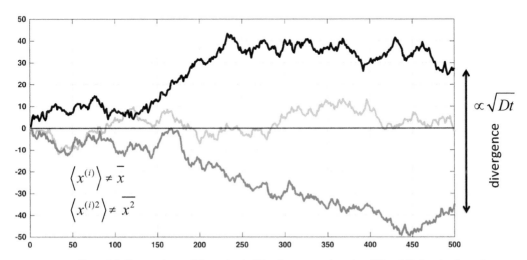

Figure 4.7 Comparison of the output of the three experiments of Fig. 4.6 showing how the divergence from the initial point increases (in a probabilistic sense) as time goes on.

Therefore, it is neither stationary nor ergodic. In this case, *to characterize the random process, we should take the ensemble set* statistical estimators instead of the time set ones.

The integration of noise is a very common example of a nonergodic process in electronic devices, and this is one reason why it is common to use the statistical averages notation $(\overline{x^n})$ instead of the time averages $(\langle x^n \rangle)$ for random electronic processes.

Turning back to the properties of ergodic processes, the exchange of time and ensemble averages allows another very important relationship identical to Eq. (4.21):

$$S_X(f) = F[R_{XX}(\tau)] = \int_{-\infty}^{\infty} R_{XX}(\tau)e^{-j2\pi f \tau} d\tau, \tag{4.77}$$

where $S_{XX}(f)$ is the *power spectrum* of the random process. This is also the *Wiener–Kinchine theorem* version for a random process. Why power spectrum and not energy spectrum? First, stationary random processes are power signals, as already explained. Second, the statistical average implies a division by N that is equivalent to the division by T of time averages. See also discussion on Fig. 4.2 for the duality between power and energy spectra. Note how the power spectrum is now defined based not on the signal's temporal characteristics but also on statistical characteristics. *Therefore, PSD is a key function to characterize a random process.*

Similar to Eq. (4.32), we can define the cross-spectral density (for ensemble averages) as

$$S_{XY}(f) = \int_{-\infty}^{\infty} R_{XY}(\tau)e^{-j2\pi ft} dt. \tag{4.78}$$

Now let us reconsider the relationship (4.74) indicating the relationship between power and variance for ergodic signals. We have again

$$P_X = \langle x^2 \rangle = \overline{x^2} = \int_{-\infty}^{\infty} S_X(f) df. \tag{4.79}$$

Furthermore, from Eq. (4.24), we have for an LTI system

$$P_Y = \int_{0}^{\infty} |H(f)|^2 S_X(f) df. \tag{4.80}$$

Therefore, for zero-mean, uniform Gaussian noise, we have that, in a low-pass LTI system

$$\sigma_Y^2 \approx |H(0)|^2 \sigma_X^2, \tag{4.81}$$

where $|H(0)|$ is the "in-bandwidth" gain of the system. Note that since $|H(0)|$ is the gain, that is, the slope of the input–output quasistationary characteristic, it is thus a revisited view of the *law of propagation of errors* discussed in Chapter 2.

The relationship (4.77) is important because *we can get the distribution of the power of a random signal based on the statistical characterization of the process and without the need for the Fourier transform defined only on deterministic signals.*

However, from an experimental point of view, let us imagine that we collect many data samples from an *ergodic* random process to store the waveform for a determined amount of time. In that case, we have the total information of the evolution of the signal for that amount of time to apply the Fourier transform. In that case, which is the relationship with the ensemble correlations function? For this task, we can take a time-truncated sample of one waveform of the ensemble

$$x(t) = \begin{cases} 0 \text{ for } |t| > T/2 \\ x(t) \text{ for } |t| < T/2. \end{cases} \tag{4.82}$$

Since it is an energy signal, we can calculate its Fourier transform

$$X_T(f) = \int_{-\infty}^{\infty} x(t)e^{-j2\pi ft} dt = \int_{-T/2}^{T/2} x(t)e^{-j2\pi ft} dt. \tag{4.83}$$

Therefore, $\left|X_T^{(k)}(f)\right|^2$ is the energy spectral density for that particular gated waveform. Thus, using Rayleigh's theorem, we get

$$\int_{-T/2}^{T/2} x^2(t) dt = \int_{-\infty}^{\infty} \left|X_T^{(k)}(f)\right|^2 df, \tag{4.84}$$

and similar to Eq. (4.70), we can estimate the average power as

$$\overline{P} = \lim_{T \to \infty} \frac{1}{T} \int_{-\infty}^{\infty} \overline{\left|X_T(f)\right|^2} dt = \int_{-\infty}^{\infty} \left(\lim_{T \to \infty} \frac{1}{T} \overline{\left|X_T(f)\right|^2} \right) dt, \tag{4.85}$$

whose integrand could be interpreted as the *power spectral density function* of the random signal. In other words, we can *define* the integrand as the PSD of the random signal

$$S(f) \triangleq \lim_{T \to \infty} \frac{1}{T} \overline{\left|X_T(f)\right|^2}, \tag{4.86}$$

and it could be shown that such definition satisfies the Wiener–Kinchine theorem. From the experimental standpoint, relationship (4.86) suggests that we can *estimate* the *single-sided* power spectral density of a random function by using

$$\hat{S}_X(f) = \lim_{T \to \infty} \frac{2}{T} \overline{\left| X_T(f) \right|^2}, \tag{4.87}$$

called *periodogram*. This means that we can take samples of a random process in several timeframes of a period T and average them to estimate the PSD.

4.4   Convergence of Concepts Between Deterministic and Random Variables

To recall, $x(t_i)$ are acquired samples of $x(t)$, where $i = 1, 2, \ldots N$. Let assume to have just four possible discrete levels A, B, C, D of the variables, corresponding to the signal values x_A, x_B, x_C and x_D. The *sample mean* of the N values is also the time mean of acquired data:

$$\bar{x} = \frac{1}{N} \sum_{i=1}^{N} x(t_i) = \langle x \rangle. \tag{4.88}$$

However, since the number of levels is finite, we can collect them into bins and rearrange the summation in the following way:

$$\bar{x} = \frac{1}{N} \sum_{i=1}^{N} x(t_i) = \frac{N_A x_A + N_B x_B + N_C x_C + N_D x_D}{N} = \sum_{j=A}^{D} \frac{N_j}{N} x_j. \tag{4.89}$$

Since the fraction N_j/N for a large number of samples tends to the probability $Pr(x_j)$ of getting the value x_j; we have

$$\bar{x} = \frac{1}{N} \sum_{i=1}^{N} x(t_i) = \langle x \rangle \overset{N \to \infty}{=} \sum_{j=A}^{D} \Pr(x_j) x_j = E[X]; \tag{4.90}$$

thus the statistical mean of the acquired data is the time mean of the process and an estimator of the expected value.

We can derive similar conclusions also for the *correlation*. Assume that we want to know how two signals are similar to each other or, in other words, correlated. Let us take, for example, the signals shown in Fig. 4.8 and assume, as in the previous example, that the resolution of our readout is limited to two levels: upper, H, and lower, L. We will call x_H and x_L the values of the x signal when it is above or below the threshold and similarly for the y. We will also assume that the two signals have zero mean value for simplicity.

To understand if they are correlated, we collect into bins the outcomes from both sampled signals $x(t_i)$ and $y(t_i)$. Intuitively, if the two signals are related, they could be on-phase (prevalence of the HH and LL bins) or counter phase (prevalence of the HL and LH bins). Thus, there is a correlation if the weight bins are clustered along the square's diagonals. Conversely, if the weights are equally distributed, we can say that

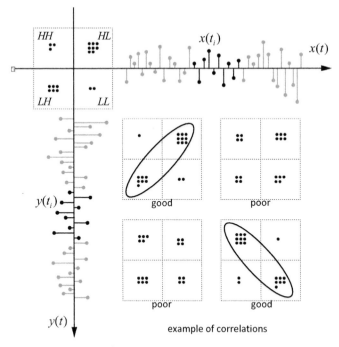

Figure 4.8 Example of cross-correlation concept.

the two signals are independent or not correlated. The best way to check clustering around the diagonals is to make an averaged sample correlation of the two signals that also corresponds to the time correlation

$$\overline{xy} = \frac{1}{N}\sum_{i=1}^{N} x(t_i)y(t_i) = \langle xy \rangle. \tag{4.91}$$

Again, since the resolution is limited, we can rearrange the summation of Eq. (4.91) in the following way:

$$\overline{xy} = \frac{1}{N}\sum_{i=1}^{N} x(t_i)y(t_i) = \langle xy \rangle = \frac{N_{HH}x_Hy_H + N_{HL}x_Hy_L + N_{LH}x_Ly_H + N_{LL}x_Ly_L}{N}$$

$$= \sum_{j=L}^{H}\sum_{k=L}^{H} \frac{N_{jk}}{N}x_j x_k \stackrel{N\to\infty}{=} \sum_{j=L}^{H}\sum_{k=L}^{H} Pr(x_j, x_k)x_j x_k = E[XY], \tag{4.92}$$

where N_{jk}/N corresponds, for a large number of samples, to the joint probability $Pr(x_j, x_k)$ of the events. Note how if we correlate the signal with itself, we get the *autocorrelation* that is

$$\overline{x^2} = \langle x^2 \rangle = P, \tag{4.93}$$

where P is the power of the signal. For signals with nonzero means, it is easy to show that

$$\overline{xy} = \langle xy \rangle = \hat{\sigma}_{XY}^2 + \overline{x} \cdot \overline{y} \stackrel{N \to \infty}{=} \sigma_{xy}^2 + E[X]E[Y] = .\tag{4.94}$$

So far, we have dealt with acquired sampled data of a single device signal. However, if we assume the process ergodic, we can extend the statistical averages no longer to past data and the entire ensemble. Therefore, the expected values are extended not only to a single device but to any device having the same model.

4.5   Low-Pass Filtering of White Noise

Now we can investigate the relationship between input and output spectral density and the autocorrelation function of the random process. Let us take an extreme case of white noise, that is, a random signal having a completely uniform power spectral density:

$$S(f) = \alpha^2,\tag{4.95}$$

whose autocorrelation function is

$$R_X(\tau) = \int_{-\infty}^{\infty} S(f)e^{j2\pi f\tau}df = \alpha^2 \cdot \delta(\tau).\tag{4.96}$$

Thus, white noise is a signal where we cannot forecast absolutely anything with any degree of confidence away from the sample itself.

Now consider white noise applied to an LTI system input whose transfer function is $H(f)$. Since $S_y = |H(f)|^2 S_x(f)$, from Eq. (4.24), we have

$$S_Y(f) = |H(f)|^2 \alpha^2.\tag{4.97}$$

Therefore, the output spectrum takes the same shape as the transfer function of the filter introduced by the LTI system.

A widespread class of LTI systems is modeled by the *first-order low-pass filter* and described using the *canonical form*

$$H(f) = \frac{k'}{1+j(f/f_0)} \to H(\omega) = \frac{k}{1+j(\omega/\omega_0)},\tag{4.98}$$

where k is the DC gain and ω_0 is its cut-off frequency. If we apply white noise to the input of the system, we get a random signal to the output, whose autocorrelation is[14]

---

[14] From the identity $F^{-1}\left[\dfrac{2/\lambda}{1+\omega^2/\lambda^2}\right] = e^{-\lambda|t|}$.

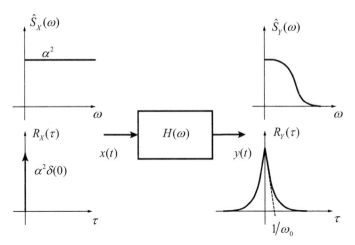

Figure 4.9 Effect of filtering on white noise.

$$R_Y(\tau) = \alpha^2 F^{-1}\left[|H(f)|^2\right] = \alpha^2 F^{-1}\left[\frac{k^2}{1 + \omega^2/\omega_0^2}\right] = \frac{\alpha^2 k^2 \omega_0}{2} e^{-\omega_0|\tau|}. \tag{4.99}$$

Thus, a random signal whose autocorrelation is a delta function is mapped into a random signal whose autocorrelation function is an exponentially decreasing function, as shown in Fig. 4.9.

Also, the single-sided power spectrum density could be found as

$$\hat{S}(\omega) = 2F[R_Y(\tau)] = \frac{2\alpha^2 k^2}{1 + \omega^2/\omega_0^2}, \tag{4.100}$$

which is referred to as a *Lorentzian form*.

What does it mean? The low-pass filtering increases the probability of forecasting the behavior at timescales close to the sample. In other words, filtering white noise slows down the output so as we can better foresee its behavior in the timeframe of the time-constant of the filter, since *the larger the time-constant, the larger the autocorrelation.*

The example shows a common trend: the larger the power bandwidth, the larger the random signal variance, and the narrower the autocorrelation, as shown in Fig. 4.10, where $B \approx f_0$ is the bandwidth of the filter.

Here we can see that the larger the bandwidth, the higher is the area below the PSD curve so that the higher the respective variance. Therefore, bandwidth reduction (e.g., using a moving average filter; see Chapter 3) means a decrease of the output noise variance and thus an increase of the signal-to-noise ratio (Chapter 2).

4.6 The Equivalent Noise Bandwidth

The purpose of this section is to characterize a linear system from the noise point of view on the base of its transfer function. If we apply white noise to a bandwidth-limited

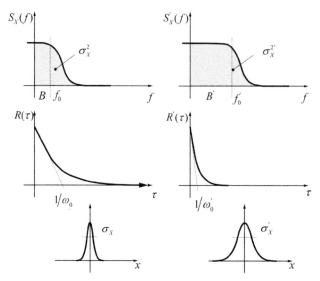

Figure 4.10 Relationship between spectral density function, autocorrelation, and variance (negative side of autocorrelation not shown).

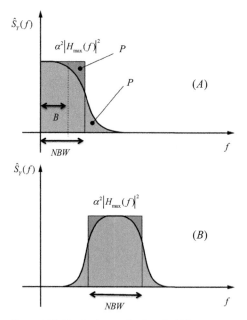

Figure 4.11 Equivalent noise bandwidth concept.

LTI system, we have seen that the output PSD $\hat{S}_Y(f) = \alpha^2|H(f)|^2$ has the same shape of $|H(f)|^2$. Looking to Fig. 4.11A for a low-pass filter, we can see that the total noise power (integral of the PSD) is equal to the area of a rectangle whose height is equal to the in-bandwidth value $|H_{max}(f)|^2$. We can calculate the base size of the rectangle NBW using the total power equivalence

$$P = \overline{y^2} = \int_0^\infty \hat{S}_Y(f)df = \int_0^\infty \alpha^2 |H(f)|^2 df = NBW \cdot \alpha^2 |H_{max}(f)|^2$$

$$\rightarrow NBW = \frac{1}{\hat{S}_Y(f)_{max}} \int_0^\infty \hat{S}_Y(f)df = \frac{1}{|H_{max}(f)|^2} \int_0^\infty |H(f)|^2 df,$$

(4.101)

where NBW is called *noise equivalent bandwidth* and $\hat{S}_Y(f)$ is the unilateral output PSD. Note how the total noise power at the system's output could be modeled by two contributions: the in-bandwidth power gain and the NBW. Therefore, the NBW model greatly simplifies design and characterization on noisy systems.

In the case where we have a first-order low-pass filter, NBW could be calculated as[15]

$$NBW = \frac{\displaystyle\int_0^\infty \alpha^2 |H(f)|^2 df}{\alpha^2} = \int_0^\infty \frac{1}{1 + (f/f_0)^2} df = \frac{\pi}{2}f_0 = \frac{\pi}{2}B.$$

(4.102)

Note that NBW is slightly larger than bandwidth B. It could be shown that the difference between NBW and B decreases as the order (selectivity) of the filter increases. Note how the concept could also be applied to band-pass shape figures such as in Fig. 4.11B.

4.7 Sum/Subtraction of Random Signals

The relationship between the energy of signals allows us to compare them. Let us consider the case where two random signals are weighted and summed together $z(t) = ax(t) + by(t)$, as illustrated in Fig. 4.12.

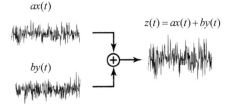

Figure 4.12 Combination of sources of random signals.

[15] From the identity $\displaystyle\int_{-\infty}^\infty \frac{1}{1 + (f/f_0)^2} df = \frac{\pi}{2}f_0.$ See the appendix on similar forms.

Assuming signals to be jointly stationary, we can use either time or ensemble averages to get cross-correlation. In the first case, we have

$$R_z(\tau) = \langle((ax(t) + by(t))(ax(t + \tau) + by(t + \tau)))\rangle =$$
$$\langle(ax(t)ax(t + \tau)) + (by(t + \tau)by(t)) + (ax(t + \tau)by(t)) + (by(t + \tau)ax(t))\rangle$$
$$= a^2 R_x(\tau) + b^2 R_y(\tau) + 2ab R_{xy}(\tau),$$

(4.103)

where we have used the identity $R_{xy}(\tau) = R_{yx}(\tau)$ since signals are real.

Eq. (4.103) for $\tau = 0$ becomes (4.30),

$$P_z = a^2 \langle x^2 \rangle + b^2 \langle y^2 \rangle + 2ab \langle xy \rangle = a^2 P_x + b^2 P_y + 2ab \langle xy \rangle.$$

(4.104)

Therefore, under the further assumption of ergodicity, the same relationship is expressed by using ensemble averages

$$P_z = a^2 \overline{x^2} + b^2 \overline{x^2} + 2ab \cdot \overline{xy}.$$

(4.105)

We can distinguish between the AC and DC components of the power; recalling that $\overline{x^2} = \sigma_X^2 + \overline{x}^2$, etc., so that we have

$$P_Z = a^2 \sigma_X^2 + b^2 \sigma_Y^2 + 2ab \cdot \sigma_{XY}^2 + (a\overline{x} + b\overline{y})^2 = P_{ZAC} + P_{ZDC}; \text{ where}$$
$$P_{ZAC} = a^2 \sigma_X^2 + b^2 \sigma_Y^2 + 2ab \cdot \sigma_{XY}^2.$$

(4.106)

We can also use the correlation coefficient as in (4.64) so that the AC power in (4.108) is expressed as

$$P_{ZAC} = a^2 \sigma_X^2 + b^2 \sigma_Y^2 + 2ab\rho_{XY} \cdot \sigma_X \sigma_Y.$$

(4.107)

For signals that are *uncorrelated*, the output becomes

$$\langle z^2 \rangle = a^2 \langle x^2 \rangle + b^2 \langle y^2 \rangle \text{ or:}$$
$$\overline{z^2} = a^2 \cdot \overline{x^2} + b^2 \cdot \overline{y^2},$$

(4.108)

which is a formalization of the *principle of superposition of powers* for uncorrelated random signals.

Assume now that we would like to convey information using the difference of two signals. For instance, as we will see in Chapter 8, instead of using as the signal a single voltage referred to the common ground, for example V_A, we use the difference between two potentials of the circuit, for example, $V_A - V_B$. The aim is to reduce the correlated noise that could concurrently affect both channels (fully differential architectures). Therefore, using Eq. (4.106) or (4.107) with a minus on the third term, we can reduce noise's influence depending on their correlation. Note that in the case of completely uncorrelated noise, that is $\rho_{XY} = 0$, the differential signal noise variance is exactly twice (assuming the same model on the channels). In contrast, for completely correlated noise (i.e., the perturbation is deterministic, $\rho_{XY} = 1$) the perturbation is thoroughly washed out. In the latter case, instead of *noise*, it is frequently referred to as *interference*.

Therefore, we can now use the concept mentioned earlier in the time domain (differentiation) by comparing two consecutive samples $z = x_2 - x_1$ at different times of the same random variable $x_1 = X(t_1)$; $x_2 = X(t_2)$. Therefore, considering (4.105) with the WSS assumption, so that $\tau = t_2 - t_1$ and $\overline{x_1^2} = \overline{x_2^2} = \overline{x^2}$, and the autocorrelation $R(t_1, t_2) = R(\tau)$, we have using Eq. (4.58)

$$P_z = \overline{x_1^2} + \overline{x_2^2} - 2R_{XX}(t_1, t_2) = 2\overline{x^2} - 2R_{XX}(\tau). \tag{4.109}$$

Equation (4.109) is at the base of techniques that will be used in Chapter 8: if the two samples are completely uncorrelated (as in white noise), then the power of the difference is double the power of a single signal. Conversely, if the samples are correlated, the power of the result is reduced by the ensemble autocorrelation $R(\tau)$ factor. Note that since $\overline{x_1} = \overline{x_2} = \overline{x}$ similarly to (4.106), we can rewrite (4.109), using Eq. (4.62):

$$P_z = P_{zAC} = \left(\sigma_X^2(t_1) + \overline{x_1}^2\right) + \left(\sigma_X^2(t_2) + \overline{x_2}^2\right) - \left(2\sigma_{XX}^2(t_1, t_2) + 2\overline{x_1}\,\overline{x_2}\right)$$
$$= 2\sigma_X^2 - 2\sigma_\tau^2, \tag{4.110}$$

where, due to WSS assumption, $\sigma_{XX}^2(t_1, t_2) = \sigma_{XX}^2(\tau)$ is the *autocovariance* and $\sigma_X^2(t_1) = \sigma_X^2(t_2) = \sigma_X^2$. That is why the differentiation of sampled signal is acting as a high-pass filter getting rid of common offset. In practice making the difference between two samples doubles the variance of the signal minus the "correlation" (covariance) between such samples.

We can now apply the same concepts in the spectrum domain. Using (4.32) in (4.103) and considering real signals, we have

$$S_z(f) = a^2 S_x(f) + b^2 S_y(f) + abS_{xy}(f) + abS_{yx}(f)$$
$$= a^2 S_x(f) + b^2 S_y(f) + abS_{xy}(f) + abS^*_{xy}(f) \tag{4.111}$$
$$= a^2 S_x(f) + b^2 S_y(f) + 2ab\text{Re}\{S_{xy}(f)\},$$

where $S_{xy}(f)$ takes into account the spectral density of components of signals generating power. Therefore, the power in the bandwidth Δf is

$$S_z(f)\Delta f = a^2 S_x(f)\Delta f + b^2 S_y(f)\Delta f + 2ab\text{Re}\{S_{xy}(f)\}\Delta f$$
$$= a^2 S_x(f)\Delta f + b^2 S_y(f)\Delta f + 2ab \cdot \rho_{XY}\sqrt{S_x(f)S_y(f)}\Delta f, \tag{4.112}$$

where

$$\rho_{XY} = \text{Re}\{c_{XY}\} \quad \text{where}$$
$$c_{XY} = \frac{S_{xy}(f)}{\sqrt{S_x(f)S_y(f)}}. \tag{4.113}$$

c_{XY} is referred to as the *noise correlation coefficient*, and it is a complex number since S_{XY} is complex even if the original signals are real. Here we intermixed the notation of ensemble and time averages due to the ergodic nature of the process.

4.8   Physical Interpretation of Cross-Spectral Density

Let consider the current and voltage on an impedance load under sinusoidal excitation in their complex phasor representation

$$v(t) = V\cos(\omega t + \theta_v) \leftrightarrow V(j\omega) = V/\sqrt{2}e^{j\theta_v}$$
$$i(t) = I\cos(\omega t + \theta_i) \leftrightarrow I(j\omega) = I/\sqrt{2}e^{j\theta_i}, \tag{4.114}$$

where V and I are the *peak* values of voltage and current, respectively. The time-average of the instantaneous power $v(t)i(t)$ dissipated in the normalized load of 1 Ohm is

$$P = \langle vi \rangle = \langle v(t)i(t) \rangle = \langle \frac{VI}{2}\cos(\theta_v - \theta_i) + \frac{VI}{2}\cos(2\omega t + \theta_v + \theta_i) \rangle$$
$$= \frac{VI}{2}\cos(\theta_v - \theta_i), \tag{4.115}$$

since the $\cos(2\omega t)$ averages to zero. This is maximum when the voltages and currents are in phase while it is zero when they are in 90° of phase difference (quadrature). In that case, we say that the power is not dissipated onto the load, but it is exchanged back and forth with the load. To cope with these different conditions, complex representations of voltage and current are used so that the average power could be expressed as

$$\langle V(j\omega)I^*(j\omega) \rangle = S = P + jQ, \tag{4.116}$$

where S is the *total power*, P is the *real (active) power*, and Q is the *reactive power*.

In Fig. 4.13 are represented two opposite examples. The first case is where voltage and currents are taken on a resistance, as shown in Fig. 4.13A. Here, the total power is equal to the active power dissipated onto the resistance as heat and reactive power Q is zero. In Fig. 4.13B is instead shown the case of a capacitor where voltage and current are in quadrature so that P is zero while Q is maximum. In this case, the energy is exchanged back and forth with the load without dissipation. We will see that the physical characteristics of noise are related to thermal properties, and therefore it is strictly related to P and not to Q. It could be shown that *electronic noise processes are associated with the dissipative behavior of devices*, thus when voltage and current are in-phase.

Thus, we can interpret the cross-spectral density (4.33), as illustrated in Fig. 4.14. The cross-spectral density $S_{xy}(f)$ is the Fourier transform of the cross-correlation of two real signals, x and y. Therefore, we can evaluate what part of

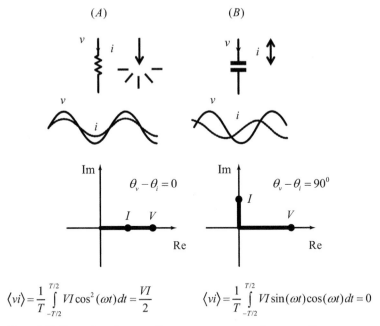

$$\langle vi \rangle = \frac{1}{T}\int_{-T/2}^{T/2} VI\cos^2(\omega t)\,dt = \frac{VI}{2}$$

$$\langle vi \rangle = \frac{1}{T}\int_{-T/2}^{T/2} VI\sin(\omega t)\cos(\omega t)\,dt = 0$$

Figure 4.13 Examples of pure active average power and pure reactive power.

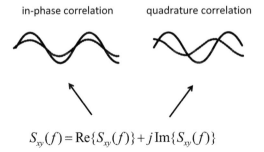

$$S_{xy}(f) = \text{Re}\{S_{xy}(f)\} + j\,\text{Im}\{S_{xy}(f)\}$$

Figure 4.14 Significance of the real and complex part of the cross-spectral density.

the cross-correlation is in-phase for each component of the spectrum and what part is in quadrature. These two contributions are the real and the imaginary parts of the $S_{xy}(f)$. Therefore, the two signals act *as if* they dissipate heat when they are in-phase. This observation is also supported by Eq. (4.111), which shows that a further contribution to the total power of two signals is the real part of $S_{xy} = (f)$ while its imaginary part is not taken into account. Therefore, what increases the power of a random process (e.g., noise power) composed of the sum of random processes is their in-phase correlation.

This observation is related to the *fluctuation-dissipation theorem*, whose informal conclusion is that the thermal noise aspects of a random process are related to the physical characteristics of heat dissipation of the same.

4.9   The Lorentzian Form

A Lorentzian *unilateral* power spectrum is given by the form

$$\hat{S}(\omega) = k^2 \frac{4\tau}{1 + \omega^2\tau^2} = k^2 \frac{4D}{D^2 + \omega^2} \rightarrow$$

$$\frac{1}{2\pi}\int_0^\infty \hat{S}(\omega)d\omega = k^2 \equiv [W], \tag{4.117}$$

shown in Fig. 4.15 in logarithmic and linear plots. We also used $D = 1/\tau$. The Lorentzian form's key feature is that it is characterized by total power k^2, as could be easily demonstrated.

This form could also be seen as the energy density spectrum $\hat{E}(\omega) = \tau \cdot \hat{S}(\omega)$ of an exponential decaying deterministic

$$x(t) = 2ke^{-\frac{2t}{\tau}}, \tag{4.118}$$

where τk^2 is the energy of the signal

$$\frac{1}{2\pi}\int_0^\infty \hat{E}(\omega)d\omega = \int_0^\infty x^2(t)dt = \tau k^2 \equiv [J]. \tag{4.119}$$

Furthermore, the Lorentzian form can be used to model the power density spectrum $\hat{S}_Y(\omega)$ of white noise $\hat{S}_X(f) = \alpha^2 = 4\tau$ filtered by a first-order low-pass filter:

$$H(\omega) = \frac{k}{1 + j\omega\tau}, \tag{4.120}$$

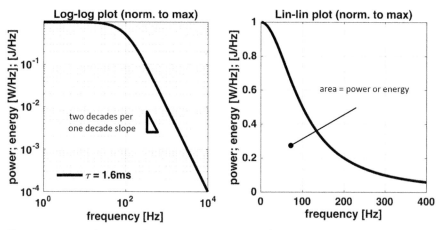

Figure 4.15 Lorentzian power/energy spectra in log and lin plots normalized to the maximum. Note that we have half energy/power (cut-off frequency) for $\omega = 1/\tau \rightarrow f = 1/(2\pi\tau) = 100$ Hz.

where the total power and the corresponding autocorrelation functions are

$$\frac{1}{2\pi}\int_0^\infty \hat{S}_Y(\omega)d\omega = k^2; \; R_Y(\tau) = k^2 e^{-\frac{|t|}{\tau}}; \; k^2 = R_Y(0). \tag{4.121}$$

4.9.1 The Squared sinc Function and Its Relationship with the Lorentzian

We now refer to a *single-sided* energy spectrum of the form

$$\hat{E}(f) = 2(k\tau)^2 |\text{sinc}(f\tau)|^2, \tag{4.122}$$

whose total energy is given by $k^2\tau$. We have shown how this energy spectrum is given by a single rectangular pulse lasting a period of time τ.

In Fig. 4.16 is shown the aspect of (4.122) in both log-log plots and lin–lin plots. Note how the decay in the log–log plot is characterized by a slope of two decades of energy per one decade of frequency. It could also be easily shown that 92% of the total energy spectrum is contained in the first lobe, up to the notch $f_0 = 1/\tau$. It is interesting to compare the Lorentzian with the squared sinc function generated by two energy signals as shown in Fig. 4.17 having the same total energy: the exponential decay and the rectangular pulse.

Also, notice how the Lorentzian envelopes the squared sinc maxima.

Another way to approximate the energy of a single square pulse is given from the observation that

$$\text{sinc}(f\tau) = \frac{1}{2} \quad \text{for } f = \frac{1}{2\tau}. \tag{4.123}$$

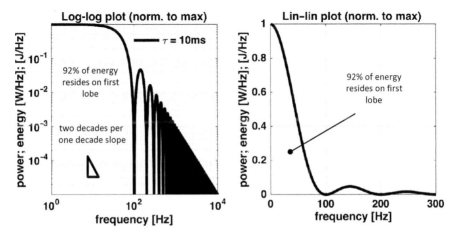

Figure 4.16 Log-log and lin–lin plots of the squared sinc function.

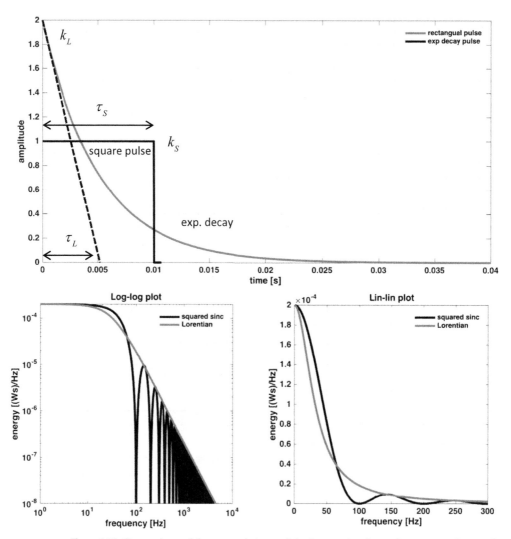

Figure 4.17 Comparison of the squared sinc and the Lorentzian forms that are transforms of signals having the same total energy. In this case, we have chosen the exponential decay amplitude twice that of the rectangular pulse and the time constant of the exponential decay half that of the rectangular pulse.

Therefore, a possible approximation of the spectral energy of a rectangular pulse could be modeled with a low-pass behavior with a cut-off frequency (-3dB) that halves the energy for that frequency

$$E(f) = \frac{k\tau^2}{1 + \dfrac{1}{f2\tau}},\qquad(4.124)$$

where the energy is half in the corner frequency and going asymptotically to $k\tau^2$ for higher bandwidth.

4.10 The Campbell and Carson Theorems

Assume we have a random train of uncorrelated pulses of the same shape

$$x(t) = \sum_{i=1}^{N} a_i g(t - t_i); \quad \text{for } -T/2 \le t \le T/2 \text{ and } 0 \text{ otherwise,} \tag{4.125}$$

where a_i is the amplitude and $g(t)$ is the function of each pulse, as shown in Fig. 4.18. We assume that this is an ergodic process.

The problem is to find the power spectrum of such a function using previous arguments. For the moment, we assume that all the pulses have unitary amplitude $a_i = 1, \forall i$.

We can express the average of the random signal as

$$\langle x(t) \rangle = \lim_{T \to \infty} \frac{1}{T} \int_{-T/2}^{T/2} \sum_{i=1}^{N} g(t - t_i) dt = \langle x \rangle = \lambda \int_{-\infty}^{\infty} g(t) dt$$

$$\text{where } \lambda = \lim_{T \to \infty} \frac{N}{T}, \tag{4.126}$$

where λ is the *average rate* of occurrence of pulses.

The autocorrelation function could be found as

$$R_X(\tau) = \lim_{T \to \infty} \frac{1}{T} \int_{-T/2}^{T/2} x(t) x(t + \tau) dt = \lim_{T \to \infty} \frac{1}{T} \int_{-T/2}^{T/2} \sum_{i=1}^{N} g(t - t_i) \sum_{i=1}^{N} g(t - t_i + \tau) dt$$

$$= \lim_{T \to \infty} \frac{1}{T} \int_{-T/2}^{T/2} \left\{ \sum_{i=1}^{N} g(t - t_i) g(t - t_i + \tau) + \sum_{i=1}^{N} \sum_{j=1 \neq i}^{N} g(t - t_i) g(t - t_j + \tau) \right\} dt.$$

$$\tag{4.127}$$

The first term is the autocorrelation function of $g(t)$, while the second term since the pulses are completely uncorrelated each other, becomes

Figure 4.18 Random train of pulses. Each pulse has the same shape $g(t)$ and variable amplitude $a(t)$.

$$\lim_{T\to\infty} \frac{1}{T} \int_{-T/2}^{T/2} \left\{ \sum_{i=1}^{N} \sum_{\substack{j=1\\ j\neq i}}^{N} g(t-t_i)g(t-t_j+\tau) \right\} dt$$

$$\lim_{T\to\infty} \frac{1}{T} \int_{-T/2}^{T/2} \left\{ \sum_{i=1}^{N} g(t-t_i) \sum_{\substack{j=1\\ j\neq i}}^{N} g(t-t_j+\tau) \right\} dt$$

$$= \langle g(t)g(t+\tau)\rangle = \langle g(t)\rangle\langle g(t+\tau)\rangle = \langle g\rangle^2 \qquad (4.128)$$

and the autocorrelation becomes

$$R_X(\tau) = \lambda \int_{-\infty}^{\infty} g(t)g(t+\tau)dt + \langle g\rangle^2. \qquad (4.129)$$

For the properties of the variance, on ergodic signals, we have $\langle x\rangle = \bar{x}$, and by (4.126),

$$R_X(0) = \overline{x^2} = \sigma_X^2 + \bar{x}^2 \qquad (4.130)$$

and therefore

$$\sigma_X^2 = \lambda \int_{-\infty}^{\infty} g^2(t)dt; \ and \ \bar{x} = \lambda \int_{-\infty}^{\infty} g(t)dt. \qquad (4.131)$$

This result is called the *Campbell's theorem*, saying that a signal with repeated equal random pulses has average and variance equal to the average and energy of the pulse function ($g(t)$ is an energy signal) times the average rate λ.

Now, we can determine the power spectral density as

$$S(f) = R_X(\tau) = \int_{-\infty}^{\infty} \left[\lambda \int_{-\infty}^{\infty} g(t)g(t+\tau)dt + \bar{g}^2 \right] e^{-j2\pi f\tau}d\tau$$

$$= \int_{-\infty}^{\infty} [\lambda R_g(\tau) + \bar{g}^2]e^{-j2\pi f\tau}d\tau = \lambda|G(f)|^2 + \delta(f)\bar{g}^2. \qquad (4.132)$$

A more formal derivation of this relationship in the case each pulse has a different amplitude a_i(so far we assumed $a_i = 1$) gives a single-sided power spectrum

$$\hat{S}(f) = 2\lambda\langle a^2\rangle|G(f)|^2 + 2\delta(f)\bar{g}^2. \qquad (4.133)$$

This is referred to as *Carson's theorem*. The result is not surprising because it says that power density is given by the energy spectrum of the single pulse $2|G(f)|^2$ multiplied by the average occurrence rate, to which is added the DC power of the average.

It should be noted that Carson's theorem *could not* be applied to periodic deterministic signals since (4.128) implies uncorrelation between pulses.

4.11 Power Spectral Density and Noise Density Notations

In electronic circuits, powers are sometimes referred (normalized) to a unitary resistor $R = 1\Omega$; thus, electrical power of deterministic or random signals are expressed as the *mean square* of voltage and current, for example,

$$\overline{v^2} = \langle v^2 \rangle \equiv [\mathrm{V}^2]; \ \overline{i^2} = \langle i^2 \rangle \equiv [\mathrm{A}^2], \tag{4.134}$$

while in the case of *power spectral densities*, the units are expressed in volts squared or ampere squared per hertz unit:

$$\overline{v^2}(f) = S_V(f) \equiv \left[\frac{\mathrm{V}^2}{\mathrm{Hz}}\right]; \ \overline{i^2}(f) = S_I(f) \equiv \left[\frac{\mathrm{A}^2}{\mathrm{Hz}}\right]. \tag{4.135}$$

Hint As a notation; we will use for PSDs the (f) term after the mean square symbol, which is absent for the power. Therefore $\overline{x^2}$ is power, while $\overline{x^2}(f)$ is power spectral density.

Very often, in datasheets and papers, to have a handy and fast reference to experimental data, it is preferred to express the random process measure in the units of the original signal instead of the square. Therefore, it is used the square root of PSD (rms density) of the random process as

$$v_{rms}(f) = \sqrt{\overline{v^2}(f)} \equiv \left[\frac{\mathrm{V}}{\sqrt{\mathrm{Hz}}}\right]; \ i_{rms}(f) = \sqrt{\overline{i^2}(f)} \equiv \left[\frac{\mathrm{A}}{\sqrt{\mathrm{Hz}}}\right]. \tag{4.136}$$

When the random process is referred to as noise, it is called with the generic term of *noise density*. Note that the area subtended by a noise density graph is *not* the rms value, but the graph is simply obtained by the power value's square root.

In Fig. 4.19 is shown the two cases where the same signal is a plot in the power spectral density and the square root of the power spectral density. Note that the slope of the function is decreased by one decade.

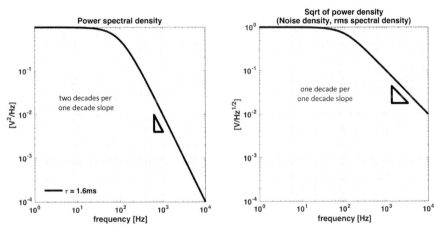

Figure 4.19 Example of power spectral density of a Lorentzian form (left) and the square root of the spectral density (right) or noise density.

As introduced above, we must be careful to calculate the rms value from the noise density graph. We have to first go into the power plot domain and then go back into the noise density plot. For example, if the spectral density is *uniform*, we can calculate the rms value as

$$\overline{v_N^2}(f) = k^2; \quad \overline{v_N^2} = k^2 \cdot B \rightarrow$$
$$v_{Nrms} = k \cdot \sqrt{B}; \quad k \equiv [\mathrm{V}/\sqrt{\mathrm{Hz}}],$$

(4.137)

where $B = f_H - f_L$ is the bandwidth under which we want to calculate the rms value of the random process. We will be more specific on other cases in the next sections.

4.12   The Sampling Process

The time sampling technique could be seen as a process under which the analog continuous signal value is multiplied by a function at fixed time intervals, T_S as illustrated in Fig. 4.20.

Let assume to multiply the analog signal $x(t)$ with a very narrow signal so that the resulting signal area is proportional to the input one sampled at the corresponding time. The best function to start with is a Dirac function. In this case, the sampling process is referred to as *ideal sampling*. The Nyquist–Shannon sampling theorem ensures that if ideal sampling is performed at a frequency that is twice the maximum signal bandwidth, all signal features are preserved. To show this, we can see that the sampled function $x_S(t)$ is given by

$$\left. \begin{array}{l} x(t) \\ s(t) = \displaystyle\sum_{n=-\infty}^{\infty} \delta(t - nT_S) \end{array} \right\} \rightarrow x_S(t) = x(t) \cdot s(t) = \sum_{n=-\infty}^{\infty} x(t)\delta(t - nT_S)$$

$$= \sum_{n=-\infty}^{\infty} x(nT_S)\delta(t - nT_S),$$

(4.138)

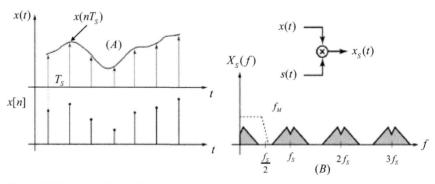

Figure 4.20 Ideal sampling of the analog signal.

where $s(t)$ is a periodic Dirac's comb (Shah) function. Since it is periodic, this expression could be expressed in Fourier series (4.5), with periodicity $T_S = \dfrac{1}{f_S}$

$$s(t) = \frac{1}{T_S} \sum_{n=-\infty}^{\infty} e^{jn2\pi f_S t}; \text{ since, } c_k = \frac{1}{T_S} \int_{-T_S/2}^{T_S/2} \delta(t) e^{-jn2\pi f_S t} dt = \frac{1}{T_S}; \quad (4.139)$$

thus, the sampled function and its Fourier transform are

$$x_S(t) = \frac{1}{T_S} \sum_{n=-\infty}^{\infty} x(t) \cdot e^{jn2\pi f_S t} \xrightarrow{F} X_S(f) = \frac{1}{T_S} \sum_{k=-\infty}^{\infty} X(f - kf_S), \quad (4.140)$$

showing that if we have a bandwidth-limited signal with $f \leq f_M$, the sampled spectrum $X_S(f)$ is composed of replicas of the baseband spectrum displaced in symmetrical fashion sidebands, as shown in Fig. 4.20B. Thus, the reconstruction of the signal could be performed by a low-pass filter that selects the first replica disregarding others. However, perfect reconstruction of the original signal could be performed only for $f_S \geq 2f_M$, as Shannon's theorem states. If the above condition is violated, replica intersect each other, thus avoiding an accurate reconstruction of the original signal, which is subject to deformation. This effect is called *aliasing*. If the signal bandwidth could not achieve the Shannon theorem condition, a low-pass filter, called the *anti-aliasing filter*, should be applied before sampling, as discussed in Chapter 3. The replication of the baseband occurs not only for signal but also for noise, giving rise to the *noise folding*.

Another important point is that the sampled signal, according to Poisson's formula, could be expressed as

$$x_S(t) = \sum_{n=-\infty}^{\infty} x(nT_S) \cdot \delta(t - nT_S) \xrightarrow{F} X_S(f) = \sum_{n=-\infty}^{\infty} T_S \cdot x(nT_S) \cdot e^{jn2\pi f T_S} \text{ and}$$

$$X_{1/T_S}(f) = \sum_{n=-\infty}^{\infty} x[n] \cdot e^{jn2\pi f T_S}, \quad (4.141)$$

where $x(nT_S)$ are the sampled values as shown in Fig. 4.20A. Therefore, the sampled signal spectrum $X_S(f)$ could be uniquely determined by a *sequence* of numbers $x[n] = T_S \cdot x(nT_S)$. Thus, Eq. (4.141) could also be interpreted as the *discrete-time Fourier transform* (DTFT) $X_{1/T_S}(f)$ of the sequence $x[n]$. In the case of A/D conversion, the above sequence could constitute the *digital data stream* of the sampled input quantity, converted in binary notation.

We have referred to an ideal sampling where a Dirac's function characterizes the sampling function. However, real sampling acts differently. More specifically, the original signal could be sampled at a specific time, and the sampling function keeps its value until the next period, as illustrated in Fig. 4.21. This kind of sampling is called *sample and hold* (S/H). It can be shown that the spectrum of the sampled and held signal is

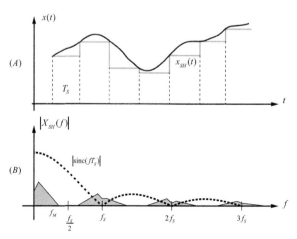

Figure 4.21 Sample-and-hold approach and related spectrum.

$$X_{SH}(f) = e^{-j\pi f T_S} \cdot \mathrm{sinc}(fT_S) \sum_{n=-\infty}^{\infty} X(f - nf_S).\tag{4.142}$$

This spectrum could be seen as one of an ideally sampled signal $x(t)$ then modulated by a $|\mathrm{sinc}(x)|$ function. This effect, on the one hand, helps the reconstruction of the signal, since it attenuates high-order replicas; however, on the other hand, it gives an undesired roll-off shape to the spectrum.

4.13   Appendix A: The Random Walk Process

Observing experimental results of random processes could lead to weird conclusions. For example, assume that we have a set of coins, and we do the experiment of flipping each of them and then stacking in distinct piles those landing on heads and the others landing on tails. Even if the coins are perfectly symmetric and assume the two events have *exactly the same probability*, the absolute *difference* of height between the two stacks could increase with the number of experiments. We do not know which of the two stacks will be higher, but we know that the higher the number of trials, the higher the probability of such a difference. This aspect is, at first sight, counterintuitive, but it is very important to understand some aspects of random processes, especially in electronic systems.

To better understand this behavior, assume that we have to deal with a binary time-dependent white noise, meaning that the signal outcome has only two possible values, $+a$ and $-a$, as shown in Fig. 4.22A with equal probability of occurrence. Now, the signal variations are accumulated in a binary integration. We can see this as a memory system, where the increment/decrement is stored forever. The integration of the binary noise of Fig. 4.22A is shown in Fig. 4.22B. We may think to repeat the experiment for

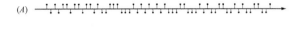

Figure 4.22 Binary white noise (A); integration of white noise (B); different integration trials (C).

different noises, as shown in Fig. 4.22C, and we may wonder what the average distance obtained by the integration function away from the starting point is. At first sight, we may think that, since we have an equal probability to be hit up and down, on average, the result will not be away from the starting point. However, this is not true.

Let us define the "distance" L_N traveled by the function after N steps. Actually, since we do want to be independent of the direction, we prefer to use the square of the distance as a measure L_N^2. Therefore, we want to evaluate the mean square of the distance after N steps: $\overline{L_N^2}$ in the population of all the possible experiments.

We can observe that after one step, we have $\overline{L_1^2} = a^2$ and that after N steps, we can have either $L_N = L_{N-1} + a$ or $L_N = L_{N-1} - a$ with equal probability. Therefore, using the mean square, we have

$$L_N^2 = \begin{cases} L_{N-1}^2 + 2aL_{N-1} + a^2 \ 50\% \text{ chances} \\ L_{N-1}^2 - 2aL_{N-1} + a^2 \ 50\% \text{ chances} \end{cases} \rightarrow \overline{L_N^2} \approx \overline{L_{N-1}^2} + a^2. \quad (4.143)$$

Therefore, by induction:

$$\overline{L_N^2} = a^2 N \rightarrow$$
$$L_{Nrms} = a\sqrt{N}. \quad (4.144)$$

This is a very important result, since we can see that, even if we have equal probability to get from the signal a positive or negative output, on average, the integration of signal departures in probability from the starting point after N steps of $\sqrt{N}$ times the unitary contribution, a.

Turning back to the example of the coin toss, we note that in any case, the law of large numbers (LNN) is not violated. Therefore, even if the difference of the two stacks increases in probability as the square root of the trials, the number of trials increases as N. Therefore, by making the ratio of one stack with the total number of trials converges to $1/2$ for both stacks.

This behavior is similar to what happens with a particle immersed in a fluid where other surrounding particles from every direction bounce it. If we memorize its starting point, we can experimentally see that the particle's position diverges from the initial point by a standard deviation of $2D\sqrt{t}$ in a random direction, where D is called the *diffusion coefficient*. This is referred to as a *random walk*. Random walk is a *nonstationary* random process since its average is not constant over time. Thus, the electronic integration of white noise could be seen as a nonstationary random walk.

This is the reason why, referring to the discussion of Section 4.3, if we do not reset an electronic charge integrator, its output goes to saturation after a given amount of time. Integration of white noise belongs to the class of Wiener random processes.

We can calculate the power spectrum of Brownian motion from that of white noise $S_W(\omega) = \alpha^2$ simply using Eq. (4.24) where the transfer function of the ideal integrator is $1/(j\omega)$; thus

$$S_B(\omega) = \frac{1}{\omega^2} S_W(\omega) = \frac{\alpha^2}{\omega^2}. \tag{4.145}$$

Therefore, the $1/f^2$ appearance of the power spectrum indicates an integration of white noise or Brownian motion behavior.

4.14   Appendix B: Summary of Important Relationships

In the following table are some useful relationships used in the book.

$x(t) = e^{-\frac{t}{\tau}}u(t) \rightarrow X(\omega) = \dfrac{\tau}{1+j\omega\tau}$	$x(t) = e^{-\frac{\lvert t \rvert}{\tau}} \rightarrow X(\omega) = \dfrac{2\tau}{1+\omega^2\tau^2}$
$\displaystyle\int_0^{\infty} e^{-bt} \cdot \cos(at)dt = \dfrac{b}{a^2+b^2}$	$x(t) = \mathrm{rect}(at) \rightarrow$ $\rightarrow F(f) = \dfrac{1}{\lvert a \rvert}\mathrm{sinc}\left(\dfrac{f}{a}\right); \ \mathrm{sinc}(\xi) = \dfrac{\sin(\pi\xi)}{\pi\xi}$
$\displaystyle\int_{\omega_1}^{\omega_2} \dfrac{1}{1+\omega^2\tau^2}d\omega = \dfrac{\tan^{-1}(\omega_2\tau) - \tan^{-1}(\omega_1\tau)}{\tau}$	$\displaystyle\int_{\tau_1}^{\tau_2} \dfrac{1}{1+\omega^2\tau^2}d\tau = \dfrac{\tan^{-1}(\omega\tau_2) - \tan^{-1}(\omega\tau_1)}{\omega}$

Time function	Single-sided energy/power spectrum		Energy/power	Notes
$R(\tau) = k^2 e^{-\lambda\lvert \tau \rvert}$	$\hat{S}(\omega) = 2\lvert F[R(\tau)]\rvert^2 = k^2\dfrac{4\omega_0}{\omega_0^2+\omega^2}$		$P = \dfrac{1}{2\pi}\displaystyle\int_0^{\infty}\hat{S}(\omega)d\omega = k^2$	Power spectrum of a signal whose autocorrelation is $R(\tau)$
$x(t) = 2ke^{-\frac{2t}{\tau}}u(t)$	$\hat{E}(\omega) = 2\lvert F[x(t)]\rvert^2 = k^2\dfrac{4\tau^2}{1+\omega^2\tau^2}$		$E = \dfrac{1}{2\pi}\displaystyle\int_0^{\infty}\hat{E}(\omega)d\omega = k^2\tau$	Energy spectrum of an exponentially decaying pulse

(cont.)

Time function	Single-sided energy/power spectrum	Energy/power	Notes
$x(t) = k \cdot \text{rect}\left(\dfrac{t}{\tau}\right)$	$\hat{E}(f) = 2(k\tau)^2 \lvert \text{sinc}(f\tau) \rvert^2$	$E = \dfrac{1}{2\pi} \displaystyle\int_0^\infty \hat{E}(\omega)\,d\omega = k^2\tau$	Energy spectrum of a rectangular pulse

$$\frac{\tau}{1+\omega^2\tau^2} = \frac{1/\tau}{(1/\tau)^2 + \omega^2} = \frac{D}{D^2 + \omega^2} = \frac{\tau}{1+(f/f_0)^2} = \frac{1}{2\pi}\frac{f_0}{f_0^2 + f^2}$$

$$\int_0^\infty \frac{1}{1+\omega^2\tau^2}\,d\omega = \frac{\pi}{2\tau}; \quad \int_0^\infty \frac{1}{1+(f/f_0)^2}\,df = \frac{\pi}{2}f_0; \quad \int_0^\infty \frac{D}{D^2+\omega^2}\,d\omega = \frac{\pi}{2}$$

Further Reading

Carlson, A. B., *Communication Systems: An Introduction to Signal and Noise in Electrical Communication.* New York: McGraw-Hill, 1986.

Gardner, W. A., *An Introduction to Random Processes with Application to Signal and Systems.* New York: McGraw-Hill, 1990.

Oppenheim, A. V., and Schafer, R. W., *Discrete-Time Signal Processing.* Upper Saddle River, NJ: Pearson Education, 2011.

Oppenheim, A. V., and Willsky, A. S., *Signals and Systems.* Upper Saddle River, NJ: Pearson Education, 2013.

Papoulis, A., and Pillai, S. U., *Probability, Random Variables, and Stochastic Processes.* New York: McGraw-Hill, 2002.

Proakis, J. G., and Manolakis, D. G., *Digital Signal Processing.* Upper Saddle River, NJ: Pearson Prentice Hall, 2007.

5 Compressive Sensing

Marco Chiani[1]

This chapter provides the essential concepts of compressive sensing (CS) – also called compressed sensing, compressive sampling, or sparse sampling. A basic knowledge of signal processing is assumed. The treatment is rigorous but limited: more details can be found in the recommended textbooks listed at the end of the chapter.

Notation Vectors and matrices are indicated in bold, $(\cdot)^T$ indicates transposition, $\|\mathbf{x}\|_2 = \sqrt{\sum_i |x_i|^2}$ is the Euclidean norm, also indicated as ℓ_2-norm.

5.1 Introduction

5.1.1 Sampling Bandlimited Signals

Assume a real time-continuous signal $x(t)$, with no frequency components for $|f| > B$ (bandlimited). The sampling theorem says that $x(t)$ can be reconstructed from its samples (Fig. 5.1) taken at sampling frequency f_s, if $f_s \geq 2B$.[2] The minimum sampling frequency $f_s = 2B$ is sometimes indicated as the Nyquist rate.

In other words, the signal $x(t)$ over a time window $t \in (0, \widetilde{T})$ is represented by the $n = f_s \widetilde{T}$ samples $x(1/f_s), x(2/f_s), \ldots, x(n/f_s)$. The minimum number of samples to represent a portion $\widetilde{T}$ of the signal $x(t)$, which we can call *degrees of freedom* of the signal over $\widetilde{T}$, is thus $n = 2B\widetilde{T}$, obtained by sampling at the Nyquist rate. These n samples $\mathbf{x} = [x_1, x_2, \ldots, x_n]^T$ can be seen as a vector that can take arbitrary values in $\mathbb{R}^n$, so we need $n = 2B\widetilde{T}$ real numbers to represent a bandlimited signal $x(t)$ over $\widetilde{T}$.

Formally, the ℓth sample can be computed as $x_\ell = \int_{-\infty}^{\infty} x(t)\delta(t - \ell/f_s)dt = x(t) * \delta(t)|_{t=\ell/f_s}$, where $\delta(t)$ is the Dirac delta function. Thus, time sampling can be interpreted as computing the scalar products between the signal $x(t)$ and some basis functions $\varphi_\ell(t) = \varphi(t - \ell/f_s)$ approximating the shifted Dirac delta function, as depicted in Fig. 5.2.[3]

In several situations we have measurements taken in the analog domain by projecting the signal $x(t)$ along $m < n$ functions $\varphi_\ell(t)$ that are quite different from the Dirac

---

[1] Professor, Alma Mater Studiorum, University of Bologna

[2] Note that this is a sufficient condition.

[3] Using a function $\varphi(t)$ in the scalar product is like to ideally sample the function $x(t) * \varphi(t)$, which means a multiplication in the frequency domain $X(f)\Phi(f)$. Therefore, any function $\varphi(t)$ with Fourier transform $\Phi(f)$ that is nonvanishing in the band $-B < f < B$ permits reconstruction of $x(t)$ from the samples $x(t) * \varphi(t)|_{t=\ell/f_s}$.

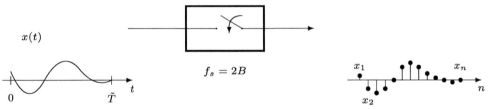

Figure 5.1 A vector $\mathbf{x} = [x_1, x_2, \ldots, x_n]^{\mathrm{T}}$ obtained by time sampling a continuous time signal $x(t)$ over a time window $\tilde{T}$ at the Nyquist rate.

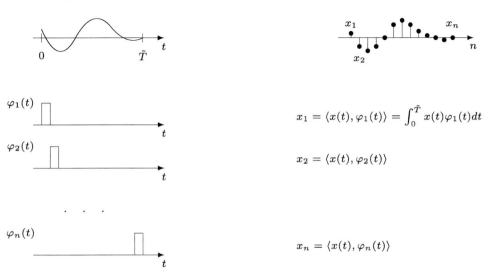

Figure 5.2 Sampling as the scalar products between the signal x(t) and the basis functions $\varphi_\ell(t) = \varphi(t - \ell/f_s)$.

delta, as represented in Fig. 5.3. For example, in magnetic resonance imaging (MRI) the $\varphi_\ell(t)$ are sinusoids and therefore each projection is a Fourier coefficient.

For simplicity, in the following we will focus on discrete-time signals, and assume that the measurement vector is the linear combination

$$
\begin{aligned}
y_1 &= a_{1,1}x_1 + \cdots + a_{1,n}x_n \\
y_2 &= a_{2,1}x_1 + \cdots + a_{2,n}x_n \\
&\;\;\cdots\cdots \\
y_m &= a_{m,1}x_1 + \cdots + a_{m,n}x_n
\end{aligned}
\tag{5.1}
$$

of the samples through known coefficients $a_{1,1}, \ldots, a_{m,n}$. Mathematically, we can think of $y_\ell = \langle \mathbf{a}_\ell, \mathbf{x} \rangle = \mathbf{a}_\ell^T \mathbf{x}$ as the scalar product between the signal vector $\mathbf{x}$ and the vector $\mathbf{a}_\ell = [a_{\ell,1}, \ldots, a_{\ell,n}]^T$.

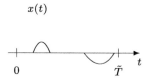

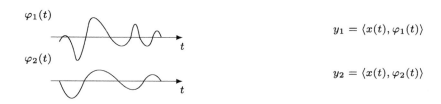

$$y_1 = \langle x(t), \varphi_1(t) \rangle$$

$$y_2 = \langle x(t), \varphi_2(t) \rangle$$

$$y_m = \langle x(t), \varphi_m(t) \rangle$$

Figure 5.3 Measurements on an analog signal obtained as m scalar products

$$\langle x(t), \varphi_\ell(t) \rangle = \int_0^{\tilde{T}} x(t) \varphi_\ell(t) dt$$ between the signal $x(t)$ and some known functions

$\varphi_1(t), \varphi_2(t), \ldots, \varphi_m(t)$.

In matrix form (5.1) can be written

$$\mathbf{y} = \mathbf{Ax}, \tag{5.2}$$

where $\mathbf{x} \in \mathbb{R}^n$, $\mathbf{A} \in \mathbb{R}^{m \times n}$, and $\mathbf{y} \in \mathbb{R}^m$. The matrix $\mathbf{A} = \{a\}_{i,j}$ is called *measurement matrix*.

The discrete-time version of the Nyquist rate sampling in Fig. 5.2 is obtained with $m = n$ and measurement matrix $\mathbf{A} = \mathbf{I}$, where $\mathbf{I}$ is the identity matrix.

5.1.2   Sparse Signals

Many signals of interest are bandlimited, but there are also several practical examples of signals with different intrinsic structure.

For example, in some cases we sense signal vectors formed by n samples, $(x_1, x_2, \ldots, x_n)$, where we know in advance that at most s components, in unknown positions, can be nonzero, the remaining $n - s$ being zero. If $s \ll n$ we say that the vector is *sparse* or s-sparse (see examples in Fig. 5.4). We denote the indexes of the nonzero elements of $\mathbf{x}$ as *signal support*.

We call also *compressible signals* those vectors having $n - s$ negligible elements: these signals can be well approximated by an s-sparse vector.

What is the amount of information in an s-sparse vector? This question is related to the possibility to compress the signal: can we represent $\mathbf{x}$ with fewer than n real numbers? Indeed, this is clearly the case: for example, a direct approach would be to

Figure 5.4 Two examples of 3-sparse signals in $\mathbb{R}^{13}$.

describe the s positions $i_1, i_2, \ldots, i_s$, followed by the values $(x_{i_1}, \ldots, x_{i_s})$. With fixed-length encoding we need $\log_2 n$ bits to represent each position, so a total of $s\log_2 n$ bits for all positions,[4] plus s real numbers for the values. If we use quantization with b bits/real number, then the total number of bits for representing an arbitrary s-sparse vector is $s\log_2 n + sb$.

However, often we do not have access to the signal, but, to $m << n$ linear projections $\mathbf{y} = \mathbf{Ax}$ of the signal (the discrete-time version of Fig. 5.3).

Let us consider a known measurement matrix $\mathbf{A}$ and a measurement vector $\mathbf{y} \in \mathbb{R}^m$. If $\mathbf{x}$ is arbitrary in $\mathbb{R}^n$, the linear system (5.2) admits a unique solution if and only if there are $m = n$ linearly independent equations. Thus, it is possible to recover $\mathbf{x}$ only if the number of measurements m is at least equal to the sample dimension n. However, we could expect that a smaller number $m < n$ of properly chosen projections is sufficient to recover the signal vector, if we know in advance that $\mathbf{x}$ is s-sparse.

Example Assume we have five sensors producing at a given time the vector $\mathbf{x} = [x_1, x_2, x_3, x_4, x_5]^{\mathrm{T}}$. Assume that, for the problem at hand, we know in advance that at most one sensor (one of the five, but we don't know which one) produces a nonzero real output (of unknown value). Thus, we have $n = 5, s = 1$. It is simple to verify that we don't need five real numbers to describe the vector $\mathbf{x} \in \mathbb{R}^5$. In fact, consider the two projections y_1, y_2 given by

$$\mathbf{y} = \begin{bmatrix} y_1 \\ y_2 \end{bmatrix} = \begin{bmatrix} 1 & 1 & 1 & 1 & 1 \\ 1 & 2 & 3 & 4 & 5 \end{bmatrix} \begin{bmatrix} x_1 \\ x_2 \\ x_3 \\ x_4 \\ x_5 \end{bmatrix}. \tag{5.3}$$

It is clear that y_1 is zero if all sensor outputs are zero. If y_1 is not zero, its value coincides with that of the only nonzero element of $\mathbf{x}$. Then, to identify $\mathbf{x}$ we need to locate the position of the nonzero element. It is easy to check that by comparing y_1 with y_2 we can find the position of the only nonzero element of $\mathbf{x}$ (in particular, with this matrix the ratio y_2/y_1 gives the position).

For example, if we read $\mathbf{y} = [-0.7, -2.1]^T$ we are able to recover the only possible 1-sparse signal vector as $\mathbf{x} = [0, 0, -0.7, 0, 0]^T$.

---

[4] Better compression can be achieved: in fact, the number of possible positions is $\binom{n}{s}$. So, with proper encoding, no more than $\log_2 \binom{n}{s} \approx s\log_2(ne/s)$ bits are required to encode the signal support.

More in general, any 1-sparse vector $\mathbf{x}$ with arbitrary dimension n can be recovered by just two projections $\mathbf{y} = \mathbf{Ax}$ by using as measurement matrix

$$\mathbf{A} = \begin{bmatrix} 1 & 1 & 1 & \cdots & 1 \\ a_1 & a_2 & a_3 & \cdots & a_n \end{bmatrix} \tag{5.4}$$

provided that all the a_i's are different. The recover algorithm is straightforward: y_1 is the value of the nonzero element of $\mathbf{x}$, while comparing y_2 with y_1 allows locating unambiguously the position of the nonzero element.

Therefore, we need just $m = 2$ properly chosen projections to fully describe a 1-sparse vector $\mathbf{x}$ of dimension n. In other words, the vector $\mathbf{y} \in \mathbb{R}^2$ can be seen as a (lossless) compressed version of the 1-sparse vector $\mathbf{x} \in \mathbb{R}^n$. The compression ratio is $n/m = n/2$.

As in the example, the objective of compressive sensing is to simultaneously acquire and compress sparse signals, by designing suitable measurement matrices $\mathbf{A}$ and implementable algorithms for signal reconstruction. The preceding example treated the trivial case of 1-sparse signal compression. Designing a suitable measurement matrix $\mathbf{A}$ and a recovery algorithm (reasonably easy to implement) for arbitrary s is generally a difficult task.

5.2   Compressive Sensing

Assume we want to solve the system

$$\mathbf{y} = \mathbf{Ax}, \tag{5.5}$$

where $\mathbf{y} \in \mathbb{R}^m$ and $\mathbf{A} \in \mathbb{R}^{m \times n}$ are known, the number of equations is $m < n$, and $\mathbf{x} \in \mathbb{R}^n$ is the unknown. Since $m < n$ we can think of $\mathbf{y}$ as a compressed version of $\mathbf{x}$. Without other constraints the system is underdetermined, and there are infinitely many distinct solutions $\mathbf{x}$ satisfying (5.5).

Theorem 5.1 *If we know a priori that at most s elements of $\mathbf{x}$ are nonzero (i.e., the vector is s-sparse), then for an observed $\mathbf{y}$ there is only one s-sparse solution (the right one) of (5.5), provided that all possible collections of 2s columns of $\mathbf{A}$ are linearly independent.*

Proof Suppose by contradiction that the linear system has two distinct s-sparse solutions $\mathbf{x}_1$ and $\mathbf{x}_2$. Then it should be that $\mathbf{y} = \mathbf{Ax}_1 = \mathbf{Ax}_2$, which implies $\mathbf{A}(\mathbf{x}_1 - \mathbf{x}_2) = \mathbf{0}$. The vector $\mathbf{x}_1 - \mathbf{x}_2$ has at most $2s$ nonzero elements; thus $\mathbf{A}(\mathbf{x}_1 - \mathbf{x}_2)$ is a linear combination of $2s$ columns of $\mathbf{A}$, which cannot be zero if these columns are linearly independent.

The theorem implies that $m \geq 2s$ must hold, or, in other words, that the number of measurements must be at least $2s$.[5]

Under the conditions of Theorem 5.1, for an observed $\mathbf{y}$ the unique sparse solution of the system (5.5) can be found by looking for the sparsest vector $\mathbf{x}$ satisfying $\mathbf{y} = \mathbf{A}\mathbf{x}$. This can be formalized as the optimization problem

$$\begin{aligned} &\text{minimize} \quad \|\mathbf{x}\|_0 \\ &\text{subject to:} \quad \mathbf{A}\mathbf{x} = \mathbf{y} \end{aligned} \tag{5.6}$$

where the ℓ_0-norm $\|\cdot\|_0$ is the number of nonzero elements.[6] A naive brute-force approach for solving (5.6) consists in testing all $\binom{n}{s}$ possible signal supports, and the corresponding $m \times s$ submatrices of $\mathbf{A}$. This approach is of combinatorial complexity. More generally, (5.6) can be reformulated as an integer programming problem, which is known to be NP-hard. Therefore, solving problem (5.6) is computationally prohibitive for nontrivial s. One key ingredient of compressive sensing is to use the ℓ_1-norm $\|\mathbf{x}\|_1 = \sum_{\ell=1}^{n} |x_\ell|$ instead of the ℓ_0-norm in (5.6), that is, solving the following convex optimization problem, known as basis pursuit[7]:

$$\begin{aligned} &\text{minimize} \quad \sum_{\ell=1}^{n} |x_\ell| \\ &\text{subject to:} \quad \mathbf{A}\mathbf{x} = \mathbf{y}. \end{aligned} \tag{5.7}$$

The problem (5.7), which can be reformulated as a linear-programming (LP) problem, is computationally much easier to solve than (5.6). It has been proved that, under some more strict conditions on $\mathbf{A}$ than that in Theorem 5.1, the solution provided by ℓ_1-norm minimization is the same as that of the ℓ_0-norm minimization. Generally speaking, the condition is that $\mathbf{A}$ must behave to some extent like an isometry for s-sparse vectors.

More precisely, for integer s define the restricted isometry constant (RIC) of a matrix $\mathbf{A}$ as the smallest number $\delta_s = \delta_s(\mathbf{A})$ such that

$$(1 - \delta_s)\|\mathbf{x}\|_2^2 \leq \|\mathbf{A}\mathbf{x}\|_2^2 \leq (1 + \delta_s)\|\mathbf{x}\|_2^2 \tag{5.8}$$

holds for all s-sparse vectors $\mathbf{x}$ [4]. The possibility to use ℓ_1 minimization instead of the impractical ℓ_0 minimization is, for a given matrix $\mathbf{A}$, related to the restricted isometry constant [4]. For example, in [5] it is shown that the ℓ_0 and the ℓ_1 solutions are coincident for s-sparse vectors $\mathbf{x}$ if $\delta_s < 1/3$. For a discussion about RIC-based and other criteria for signal recovery see [6–9].

Then, the next question is how to design a measurement matrix with prescribed characteristics that allow use of (5.7). In this regard, the key idea in compressive

---

[5] Note that the case presented in the example for $s = 1$ is optimum: any two columns of the matrix in (4) are linearly independent, and we compress any 1-sparse vector with just $m = 2$ measurements (which is the minimum possible).

[6] Strictly speaking ℓ_0 is not a norm.

[7] The use of the ℓ_1-norm to find sparse solutions precedes compressive sensing [1–3].

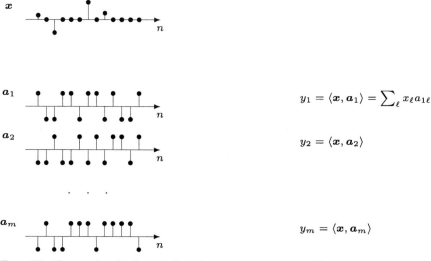

Figure. 5.5 Discrete-time implementation of compressed sensing: taking m scalar products between the signal **x** and some functions $\mathbf{a}_\ell$ designed by randomly generating ± 1.

sensing is to design **A** simply by randomly generating its entries according to some statistical distribution. In fact, it is possible to show that with high probability a randomly generated matrix allows compressed sensing with algorithms of the type (5.7), provided that m and n are large [4, 10–14].

For example, the entries of **A** can be simply independent identically distributed (i.i.d.) random numbers according to a Gaussian distribution (Gaussian random matrix).

Another possible approach is to use random ± 1 for the entries of **A**, as depicted in Fig. 5.5 (Rademacher random matrix). In this case no multiplications are needed for acquisition: the projections are simply the sum or difference of the elements of **x**.

We have seen that for the optimization problem (5.6) we can in principle design measurement matrices with $m \geq 2s$, but with a recovery algorithm of combinatorial complexity.

In contrast, assuming a random generation of the measurement matrix, (5.7) needs a larger number of measurements, in the order of $m \geq 2s\log(n/m)$ for (s, n, m) large and $s \ll n$ [7, 15].

5.2.1   Signals that are Sparse in a Transformed Domain

Assume that the vector **x** is not sparse, but its transformed version in a different basis $\tilde{\mathbf{x}} = \mathbf{Y}^{-1}\mathbf{x}$ (e.g., its Fourier transform) is sparse, where **Y** is the $n \times n$ transformation matrix.

Example The signal $A_\ell \cos(2\pi \ell i/n + \psi_\ell), i = 0, \dots, n - 1$, which is a discrete sinusoid at frequency ℓ/n, is not sparse in the time domain, but its n-point discrete Fourier transform (DFT) has just two nonzero elements. The sum of K sinusoids of this type is $2K$-sparse in the frequency domain (Fig. 5.6).

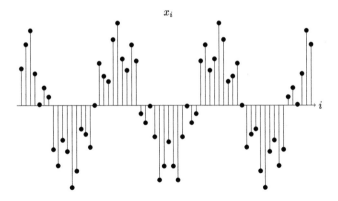

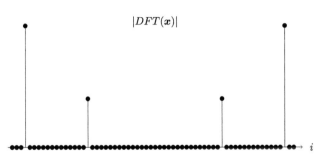

Figure 5.6 The signal $x_i = \cos(2\pi 3i/n) - 0.4 \cos(2\pi 17i/n), i = 0, \ldots, n - 1, n = 64$ is not sparse in the time domain (upper graph), but it is sparse in the Fourier transform domain (DFT magnitude, lower graph).

For signals that are sparse in a transformed domain the measurement values can be written

$$\mathbf{y} = \mathbf{Ax} = \mathbf{AY\tilde{x}},$$

where $\mathbf{\tilde{x}}$ is sparse. Therefore, signal reconstruction is simply as in (5.6) or (5.7), by using $\mathbf{AY}$ instead of $\mathbf{A}$. For example, (5.7) becomes

$$\text{minimize} \quad \sum_{\ell=1}^{n} |\tilde{x}_\ell| \tag{5.9}$$
$$\text{subject to:} \quad \mathbf{AY\tilde{x}} = \mathbf{y}.$$

This would produce the sparse vector $\mathbf{\tilde{x}}$, from which we then compute $\mathbf{x} = \mathbf{Y\tilde{x}}$. Clearly, in this case the conditions allowing the signal reconstruction, for example, the conditions on the RIP for (5.7), or Theorem 5.1 for (5.6), must be fulfilled by $\mathbf{AY}$.

Therefore, in principle **A** should be designed for the specific transformation **Y** defining the domain of sparsity of the signal at hand. However, in CS the measurement matrix **A** is usually randomly generated. Since **Y** is an orthogonal matrix, if we randomly generate the entries of **A** the resulting matrix **AY** has, roughly speaking, the same structure as **A**. For example, columns of **A** that are linearly independent remain independent once multiplied by **Y**.

Therefore, an important advantage in compressive sensing is that the sensing is completely nonadaptive: a randomly generated **A** can be used for collecting measurements irrespective of the basis over which the signal is sparse. Thus, the same matrix can be used for signals that are sparse in different transformed domains. The sparsity basis **Y** is needed only for signal reconstruction as in (5.9).

5.2.2 Compressive Sensing in the Presence of Noise for Compressible Signals

Often the signals of interest are not exactly sparse, but rather compressible (i.e., well approximated by a sparse signal). Moreover, there could be noise affecting the measurement, so that the measured vector can be written as

$$\mathbf{y} = \mathbf{Ax} + \mathbf{z},$$

where $\mathbf{z}$ is the noise and $\mathbf{x} \in \mathbb{R}^n$ is a real vector (not necessarily exactly sparse). In this case we can look for a sparse approximation of $\mathbf{x}$ with maximum representation error κ, by solving the optimization problem

$$\text{minimize} \|\hat{x}\|_0 \text{ subject to } \|\mathbf{A}\hat{\mathbf{x}} - \mathbf{y}\|_2 \leq \kappa. \tag{5.10}$$

The resulting $\hat{x}$ is a sparse approximation of $\mathbf{x}$. Under some circumstances the recovery error $\|\hat{x} - \mathbf{x}\|_2$ can be bounded, based on the signal, noise, and measurement matrix **A** characteristics [8, 9].

5.2.3 Sparse Recovery Algorithms

A central problem in compressive sensing is to recover a sparse signal $\mathbf{x} \in \mathbb{R}^n$ from the $m < n$, possibly noisy, measurements $\mathbf{y}$. Several algorithms have been proposed, with the aim to solve (5.6), (5.7), or (5.10). The main characteristics of these algorithms are capability to deal with large dimensional problems, speed, robustness to model mismatch and noise, and performance guarantees. The recovery algorithms can be classified as follows:

- Convex optimization-based methods. For example, interior-point methods can be used to solve the ℓ_1-norm minimization problem in polynomial time, $O(n^3)$.
- Greedy algorithms. In general, these algorithms try to iteratively find the signal support and then estimate the sparse vector that better matches the measurements.

Matching pursuit (MP), orthogonal matching pursuit (OMP), and compressive sampling matching pursuit (CoSaMP) are in this category. Thresholding methods are also in this category.

The performance and complexity of these algorithms depend on the problem dimension, measurement matrix, and required accuracy. Other algorithms based on combinatorial techniques and on Bayesian methods have been studied for CS. For a detailed discussion see [8, 9, 16, 17].

5.3   Summary of Compressive Sensing

In summary, compressive sensing for a time sequence of n-dimensional signals $\mathbf{x}_1, \mathbf{x}_2, \ldots$, each of which is s-sparse, consists of the following steps:

1. Choose m sufficiently greater than $2s\log(n/m)$.
2. Generate randomly $m \times n$ numbers, according to a Gaussian (reals) or a Rademacher (± 1) distribution and put them in a matrix $\mathbf{A}$.
3. Signal acquisition:
   – Collect the compressed m-dimensional measurements $\mathbf{y}_1 = \mathbf{A}\mathbf{x}_1, \mathbf{y}_2 = \mathbf{A}\mathbf{x}_2, \ldots$
   – Transmit or store $\mathbf{y}_1, \mathbf{y}_2, \ldots$
4. Signal recovery. Given the stored or received $\mathbf{y}_1, \mathbf{y}_2, \ldots$
   – Apply a sparse recovery algorithm (previous section) to find the sparsest solutions $\hat{x}_k$ of $\mathbf{y}_k = \mathbf{A}\hat{x}_k$, $k = 1, 2, \ldots$.

5.4   Applications

There are several applications of CS. In the following some of these are highlighted, together with some references.

5.4.1   Analog to Information Conversion

Here CS is used to acquire a continuous-time signal $x(t)$, assumed sparse, without sampling at the Nyquist rate [18–23]. The scheme is that depicted in Fig. 5.7, where m samples are collected every $\widetilde{T}$ seconds, by using parallel correlators. For the analog implementation we therefore need (see, e.g., [19]):

- A circuit to generate the measurement waveforms $\varphi_\ell(t)$, $\ell = 1, \ldots, m$
- m multipliers to compute $x(t)\varphi_\ell(t)$
- m integrators.

If the measurement waveforms $\varphi_\ell(t)$ can take only the values ± 1, then the multipliers are not necessary (multiplication by -1 is just a polarity change).

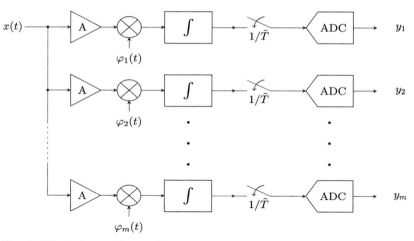

Figure 5.7 Block diagram for an of the CS linear transformation.

For example, in [18, 24] implementations based on compressive sensing are reported for acquisition of sparse one-dimensional wideband signals.

5.4.2 Compressive Sensing for Image Acquisition: Single-Pixel Camera

It is well known that images are generally compressible in the wavelet domain. In single-pixel camera, an image is correlated with a matrix composed of just 0's and 1's, that physically is realized with a microarray consisting of a large number of micromirrors. These micromirrors can be turned on or off individually, while the light reflected from the image on this array is combined (added) at the only present sensor [25]. Depending on the micromirror status, the light from that part of the image is added or not. Therefore, by randomly changing the micromirrors status we compute random linear measurements of the scene under view [25].

5.4.3 Compressive Sensing for Magnetic Resonance Imaging and for Biomedical Signal Processing Applications

In magnetic resonance imaging (MRI) for biomedical applications the acquisition speed is of fundamental importance. This speed is fundamentally limited by physical (gradient amplitude and slew-rate) and physiological (nerve stimulation) constraints. MR images are compressible in the wavelet domain and are reconstructed from incomplete frequency information. The use of compressed sensing techniques is therefore investigated to reduce the amount of acquired data without degrading the image quality [26]. Many other biomedical signals are sparse or quasisparse in some domain and can therefore be processed by using CS methods. A comprehensive review of the CS literature, focused on biomedical applications and in particular on the compression of biosignals commonly of interest in ambulatory monitoring applications, is presented in [27].

References

[1] B. F. Logan, *Properties of High-Pass Signals*. PhD thesis, Columbia University, New York, 1965.

[2] D. L. Donoho and B. F. Logan, Signal recovery and the large sieve, *SIAM J. Appl. Math.*, vol. 52, no. 2, pp. 577–591, 1992.

[3] D. L. Donoho and X. Huo, Uncertainty principles and ideal atomic decomposition, *IEEE Trans. Inf. Theory*, vol. 47, no. 7, pp. 2845–2862, Nov. 2001.

[4] E. Candes and T. Tao, Decoding by linear programming, *IEEE Trans. Inf. Theory*, vol. 51, no. 12, pp. 4203–4215, Dec. 2005.

[5] T. Cai and A. Zhang, Sharp RIP bound for sparse signal and low-rank matrix recovery, *Appl. Comput. Harmon. Anal.*, vol. 35, no. 1, pp. 74–93, Aug. 2013.

[6] E. J. Candès and M. B. Wakin, An introduction to compressive sampling, *IEEE Signal Proc. Magazine*, vol. 25, no. 2, pp. 21–30, 2008.

[7] D. L. Donoho and J. Tanner, Precise undersampling theorems, *Proc. IEEE*, vol. 98, no. 6, pp. 913–924, May 2010.

[8] R. Baraniuk, M. A. Davenport, and M. F. Duarte, *An introduction to compressive sensing*. Connexions e-textbook, 2011.

[9] S. Foucart and H. Rauhut, *A Mathematical Introduction to Compressive Sensing*. New York: Springer Science+Business Media, 2013.

[10] D. Donoho, Compressed sensing, *IEEE Trans. Inf. Theory*, vol. 52, no. 4, pp. 1289–1306, April 2006.

[11] E. J. Candes, J. Romberg, and T. Tao, Robust uncertainty principles: exact signal reconstruction from highly incomplete frequency information, *IEEE Trans. Inf. Theory*, vol. 52, no. 2, pp. 489–509, Feb 2006.

[12] E. J. Candes and T. Tao, Near-optimal signal recovery from random projections: Universal encoding strategies? *IEEE Trans. Inf. Theory*, vol. 52, no. 12, pp. 5406–5425, Dec. 2006.

[13] A. Elzanaty, A. Giorgetti, and M. Chiani, Limits on sparse data acquisition: RIC analysis of finite Gaussian matrices, *IEEE Trans. Inf. Theory*, vol. 65, no. 3, pp. 1578–1588, Mar. 2019.

[14] A. Elzanaty, A. Giorgetti, and M. Chiani, Weak RIC analysis of finite Gaussian matrices for joint sparse recovery, *IEEE Signal Proc. Lett.*, vol. 24, no. 10, pp. 1473–1477, Oct. 2017.

[15] E. J. Candès, The restricted isometry property and its implications for compressed sensing, *Comptes Rendus Math.*, vol. 346, no. 9, pp. 589–592, May 2008.

[16] A. Maleki and D. L. Donoho, Optimally tuned iterative reconstruction algorithms for compressed sensing, *IEEE J. Sel. Topics Signal Proc.*, vol. 4, no. 2, pp. 330–341, Apr. 2010.

[17] A. Y. Yang, Z. Zhou, A. G. Balasubramanian, S. S. Sastry, and Y. Ma, Fast ℓ_1 minimization algorithms for robust face recognition, *IEEE Trans. Image Proc.*, vol. 22, no. 8, pp. 3234–3246, Aug. 2013.

[18] M. Wakin, S. Becker, E. Nakamura, M. Grant, E. Sovero, D. Ching, et al. A nonuniform sampler for wideband spectrally-sparse environments, *IEEE J. Emerg. Sel. Topics Circuits Syst.*, vol. 2, no. 3, pp. 516–529, 2012.

[19] F. Chen, A. P. Chandrakasan, and V. M. Stojanovic, Design and analysis of a hardware-efficient compressed sensing architecture for data compression in wireless sensors, *IEEE J. Solid-State Circuits*, vol. 47, no. 3, pp. 744–756, Mar. 2012.

[20] J. Haboba, M. Mangia, F. Pareschi, R. Rovatti, and G. Setti, A pragmatic look at some compressive sensing architectures with saturation and quantization, *IEEE J. Emerg. Sel. Topics Circuits Syst.*, vol. 2, no. 3, pp. 443–459, Sept. 2012.

[21] D. Gangopadhyay, E. G. Allstot, A. M. R. Dixon, K. Natarajan, S. Gupta, and D. J. Allstot, Compressed sensing analog front-end for bio-sensor applications, *IEEE J. Solid-State Circuits*, vol. 49, no. 2, pp. 426–438, Feb 2014.

[22] F. Chen, F. Lim, O. Abari, A. Chandrakasan, and V. Stojanovic, Energy-aware design of compressed sensing systems for wireless sensors under performance and reliability constraints, *IEEE Trans. Circuits Syst.*, vol. 60, no. 3, pp. 650–661, March 2013.

[23] A. Elzanaty, A. Giorgetti, and M. Chiani, Lossy compression of noisy sparse sources based on syndrome encoding, *IEEE Trans. Commun.*, vol. 67, no. 10, pp. 7073–7087, Oct. 2019.

[24] J. Yoo, S. Becker, M. Loh, M. Monge, E. Candes, and A. Emami-Neyestanak, A 100 MHz–2GHz 12.5 x sub-Nyquist rate receiver in 90 nm CMOS," in *Proc. of 2012 IEEE Radio Frequency Integrated Circuits Symposium*, June 2012, pp. 31–34.

[25] M. F. Duarte, M. A. Davenport, D. Takhar, J. N. Laska, T. Sun, K. E. Kelly, *et al.*, Single-pixel imaging via compressive sampling, *IEEE Signal Process. Mag.*, vol. 25, no. 2, p. 83, March 2008.

[26] M. Lustig, D. L. Donoho, J. M. Santos, and J. M. Pauly, Compressed sensing MRI, *IEEE Signal Process. Mag.*, vol. 25, no. 2, pp. 72–82, March 2008.

[27] D. Craven, B. McGinley, L. Kilmartin, M. Glavin, and E. Jones, Compressed sensing for bioelectric signals: a review, *IEEE J. Biomed. Health Inform.*, vol. 19, no. 2, pp. 529–540, 2015.

Part II

Noise and Electronic Interfaces

6 The Origin of Noise

Understanding the origin of the noise is essential, as it gives hints on how to reduce its effects even from the electronic point of view. This chapter will shortly analyze the physics background of some sources of random processes that are limiting sensing systems referred to as "thermal," "shot," and "flicker" noises. It will also show how thermal and shot noises are at the base of other observed electronic effects such as "kTC," "phase," and "current" noises. The discussion will use analogies between mechanical and electronic effects of thermal agitation. This is important not only for understanding the process but also to unify the noise model in microelectromechanical sensors systems to use the same analytical framework.

6.1 Thermal Noise

Thermal noise is the most common source of noise found in electronics, and it will be introduced starting from some aspects of thermal agitation in classical statistical mechanics. This is because noise is a random process resulting from the sum of a huge number of single contributions that could be modeled only from a statistical perspective.

6.1.1 A Simplified Mechanical Model

A pressure sensor is illustrated in Fig. 6.1, where a gas whose pressure has to be measured is enclosed in a container by a lid/piston having 1 degree of freedom. The piston is kept in mechanical equilibrium by a spring. The increase of the gas's pressure moves the piston along the degree of freedom with respect to the initial equilibrium point so that its displacement gives the output of the system. The gas is composed of *identical N* single particles colliding with each other and against the container's borders by elastic impacts. Instead of using the macroscopic definition of "pressure," we will describe that as the average effect of the force exerted by collisions of particles with the piston, assuming the entire system is in thermal equilibrium.[1]

Each particle's velocities $v_1, v_2 \ldots v_N$ are defined by their components along each reference axis (at the moment, only two for simplicity) as illustrated by the enlargement of Fig. 6.1 so that for each particle we have $v_1^2 = v_{1x}^2 + v_{1y}^2$, $v_2^2 = v_{2x}^2 + v_{2y}^2 \ldots$ and so forth.

Considering the mean square of the velocities along each axis,

---

[1] Thermal equilibrium means that there is no heat transfer inside and outside the system; that is, all parts are at the same temperature of the environment.

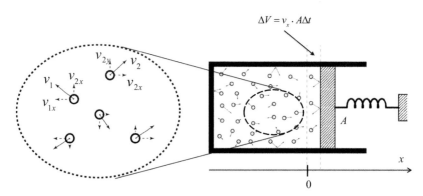

Figure 6.1 Schematic view of an absolute mechanical pressure sensor. The enlargement shows the instant velocities of the particles.

$$\overline{v_x^2} = \frac{v_{1x}^2 + v_{2x}^2 + v_{3x}^2 + \cdots}{N} \; ; \; \overline{v_y^2} = \frac{v_{1y}^2 + v_{2y}^2 + v_{3y}^2 + \cdots}{N} , \tag{6.1}$$

we can determine the mean square of the velocities of all particles as

$$\overline{v^2} = \frac{v_1^2 + v_2^2 + v_3^2 + \cdots}{N} = \frac{v_{1x}^2 + v_{2x}^2 + v_{3x}^2 + \cdots}{N}$$
$$+ \frac{v_{1y}^2 + v_{2y}^2 + v_{3y}^2 + \cdots}{N} = \overline{v_x^2} + \overline{v_y^2} . \tag{6.2}$$

From a classical probability point of view (principle of indifference), there is no reason why the mean square along the x-axis should be different from that calculated along the y-axis; thus $\overline{v_x^2} = \overline{v_y^2}$. Therefore, we can expand (6.2) on three dimensions so that

$$\overline{v_{x,y}^2} = \frac{1}{2}\overline{v^2} \quad \text{for 2 dimensions}$$
$$\overline{v_{x,y,z}^2} = \frac{1}{3}\overline{v^2} \quad \text{for 3 dimensions,} \tag{6.3}$$

where $v_{x,y,z}$ stands for particle velocity on x or y or z.

Therefore, the total mean kinetic energy and the kinetic energy along an axis of a particle are given by

$$\overline{E_C} = \frac{1}{2}m\overline{v^2}$$
$$\overline{E_{Cx,y,z}} = \frac{1}{2}m\overline{v_{x,y,z}^2} . \tag{6.4}$$

The following relationships apply

$$\overline{E_{Cx,y}} = \frac{1}{2}\overline{E_C} \quad \text{for 2 dimensions}$$
$$\overline{E_{Cx,y,z}} = \frac{1}{3}\overline{E_C} \quad \text{for 3 dimensions.} \tag{6.5}$$

This means that the total *mean* kinetic energy of a gas is equally divided along the *translational* degrees of freedom. It is one aspect of the *equipartition theorem of energy*.

Since the piston has a degree of freedom along the x-axis, each particle delivers a single momentum[2] $2m|v_x|$. The pressure is the action of the force exerted by particles over the area A of the piston. It is given by the momentum per time delivered by colliding particles to the piston[3]

$$P = \frac{F}{A} = \frac{1}{A} \frac{\text{no. of momenta delivered}}{\Delta t}$$

$$= \frac{1}{A} \frac{\text{no. of hits}}{\Delta t} \cdot (\text{sin gle momentum delivered}) \qquad (6.6)$$

$$= \frac{\text{no. of hits}}{A\Delta t} \cdot 2m|v_x|,$$

where m is the mass of the particle. The distribution of particles' velocities along the only degree of freedom x follows an unknown probability density function $p(v_x)$. Thus, taking a specific velocity $v_x = w$, the number of particles in the box having x-velocity in the interval $w < v_x < w + \Delta w$ is

$$\Delta N(w) = N \cdot p(w)\Delta w. \qquad (6.7)$$

However, just a fraction of the previous subclass of particles can hit the piston in a timeframe Δt, occupying a volume $\Delta V(w) = w \cdot A\Delta t$ that is the volume given by the area of the piston times the trip covered by a particle traveling at $v_x = w$. Thus, the number of hits with the piston given by particles having velocity $v_x = w$ in timeframe Δt is

$$\text{no. of hits}(w) = \frac{1}{2}\frac{\Delta V(w)}{V}\Delta N(w) = \frac{1}{2}\frac{w \cdot A\Delta t}{V} \cdot N \cdot p(w)\Delta w$$

$$= \frac{1}{2}n \cdot A\Delta t \cdot wp(w)\Delta w, \qquad (6.8)$$

where $n = N/V$ is the particle concentration the $1/2$ factor derives from the fact that, on average, only 50% of particles directed toward the piston can hit it. Thus, the fraction of pressure given by the subclass of particles hitting the piston and having velocity $v_x = w$ is

$$\Delta P(w) = \frac{\text{no. of hits}(w)}{A\Delta t} \cdot 2mw = m \cdot n \cdot w^2 p(w)\Delta w. \qquad (6.9)$$

Now, we can get the value of the pressure P by going from finite variations to differentials and by integrating both terms of (6.9)

---

[2] The variation of momentum of a particle bounced back is $-2m|v_x|$; therefore what is *delivered* to the piston is $2m|v_x|$.

[3] $F = m \cdot a = m \cdot \left(\frac{dv}{dt}\right) = \frac{d}{dt} \cdot (mv) = \frac{d}{dt}M \approx \frac{\Delta M}{\Delta t} = \frac{\text{no. of momentum delivered}}{\Delta t}$.

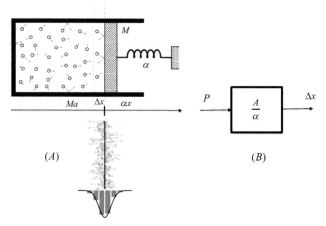

Figure 6.2 (A) Noise in a mechanical absolute pressure sensor. (B) System representation of the sensor.

$$P = \int_0^\infty dP = m \cdot n \cdot \int_0^\infty w^2 p(w)dw = m \cdot n \cdot \overline{v_x^2}, \tag{6.10}$$

where v_x^2 is the mean square of the particle velocities along the x-axis. Note that the integral is taken on the positive axis since we assumed only positive velocities.

Until now, we have not made any assumption on the kind of particles except that they are single and subject to elastic collisions. Therefore, the same conclusion could be made on "particles" such as electrons, ions, or billiard balls. A more sophisticated analysis shows that the result of (6.10) is also valid for mixed gases, composed of particles of different masses, and for multiatomic particles such as molecules considering more degree of freedoms such as the rotational ones. It should be pointed out that even if particles could have more degrees of freedom, what is effective concerning the piston is the translational motion. Therefore, the mean kinetic energy is a fundamental "status" of the matter that can be associated with one of the main characteristics of the matter, which is called *kinetic temperature*. In other words, we can define the temperature as something that is proportional to the mean kinetic energy of the gas, or we can simply state that the temperature *is* the mean kinetic energy of the gas.

This consideration led to defining a relationship between mean kinetic energy and temperature according to the relationship

$$\overline{E_C} \triangleq \frac{3}{2}kT, \tag{6.11}$$

where k is the Boltzmann constant. The relationship (6.11) has two main consequences. First, it sets the mean kinetic energy along 1 degree of freedom as

$$\overline{E_{C_{x,y,z}}} = \frac{1}{2}kT. \tag{6.12}$$

Second, putting (6.11) into (6.10), we get

$$P = nkT = \frac{N}{V}kT \rightarrow PV = NkT,$$ (6.13)

which is the *ideal gas law* also derived from the classical statistical mechanical point of view.[4]

Now, consider the mechanical pressure sensor as illustrated in Fig. 6.2A. The sensor is based on a mechanical equilibrium between the dynamic and the elastic force

$$F = \alpha \Delta x \rightarrow P = \frac{\alpha}{A}\Delta x,$$ (6.14)

where $\alpha \equiv [N/m]$ is the stiffness constant and α/A is the *sensitivity* of the sensor as shown in Fig. 6.2B. Furthermore, the whole system is in *thermal equilibrium*. Thus, the pressure amount is measured by a displacement of the piston with respect to the reference.

The problem is that the pressure is not related to a continuous entity but to a stochastic number of collisions of particles with the piston's surface, giving rise to an erratic movement of the position around the equilibrium point. From another viewpoint, we can refer to this behavior as *noise*.

Can we estimate the "strength" of that noise? Let's derive an estimate of such erratic movement using an energetic equivalence. We first derive the energy of the deterministic *free movement* of the piston, which underlies the harmonic solution of the derivative equation

$$M\frac{d^2x}{dt^2} + \alpha x = 0$$
$$\rightarrow x(t) = \Delta x \cos(\omega_0 t + \varphi); \quad \omega_0 = \sqrt{\frac{\alpha}{M}},$$ (6.15)

where ω_0 is the resonance frequency of the system, M is the piston's mass, and Δx is the displacement of the piston (see Fig. 6.2). Therefore, the total energy of the harmonic system is calculated as the potential and the kinetic energies sum

$$E_{TOT} = E_P + E_C = \frac{1}{2}\alpha x^2(t) + \frac{1}{2}M\left(\frac{d}{dt}x(t)\right)^2$$
$$= \frac{1}{2}\alpha \Delta x^2 \cos^2(\omega_0 t + \varphi) + \frac{1}{2}M\omega_0^2 \Delta x^2 \sin^2(\omega_0 t + \varphi)$$ (6.16)
$$= \frac{1}{2}M\omega_0^2 \Delta x^2 = \frac{1}{2}\alpha \Delta x^2.$$

Now we can assume that the energy related to the random process is equivalent to the energy of the deterministic mechanical harmonic oscillator, that is, that in 1 degree of freedom (d.o.f.)

mean kinetic energy on 1 d.o.f. = total harmonic energy

$$\Delta x^2 \Leftrightarrow \overline{\Delta x_N^2},$$ (6.17)

---

[4] The relation is equivalent to $PV = \tilde{n}RT$, where $\tilde{n}$ is the number of moles and R is the gas constant.

where $\overline{\Delta x_N^2}$ is the mean square of the *random variable* associated with the stochastic process due to the collisions of the particles with the piston. Therefore

$$\frac{1}{2}a\Delta x^2 = \frac{1}{2}a\overline{\Delta x_N^2} = \frac{1}{2}kT$$
$$\rightarrow \overline{\Delta x_N^2} = \frac{kT}{\alpha} \tag{6.18}$$
$$\rightarrow x_{Nrms} = \sqrt{\overline{\Delta x_N^2}} = \sqrt{\frac{kT}{\alpha}},$$

where x_{Nrms} is the root mean square value of *thermomechanical noise*.

Equation (6.18) shows that

- The amount of noise is proportional to the temperature.
- The amount of noise is independent with respect to the piston's mass M.
- The higher the stiffness constant, the lower the noise and the sensitivity of the pressure sensor.

The equivalence (6.17) is away from a formal mathematical rigor; however, it leads to conclusions consistent with experimental results. More specifically, we made a shortcut by equating the energy of an ideal oscillator (a Dirac function centered in ω_0) with the energy of the random variable describing the position of the piston whose PSD is spread around the resonance frequency. In a more formal treatment, the general equation describing the piston's position should include a term considering its collision with particles that could be modeled with viscosity, friction, or resistance (Langevin equation). Moreover, it could be shown that any noise process is strictly related to the dissipative behavior of the system (fluctuation-dissipation theorem). Surprisingly, even if mechanical resistance is the "cause" of noise, Eq. (6.18) does not show this because this is the result of a mathematical integration that hides the relationship with respect to the dissipative variables. This will be shown in a similar (but not analogous) case, such as the kT/C noise.

6.1.2   Electronic Thermal Noise from the Experimental Viewpoint

Thermal noise in electronic devices is a random process due to the thermal agitation of electrical free carriers in thermal equilibrium with conductor lattice and with the surroundings, thus causing random fluctuations of voltage/currents at device terminals. We characterize here electronic thermal noise first *from experimental evidence*.

The tested circuit is shown in Fig. 6.3, where we measure the voltage across a resistor in open-circuit conditions. No static current is flowing through the resistor, and the whole system is at thermal equilibrium.

If we measure (through an instrument with sufficient accuracy with respect to our task) the voltage across the circuit biased by an ideal generator V_0 as illustrated in Fig. 6.3A, we observe the following behavior:

- The output voltage $V(t)$ is affected by random fluctuations.

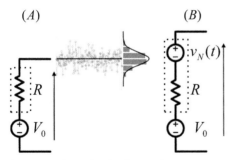

(A)   (B)

Figure 6.3 Thermal noise in a resistor.

- Its experimental mean tends toward the bias value V_0, that is, $\langle V(t) \rangle \to V_0$ as the sample space increases; therefore, associating the random variable $v_N(t)$ with the perturbations around the bias we have $\langle V(t) \rangle = \langle V_0 + v_N(t) \rangle = V_0 + \langle v_N(t) \rangle \to V_0; \langle v_N(t) \rangle \to 0$. Therefore, we assume that the perturbation has an *expected value* equal to zero.[5]
- $v_N(t)$ is normally distributed, and its mean square $\langle v^2(t) \rangle = \sigma_N^2$ increases proportionally to the *temperature* and *resistance* values.
- The corresponding power density $S_N(f)$ is *uniform* in a very large bandwidth.

We will refer to $v_N(t)$ as the random variable modeling a stochastic process called *thermal or Johnson noise* since it was first observed and analyzed by Johnson in 1927. Following the previous experimental results, and assuming the ergodicity of the process, we can model the thermal noise of a noisy resistor as shown in Fig. 6.4A as a random voltage generator $v_N(t)$ whose noise power is given by the mean square value $\overline{v_N^2}$ or by the spectral power density $\overline{v_N^2}(f)$ and placed in series to a *noiseless* resistor R as shown in Fig. 6.4B.

Following Norton's theorem, an equivalent representation of the model mentioned above is represented in Fig. 6.4C, where the thermal noise of a resistor R is characterized by a random current generator $i_N(t)$ whose power is characterized by the mean value is $\overline{i_N^2}$ or by the spectral power density $\overline{i_N^2}(f)$ and placed in parallel to a *noiseless* resistor R. Using Ohm's law $v_N(t) = R \cdot i_N(t)$ we have that

$$\overline{i_N^2}(f) = \frac{\overline{v_N^2}(f)}{R^2} \text{ and } \overline{i_N^2} = \frac{\overline{v_N^2}}{R^2}. \tag{6.19}$$

6.1.3   Thermal Noise Power Spectra Density Calculation: The Nyquist Approach

The calculation of the power spectrum density (PSD) was determined by Nyquist one year after Johnson's paper (for this reason, it is also referred to as the Johnson–Nyquist noise). Even if there are many other derivations of the same expression (using

---

[5] This comes from the law of large numbers (LNN). The expected value of random noise equal to zero also comes from thermodynamic considerations.

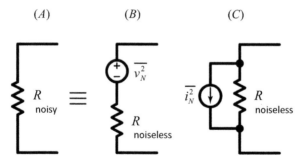

Figure 6.4 Schematic representation of the thermal noise of a noisy resistor R (A) as a voltage (B) or current (C) generator in series or parallel configuration with a noiseless resistor, respectively.

both microscopic and macroscopic arguments), the Nyquist method remains one of the most elegant macroscopic approach.

Referring to Fig. 6.5A, we assume that two noisy resistances R_1 and R_2 are connected in thermal equilibrium. The noise power $\overline{v_1^2}$ generated by the first resistance and adsorbed by the second one is $R_2\overline{v_1^2}/(R_1 + R_2)^2$, and the noise power $\overline{v_2^2}$ generated by the second and absorbed by the first is $R_1\overline{v_2^2}/(R_1 + R_2)^2$. Because the system is at thermal equilibrium, according to the second principle of thermodynamics, it is not possible to transfer heat (power) between different locations; therefore, the two noise power flows must be equal. Furthermore, this equilibrium should be valid for any frequency band; otherwise, the second law could be violated by simply using a filter, which is not possible.

From this observation, we can observe that if $R_1 = R_2 = R$ the power spectral density of spectrum $\overline{v^2}(f)$ should be

1. Independent of the structure and material of the resistors
2. A universal function of T, R, f, that is, temperature, resistance value, and frequency

If $R_1 \neq R_2$, due to the respective powers equivalence, for a small amount of the spectrum, the following should be valid:

$$\frac{R_2}{(R_1 + R_2)^2}\overline{v_1^2}(f)\Delta f = \frac{R_1}{(R_1 + R_2)^2}\overline{v_2^2}(f)\Delta f. \tag{6.20}$$

Therefore,

3. The noise power in a bandwidth Δf $\overline{v^2}(f)\Delta f$ is proportional to the resistance value that is generating the power.

If $R_1 = R_2 = R$ but they are at different temperatures $T_1 \neq T_2$, the power flows do not cancel out, and since the temperature is the mean kinetic energy of the electron gas, the net heat transfer (kinetic energy per time) is[6]

---

[6] Remember that $V^2/4R$ is the maximum power transfer between a source of internal resistance R_S with a load R_L and it occurs when the two resistances are matched: $R_S = R_L = R$.

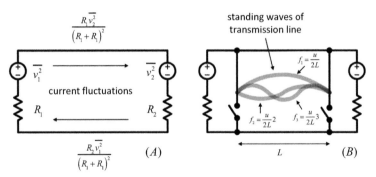

Figure 6.5 Nyquist's model of PSD calculation of thermal noise. (A) Circuit model. (B) Connection with the transmission line.

$$\frac{1}{4R}\left(\overline{v_1^2}(f)\Delta f - \overline{v_2^2}(f)\Delta f\right) \propto (T_1 - T_2). \tag{6.21}$$

Therefore

4. The noise power $\overline{v^2}(f)\Delta f$ in a bandwidth Δf should be proportional to the temperature.[7]

We found a relationship on a bandwidth, but not on the whole spectrum. Referring back to Fig. 6.5B, let us assume the particular case in which a lossless transmission line connects two equal value resistances with matched impedance $R = Z_0 = \sqrt{l/c}$, where l and c are the inductance and capacitance per unit lengths.

The power delivered by resistors to the transmission line in a small amount of the spectrum is

$$\Delta P = \frac{1}{4R}\overline{v^2}(f)\Delta f. \tag{6.22}$$

Therefore, the total energy stored in the transmission line is

$$\Delta E = \Delta P \cdot t \cdot 2 = \Delta P \frac{2L}{u} = \frac{1}{2R}\frac{L}{u}\overline{v^2}(f)\Delta f, \tag{6.23}$$

where L is the length of the transmission line, and u is the propagation's velocity and $t = L/u$ the propagation time-of-flight.

Now, if we suddenly shortcut the transmission line as shown in Fig. 6.5B, the energy trapped in the line is shared between its N standing waves (considering both in-phase and counter-phase waveforms) whose frequencies are $f_N = u/(2L) \cdot N$. Therefore, the number of modes (related to the degrees of freedom) for each very small spectrum partition Δf is

---

[7] Based on the preceding observation, one may wonder if it could be possible to transfer energy by noise in the presence of temperature gradients. This is possible, but the amount of transferred power is very low for practical uses, on the order of pWs.

$$M = \frac{\Delta f}{(u/2L)} = \frac{2\Delta fL}{u}, \tag{6.24}$$

where $M = 1/2N$ since we have two stationary waves per mode due to phase shift. Due to the equipartition theorem in thermal equilibrium, an amount of energy $1/2kT$ is exchanged for each degree of freedom, thus considering (6.23) and (6.24)

$$\Delta E = M \cdot kT = \frac{2\Delta fL}{u} kT = \frac{1}{2R} \frac{L}{u} \overline{v^2}(f)\Delta f$$
$$\rightarrow \overline{v^2}(f) = 4kTR, \tag{6.25}$$

where we have considered double $1/2kT$ energy per mode because there are 2 degrees of freedom per mode due to the phase. The previous PSD expression is referred to as the *Nyquist formula* for thermal noise, where $4kTR$ is the expression of the *unilateral spectral power density* of the thermal noise.

If instead of a transmission line, we couple to the noisy resistor a generic impedance Z,

$$Z = R + jX, \tag{6.26}$$

composed by real (resistance) and imaginary (reactance) parts, using similar arguments, it could be shown that noise is *due only to the resistance*:

$$\overline{v^2}(f) = 4kT\text{Re}\{Z\}. \tag{6.27}$$

In other terms, noise is present only where voltage and current are in-phase and in the component of the impedance responsible for heat dissipation. This is also called the *generalization of the Nyquist formula*.

An apparent problem with the Nyquist formula's conclusion is that the noise power energy becomes infinite as $f \to \infty$. This paradox could be solved using quantum physics mechanics because for $hf > kT$ (where h is Plank's constant), the *granular effect* of energy exchanges becomes nonnegligible. However, for the scope of this book, we can safely assume that the (single-sided) power spectral density of thermal noise is

$$\overline{v_N^2}(f) = 4kTR \text{ for } f \leq \frac{kT}{h} \approx THz. \tag{6.28}$$

6.1.4   Thermal Noise PSD Calculation Using an Energy Tank

There are other ways to calculate thermal noise PSD in macroscopic systems using arguments more familiar to electrical engineers. One of them is based on using an energy tank assuming that the PSD is uniform (white) by experimental evidence.

We can assume that the noisy resistor is connected to an "energy tank" such as a generic capacitor C in thermal equilibrium, as shown in Fig. 6.6. Actually, we can always assume that a *parasitic* capacitance is present between the terminals of a real resistor[8]. Using the model of Fig. 6.4, we represent the resistor as a noiseless one with a current noise generator in parallel.

The random fluctuations of the current induce random fluctuations on the capacitor voltage, whose mean value $\overline{V_C^2}$ is

$$\overline{V_C^2} = \int_0^\infty \overline{i_N^2}(f)|H|^2 df = \int_0^\infty \overline{i_N^2}(f)\left|\frac{R}{1+j\cdot 2\pi f\cdot RC}\right|^2 df$$

$$= \int_0^\infty \overline{i_N^2}(f)R^2\frac{1}{1+(f/f_0)^2} df \text{ where } f_0 = \frac{1}{2\pi RC}. \tag{6.29}$$

However, since the spectral density of the noise is uniform over the bandwidth, we can write

$$\overline{V_C^2} = \overline{i_N^2}(f)R^2\cdot\int_0^\infty\frac{1}{1+(f/f_0)^2} df = \overline{i_N^2}(f)R^2\cdot\frac{\pi}{2}f_0 = \overline{i_N^2}(f)\frac{R}{4C}. \tag{6.30}$$

However, from the Boltzmann statistics, the mean kinetic energy on one degree of freedom (see Fig. 6.6) should be

$$\frac{1}{2}C\overline{V_C^2} = \frac{1}{2}kT;$$

$$\overline{i_N^2}(f) = \frac{4kT}{R}; \rightarrow \overline{v_N^2}(f) = 4kTR, \tag{6.31}$$

where $4kTR$ is the *unilateral spectral power density* of the thermal noise corresponding to the *Nyquist formula*.

We could have been using any other energy reservoirs such as an inductor L or even an LC circuit and obtained the same result.

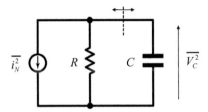

Figure 6.6 Characterization of the thermal noise by means of an RC circuit in thermal equilibrium.

---

[8] A more general derivation could also be done considering an inductance in series. However, the result leads to the same outcome, and we will omit this more general approach for simplicity.

6.1.5   The *kTC* Noise

Let us assume to connect a noisy resistor to a capacitor, similar to that described in Section 6.1.4. This is one of the most frequent conditions in electronic systems. What is the characterization of the random signal $\overline{v_{ON}^2}$ across the capacitor induced by thermal noise?

A few items should be pointed out before the discussion. First, an ideal capacitor is not noisy by definition by the generalized Nyquist formula. Second, the voltage fluctuations are due only to the resistance, which is inducing its noise on the capacitor plates. To calculate this effect, called *kTC noise*, we can refer to Fig. 6.7. A noisy resistor, as illustrated in Fig. 6.7A, could be modeled as a noiseless resistor in series to a power noise generator as in Fig. 6.7B. Therefore, the resistor noise power spectral density $\overline{v_N^2}(f) = 4kTR$ is mapped as shown in Fig. 6.7C into a voltage noise $\overline{v_{ON}^2}(f)$ according to

$$\overline{v_{ON}^2}(f) = |H(f)|^2\overline{v_N^2}(f) = |H(f)|^2 4kTR, \tag{6.32}$$

from which we can calculate the variance of the process by integrating over the whole spectrum:

$$\overline{v_{ON}^2} = \int_0^\infty |H(f)|^2\overline{v_N^2}(f)df = \int_0^\infty |H(f)|^2 4kTRdf = 4kTR\int_0^\infty \frac{1}{1+(f/f_0)^2}df$$
$$= 4kTR \cdot \frac{\pi}{2}f_0 = 4kTR \cdot \frac{\pi}{2} \cdot \frac{1}{2\pi RC} = \frac{kT}{C}. \tag{6.33}$$

Thus, the *kTC* noise (also referred to as kT/C) on the capacitor is not dependent on the value of the resistor R[9]. How is it possible, since resistance is where noise is originated? As illustrated in Fig. 6.8, a reduction of the value from R_1 to R_2 implies a reduction of the unilateral noise power spectrum.

However, the noise bandwidth B_N is increased by the same amount from $1/R_1C$ to $1/R_2C$ to keep the area at the same value kT/C.

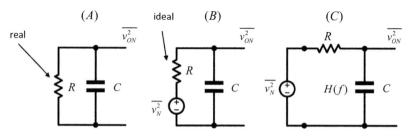

Figure 6.7 Determining the *kTC* noise model.

[9] The similarity between the kT/C and the kT/α between the mechanical and the electronic examples should be carefully evaluated because even if the result is similar, in one case, it is the derivation of a second-order ideal system, and in the other, it is the derivation of a first-order system. However, if we had considered viscosity in the mechanical system and an RLC in the electronic system, respectively, we could have obtained the same final result.

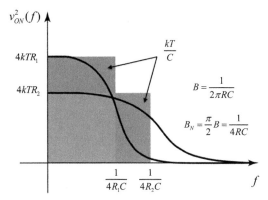

Figure 6.8 Tradeoff between the amount of noise power and the noise bandwidth in kTC noise.[10]

Hint The kT/C noise is an example of a situation in which *correlation does not imply causation*. If we experimentally change a resistance value across a capacitor, we get the same noise standard deviation (i.e., no correlation), and we might think that there is no causation between resistance and noise. However, the "cause" of the effect is the resistance itself.

Another interesting aspect of kTC noise is related to the signal-to-noise-ratio of information stored in a capacitor. This is also called an *analog sampler*. Let us assume that the information is stored in a capacitor C as a voltage. For example, we can charge the capacitor by means of a switch up to the desired voltage V_0. Whatever the switch's resistance and the wiring is, the voltage standard deviation on the capacitor is $v_{Nrms} = \sqrt{kT/C}$. Once the switch is opened to store V_0, an uncertainty due to the kTC noise (see Fig. 6.9) should be considered.

Therefore, we can calculate the SNR of the storage process as

$$SNR = \frac{V_0^2}{v_N^2} = \frac{V_0^2}{\left(\dfrac{kT}{C}\right)}. \tag{6.34}$$

However, since the energy stored into a capacitor has the form $1/2 \cdot CV^2$, the following expression can be determined:

$$SNR = \frac{V^2}{\left(\dfrac{kT}{C}\right)} = \frac{CV^2}{kT} = \frac{\dfrac{1}{2}CV^2}{\dfrac{1}{2}kT} = \frac{\text{energy stored on } C}{\text{noise energy exerted on 1 d.o.f.}}. \tag{6.35}$$

---

[10] It should be pointed out that the graphical representation is valid on a *lin–lin* plot and not on a *log–log* plot as frequently represented.

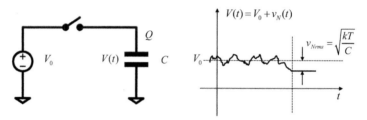

Figure 6.9 Sampling information over a capacitor with kTC noise.

Equation (6.35) shows how the SNR is strictly related to the energetic properties of the system, giving the ratio between the total stored energy over the mean kinetic energy of electron gas over 1 degree of freedom.

The same result could be obtained referring to the variable charge Q stored into the capacitance instead of the voltage V. In fact, we can see that thermal noise is also mapped into a charge variance $\overline{q_N^2}$:

$$Q = CV \rightarrow \overline{q_N^2} = C^2 \overline{v_N^2} \rightarrow \overline{q_N^2} = kTC, \tag{6.36}$$

so that the SNR could be calculated as

$$SNR = \frac{Q^2}{\overline{q_N^2}} = \frac{\frac{1}{2}CV^2}{\frac{1}{2}kT}, \tag{6.37}$$

which is the same expression found before.

6.1.6   Thermal Noise in Resistor–Capacitor Transients

Whenever a noisy resistor drives a capacitor, we have a variance in the capacitor voltage that changes with time. To calculate it, we consider the basic equations of the RC circuit. Referring to Fig. 6.10, we have

$$\frac{dQ(t)}{dt} + \frac{Q(t)}{RC} = \frac{E(t)}{R}, \tag{6.38}$$

where $E(t)$ could be considered as the sum of *both* a driving potential *and* the voltage noise model and $Q(0) = 0$ so that the voltage across the capacitance at the starting point $V_C(0) = 0$. The equation is not separable and could be rewritten[11] as

---

[11] To make it separable, we multiply each term by a factor $\alpha(t)$ so that $\alpha(t)\dfrac{dQ(t)}{dt} + \alpha(t)\dfrac{Q(t)}{RC} = \alpha(t)\dfrac{E(t)}{R}$ and the first left term could be considered equal to $d/dt\big(\alpha(t)Q(t)\big)$ if and only if $\alpha(t) = e^{t/RC}$ from which (6.38) is now separable.

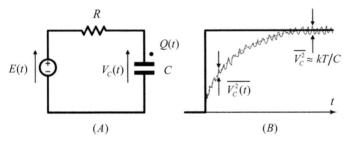

Figure 6.10 Variance of the capacitance voltage versus time.

$$\frac{d}{dt}\left(Q(t)e^{t/RC}\right) = \frac{1}{R}\left(E(t)e^{t/RC}\right),\qquad(6.39)$$

which can now be separated and solved as

$$Q(t) = \frac{1}{R}e^{-t/RC}\cdot\int_0^t E(\tau)e^{\tau/RC}d\tau.\qquad(6.40)$$

If we apply a deterministic DC voltage step equal to E_0, we have $Q(t) = CE_0(1 - e^{-t/RC})$, which is the typical exponential behavior of a capacitor C whose charge converges to CE_0.

Now assuming that $Q(t)$ has zero mean value, the mean square value could be written as

$$\overline{Q^2(t)} = \frac{1}{R^2}e^{-2t/RC}\int_0^t\int_0^t \langle E(\tau_1)E(\tau_2)\rangle e^{-(\tau_1+\tau_2)/RC}d\tau_1 d\tau_2 \text{ where}$$

$$\langle E(\tau_1)E(\tau_2)\rangle = 2kTR\delta(\tau_1 - \tau_2),\qquad(6.41)$$

where we used a double-sided spectrum for thermal noise.[12] The Dirac function is given because it comes from the autocorrelation of white noise. Now by inserting a change of variables into the integrals, we obtain

$$\overline{Q^2(t)} = kTC(1 - e^{-2t/RC})$$

$$\rightarrow \overline{V_C^2(t)} = \frac{kT}{C}(1 - e^{-2t/RC}).\qquad(6.42)$$

This result is very important because it states that in a transition where we assume initial zero charge on the capacitor, we have a variance that increases from zero to the asymptotic value of kT/C as previously determined with a correlation time equal to $RC/2$. In other terms, the variance of the voltage across the capacitor charged through a noisy resistor is negligible for $t << RC/2$ and equal to kT/C for $t >> RC/2$.

---

[12] Here, we use $2kTR$ since it should be considered as a double-sided power density function from the Wiener–Khinchin theorem.

6.2   Current Noise (Shot Noise)

Current noise occurs whenever discrete charges, that is, electrons, pass through a barrier potential as in vacuum tubes and semiconductor junctions, as shown in Fig. 6.11. In such conditions, the current $I(t)$ exhibits fluctuations $i_N(t)$ around a mean value I_0 that is referred to as *current noise*.

6.2.1   Shot Noise from the Experimental Viewpoint

A device characterized by an electrical potential barrier (such as a junction diode) is biased in a DC point to set a mean current I_0 flowing across it. The following points summarize the experimental evidence of current noise:

- The measured current $I(t)$ is not constant but is affected by random fluctuations.
- Its experimental mean tends toward I_0, that is, $\langle I(t) \rangle \to I_0$ as the sample space increases, therefore associating random variable $i_N(t)$ with fluctuations we have $\langle I(t) \rangle = \langle I_0 + i_N(t) \rangle = I_0 + \langle i_N(t) \rangle \to I_0$; $\langle i_N(t) \rangle \to 0$. Therefore, we assume that the perturbation has an *expected value* equal to zero.
- $i_N(t)$ is normally distributed, and its mean square $\langle i_N{}^2(t) \rangle = \sigma_N^2$ does not significantly increase with the temperature.
- The spectral power density $S_N(f)$ is *uniform* over a large bandwidth.

We will refer to $i_N(t)$ as the random variable modeling a random process called *current (or shot) noise*.

As shown in Fig. 6.12, we can consider the current as the result of many charges passing on a section on a fixed amount of time $I = qN/\Delta t$, where N is the number of charges counted over the period time Δt. However, because we should consider an electron's passage through the barrier as a random process, the number of charges may differ for subsequent equal periods of time. For instance, as shown in Fig. 6.12, the number of charges is N_1 in the first and N_2 in the second periods. If we collect a large number of period samples, we can calculate an average $\overline{N}$ of charges

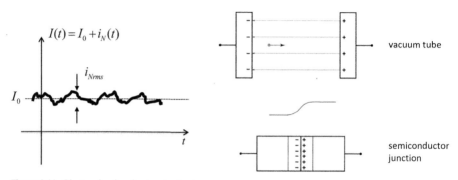

Figure 6.11 Shot noise in electronic devices.

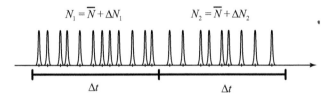

$$N_1 = \overline{N} + \Delta N_1 \qquad\qquad N_2 = \overline{N} + \Delta N_2$$

Figure 6.12 Representation of the electronic current as composed of single charge events passing through a section.

per period that accounts for I_0. However, we can also derive statistics regarding the variation of events ΔN around $\overline{N}$ to characterize the current noise i_N.

The differences between shot noise and thermal noise should be pointed out. In thermal noise, the random process's uncertainty is due to the migration of electrons in both directions of a degree of freedom. It assumes thermal equilibrium, thus no current flowing through a resistance. Therefore, evaluating thermal noise in nonequilibrium (e.g., a resistance is dissipating heat through a current) is an approximation. Conversely, shot noise does not assume thermal equilibrium and is based on the granularity of charges passing through a barrier in one direction.

6.2.2   Characteristics of Current (Shot) Noise as a Poisson Process

Now let us consider the situation of Fig. 6.13, where pulses are shown describing the passage of electrons through the barrier potential as a random process.

We will call $X(0, t)$ the discrete random variable corresponding to the event: "number of charges crossing a section in a period of time $[0, t)$ " where its probability distribution depends on t. We will call the probability mass function

$$p_k(0, t) = \Pr[X(0, t) = k]; \; k = 1, 2, 3, \ldots \tag{6.43}$$

the probability that k events occur in the interval $[0, t)$. We also refer to Δt as a small increment of the period mentioned above, as shown in Fig. 6.13. We assume the following hypotheses:

Assumption 1: The random variables

$$X(t_1, t_2); \; X(t_2, t_3); \; X(t_3, t_4) \; \ldots, \; \text{with } t_1 < t_2 < t_3 \ldots \tag{6.44}$$

are *mutually independent*.

Assumption 2: The probability of having exactly one arrival is proportional to the length of Δt, that is

$$p_1(t, t + \Delta t) = \lambda \Delta t + o(\Delta t) \text{ with } \lambda = \text{average density or mean rate} \equiv [1/s]. \tag{6.45}$$

Assumption 3: The probability of having two or more passages during a sufficiently small period is negligible

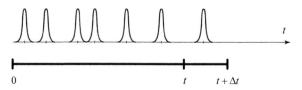

Figure 6.13 Definitions of the current noise process. Each pulse is the random passage of an electron through the potential barrier.

$$\sum_{k=2}^{\infty} p_k(t, t + \Delta t) = o(\Delta t). \tag{6.46}$$

The problem now is to calculate over a small increase of the period Δt the differential properties of the probability distributions to determine the analytical expressions of $p_k(0, t)$.

From the second and third assumptions, we get that the probability of having *zero* events in the whole interval is

$$p_0(t, t + \Delta t) = 1 - \sum_{k=1}^{\infty} p_k(t, t + \Delta t) = 1 - \lambda \Delta t + o(\Delta t). \tag{6.47}$$

Since the two intervals are nonoverlapping, the probability of having no events on the whole segment means we must have no passages on both subintervals

$$p_0(0, t + \Delta t) = p_0(0, t) \cdot p_0(t, t + \Delta t) = p_0(0, t) \cdot (1 - \lambda \Delta t + o(\Delta t)). \tag{6.48}$$

Rearranging Eq. (6.50), we get the incremental variation of p_0 with respect to Δt

$$\frac{p_0(0, t + \Delta t) - p_0(0, t)}{\Delta t} = -p_0(0, t) \cdot (\lambda - \frac{o(\Delta t)}{\Delta t}), \tag{6.49}$$

which for $\Delta t \to 0$ becomes

$$\frac{dp_0(0, t)}{dt} = -p_0(0, t) \cdot \lambda. \tag{6.50}$$

Using the boundary condition, $p_0(0, 0) = 1$ we get the solution of Eq. (6.50) as

$$p_0(0, t) = e^{-\lambda t}. \tag{6.51}$$

To calculate $p_1(0, t)$ the process is very similar. However, the case of having one passage over the whole interval implies taking into consideration that we have only two possibilities: either the event occurs in the first interval and *not* in the second, *or* the event occurs in the second interval and *not* in the first. Thus, using the probability properties, we have

$$p_1(0, t + \Delta t) = p_1(0, t) \cdot p_0(t, t + \Delta t) + p_0(0, t) \cdot p_1(t, t + \Delta t). \tag{6.52}$$

Using the second assumption, Eqs. (6.48) and (6.51) for $\Delta t \to 0$ we get

$$\frac{dp_1(0,\ t)}{dt} = -\lambda p_1(0,\ t) + \lambda e^{-\lambda t}; \tag{6.53}$$

then using the boundary condition, $p_1(0,\ 0) = 0$ we get the solution of Eq. (6.53) as

$$p_1(0,\ t) = \lambda t e^{-\lambda t}. \tag{6.54}$$

We can determine higher orders considering that for k events, we have $k!$ possibilities on the two intervals; thus, by using mathematical induction, we have

$$p_k(0,\ t) = \frac{(\lambda t)^k e^{-\lambda t}}{k!}. \tag{6.55}$$

Equation (6.55) is the *Poisson probability mass function* of a k-event process defined by a mean rate λ on an interval $[0,\ t]$. The phenomena modeled by Poisson distributions are called *Poisson processes*, and shot noise is a Poisson process.

A property of the Poisson distribution is the expected value that can be evaluated using known properties of series expansion[13]:

$$\begin{aligned}
E[X(0,\ t)] &= \sum_{k=0}^{\infty} k \cdot p_k(0,\ t) = e^{-\lambda t} \sum_{k=0}^{\infty} k \cdot \frac{(\lambda t)^k}{k!} \\
&= (\lambda t) e^{-\lambda t} \sum_{k=1}^{\infty} \frac{(\lambda t)^{k-1}}{(k-1)!} = (\lambda t) e^{-\lambda t} e^{\lambda t} = \lambda t.
\end{aligned} \tag{6.56}$$

The result is very interesting for its simplicity but is not surprising because it states that the mean value of the process is given by the mean rate times the observation period.

Using the same approach as Eq. (6.56), a more unexpected result could be found; that is, the variance of the process is

$$\sigma^2_{X(0,\ t)} = \lambda t, \tag{6.57}$$

which is the same as the mean value.

To summarize, considering the charge transit process as a generic "event" occurring in a time interval $[0, t]$, it is

λ = mean rate or average density of events
λt = mean value = expected number of events in the time interval   (6.58)
$\sqrt{\lambda t}$ = standard deviations of events in the time interval.

When the number of events in the period becomes relevant, the Poisson process is subject to the central limit theorem, and the distribution becomes a Gaussian one. As shown in Fig. 6.14, the probability of having k events is

[13] $e^x = \sum_{n=0}^{\infty} \frac{x^n}{n!} = \sum_{n=1}^{\infty} \frac{x^{n-1}}{(n-1)!} = 1 + x + \frac{x^2}{2} + \frac{x^3}{6} + \dots.$

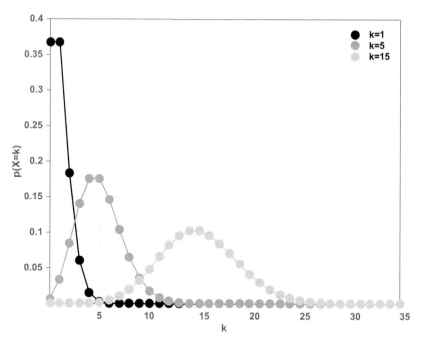

Figure 6.14 Plot showing Poisson distribution versus expected events k that closely resembles a Gaussian as the mean increases (here $t = 1$ for simplicity).

plotted versus the number of events. The distribution is asymmetric for low values of the expected number of events λt and most prominent for low mean. However, as the mean gets large, the distribution closely resembles a Gaussian, as expected by the central limit theorem.

6.2.3   Calculation of Current (Shot) Noise Power Spectral Density

To calculate the noise power spectrum, we apply the *Carson theorem*. It states that a signal $x(t)$ composed of a large number of randomly occurring pulse-shaped functions $g(t)$(such as those shown in Fig. 6.12) modulated by a random value a_i

$$x(t) = \sum_{i=1}^{N} a_i g(t - t_i) \tag{6.59}$$

is characterized by the unilateral power spectral density given by

$$S_X(f) = 2\lambda \overline{a^2} |G(f)|^2 + 2\langle g(t) \rangle^2 \delta(f), \tag{6.60}$$

where $G(f)$ is the Fourier's transform of the energy-limited function $g(t)$.

Now, considering that in our case $a = q$ is the electron charge and that

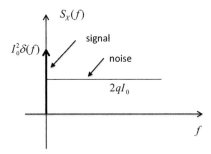

Figure 6.15 Graphical representation of the current noise power spectrum.

$$\overline{a^2} = q^2$$
$$g(t) = \delta(t)$$
$$\lambda q = \frac{\text{average of charges}}{\text{interval}} = I_0 \tag{6.61}$$

we get

$$S_X(f) = 2qI_0 + 2I_0{}^2\delta(f), \tag{6.62}$$

which is the power spectral density of the current noise *and* signal. More specifically, the first term of the right-hand side is related to the power of noise, and the second one to the power of the DC signal, as shown graphically in Fig. 6.16.

Figure 6.15 shows how current noise (or shot noise) could not be independent of signal, but it is dependent on the signal itself. Furthermore, since current noise is a Poisson's process, we have $\langle I(t) \rangle = \langle I_0 + i_N(t) \rangle = \lambda q = I_0$ and $\langle i_N^2(t) \rangle = \sigma_N^2 = \lambda q = I_0$.

The electronic circuit model of current noise is similar to thermal noise. For example, for an electronic device such as the junction diode (see Chapter 9), the current noise could be modeled as in Fig. 6.16: a noisy device is shown as equivalent to a noiseless diode with a superposed noise generator of the power spectrum $\overline{i_N^2}(f) = 2qI_0$.

Therefore, the current noise modeling is identical to that of thermal noise, even if the physical origin is different. Whenever noise is characterized by uniform PSD, it is called "white."

6.2.4 Relationship Between Shot Noise and Thermal Noise

It is useful to point out the main differences between thermal noise and shot noise:

- Thermal noise formulas and relationships have been calculated in thermal equilibrium and thus are related to this situation or small deviations from equilibrium.
- Shot noise is intrinsically calculated in a nonequilibrium condition because it is due to the current fluctuation of charges from one side to the other in an electrical potential barrier.

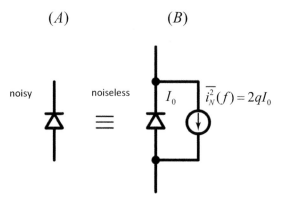

Figure 6.16 Current noise model in a junction diode.

However, it can be seen that thermal noise can be modeled as the combination of the shot noise of two almost equal forward and reverse currents. Therefore, shot noise and thermal noise can be represented by a unique model where thermal noise is a particular case of shot noise when the system is in almost thermal equilibrium.

6.3   Noise in Optical Detectors

An optical detector is a device that transduces light into an electrical signal. For this reason, it is convenient to treat optical detectors as *a device converting a flux of photons into a number of electrical charges*. Modern solid-state optical sensors could deal with very low light levels into very reduced dimensions. Thus, the devices' analysis should take into account the granularity of the signal as a necessary step for describing noise properties. The specific structures of possible implementations of optical detectors are discussed in Chapter 9, so we will discuss concepts related to noise modeling in such devices.

6.3.1   Noise Photocurrents

Photodiodes operating in continuous time mode should be modeled by adding more noise sources to the simplified model of Fig. 6.16. More specifically, we have to add the contribution of the noise of the resistance of the device and that of the *dark current*.

The *dark current* is essentially the junction's leakage current when no light is acting on the photodiode, that is, at dark conditions. The resistance of the device R_P is *not* the equivalent model resistance of the photodiode (because it is the result of a model, it does not create noise), but it is the real series resistance of the photodiode contacts and substrates.

Therefore, in the bandwidth Δf of the system, the main noise sources are

• Shot noise due to the photocurrent $\overline{i^2}_O = 2qI_O\Delta f$, where I_O is the photocurrent

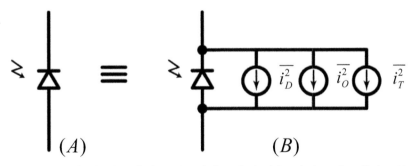

Figure 6.17 Noise in photodiodes, (A) real photodiode. (B) Noiseless photodiode with noise sources.

- Shot noise due to the dark current $\overline{i^2}_D = 2qI_D\Delta f$, where I_D is the dark current
- Thermal noise due to the photodiode's resistance $\overline{i^2}_T = \dfrac{4kT\Delta f}{R_P}$, where R_P is the series resistance of the photodiode.

A complete model of the noisy photodiode is shown in Fig. 6.17.

6.3.2   Shot Noise in Photosites

So far, we have dealt with continuous-time signals from photodiodes. However, photodiodes and charge-coupled devices (CCDs) could operate in discrete-time mode (Chapter 9). The operation mode is very similar in the two cases. For a fixed amount of time, called *integration time*, the light intensity (i.e., energy) is detected by evaluating the photon flux by measuring the photogenerated electrical charges. Then, the information is reset, and the operation is repeated periodically.

Unfortunately, the operation is affected by some degree of uncertainty due to the random process. One of these is the shot noise due to the granularity of the signal itself that Poisson's statistic could describe. It is another way to look at the current noise.

Assume that we operate in discrete time for a fixed amount of light signal and repeat the experiment of collecting and counting the number of photocharges. We expect to count the same number of charges in an ideal case. However, this is not true due to the granularity of the signal and the randomness of each photogeneration event's occurrence.

Therefore, the process follows Poisson statistics, as we already discussed. Assume we perform K independent experiments as shown in Fig. 6.18. We can define the mean collected photocharges N as

$$\overline{n} = \lambda T_i = \frac{1}{K}\sum_{i=1}^{K} n[i] = N \equiv [\cdot], \qquad (6.63)$$

where λ is the mean rate of collected charges, T_i is the integration time, and $n[i]$ is a sequence of numbers. Now, we can define the variance of the process following a Poisson statistics process that is equal to the mean:

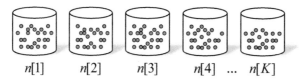

$$n[1] \qquad n[2] \qquad n[3] \qquad n[4] \;\; \ldots \;\; n[K]$$

Figure 6.18 Operation in discrete time of photodetectors.

$$\sigma^2 = \lambda T_i = \overline{(n - N)^2} = \frac{1}{K}\sum_{i=1}^{K}(n[i] - N)^2 = N \equiv [\cdot^2]$$

$$\sigma = \sqrt{N} \equiv [\cdot]. \tag{6.64}$$

Note how the variance is again N; however, the dimension is the square of a dimensionless quantity in this case.

6.4   Flicker or 1/*f* Noise

The term "flicker noise" or "1/*f* noise" refers to any kind of noise having power spectral density in the form

$$S(f) = \frac{k^2}{f^\alpha}; \; \text{with } 0 < \alpha < 2, \tag{6.65}$$

where k is a constant. Therefore, the characterization of such noise is essentially made from the behavioral point of view and not from its physical origin. Flicker noise has been observed in a large variety of electrical process fluctuations such as voltage/current in electronic devices (vacuum tubes, diodes, transistors, resistors), resistance in semiconductors, metallic films, and so forth and current/voltage across natural and synthetic membranes and nerves.

Furthermore, 1/*f* PSD fluctuation spectra are found in a large variety of observed phenomena: geological, astronomical, physiological, chemical, and biological, and a large amount of literature could be found on this matter. Of course, the physical model underlying those aspects could be very different, and no unified theory on such noise has been found even though several attempts have been made in this direction. In this book, we will restrict the argument only to electronic devices. In this respect, the physical phenomena and the models are much more defined; however, there is not yet agreement on whether all observed electrical 1/*f* noises belong to the same physical phenomenon. We will present models evolved on the *sum-of-distributed-time-constants* approach, proposed since 1937 for vacuum tubes.

Figure 6.19 shows the behavior in both the time and power domains of Eq. (6.65) for different values of α. Note how the time behavior becomes increasingly "bumpy" and seemingly nonstationary as α increases. Note also how for $\alpha = 0$ the behavior is the same as thermal and current noise referred to as "white."

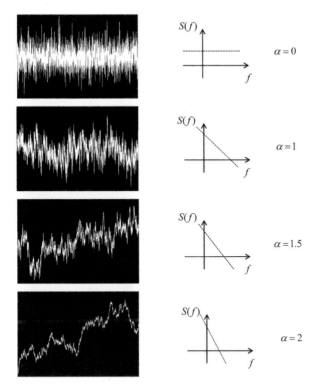

Figure 6.19 Temporal and spectral representations of noises with PSDs defined as in (6.65). Note how for $\alpha = 0$ the noise is "white."

Frequently, the investigation on $1/f$ noise follows a reverse direction with respect to other noise source analyses. Instead of starting from the physical model, there are attempts to understand the physical mechanism from a mathematical point of view. In any case, the exact derivation of the PSD for some physical phenomena is still an open issue. For instance, for 1/f noise to be stationary, its variance

$$\sigma^2 = \int_0^\infty S(f)df \qquad (6.66)$$

must be finite, which is not the case with Eq (6.65). Thus, we can either assume flicker noise as nonstationary or $S(f)$ defined in a limited frequency range. We will discuss this better later.

Referring to electronic devices, we can find a spectral current power density of the form (together with other noise sources)

$$S_I(f) = k^2 \frac{I^2}{f^\alpha}. \qquad (6.67)$$

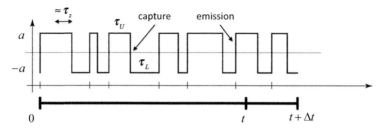

Figure 6.20 RTS model.

A persuasive model that has been formulated for $1/f$ noise in electronic devices is based on the concept of trapping. In small solid-state electronic devices, the alternate trapping and emission of carriers at an individual defect site induces discrete switching in the device resistance, referred to as a random telegraph signal (RTS) for a single charge carrier.

When in a conducting channel, a free carrier is frozen, falling into a trap; it gives no more contribution to the current. Therefore, the behavior of a carrier could be modeled by a random telegraph signal as shown in Fig. 6.20, going from a lower state "-a" to an upper level, "a." We can treat the probability of having k events in a given period of time according to the Poisson process as

$$p_k(0,\ t) = \frac{(\lambda t)^k e^{-\lambda t}}{k!} \quad \text{where } \lambda = \text{mean transition between states.} \tag{6.68}$$

Here, for simplicity, we assume that λ is the mean rate of transitions, that is, the changes of sign.

Thus, for the computation of the autocorrelation function, we can refer to Fig. 6.21, where is shown how the product $x(t)x(t+\tau)$ can give only two results: a^2 in the case of even number of transitions in the interval $[t, t+\tau]$ and $-a^2$ in the case of odd transitions in the same period. Then, we must multiply such value for the probability given by the Poisson distribution, giving

$$R_x(\tau) = \overline{x(t)x(t+\tau)}$$

$$= a^2[p_0(0,\tau) + p_2(0,\tau) + \ldots] - a^2[p_1(0,\tau) + p_3(0,\tau) + \ldots]$$

$$= a^2 e^{-\lambda\tau}\left[1 - \lambda\tau + \frac{(\lambda\tau)^2}{2!} - \frac{(\lambda\tau)^3}{3!} + \ldots\right] = a^2 e^{-\lambda\tau}e^{-\lambda\tau} = a^2 e^{-2\lambda\tau} = a^2 e^{-\frac{\tau}{\tau_z}}.$$

$$\tag{6.69}$$

Therefore, the autocorrelation function is an exponentially decaying function with a relaxation time constant equal to $\tau_z = 1/(2\lambda)$ equal to the average trap time. Now the power spectral density could be found using the usual relationship

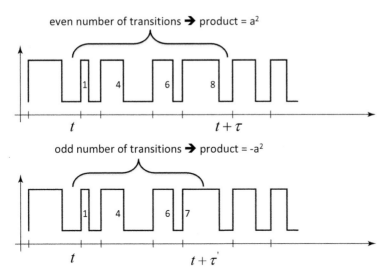

even number of transitions ➜ product = a²

odd number of transitions ➜ product = -a²

Figure 6.21 Calculation of the correlation of an RTS signal.

$$S_{\tau_z}(f) = 2\int\limits_{-\infty}^{\infty} R_x(\tau)e^{-j\omega\tau}d\tau = 2a^2\int\limits_{-\infty}^{\infty} e^{-2\lambda|\tau|}e^{-j\omega\tau}d\tau = 4a^2\int\limits_{0}^{\infty} e^{-2\lambda\tau}\cos(\omega\tau)d\tau$$

$$= 4a^2\frac{2\lambda}{4\lambda^2 + \omega^2} = a^2\frac{4\tau_z}{1 + \omega^2\tau_z^2}.$$

(6.70)

The result shows how an exponential decreasing autocorrelation function provides a power spectral density of the form given in (6.70), which is a *Lorentzian spectrum* as discussed in Chapter 4.

One may wonder why the "average trap time" is half the up and down average state times. The answer is that the energy of a rectangular pulse is equal to an exponentially decaying pulse with half characteristic time and double amplitude (Chapter 4). More specifically, the same Lorentzian spectrum is given if we consider that the current of a device is subject to a stochastic process where random events, such as the release of charges, induces perturbations that are "absorbed" by the system in a time-constant $\tau_z/2$ as shown in Fig. 6.22. This kind of process is called a *relaxation process*.

This could be easily shown that if we apply Carson's theorem to the function $g(t) = 4a\exp(2t/\tau_z)$ occurring at the averaging time $\lambda_C = 1/(4\tau_z)$ and amplitude $4a$ we get the same Lorentzian spectrum of (6.70). This means that an RTS signal could model the representation of the trap-and-release charge process, but from the physical point of view, this model is equivalent to a relaxation process (occurring at positive edges of the RTS waveform) with a characteristic time constant equal to $\tau_z/2$. Note that the RTS signal could not be calculated with Carson's theorem since the shape of the single waveform is not fixed but stochastically varying.

A more detailed analysis shows that if we characterize as τ_U and τ_L the average time of the upper and lower states of the RTS signal, respectively (see Fig. 6.20), we get a Lorentzian spectrum whose frequency of transitions between states is given by

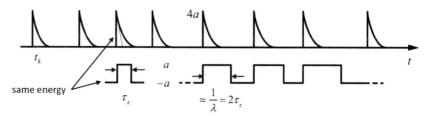

Figure 6.22 Random pulse train of exponentially decaying functions.

$1/\tau_z = 1/\tau_U + 1/\tau_L$, where for $\tau_L = \tau_U = \tau$ we get $\tau_z = \tau/2$, giving the same expression as (6.70).

Now we can assume that the overall current is composed of a very large number of single random contributions subject to trapping. However, we must consider that there is not only a kind of trap but a distribution of traps characterized by different trapping times. Therefore, noise is constructed by the linear superposition of random processes whose time-constant τ are distributed between a lower value τ_1 and an upper level τ_2 according to a probability density function $p(\tau)$ of trapping times so as the overall spectra is given by

$$S_x(f) = \int_0^\infty S_\tau(f)p(\tau_z)d\tau = a^2 \int_0^\infty \frac{4\tau p(\tau_z)}{1+\omega^2\tau_z^2}d\tau, \qquad (6.71)$$

where of course it should be

$$\int_0^\infty p(\tau_z)d\tau_z = 1. \qquad (6.72)$$

Now we can assume that the traps are distributed according to the following law:

$$p(\tau_z) \cdot \tau_z = k, \qquad (6.73)$$

assuming that large relaxation-time traps are less probable than that related to short relaxation time. It should be pointed out that it is assumed that there should be *no interaction or transition between traps* (isolated traps condition), meaning that trapping events are statistically independent.

The Lorentian power spectra superposition effect is shown in Fig. 6.23, where the envelope of the corners gives rise to a $1/f$ slope figure.

Thus, under the assumption of (6.73), we can write

$$S(\omega) = k(2a)^2 \int_{\tau_1}^{\tau_2} \frac{1}{1+\omega^2\tau^2}d\tau = kI^2 \frac{\tan^{-1}(\omega\tau_2) - \tan^{-1}(\omega\tau_1)}{\omega}$$

$$\text{If } \begin{cases} \omega\tau >> 1 \rightarrow \tan^{-1}(\omega\tau) \approx \dfrac{\pi}{2} \\ \omega\tau << 1 \rightarrow \tan^{-1}(\omega\tau) \approx 0. \end{cases} \qquad (6.74)$$

Therefore, for the *mid*-frequency, we have

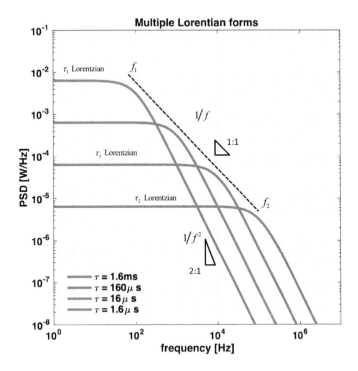

Figure 6.23 Effect of superposition of relaxation processes in the composition of 1/f noise.

$$\frac{1}{\tau_1} < \omega < \frac{1}{\tau_2}$$

$$S(f) = kI^2 \frac{\pi/2 - 0}{2\pi f} = \frac{kI^2}{4} \frac{1}{f}, \tag{6.75}$$

where the 1/f slope is mathematically proven.

For the frequency lower than that related to the longest time constant

$$\omega << \frac{1}{\tau_1}$$

$$S(\omega) = k(2a)^2 \int_{\tau_1}^{\tau_2} d\tau = kI^2(\tau_2 - \tau_1). \tag{6.76}$$

And for frequency higher than that related to the shortest time constant

$$\omega >> \frac{1}{\tau_2}$$

$$S(\omega) = k(2a)^2 \int_{\tau_1}^{\tau_2} \frac{1}{\omega^2 \tau_z^2} d\tau_z = kI^2 \left(\frac{1}{\tau_1} - \frac{1}{\tau_2} \right) \frac{1}{\omega^2}. \tag{6.77}$$

The effect is shown in Fig. 6.23, where 1/f behavior is apparent in a midpoint between the two relaxation-time frequencies and the 1/f² behavior at high frequency.

An interesting and similar conclusion is given for different trap distributions. For example, it can be shown that

$$\text{If}[\tau_z p(\tau)] \propto \tau_z^{\alpha-1} \rightarrow S(f) \propto \frac{1}{f^\alpha}. \tag{6.78}$$

The derivation of flicker noise from the summation of RTS signals with different time constants *is just an interpretation* of the origin of the $1/f$ spectrum in electronic devices, even if it is one of the most accredited.

6.4.1   The Issue of Stationary in Flicker Noise and Its Memory

One of the main problems related to the flicker noise framework, on both conceptual and experimental sides, is that the analytical function $1/f$ does not converge for a frequency approaching zero, and we cannot experimentally validate this model unless waiting an infinite time. This aspect has been extensively analyzed, and there are still open issues and debates on the matter. There are two ways to look at this issue:

1. We assume the flicker noise is a stationary process due to its physical origins. For example, according to the trap model discussed earlier, we assume a lower-bound trap time so that a Lorentzian PSD flattens the spectral power density behavior of $1/f$ noise at lower frequencies. Therefore, the integral does not diverge, and thus the total noise power is limited.
2. We assume that the flicker noise is nonstationary. However, it shows a stationary PSD down to the frequency given by the reciprocal of the observation time.

The latter hypothesis seems to conflict with the general treatment of non-stationary processes. It could be shown that in most cases, the PSD and the autocorrelation functions of a nonstationary process are dependent on the observation time. For example, the power and the variance of a nonstationary process increase with time. However, Keshner showed that even if the $1/f$ noise model is assumed as a nonstationary process, it shows a stationary PSD down to a lower-bound frequency given by the reciprocal of the observation time. The time-dependent PSD part is for frequencies that are not measurable within the observation time. In conclusion, the only way to validate the assumptions (if the process is stationary or nonstationary) is to increase the observation time, since for defined timeframes, independently from assumptions, the PSD is experimentally stationary and analytically defined.

Independent from the assumptions mentioned above, some observations allow us to overlook the stationarity hypotheses from the application point of view. Flinn pointed out that the $1/f$ *function* divergence is quite slow versus frequency. For example, if we take a lower limit of 10^{-17} Hz corresponding to the estimated reciprocal of the "age" of the universe and 10^{23} Hz corresponding to the reciprocal of the time occurring for the light to traverse the electron radius, we have 40 decades. It is surprising that for this huge range, the rms fluctuations would be only six times as great[14] as those taken in a bandwidth 1–10 Hz. Therefore, a lower bound choice has a weak effect on the total power calculus. Therefore, we can set a lower-bound reference frequency in noise calculation (e.g., the reciprocal of the observation or lifetimes of the system) without any sensible change of the estimated noise variance. We will see this aspect better in Chapter 7.

Another important issue is understanding the level of memory of the $1/f$ noise. As shown by Keshner, we can model the $1/f^\alpha$ noises by starting from the white noise using a lumped model of an infinite chain of low-pass filters, as shown in Fig. 6.24A. In this model, a source of white noise is fed to the chain's input to get flicker noise at the output. We can use an infinite chain of integrators in a similar approach, as shown in Fig. 6.24B. The PSD response of a single stage is illustrated in Fig. 6.24C. However, we can approximate the behavior of $1/f^\alpha$ noises using a finite number of integrators for a given spectrum.

The choice of the resistances sets the pole and the zero of the transfer function of a single segment of the chain. If the fluctuations are slower than the frequency indicated by the function's pole, the memory modeled by the capacitor loses the information, and the response is uniform, as in the white noise. Conversely, if the frequency is higher than that indicated by the function's zero, the memory has no time to store the information, and the response is again as for white noise. In between those boundaries, we have the full function of the memory, and the response follows the $1/f^2$ function. It can be shown that this behavior follows the same model of a particle motion in thermal agitation in a viscous medium frequently referred to as *Brownian motion* (see Chapter 4). Note also that the first part of the function is a Lorentzian function, and this approach is similar to the one previously used for the trap model. Therefore, by making an infinite chain of single-lumped networks and using the right choice of time constants, we can compose $1/f$ noise, as already discussed in this section.

This is very useful to understand the memory behavior of flicker noise. In principle, white noise is memoryless, as evidenced by its autocorrelation function, a Dirac function. Brownian or $1/f^2$ noise is a single-memory process. We can get Brownian noise by filtering white noise with an ideal integrator. In between those two examples, we have the $1/f$ noise.

It can be shown that the circuit of Fig. 6.24B models the $1/f$ process if we place a time constant for each decade in the frequency domain. This means that the flicker noise determines future states by summing up equally (by means of the state variables relaxation times) the memory (making an average) of the past 1, 10, 100, 1,000, and so forth seconds.

Furthermore, it could be seen that, for a given approximation, $1/f$ noise has the largest number of time constants within an observation time with respect to other $1/f^\alpha$ noise sources. This means that we can approximate the power spectral density behavior of $1/f^\alpha$ noises with a limited number of integrators, but the $1/f$ noise needs the most significant number of them among other flicker noises ($\alpha \neq 1$) for an equal degree of approximation.

---

[14] $\sqrt{\int_{10^{-17}}^{10^{23}} 1/f\, df} = \sqrt{\ln(10^{40})} = 9.6$; whilst $\sqrt{\ln(10)} = 1.5$.

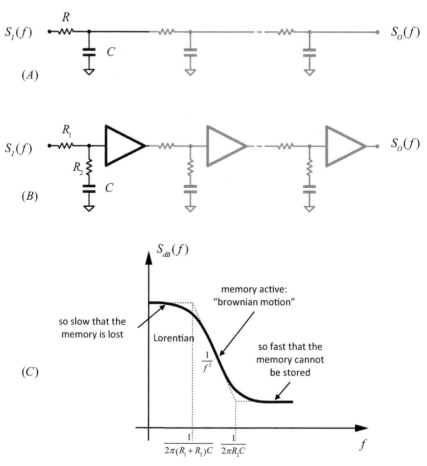

Figure 6.24 Electronic lumped model of flicker noise and memory states. (A) Generation of $1/f$ noise by infinite low-pass filters. (B) Generation of $1/f$ noise by integrators (single memory) cascade. (C) PSD response of single-integrator stage.

6.5   The Colors of Noise

We have classified noise according to physical origins. However, we can classify them according to the shape of the PSD.

As shown in Fig. 6.25, a uniform PSD is called *white noise*. Thus, both thermal and shot noises could be classified as white noise. Conversely, $1/f$ noise is frequently referred to as pink noise. Finally, PSDs following behavior proportional to $1/f^2$ and f are referred to as *red noise* and *blue noise*. If we map such noises in the audio bandwidth, we can sense that white noise is similar to that of the wind or to the sound of a spilling gas valve. On the other hand, pink and red noises are typical of a waterfall due to the predominance of low frequencies.

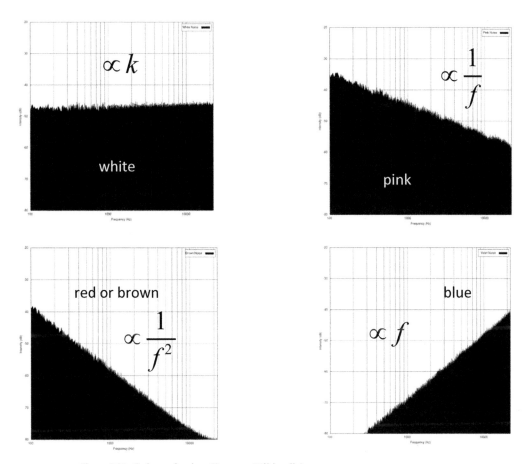

Figure 6.25 Colors of noise. (Source: Wikipedia)

6.5.1 Autocorrelation Functions of Noises

An important aspect of the different kinds of noise is knowing their autocorrelation function. This is fundamental to know the effect of subtracting two samples of the same process, as will be discussed in Chapter 8. The basic relationship is the Wiener–Khinchin theorem. In the white noise case, since the power spectrum is uniform, the autocorrelation function is a Dirac function. Therefore, it is a perfectly uncorrelated noise, and the standard deviation of the difference of two samples results in the sum of the two standard deviations.

This is not the case with other noise sources. This is because if the PSD is not uniform, this necessarily indicates an autocorrelation function that is different from a Dirac function. For instance, for the $1/f$ noise, Keshner showed that for a given window of observation $[t_1, t_2]$ the autocorrelation function is

$$R(t_2, \tau) = \ln(4t_2) - \ln(\tau) \, for : \, \tau << t_2, \tag{6.79}$$

where $\tau = t_2 - t_1$. This result shows that the autocorrelation function (which is dependent on time as nonstationary processes) is a decreasing function versus τ (once t_2 is fixed), and therefore we can use this observation to reduce the noise standard deviation by making the difference of two samples (see Chapter 4 and the *correlation double sampling* technique as discussed in Chapter 8).

6.6   Thermomechanical Noise

6.6.1   Quick Review of Second-Order Systems

Before going into the description of mechanical sensing systems and related noise, the basic concepts of second-order transfer functions will be summarized here.

If $x(t)$ and $y(t)$ are the input and output of a *linear time-invariant (LTI)* system, the ratio of their complex transform $X(s)$ and $Y(s)$ is referred to as the *transfer function* of the system (see Fig. 6.26). A linear system's transfer function describes its behavior completely in *both* the time domain and frequency domain.

A transfer function whose denominator is described by a second-order polynomial function of the complex variable s is called a *second-order transfer function*. The following *canonical expression* characterizes a *second-order low-pass system:*

$$H(s) = \frac{Y(s)}{X(s)} = \frac{k\omega_0^2}{s^2 + \dfrac{\omega_0}{Q}s + \omega_0^2}, \tag{6.80}$$

where ω_0 is the *resonance frequency*, Q is called a *quality factor* and k is the *gain*. In the mathematical representation, Q is very useful to understand how the poles of the function are displaced. It is straightforward to show that if $Q < 1/2$, $Q = 1/2$, $Q > 1/2$ the two poles (solutions of the denominator quadratic equation) are real, overlapped, and complex conjugate, respectively. For $Q = 1/\sqrt{2}$ we have a particular condition, referred to as *quadrature*, where the real and imaginary components of the poles are equal. We will see that the Q factor will also have several *physical* interpretations. If we make the usual substitution $s \leftarrow j\omega$ for the harmonic regime, it becomes

$$H(j\omega) = \frac{Y(j\omega)}{X(j\omega)} = k\frac{\omega_0 Q}{j\omega} \frac{1}{1 + jQ\left(\dfrac{\omega}{\omega_0} - \dfrac{\omega_0}{\omega}\right)}, \tag{6.81}$$

which clearly shows a low-pass behavior, whose module is

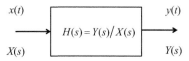

$$H(s) = Y(s)/X(s)$$

Figure 6.26 Linear system and related transfer function.

Table 6.1 Time response of a second-order linear system.

Quality factor $Q = \tau\omega_0/2$	Impulse response	Decay time constant	Notes
$Q \leq 1/2$	$y(t) = Ae^{\frac{t}{\tau_D}}$	$1/\tau_D = 1/\tau + \sqrt{1/\tau - \omega_0^2}$	$\tau = 2Q/\omega_0$
$Q = 1/2$	$y(t) = Ae^{\frac{t}{\tau_D}}$	$1/\tau_D = 1/\tau$	$\tau = 2Q/\omega_0$
$Q > 1/2$	$y(t) = Ae^{\frac{t}{\tau_D}}\cos(\omega t)$	$1/\tau_D = 1/\tau$	$\omega = \sqrt{\omega_0^2 - 1/\tau}$

$$|H(j\omega)| = kQ\frac{\omega_0}{\omega}\frac{1}{\sqrt{1 + Q^2\left(\frac{\omega}{\omega_0} - \frac{\omega_0}{\omega}\right)^2}}, \tag{6.82}$$

whose *Bode plot* is illustrated in Fig. 6.27. The behavior of the plot is strongly related to the value of Q. If the poles are real (ω_A and ω_B) and fairly separated, they are clearly visible in the plot. If they are in quadrature, the curve is maximally flat until ω_0. Finally, if the poles are beyond quadrature, the plot shows a resonance peak whose amplitude is about the value of Q itself.

This behavior is reflected in the time domain. Since the *transfer function* is the complex transform of the *impulse response*, we can easily calculate the time response of the system that has different behaviors according to the value of Q as summarized in Table 6.1, where τ_D is the *decay time* of the response.

To better understand the behavior, three plots with different quality factors are shown in Fig. 6.28. As clearly shown, if the poles are complex ($Q > 1/2$), we have a "ringing" behavior with "overshoot" bumps. This is called *underdamped* behavior. On the other hand, if the poles are real ($Q \leq 1/2$), we have no ringing, and the behavior is called *overdamped*. The transition point where the poles are real and coincident ($Q = 1/2$) is called *critically damped*.

The summary of the time and frequency behavior of the second-order systems is shown in Table 6.2. Note how the system should be beyond quadrature ($Q > 1/\sqrt{2}$) to have a peak in the frequency domain while it is enough to have complex poles ($Q > 1/2$) to have an overshoot in the time domain.

A band-pass behavior characterizes another second-order system. The following canonical expression characterizes a *second-order band-pass* system

$$H(s) = \frac{k\frac{\omega_0}{Q}s}{s^2 + \frac{\omega_0}{Q}s + \omega_0^2}, \tag{6.83}$$

Table 6.2 Summary of the frequency and time behavior of second-order systems.

Quality factor Q	Poles	Referred to as	Frequency domain	Time-domain
$Q < 1/2$	Real	Overdamped	No resonance peak	No overshoot
$Q = 1/2$	Real coincident	Critically damped	No resonance peak	One overshoot bump
$1/2 < Q < 1/\sqrt{2}$	Complex conjugates, before quadrature	Moderately underdamped	No resonance peak	More than one overshoot bump
$Q = 1/\sqrt{2}$	Complex conjugates, quadrature	Strongly underdamped	Resonance peak	Damped ringing
$Q > 1/\sqrt{2}$	Complex conjugates, beyond quadrature	Strongly underdamped	Resonance peak	Damped ringing

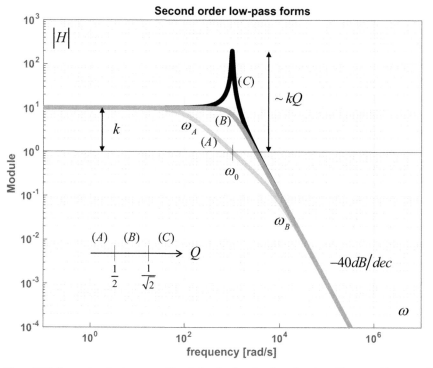

Figure 6.27 Second-order low-pass filter behavior in a log-log plot. $k = 10$, $\omega_0 = 1k$, $\omega_A = 100$, $\omega_B = 100k$, $Q = 0.1$ (A), $Q = 1/\sqrt{2}$ (B), $Q = 20$ (C).

where again ω_0 is the *resonance frequency*, Q is called the *quality factor* and k is the *gain*. Q is very useful to understand how the poles of the function are displaced. If we make the usual substitution $s \leftarrow j\omega$, it becomes

$$H(j\omega) = \frac{k}{1 + jQ\left(\dfrac{\omega}{\omega_0} - \dfrac{\omega_0}{\omega}\right)}, \qquad (6.84)$$

whose module is

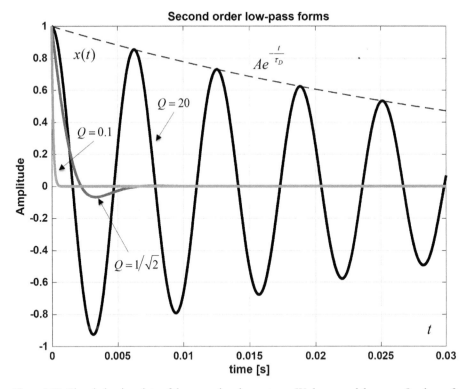

Figure 6.28 Time behavior plots of the second-order systems. We have used the same Q values of the frequency-response plot.

$$|H(j\omega)| = \frac{k}{\sqrt{1 + Q^2 \left(\dfrac{\omega}{\omega_0} - \dfrac{\omega_0}{\omega} \right)^2}} \qquad (6.85)$$

and could be plotted as in Fig. 6.29, which clearly exhibits a band-pass behavior.

Again, the behavior of the plot is strongly related to the value of Q. If the poles are real (ω_A and ω_B) and their values fairly separated; they are clearly visible in the plot. If they are in quadrature, the curve is smoothly bent in ω_0. Finally, if the poles are beyond quadrature, the plot shows a resonance peak whose amplitude is about the value of Q itself. Note how the maximum value of the function is k.

6.6.2 Bandwidth and Noise Bandwidth of the Bandpass Function

Using (6.85), it is easy to show that variation of the frequency $\Delta\omega$ with respect to the resonant frequency ω_0 so that the functions decay to $k/\sqrt{2}$ (i.e., where the power of input signal is halved, -3 dB) is given by

$$\Delta_{1/2}\omega_{-3dB} \simeq \frac{\omega_0}{2Q}. \qquad (6.86)$$

This is the cutoff frequency on one side of the function; therefore, considering the symmetry of the band-pass function, its bandwidth is

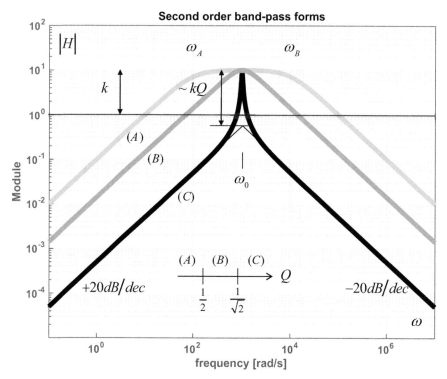

Figure 6.29 Second-order band-pass filter behavior in a log-log plot. $k = 10$, $\omega_0 = 1k$, $\omega_A = 100$, $\omega_B = 100k$, $Q = 0.1$ (A), $Q = 1/\sqrt{2}$ (B), $Q = 20$ (C).

$$BW_E(\omega) = \Delta\omega_{-3dB} = \frac{\omega_0}{Q} \text{ and}$$

$$BW_E(f) = \Delta f_{-3dB} = \frac{f_0}{Q}.$$ (6.87)

Our goal is to calculate the equivalent noise bandwidth (NBW) for a second-order bandpass filter.

The same function as in Fig. 6.29 for $Q = 20$ is illustrated in Fig. 6.30 in a lin–lin plot for the square of the module. The gray shape illustrates the area covered by the function and is centered on the maximum.

As known from Chapter 2, Section 2.6, the equivalent noise bandwidth is calculated on the square of the transfer function module, since it has to transform the power spectra

$$NBW = \frac{1}{|H(\omega)|^2_{\text{max}}} \frac{1}{2\pi} \int_0^\infty |H(\omega)|^2 d\omega.$$ (6.88)

Therefore, for the special case of the bandpass second-order filter, it becomes[15]

---

[15] Recall that $\displaystyle\int_0^\infty \frac{1}{1 + Q^2\left(\dfrac{\omega}{\omega_0} - \dfrac{\omega_0}{\omega}\right)^2} d\omega = \frac{\pi}{2}\frac{\omega_0}{Q}.$

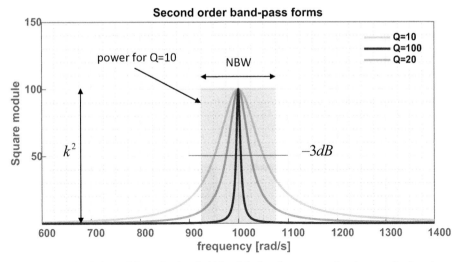

Figure 6.30 Calculation of the noise bandwidth of the bandpass second-order transfer function, where $k = 10$, $\omega_0 = 1k$, $Q = 20$. The graph is illustrated in a lin–lin plot style.

$$NBW = \frac{1}{k^2}\frac{1}{2\pi}\int_0^\infty \frac{k^2}{1 + Q^2\left(\dfrac{\omega}{\omega_0} - \dfrac{\omega_0}{\omega}\right)^2}d\omega = \frac{1}{4}\frac{\omega_0}{Q} = \frac{\pi f_0}{2Q}, \qquad (6.89)$$

which is very similar to the case of first-order systems where NBW is equal to the reciprocal of four times the time constant. We will see that Q/ω_0 is the energy time constant; therefore, NBW is reciprocal to four times the energy time constant.

6.6.3   Physical Models

The second-order transfer function could be very useful to model resonating physical systems such as electronic or mechanical, as illustrated in Fig. 6.31.

In these systems, we assume that some dissipative mean (transforming resonant energy in heat) is acting, such as the resistance in the electronic oscillator and the gas's viscosity in the mechanical one. In the electronic oscillator case, we have a driving voltage $v(t)$ that is forcing a potential on a resistor-inductor-capacitor (RLC) loop. Therefore, applying the Kirchhoff voltage law on the loop, we have

$$v(t) = v_L(t) + v_R(t) + v_C(t)$$

$$v(t) = L\frac{d^2q}{dt^2} + R\frac{dq}{dt} + \frac{q}{C}, \qquad (6.90)$$

where $q = q(t) \equiv [C]$ is the charge on the capacitor, $R \equiv [\Omega]$ is the resistance, and $L \equiv [\Omega \cdot s]$ is the inductance. Therefore, using the Laplace transform, we have

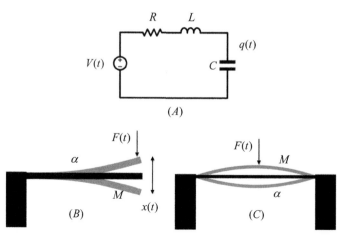

Figure 6.31 Electronic (A) and mechanical resonators as the cantilever (B) and membrane (C).

$$V(s) = Ls^2Q(s) + RsQ(s) + \frac{Q(s)}{C}$$
$$\rightarrow \frac{Q(s)}{V(s)} = H(s) = \frac{1/L}{s^2 + \frac{R}{L}s + \frac{1}{LC}}, \tag{6.91}$$

which describes the behavior of the charge with respect to the forcing voltage in second-order canonical form, where R is the dissipative element.

Similarly, the mechanical system could be described as a typical *spring–mass–damper system* subject to an external force $F(t)$ by means of the sum of dynamic, damping, and elastic forces:

$$F(t) = F_A(t) + F_D(t) + F_E(t)$$
$$F(t) = M\frac{d^2x}{dt^2} + b\frac{dx}{dt} + \alpha x, \tag{6.92}$$

where $x = x(t) \equiv [m]$ is the displacement, $M \equiv [kg]$ is the mass of the object, $b \equiv [F/v]$ is the damping coefficient, and $\alpha \equiv [F/m]$ is the stiffness constant.[16] Therefore, the behavior could be described as

$$F(s) = Ms^2X(s) + bsX(s) + \alpha X(s)$$
$$\rightarrow \frac{X(s)}{F(s)} = H(s) = \frac{1/M}{s^2 + \frac{b}{M}s + \frac{\alpha}{M}}, \tag{6.93}$$

which describes the behavior of the displacement with respect to the acting force.

A summary of the mechanical and electronic system behaviors is illustrated in Tables 6.3 and 6.4. For a physical explanation of the quality factor in Table 6.3, see the next paragraph.

---

[16] Note that $b \equiv [F/v]$ has the units of the ratio of the force over the velocity, that is, the reciprocal of physical *mobility* (see Chapter 10).

Table 6.3 Relationships between the parameters of the second-order equations (6.90) and (6.92).

	Quality factor Q	Resonant frequency ω_0	Decay time $\tau = 2Q/\omega_0$
Electronic	$Q = \dfrac{1}{R}\sqrt{\dfrac{L}{C}}$	$\omega_0 = 1/LC$	$\tau = \dfrac{2L}{R}$
Mechanical	$Q = \dfrac{\sqrt{\alpha M}}{b}$	$\omega_0 = \sqrt{\dfrac{\alpha}{M}}$	$\tau = \dfrac{2M}{b}$

Table 6.4 Free behavior of perturbed electronic and mechanical resonators.

	Damping frequency	Damping frequency ω_D
Electronic	$q(t) = q(0)e^{-\frac{t}{\tau}}\cos(\omega_D t)$	$\omega_D = \sqrt{\omega_0 - \dfrac{1}{\tau^2}} = \omega_0\sqrt{1 - \dfrac{1}{4Q^2}}$
Mechanical	$x(t) = x(0)e^{-\frac{t}{\tau}}\cos(\omega_D t)$	

In general, a resonator exchanges its energy between two forms. More specifically, in an electronic oscillator, the total energy switches between that one stored in a capacitor (electric field) with that one stored in the inductor (magnetic field). Similarly, this happens in a mechanical resonator between the kinetic and potential energy forms, such as in a pendulum. However, any time the energy is exchanged, a part is dissipated in heat by the damping factor. It could be shown that the mean energy is decaying with a constant time τ_E that is half of the signal decay time τ. The behavior is shown in Fig. 6.32.

In Table 6.5 is summarized the energy components of the various resonators.

Therefore, the fact that energy decay time is half of the signal decay time $\tau_E = \tau/2$ allows showing important physical characteristics of the resonator referred to as the *quality factor Q*. More specifically, the *dissipation rate* could be calculated as the energy loss in an energy decay time constant[17]

$$P_{loss} = \frac{\langle U(0)\rangle}{\tau_E} \equiv [W].$$ (6.94)

Therefore, we can define a Q factor as

$$Q \triangleq 2\pi \cdot \frac{\text{energy stored per cycle}}{\text{energy dissipated per cycle}} = 2\pi \cdot \frac{\langle U\rangle}{P_{loss}T_0}$$

$$= \omega_0 \cdot \frac{\langle U\rangle}{P_{loss}} = \omega_0 \cdot \frac{\text{energy stored per cycle}}{\text{dissipation rate}} = \omega_0 \cdot \tau_E \equiv [\text{rad}] \equiv [\cdot].$$ (6.95)

---

[17] This is a typical first-order approximation of exponential decays where the average slope of the exponential decay is given by the slope of the tangent drawn at the initial point of the function.

Table 6.5 Main characteristics of the electronic and mechanical resonators.

	Energy form 1	Energy form 2	Energy decay time $\tau_E = Q/\omega_0$	Energy per cycle
Electronic	$U_L = \frac{1}{2}LI^2$	$U_C = \frac{1}{2}\frac{q^2}{C}$	$\tau_E = \frac{\tau}{2} = \frac{L}{R}$	$\langle U(t)\rangle = \frac{1}{2}\frac{q^2(0)}{C}e^{-\frac{t}{\tau_E}}$
Mechanical	$U_K = \frac{1}{2}M\left(\frac{dx}{dt}\right)^2$	$U_E = \frac{1}{2}\alpha x^2$	$\tau_E = \frac{\tau}{2} = \frac{b}{M}$	$\langle U(t)\rangle = \frac{1}{2}\alpha x^2(0)e^{-\frac{t}{\tau_E}}$

$q(0)$ and $x(0)$ are the initial charge and displacement, respectively.

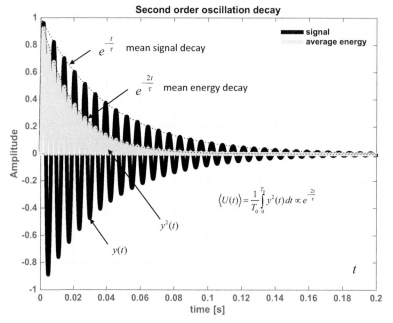

Figure 6.32 Signal and mean energy decay in a resonator. A generic waveform $y(t) = Ae^{t/\tau}\cos(\omega t)$ is plotted alongside the square of the same function. The decay of the envelope of the square is twice as fast as the original signal. It could be shown that the mean energy per cycle decay follows this time constant.

The last expression shows another view of the quality factor as *the number of radians required for the resonator's energy to decay to 1/e (37%) of its initial value.*

6.6.4   Thermomechanical Noise

Every time a mechanical part with some degree of freedom is subject to the hits of a very large number of particles, it is subject to a *fluctuation force* due to *thermomechanical noise*, as shown in Fig. 6.33.

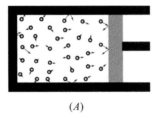

(A)

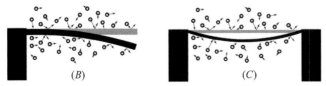

(B)   (C)

Figure 6.33 Thermomechanical noise in mechanical systems such as piston (A), cantilever (B), and membrane.

The noise spectrum could be calculated using the same arguments of Section 6.1 that use methods based on the energy of the systems. The first point is to calculate the relationship between the force F and the velocity v since all the gas's energy properties are related to the velocity distribution. Therefore, using (6.93) with respect to the velocity of the moving mechanical part, we have

$$
\begin{aligned}
M\frac{dv}{dt} + bv + a\int vdt &= F \\
\rightarrow MsV(s) + bV(s) + \frac{a}{s}V(s) &= F(s) \\
\rightarrow H_{VF}(s) = \frac{V(s)}{F(s)} &= \frac{1}{Ms + b + \dfrac{a}{s}},
\end{aligned}
\tag{6.96}
$$

where $V(s)$ and $F(s)$ are the transform of the force and velocity, respectively.

The last equation of (6.96) could be expressed in terms of angular frequency as

$$
|H_{VF}(j\omega)|^2 = \left|\frac{V(j\omega)}{F(j\omega)}\right|^2 = \frac{1}{b^2 + \left(M\omega - \dfrac{a}{\omega}\right)^2} = \frac{1/b^2}{1 + Q^2\left(\dfrac{\omega}{\omega_0} - \dfrac{\omega_0}{\omega}\right)^2},
\tag{6.97}
$$

which is a *second-order band-pass* transfer function.

The transfer function (6.97) is important to understand the relationship between the power spectral functions of force and velocity distributions. Furthermore, assuming the spectral density function of the force induced by noise $\overline{F_N^2}(\omega)$ constant in the range of interest, we have

$$
\overline{v_N^2} = \frac{1}{2\pi}\int_0^\infty |H_{VF}(j\omega)|^2 \overline{F_N^2}(\omega)d\omega = \frac{1}{b^2} \cdot NBW \cdot \overline{F_N^2}(\omega).
\tag{6.98}
$$

We can calculate the mean square of the velocity as

$$\overline{v_N^2} = \frac{1}{b^2} \frac{1}{4} \frac{\omega_0}{Q} \overline{F_N^2}(\omega) = \frac{1}{b^2} \frac{1}{4} \frac{b}{M} \overline{F_N^2}(\omega) = \frac{1}{4bM} \overline{F_N^2}(\omega). \tag{6.99}$$

Now, since we write the overall velocity noise and equate the system kinetic energy to the thermal energy for 1 degree of freedom,

$$\frac{1}{2} M \overline{v_N^2} = \frac{1}{2} kT \;\; \rightarrow \;\; v_{rms} = \sqrt{\frac{kT}{M}}, \tag{6.100}$$

from which we have

$$\overline{F_N^2}(\omega) = \overline{F_N^2}(f) = 4kTb, \tag{6.101}$$

which is very similar to the equivalent $4kTR$ of electrical circuits.

One may wonder if the result so far gained is consistent with the result of Section 6.1, where the rms deviation of the piston of the pressure sensor is $\sqrt{kT/\alpha}$. We could easily demonstrate this assuming that most of the energy is accumulated in a peak of resonance given by high Q values. For such values, the NBW of the low-pass filter is almost equivalent to that one of the band-pass and using (6.84), we get

$$\overline{x_N^2} = \text{NBW} \cdot |H|^2_{\text{max}} \cdot \overline{F_N^2}(\omega) = \frac{1}{4} \frac{\omega_0}{Q} \cdot \frac{Q^2}{\alpha^2} \cdot 4kTb = \frac{kT}{\alpha}. \tag{6.102}$$

6.7   Phase Noise

Phase noise is an effect of several kinds of noises, primarily the thermal one, on an electronic oscillator signal. Broadly speaking, phase noise is the perturbation in the movement of any physical system characterized by harmonic motion. Electronic phase noise is one of the most difficult phenomena to be predicted and modeled because of the nonlinear effects of the oscillator.

We first use a mechanical system model to show the concept: in Fig. 6.34, we have two free-running mechanical systems whose motion is described by a harmonic function. The first (Fig. 6.34A) is a rotating balance wheel, and the second is an oscillating piston. The systems are immersed in a gas of particles interacting with the moving parts. Whenever a particle hits the moving part, there is an exchange of momentum resulting in a perturbation of the trajectory. In the first case, the perturbation affects just the phase since the wheel's diameter is fixed; in the other case, the perturbation could change both phase and amplitude. In principle, if no collisions occur and there is no friction, the harmonic motion would continue indefinitely. Any collision determines a loss of the kinetic energy of the moving part. This effect is macroscopically perceived as the "viscosity" of the medium, with motion damping.

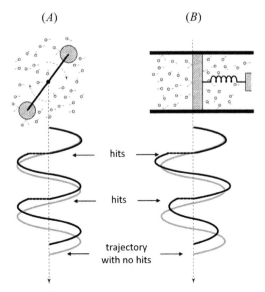

Figure 6.34 The concept of phase noise in a mechanical model. The harmonic trajectories of free rotating (A) or oscillating (B) mechanical systems are perturbed by particles' hits. The first is perturbed in phase only, while the second is perturbed in both phase and amplitude. (Phase change is emphasized.)

Figure 6.35 shows that, in the case in which we have both "degrees of freedom," the perturbation could increase (or decrease) amplitude and phase depending on the instant when the object is hit. Specifically, phase is affected more when the hit occurs during the zero-crossing of the harmonic function (in-quadrature or sine perturbation), while the amplitude is maximally affected when the hit occurs at the top of the evolution (in-phase or cosine perturbation). This means that the power of the perturbation, that is, the spectral density of this kind of noise, could be seen as composed of both amplitude and phase contributions where the kinetic energy of the particle is shared. This can be seen in the complex plane, as shown in Fig. 6.35C.

As shown in Fig. 6.35, the effect of the perturbation is to partly exert a phase change on the "signal" that, in this case, is the natural harmonic wave of the oscillator. Therefore, from the electronic systems point of view, the phase change could be represented as the phase modulation of a carrier that is the oscillator's sinusoidal signal. In conclusion, differently from the other linear models, where typically noise is added to the signal, here we have a nonlinear effect due to the modulation process, represented by a multiplication between signals as shown in Fig. 6.36.

To better understand the role of the phase, we have to consider the structure of the harmonic function

$$x(t) = A \cos(\underbrace{\omega t + \varphi}_{\phi(t)}), \tag{6.103}$$

where the "phase" $\phi(t) \equiv [\cdot] \equiv [\text{rad}]$ is simply the argument of the cosine, that is, the advancement of the phase vector (phasor) versus time in the complex plane. Its

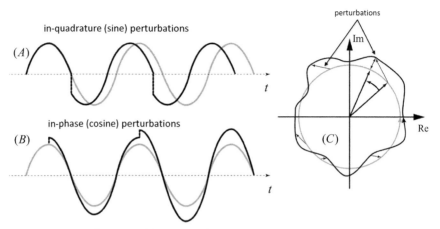

Figure 6.35 Depending on the hit time, the free trajectory's perturbation could be exerted in both phase (A) and amplitude (B). However, the perturbation could be a composition of the two cases mentioned above, as seen in the complex plane (C).

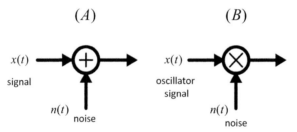

Figure 6.36 The concept of phase noise in electronic design. In linear systems, we have always considered noise as an additional perturbation to signal (A). Conversely, in phase noise, we have a modulation effect of the noise with the "carrier" of the oscillator. Therefore, phase noise should be considered a multiplication (nonlinear) effect on the signal (B). Note that the multiplier scheme models both amplitude modulation (AM) and phase modulation (PM).

unity is a ratio of the radius with the circumference, so it is dimensionless, or we can use the auxiliary unit of *radiant (rad)*. As in linear motion, we can devise a "velocity" as the derivative of the advancement of motion with respect to time, called angular frequency. Therefore, we *define* the *instantaneous frequency* and the *instantaneous phase* as

$$\omega(t) \triangleq \frac{d\phi(t)}{dt}$$

$$\phi(t) \triangleq \phi(0) + \int_0^t \omega(t)dt. \tag{6.104}$$

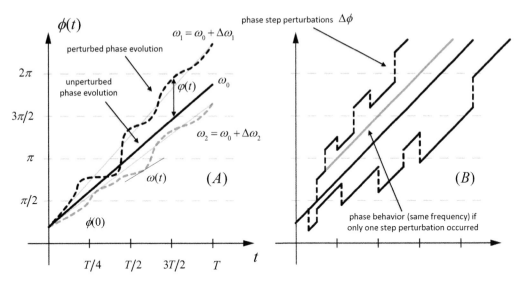

Figure 6.37 Role of the phase. (A) The straight line is related to a constant frequency (angular velocity), while the curve is the deviation from the constant frequency due to phase changes. (B) The effect of random step variations of phase. Note how the average frequency is the same, but the deviation from the unperturbed phase evolution follows a "random walk."

As shown in Fig. 6.37A, the phase evolution is a generic monotonic increasing function whose derivative in each point is $\omega(t)$. If the advancement follows a uniform circular motion, the phase function versus time is a line whose slope is the angular frequency ω_0. If there is no regular advancement, we can evaluate an "average" frequency between two points of the function by calculating the slope of the line connecting the two points.

Referring to the specific case of the oscillator, it is better to use as a reference the natural frequency ω_0 of the oscillator so that we can rewrite the harmonic function as

$$x(t) = A \, \cos(\underbrace{\omega_0 t + \varphi(t)}_{\phi(t)}), \tag{6.105}$$

where $\varphi(t)$ is the *phase error* that can be visualized in Fig. 6.37 as the distance between the dotted and the solid line related to the natural frequency.

In the presence of a perturbed phase evolution, we can calculate the mean frequency, for example (but this is only a possible choice) choosing a period T on which to calculate it:

$$\omega_1 = \frac{\phi(T)}{T} = \frac{1}{T} \int_0^T \left(\omega_0 t + \varphi(t) \right) dt = \omega_0 + \frac{1}{T} \int_0^T \varphi(t) = \omega_0 + \Delta\omega_1. \tag{6.106}$$

Now, if we assume that the perturbation of the phase is given by random step functions as shown in Fig. 6.37B, we can see that the deviation of the perturbed phase evolution with respect to the unperturbed phase evolution is increasing with time following a *random walk* behavior (see the appendix in Chapter 4). More precisely, we have the same probability of getting a perturbation in advantage or disadvantage of the phase; the system keeps the memory of the last increment or decrement (see Section 6.4.1), and thus the error is accumulated (integrated) as in the random walk. Even if we do not know if we will have a phase lag or gain at the end of some future time, we know that we will have a phase difference with respect to the original unperturbed trajectory whose mean value increases in probability as the square root of time.

A simplified electronic oscillator model is composed of an LC resonating circuit coupled with a negative-resistance dipole whose role is to sustain the power dissipation of the resonator, keeping the amplitude as constant as possible (Fig. 6.38A). In brief, a free oscillator (without the restoring part) would have a damping behavior set by the Q factor of the resonator. In the presence of the restorer, we have a virtual increase of the Q factor (since the negative resistance model of the restore tends to neglect the effect of the real resistance) to restrict all the spectral power around the natural frequency of the oscillator. In principle, an ideal persistent oscillator would have a spectral power modeled by a Dirac function centered on its natural frequency. Unfortunately, the noise power of both the real resistance $\overline{v_N^2}$ and of the restorer $\overline{v_G^2}$ contributes to perturb the oscillator trajectory by changing the spectral power characteristic following a modulation effect.

In dynamical systems such as the electronic oscillator, we can describe the time evolution of the system using the state-space diagram, as shown in Fig. 6.38B. "States" are arbitrarily chosen variables that uniquely describe the system's time evolution. An ideal oscillator has a time evolution that is described by a cyclic curve (run every natural period of the oscillator) defined in the state space called *limit cycle*, or orbit, or trajectory.

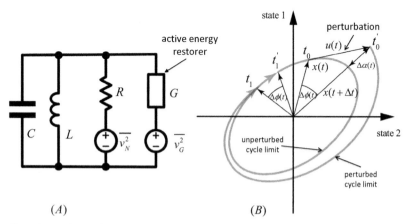

(A)   (B)

Figure 6.38 Electronic oscillator. It is composed of an RLC circuit together with a negative resistance dipole that serves to supply the energy for sustainable and constant amplitude oscillations (A). The perturbation effect could also be seen in the state space (B) with perturbed amplitude and phase.

As shown in the state space of Fig. 6.38B, the perturbation $u(t)$ could change both the phase and angle of the orbit of the state $x(t)$(composed of two components) such that

$$u(t) = \Delta x = x(t + \Delta t) - x(t) = [1 + \Delta a(t)] \cdot x(t + \frac{\Delta \phi(t)}{\omega_0}) - x(t);$$
$$\omega_0 = \frac{2\pi}{T}; \; \Delta t = \frac{\Delta \phi(t)}{\omega_0}, \tag{6.107}$$

where $\Delta a(t)$ and $\Delta \phi(t)$ are the induced perturbations in amplitude and angle and are functions of the original perturbation. To understand Eq. (6.107), refer to Fig. 6.35, where $u(t)$ is the difference between two points taken at different times whose difference is Δt. If the trajectory (or function) is smooth, the difference is very small, but if the perturbation occurs within the time-lapse Δt, that makes a difference in both amplitude and phase depending on when it occurs. In the presence of an impulse perturbation, the restorer keeps the perturbed trajectory of the oscillator as close as possible to the original one, and the amplitude component is absorbed while the phase increases/decreases following a step function. This could be seen in Fig. 6.38B, where we compare what happens with/without perturbation in the limit cycle between two times t_0 and t_1 for unperturbed and t_0' t_1' for perturbed motion. In the case of the perturbation, the orbit gets a phase lead with respect to the unperturbed orbit that cannot be deleted. While the feedback system of the oscillator limits the amplitude perturbation, the phase lead remains as time goes on. Therefore, the phase is accumulated after any perturbation. Even if the perturbation has the same probability between opposite directions, the net contribution diverges as in the random walk. Therefore, the phase keeps the memory of events and sums all impulse perturbations in step functions changes of phase. Therefore

$$\Delta a(t) \rightarrow 0 \text{ for } t \rightarrow \infty$$
$$\varphi(t) = \Delta \phi(t) \sim \int_{-\infty}^{t} u(t) dt. \tag{6.108}$$

Therefore, the phase acts as an "integrator" of perturbations, even if it is not an LTI system since the output depends on when the perturbation occurs (for this reason, we used the sign $\sim$).

The effect of the perturbations in the frequency domain is to spread the power spectral density from a Dirac function as shown in Fig. 6.39A of an ideal oscillator into a broader symmetrical spectrum around the natural frequency (Fig. 6.39B) because of the modulation effect. This could be modeled with a *complex phasor diagram* as illustrated in Fig. 6.39C: the natural frequency is represented by an arrow with suppressed rotation, while any spectral component given by a deviation $\Delta \omega$ from ω_0 could be represented by subphasors rotating at a frequency $\Delta \omega$ in opposite directions. As known from the theory of modulation, the final phasor state is given by the vector sum of the original phasor (natural frequency acting as a "carrier") with the composition of the two antirotating subphasors (perturbation spectral components acting as a "modulating signal"). Therefore, the power associated with those

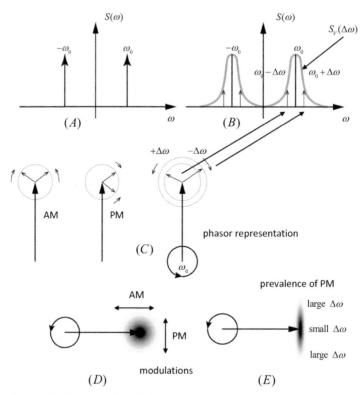

Figure 6.39 Phasor model of phase noise. (A) Ideal oscillator spectrum. (B) Effect of the modulation due to noise perturbation. (C) Phasor model of the modulation due to a given frequency perturbation $\Delta\omega$. (D) AM and PM modulation due to wide-spectrum noise in a free oscillator. (E) Effect of the modulation in an AM constrained oscillator.

subphasors could be represented as symmetrical spectral lines centered on the natural frequency and lying on the upper and lower sideband, respectively, as shown in Fig. 6.39B.

The effect of the noise on a free-running oscillator could be represented in a distribution of scattered movements, as shown in Fig. 6.39D. Any change (due to perturbations) of the phasor length is associated with amplitude modulation (AM), while the changes of phasor angle are associated with phase modulation (PM). Therefore, a generic perturbation could be decomposed of AM and PM modulations. Both AM and PM modulations contribute to the spectral sidebands such as that in Fig. 6.39B.

However, in a sustained oscillator, the AM modulation is reduced as much as possible by restoring the impedance loss. Therefore, most of the perturbations are inducing PM, as in Fig. 6.39E. Also, note that for large frequency changes, PM is predominant, while for small perturbations, AM components become important.

A noise modulated with a periodic waveform is a *cyclostationary process.* A process is cyclostationary if and only if its autocorrelation function $R(t, \tau)$ and mean are periodic with some periodicity. The spectrum of cyclostationary noise is like that in Fig. 6.39B with more replicas if the carrier is not a sinusoid. It could be shown that in the case of cyclostationary noise, the power components of the two sidebands are correlated in both AM and PM components and are symmetric (if the signal is real) and stationary. Unfortunately, oscillator phase noise is not always cyclostationary because the modulation is not synced elsewhere, but the average frequency moves with the phase deviation itself. This aspect has a consequence also in the definition of the oscillator frequency. Due to perturbations, the oscillator frequency should be calculated based on an average represented as the barycenter of the spread distribution of Fig. 6.39D and E). However, in the long-range, even the oscillator frequency has a shift due to perturbations (see Fig. 6.37); thus, the barycenter itself has slow movement in the phasor space. In any case, only for small observation times, the phase noise process could be considered stationary.

Turning back to the observation leading to (6.108), we have that

$$\varphi(t) = \Delta\phi(t) \sim \int_{-\infty}^{t} u(t)dt$$

$$\rightarrow S_\varphi(f) \sim \frac{S_u(f)}{(2\pi f)^2}, \tag{6.109}$$

where $u(t)$ is the perturbation and $S_u(f)$ its PSD.

Therefore, any components of the voltage/current power noise spectrum $S_u(f)$ are mapped into the phase power $S_\varphi(f)$ spectrum with a gain of $1/f^2$ as shown in Fig. 6.40. This is exactly the same (except the reference here is ω_0) as the effect of integration of noise occurring in Brownian motion; Furthermore, on the basis of (6.109), the presence of $1/f$ noise is mapped into phase noise as $1/f^3$ as shown in Fig. 6.40B.

However, the observations mentioned above have been discussed using linear systems arguments while we should consider the effect of the modulation as seen in Fig. 6.39A. Therefore, due to the nonlinear effect of the feedback system, the oscillator carrier is not a pure sinusoidal waveform but is composed of harmonics. This implies that also the original power components of noisy devices $S_u(f)$ around the multiples of the oscillator frequency $f_0, 2f_0, 3f_0, \ldots$ are folded back in the phase noise $S_\phi(f)$ as shown in Fig. 6.41A and B according to a weight factor c_x related to the harmonic coefficients of the Fourier's transform of the oscillation waveform.

The observed spectrum of a noisy oscillator is the mapping in the frequency domain of the modulation components, as shown in Fig. 6.39B. However, since the carrier is not a pure sinusoidal waveform, we also have a modulation effect in the carrier harmonics, as shown in Fig. 6.41C. $S_V(f) \equiv [V^2/Hz]$ is what we really measure from an oscillator using a spectrum analyzer, while $S_\phi(f) \equiv [rad^2/Hz]$ could be measured only using an ideal phase detector, so it is used more as a mathematical model. Usually,

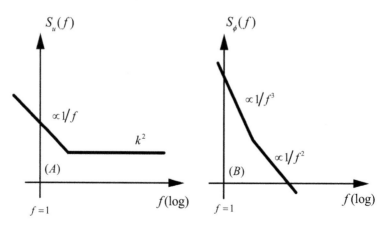

Figure 6.40 Transformation of random noise (A) into phase noise spectra (B).

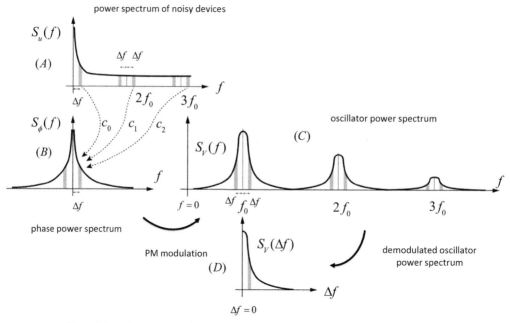

Figure 6.41 Phase noise relationships. (A) Power spectrum of the perturbation. (B) Lin–lin power spectrum of phase noise. (C) Mapping of flicker noise into phase noise. (D) Modulated power spectrum of the oscillator. (E) Translated power spectrum of the oscillator.

the oscillator signal is also treated by a coherent demodulator, obtaining $S_V(\Delta f) \equiv [V^2/\text{Hz}]$ as shown in Fig. 6.41D, also referred to as $L(\Delta f)$ if the power is referred to the carrier. $S_V(f)$ and $S_V(\Delta f)$ contains both amplitude and phase noise. Usually, the frequency components distant from the harmonics $f_0, 2f_0, 3f_0, \ldots$ are small signals and easy to be treated mathematically and with simulation tools. On the other hand, spectrum components close to harmonics are large signals affected by nonlinearities and thus difficult to model.

> **Hint** Phase noise spectrum is not expressed in terms of real power. This is because the status of a harmonic evolution (and thus its power) could be represented in the complex plane in two equivalent forms: by *Re* and *Im* components or by *module* and *phase* components. In the first case, the power components (*Re* and *Im*) are expressed with the same units (I, V, etc.), while in the second case, the module is expressed in physical units (I, V, etc.), while the phase is not. Therefore, the associated "power" is expressed in the square of dimensionless units (or rad^2). This means that to compose these two "power" components, we need a coupling factor for the phase, as we will see in Section 6.7.2.

The $1/f^2$ mapping discussed so far could also be seen in other arguments regarding electronic circuits. The RLC parallel resonator (Fig. 6.38A) acts as a two-pole band-pass filter. Using Eq. (6.84), it is easy to show that the *voltage-to-current* relationship is a transfer function that, for high-quality factor values and small deviations Δf around f_0, has the expression

$$\left| Z(f_0 + \Delta f) \right| = \left| H(\Delta f) \right| = \left| \frac{v(f_0 + \Delta f)}{i(f_0 + \Delta f)} \right| \approx R \frac{f_0}{2Q \cdot \Delta f}. \tag{6.110}$$

This relationship could also be seen as a transfer function with -3 dB bandwidth equal to f_0/Q, and where, for $f = f_0$(resonance) $Z = R$ where R is the loss of the resonator. Note that the quality factor Q should also consider the effect of the active energy restorer. Thus, on each side, there is a roll-off of the impedance equal to $f_0/2Q$ (see Section 6.6.2). In other terms, the magnitude of the impedance *decreases* of a factor $f_0/2Q$ for any variation of the frequency $\Delta f = f - f_0$. Therefore, the noise power spectral density $S_u(f) = \overline{i_n^2}(f) = 4kT/R$ is mapped into a frequency shift $\Delta f = f - f_0$ around f_0 by

$$S_V(\Delta f) = \overline{v_n^2}(\Delta f) = \left| H(\Delta f) \right|^2 \overline{i_n^2}(f) \approx \frac{4kT}{R} \left(\frac{f_0}{2Q \cdot \Delta f} \right)^2. \tag{6.111}$$

Thus, the power spectral density of the output is shaped by the resonator even if the original noise of devices might be uniform. The $1/f^2$ slope is because the RLC tank rolls off as $1/f$ on both sides of the resonating frequency in amplitude; thus, we should expect twice that slope in power terms.

We can see the system's behavior according to the equipartition theorem of thermo-dynamics (where the average kinetic energy of particles in thermodynamic equilibrium is equally shared between the degrees of freedom) so that amplitude and phase-noise power are equally balanced. However, the amplitude limiting mechanism active in any real oscillator removes half the noise given by (6.111).

The noise of the oscillator is given not only by the loss of the resonator but also by the noise of the restorer (see Fig. 6.38A); therefore, we have to take into account a multiplicative factor F for the excess noise: $S_u(f) = \overline{i_n^2}(f) = F4kT/R$. Furthermore, the measured noise $L_{dB}(\Delta f)$ is usually referred to as the power of the harmonic signal

$$L_{dB}(\Delta f) = 10\log \frac{P_N}{P_S} = 10\log \frac{S_V(f)}{P_S} = \frac{1/2|H(\Delta f)|^2 \overline{i_n^2(f)}}{1/2 V_0^2}$$
$$= 10\log \left(\frac{F}{P_S} \frac{2kT}{R} \left(\frac{f_0}{2Q \cdot \Delta f} \right)^2 \right) \equiv [\mathrm{dBc/Hz}],$$

(6.112)

which is a part (there are other empirical terms in the general expression) of what is referred to as the *Leeson formula*, where usually the F factor is calculated a posteriori as a fitting parameter of experimental data. This part of the Leeson formula could be applied only where $L_{dB}(\Delta f)$ has the $(1/\Delta f)^2$ behavior, that is, away from the carrier, as will be discussed in Section 6.7.1.

A final note on the phase noise process's stationarity should be considered. Looking back to Fig. 6.39D and E, we assumed that the barycenter of the phasor position is stable on the natural frequency. This could be true if the oscillator is constantly synchronized with a reference. However, the barycenter slowly moves for free-running oscillators due to the phase's perturbation. It could be shown that if the barycenter of the phasor is stable in time, phase noise is, for very small phase deviations, a wide sense stationary process (WSS), but not in general. However, if the observation time is shorter than natural frequency motion, we can use the WSS assumption, similarly to flicker noise.

6.7.1 The Total Oscillator Noise

The spectrum that we can experimentally observe with a spectrum analyzer is the demodulated $S_V(\Delta f)$ or $L_{dB}(\Delta f) = S_{V(\mathrm{dB})}(\Delta f)$ containing both amplitude and phase noise power components. On the other hand, it is difficult to measure phase noise directly $S_\phi(f)$ unless we use an ideal phase detector; however, we will see in Section 6.7.2 that $S_\phi(\Delta f)$ could be, under certain assumptions, estimated from $S_V(\Delta f)$. From the measurement point of view, if we directly test the oscillator's power, we always get on average a $V_0^2/2$ value. In an ideal free oscillator, PSD is a Dirac's delta function centered on resonance frequency having such a total power. On the other hand, in a noisy oscillator, the PSD impulse function spreads around f_0 by keeping the same average total power. The noisier, the larger the spread. The perturbations scatter the point with equal probability with respect to the degrees of freedom, having (for relatively small observation times) as average the natural frequency. However, since the sustained oscillator constrains the amplitude, most of the perturbations act on the phase degree of freedom, as shown in Fig. 6.39E. The phasor representation shows that perturbations act on the difference with respect to the natural frequency $\Delta f = f - f_0$, so the power spectrum should be referred to in the demodulated spectrum. Furthermore, in terms of power, we should evaluate the fluctuations with respect to the average power $V_0^2/2$, so the best representation is to refer the total spectrum to that average power "signal", as expressed by $S_{V(\mathrm{dB})}(\Delta f)$.

The "power" spectrum of the phase has an asymptote in zero since the divergence of ensembles of oscillator phases, that is, $\overline{\phi^2(t)} = 2Dt$ increases as time goes on, as in

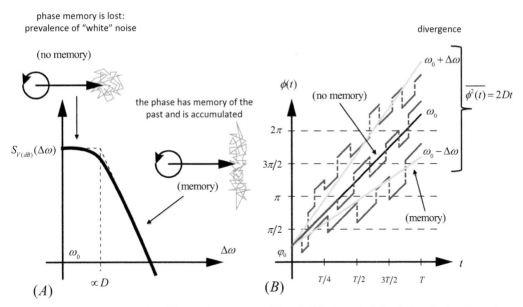

Figure 6.42 Total oscillator noise spectrum (A) and diffusion principle of phase in the phase–time diagram (B).

a diffusion process. However, this is an aspect of *one variable of the power*, not of the *total power*, which is measured by $S_{V(dB)}(\Delta f)$ whose PSD necessarily flattens out with a corner frequency since the total power should be limited.

As shown in Fig. 6.42, in an ensemble of collected waveforms where we measure the phase lag, we can see a spread around the ensemble average given by $\overline{\phi^2(t)} = 2Dt$ where D is the diffusion coefficient since it follows a random walk (Fig. 6.42B). D depends on the number of collisions, and thus it is related to dissipative characteristics (viscosity/resistance) of the system. Fig. 6.42 shows that the final phase destination sets the frequency shift $\Delta\omega$ with respect to the average frequency ω_0.

Therefore, the diffusion process sets slope boundaries of average size D where the phase's evolution follows a different behavior. The farther from the resonant frequency we are, the more constructive sum of the phase changes we get: the process keeps the memory of the variations, and they are integrated as in the random walk. The higher the deviation from the natural frequency, the clearer is the memory of the past. In other terms, the more we deviate from the original frequency, the more we can count the number of hits (memory) that caused such a deviation. This is why the power spectrum is of the $1/f^2$ form as in the diffusion process, such as the random walk. However, inside the boundaries, the memory of the past is lost since we cannot evaluate the number of perturbations that occurred, similarly to a completely uncorrelated white noise. This is the reason why in the proximity of $\Delta\omega = 0$ we have a uniform power spectrum. This is also clear from the phasor, where for small variations of frequency, we contribute to AM modulation represented by white noise.

Since the phase noise follows a diffusion process, it can be shown that the total power spectrum (oscillator + noise) has the Lorentzian form

$$S_V(\Delta\omega) = V_0^2 \frac{D}{D^2 + (\Delta\omega)^2}, \qquad (6.113)$$

where D is called the diffusion constant of the phase noise; see Chapter 4. Of course, the total integrated power has the value $V_0^2/2$, as discussed. It is difficult to estimate a priori the corner frequency, and usually, it is fit based on experimental data (together with other experimental evidence of $L(\Delta\omega)$); however, it completely defines the physical process.

If the oscillator is very much constrained to the limit cycle (very small AM modulation), then the flat part of the Lorentzian is reduced, and the corner frequency increases, but in any case, the total power should be $V_0^2/2$.

We can rewrite (6.113) in terms of $L(\Delta\omega)$ as

$$L(\Delta\omega) = \frac{S_V(\omega)}{V_0^2/2} = \frac{2D}{D^2 + (\Delta\omega)^2}, \qquad (6.114)$$

and in terms of regular frequency as

$$L_V(\Delta f) = \frac{\beta}{f_L^2 + \Delta f^2} \simeq \frac{\beta}{\Delta f^2} \equiv [V^2/Hz], \qquad (6.115)$$

where $L_V(\Delta f)$ is the double-sided power spectrum, β is a constant, and f_L is the cutoff frequency of the Lorentzian. The last approximated equivalence is given for $f \gg f_L$, giving a hint that L is an approximation of $S_\phi(\Delta f)$ away from f_0.

6.7.2   Characterization of Phase Noise from Total Noise from a Modulation Viewpoint

The measured PSD of a noisy oscillator considers both phase and amplitude noise even if most of the power corresponds to phase noise. A typical experimental phase noise spectrum is shown in Fig. 6.43A. Even if the plot is in a log scale of the ordinate variable as in Fig. 6.41D, there is a quick peak on the natural frequency due to the small number of measuring process points.

A useful way to characterize the experimental spectrum of an oscillator is to refer it to the power of the fundamental and express it in dB:

$$L_{dB}(\Delta f) = 10\log\left(\frac{S_V(\Delta f)}{P_{sig}}\right) \equiv [dBc/Hz], \qquad (6.116)$$

where P_{sig} is the power of the carrier and dBc are called *decibels below the carrier per hertz*.[18] However, the power of the carrier could be so high with respect to the noise shape as to exceed the dynamic range of the spectrum analyzer. Thus, it is preferred

---

[18] This definition could be misleading since the frequency is within the argument of the log; thus, doubling the bandwidth does not mean that the decibel value doubles.

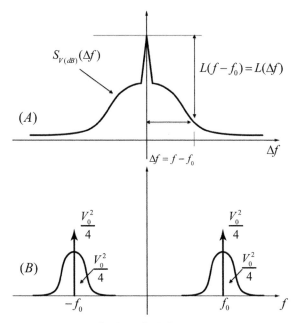

Figure 6.43 Determination of $L(\Delta f)$.

to use demodulation to observe the spectrum around the natural frequency of the oscillator.

We may consider that phase change in a period is *very small* so that we can express the signal as

$$V(t) = V_0 \sin(2\pi f_0 t + \phi(t)) \approx V_0 \sin(2\pi f_0 t) + V_0 \phi(t)\cos(2\pi f_0 t). \tag{6.117}$$

This is a representation of phase noise as additive amplitude noise whose unilateral power spectrum is

$$\widetilde{S}_V(f) = \frac{V_0^2}{2}[\delta(f - f_0)] + \frac{V_0^2}{2}[S_\phi(f - f_0)]. \tag{6.118}$$

This shows that the power is split between carrier power and the sidebands power of equal amount, as shown in Fig. 6.43B (in a double-sided spectrum). This abrupt split between the two components is not real but is due to the approximation of small phase deviation so that we can model the effect of modulation.

Now we can use a *carrier suppression technique* based on a quadrature demodulator to eliminate the first term. Therefore, *referring to the demodulated signal,* we have

$$S_V(\Delta f) \approx \frac{V_0^2}{2} S_\phi(\Delta f), \tag{6.119}$$

so that

$$L_{dB}(\Delta f) = L_{dB}(f - f_0) = 10\log\left(\frac{S_V(f)}{S_V(f_0)}\right) = 10\log\left(S_\phi(f - f_0)\right)$$

$$\to S_\phi(\Delta f) = 10^{\frac{L_{dB}(\Delta f)}{10}},$$

(6.120)

where $S_V(f_0) = V_0^2/2$. This relationship means that, for small phase changes, $L(f)$ is simply an approximation of $S_\phi(f)$.

Since historically $L(f)$ has been used as an "identity" of phase noise, in the case in which we measure phase noise directly with a phase detector, the relationship (6.116) could lead to incorrect results (e.g., values greater than 0 dBc). This happens whenever the small-angle perturbation assumption is violated. For this reason, in the case phase noise is measured directly, it is better to use (6.119) and (6.120) as a *definition* of $L(f)$.

6.7.3 Jitter and Its Estimation from Phase Noise

Jitter is the deviation of the periodicity in a harmonic signal as shown in Fig. 6.44 due to phase noise. The same phenomenon could be seen as period variations of a binary signal (e.g., a clock) whose state transitions correspond to zero crossing of the sinusoid from which it has been derived. Evaluating fluctuations of periods in a time-domain reference is very important in telecommunications and digital signal processing.

Looking at Fig. 6.44, we can set the edges of the square waveform when the sinusoidal signal $\sin(2\pi f_0 + \varphi(t))$ crosses the zero reference. Note that $\varphi(t)$ is time-variant since it is affected by noise, as illustrated in Fig. 6.37A. Therefore, the zero-crossing of a *sinusoidal* signal in 0 and N periods is

$$2\pi f_0 t_0 + \varphi(t_0) = 0$$
$$2\pi f_0 t_N + \varphi(t_N) = 2\pi N,$$

(6.121)

where t_0 is the time of the first crossing observed and t_N the crossing time after N periods. Subtracting one from the other, we have

$$\tau = t_N - t_0 = NT_0 + \Delta t$$
$$\Delta t = \frac{T_0}{2\pi}[\varphi(t_0) - \varphi(t_N)],$$

(6.122)

where Δt is the variation of the period (jitter) that we can estimate statistically.

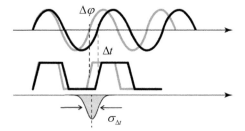

Figure 6.44 Relationship between phase and jitter.

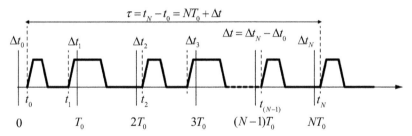

Figure 6.45 Determination of the jitter from phase noise.

Therefore

$$\overline{\Delta t^2} = \frac{T_0^2}{4\pi^2} \overline{\Delta \varphi^2}.$$ (6.123)

Looking at Fig. 6.45, we can see that the difference $\Delta t = \Delta t_N - \Delta t_0$ between the unperturbed clock and the perturbed one is taking into account all the period variations.

We can calculate a first form of the jitter, which is called *absolute jitter* or *long-term jitter* $\sigma_{abs}(\tau)$. Consider in a phase-noise-affected square wave a sample of total length τ composed of N periods where each single period variation (with respect to the average period) is Δt_n, as shown in Fig. 6.46. We take an ensemble of measurements, where the ith ensemble variation on N periods $\Delta t^{(i)}$ is

$$\Delta t^{(i)} = \sum_{n=0}^{N} \Delta t_n^{(i)}; \quad \tau \simeq NT_0.$$ (6.124)

Now we estimate the statistical properties of the ensemble in order to understand the average variation of the period in N cycles.

Following (6.123), we can first evaluate the statistical properties of the phase and then turn back to the period. Therefore

$$\overline{\varphi^2(\tau)} = E\left[(\varphi(t_N) - \varphi(t_0))^2\right] = \overline{\left(\varphi(t_N) - \varphi(t_0)\right)^2}$$
$$= \overline{\varphi^2(t_N)} - 2\overline{\varphi(t_N)\varphi(t_0)} + \overline{\varphi^2(t_0)} = 2\left(\overline{\varphi^2(t_0)} - \overline{\varphi(t_N)\varphi(t_0)}\right).$$ (6.125)

Since it is stationary[19], we have[20]

$$\overline{\varphi^2(t_N)} = \overline{\varphi^2(t_0)} = \int_0^\infty L_V(f)df = R_\varphi(0),$$ (6.126)

where $S_V(f)$ is a single-sided spectrum. It should be noted that $\overline{\varphi^2(t_N)}$ is a variance at t_N of an ensemble.

---

[19] It could be shown that for small phase perturbations, phase noise is wild sense stationary (WSS).
[20] In the following, for simplicity of notation we will use f in place of Δf.

For the autocorrelation, since $\tau = t_N - t_0$

$$\overline{\varphi(t_N)\varphi(t_0)} = R_\varphi(\tau) = 1/2 \cdot F^{-1}\{L_V(f)\} = 1/2 \int_{-\infty}^{\infty} L_V(f)e^{j2\pi f\tau}df$$

$$= \int_0^{\infty} L_V(f)\cos(2\pi f\tau)df,$$

(6.127)

where the ½ factor is required because $S_V(f)$ is a single-sided power spectrum whilst Fourier transform is defined on a double-sided (symmetric) spectrum.

Therefore[21]

$$\overline{\varphi^2(\tau)} = 2\int_0^{\infty} L_V(f)[1 - \cos(2\pi f\tau)]df = 4\int_0^{\infty} L_V(f)\sin^2(\pi f\tau)df \equiv [\text{rad}^2].$$

(6.128)

And referring back to time with (6.123),

$$\sigma^2_{abs(\Delta t)} = \overline{\Delta t^2} = \frac{T_0^2}{\pi^2}\int_0^{\infty} L_V(f)\sin^2(\pi f\tau)df \equiv [s^2],$$

(6.129)

where $\tau = t_N - t_0$. Equation (6.129) allows to calculate the absolute jitter on the basis of the phase noise spectrum, and it is dependent, of course, on the observation time τ.

Assuming the phase spectrum modeled by a Lorentzian as in (6.115) (but this is not always the case from the experimental point of view), we could have interesting conclusions expressed in closed forms.

Assuming $S_V(f)$ as Lorentzian, we have

$$\overline{\varphi^2(t_0)} = \int_{-\infty}^{\infty} L_V(f)df = \frac{\beta\pi}{f_L}$$

(6.130)

and

$$\overline{\varphi(t_N)\varphi(t_0)} = R_\varphi(\tau) = \frac{\beta\pi}{f_L}e^{-2\pi f_L\tau} \approx \frac{\beta\pi}{f_L}(1 - 2\pi f_L\tau)$$

(6.131)

because, as seen before, the Fourier transform of a Lorentzian spectrum is an exponentially decaying function.

Therefore

$$\overline{\varphi^2(\tau)} = 2\left(\overline{\varphi^2(t_0)} - \overline{\varphi(t_N)\varphi(t_0)}\right) = 4\beta\pi^2\tau$$

$$\rightarrow \sigma_{abs(\varphi)} = \sqrt{4\beta\pi^2\tau} \equiv [\text{rad}],$$

(6.132)

[21] $1 - \cos(2\pi f\tau) = 2\sin^2(\pi f\tau)$.

and referring back to time

$$\sigma_{abs(\Delta t)} = \frac{T_0}{2\pi} \sigma_{abs(\varphi)} = \frac{\sqrt{4\beta\pi^2\tau}}{2\pi f_0} = \frac{\sqrt{\beta}}{f_0}\sqrt{\tau} = \alpha\sqrt{\tau} \equiv [\text{s}], \quad (6.133)$$

where α is a constant. The above is a simple and important relationship, showing that the jitter is increased as the square root of the period because the mean square error of each cycle is independent, and the contributions sum up. This should not surprise since it is a diffusion process as a *random walk*. This jitter expression is useful whenever we use a clock for measuring time-dependent events in sensing systems.

There is another expression of the jitter, which is called *rms jitter* or *cycle jitter*:

$$\sigma_{rms} = \lim_{N \to \infty} \sqrt{\frac{1}{N} \sum_{n=1}^{N} \Delta t_n^2}. \quad (6.134)$$

Thanks to the previous discussion, we can easily find it as

$$\sigma_{rms} = \sigma_{abs(\Delta t)}(\tau = T_0) = \frac{\sqrt{\beta}}{f_0\sqrt{f_0}} \equiv [\text{s}]. \quad (6.135)$$

Therefore, from (6.133) and (6.135), we have

$$\sigma_{abs(\Delta t)} = \sigma_{rms}\sqrt{f_0}\sqrt{\tau}. \quad (6.136)$$

Now it is useful to summarize some important hints:

- The expressions (6.133) and (6.135) are calculated under the assumption of the Lorentzian model of phase noise.
- In case the spectrum is not Lorentzian (the experimental phase noise is widely variable on the basis of many other physical effects), we should use (6.129) for the absolute jitter and for the rms jitter

$$\sigma_{rms}^2 = \frac{T_0^2}{\pi^2} \int_0^\infty L_V(f)\sin^2(\pi f\tau)df \equiv [\text{s}^2], \quad (6.137)$$

where $\tau \approx NT_0$ or better, for cycle-to-cycle jitter $\tau \approx T_0$. The modulation of the $L_V(f)$ given by $\sin^2(\pi f\tau)$, with $\sin^2(\pi f\tau) \approx \pi^2 f^2/f_0^2$ a high-pass filter up to the frequency $f \le f_0/2$ and should be carefully considered.

- Frequently, for the rms jitter, the following expression is considered

$$\sigma_{rms}^2 \simeq \frac{T_0^2}{(2\pi)^2} \overline{\varphi^2(T_0)} = \frac{T_0^2}{(2\pi)^2} \cdot \int_0^\infty L_V(f)df = \frac{T_0^2}{4\pi^2} \int_0^\infty L_V(f)df; \quad (6.138)$$

however, it does not take into account the $\sin^2(\pi f \tau)$ modulation. This means that we disregard the cross-correlation (6.127), and this could sometimes give large overestimations of the jitter.

Further Reading

Baghdady, E. J., Lincoln, R. N., and Nelin, B. D., Short-term frequency stability: Characterization, theory, and measurement. *Proc. IEEE*, vol. 53, pp. 704–722, 1965.

Brillouin, L., *Science and Information Theory*, 2nd ed. Mineola, NY: Dover Publications, 1962.

Cutler, L. S. and Searle, C. L., Some aspects of the theory and measurement of frequency fluctuations in frequency standards. *Proc. IEEE*, vol. 54, pp. 136–154, 1966.

Drakhlis, B., Calculate oscillator jitter by using phase-noise analysis, *Microwaves RF*, vol. 40, no. 2, pp. 109–119, 2001.

Feynman, R. P., Robert, B. L., Sands, M., and Gottlieb, M. A., *The Feynman Lectures on Physics*. Reading, MA: Pearson/Addison-Wesley, 1963.

Gabrielson, T. B., Mechanical-thermal noise in micromachined acoustic and vibration sensors. *IEEE Trans. Electron Devices*, vol. 40, pp. 903–909, 1993.

Hajimiri, A., and Lee, T. H., A general theory of phase noise in electrical oscillators. *IEEE J. Solid-State Circuits*, vol. 33, pp. 179–194, 1998.

Hajimiri, A., and Lee, T. H., *The Design of Low Noise Oscillators*. Norwell, MA: Kluwer Academic, 2003.

Hajimiri, A., Limotyrakis, S., and Lee, T. H., Jitter and phase noise in ring oscillators. *IEEE J. Solid-State Circuits*, vol. 34, pp. 790–804, 1999.

Ham, D., and Hajimiri, A.,Virtual damping and Einstein relation in oscillators, *IEEE J. Solid-State Circuits*, vol. 38, no. 3, pp. 407–418, Mar. 2003.

Herzel, F., and Razavi, B., A study of oscillator jitter due to supply and substrate noise. *IEEE Trans. Circuits Syst. II: Analog Digital Signal Process.*, vol. 46, pp. 56–62, 1999.

Johnson, J. B., Thermal agitation of electricity in conductors. *Nature*, 119, p. 50, 1927.

Johnson, J. B., Thermal agitation of electricity in conductors. *Phys. Rev.*, 541, pp. 97–129, 1927.

Keshner, M. S., 1/F Noise. *Proc. IEEE*, vol. 70, pp. 212–218, 1982.

Lee, T. H., and Hajimiri, A., Oscillator phase noise: A tutorial. *IEEE J. Solid-State Circuits*, vol. **35**, 2000.

Leeson, D. B., A simple model of feedback oscillator noise spectrum. *Proc. IEEE*, vol. 54, pp. 329–330, 1966.

MacDonald, D. K. C., *Noise and Fluctuations*. New York: John Wiley & Sons, 1962.

McNeill, J. A., Jitter in ring oscillators, 1994.

Nyquist, H., Thermal agitation of electric charge in conductors. *Phys.Rev.*, 32, pp. 110–113, 1928.

Phillips, J., and Kundert, K., An introduction to cyclostationary noise. *Cust. Integr. Circuits Conf.* pp. 1–43, 2000.

Razavi, B., A study of phase noise in CMOS oscillators. *IEEE J. Solid-State Circuits*, vol. 31, pp. 331–343, 1996.

Rice, S. O., Mathematical analysis of random noise. *Bell System Tech. J.* 23, pp. 282–332, 1944.

Sarpeshkar, R., Delbruck, T., and Mead, C. A., White noise in MOS transistors and resistors. *IEEE Circuits Devices Mag.*, vol. 9, pp. 23–29, 1993.

7 Noise in Electronic Devices and Circuits

This chapter will treat two essential steps in electronic sensor design. The first is the passage from functional blocks to lumped model electronic circuits. In this approach, the noise will no longer be associated with behavioral blocks but with circuit topology and electronic device elements. The second step is to analyze the effects of the readout mode on noise, emphasizing the differences between continuous and discrete-time approaches. Finally, we will discuss some tradeoffs related to bandwidth and resolution in acquisition chains following the relationships of Chapter 3.

7.1 Thermal Noise Limited Signal-to-Noise Ratio and Bandwidth

Most electronic systems working at ambient conditions are thermal noise limited in resolution. A starting point could be to calculate the thermal noise given by a resistor with respect to its resistance values and compare it with a 1 V reference to have a comparison table. The signal-to-noise ratio (SNR) is therefore given by

$$\text{SNR} = 20\log\left[\frac{1\text{ V}}{\sqrt{4kTRB}}\right], \tag{7.1}$$

where B is the bandwidth of the system. Then, we will convert the above value to the effective number of bits (ENOB) to compare with the characteristics of A/D converters.

A summary plot of this behavior is shown in Fig. 7.1.

The plot is very illustrative of the constraints in which sensor design could operate. In general, sensor systems have a relatively low frequency in signal bandwidth, and we have chosen a range from 10 Hz to 10 MHz. Moreover, the input equivalent resistance rarely is below tens of ohms.

Therefore, if we need to design an interface in such boundaries, the dynamic range does not exceed 150 dB or 24 effective number of bits. There are, especially in DC measurements, techniques that cover higher dynamic ranges but are based on dividing the dynamic range (DR) into subdomains on which the interface is adapted according to the signal strength. See Sections 2.11.2 and 2.13.1 of Chapter 2 for a discussion.

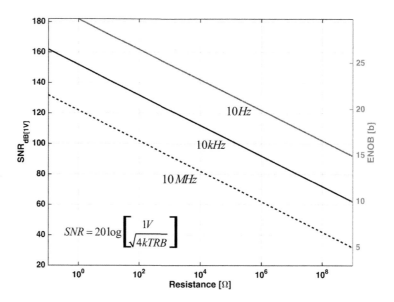

Figure 7.1 Graphical view of the signal-to-noise ratio given by a variable value resistance with respect to the reference of 1 V. The corresponding effective number of bits of an A/D converter are shown on the right.

7.2   Pink and White Noise Combination

We already introduced the classifications in "colors" of noises according to their power spectral density shape. In most electronic applications, we have to deal with white (thermal and shot) combinations with pink (flicker) noises where a first-order low-pass behavior defines the system bandwidth.

Figure 7.2A shows the shape of the power spectral density (PSD) in a noisy system characterized by white and pink noise and whose bandwidth is limited by the cutoff frequency f_0 of a first-order low-pass filter. Note that $1/f$ noise follows an asymptotic line with a 1:1 slope (i.e., one decade of the ordinate per one decade of the abscissa) while a straight line defines the white noise. The intercept point between the two asymptotic lines is referred to as *noise corner frequency* f_C, and it is the frequency where the pink and white noises contribute equally to the total noise. There is a typical dominant single-pole cutoff frequency of the system f_0 at the far side of the graph. Since we are dealing with a power plot, the slope of the graph beyond the cutoff frequency is characterized by a 2:1 slope. If we want to visualize the noise power geometrically in a specific bandwidth, we have to focus on the lin–lin plot of the PSD function shown in Fig. 7.2B (and not in the log-log plot). Note how this plot is very different from the log-log one, and we have to zoom in the very beginning part of the graph to visualize the hyperbolic behavior of the function. Therefore, the corner and cutoff frequencies are far from the enlarged linear plot. If we want to include them in the plot, the pink noise part of the plot would be squeezed as a very sharp pulse on the

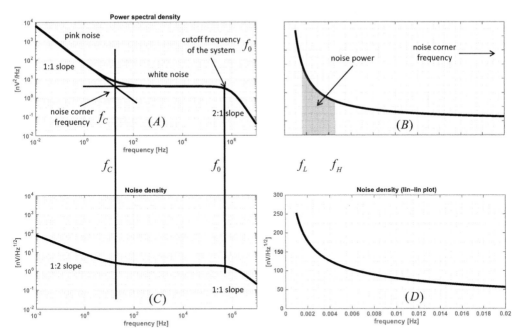

Figure 7.2 Combination of white and pink noise. (A) Power spectral density. (B) Power spectral density in lin–lin plot. (C) Noise density. (B) Noise density in lin–lin plot. [Example extrapolated from OPA209 by Texas Instrument: $k_W = 2.2\text{nV}/\sqrt{\text{Hz}}$; $k_P = 8\text{nV}/\sqrt{\text{Hz}}$]

left. This observation emphasizes the log-log plot's strength with respect to other representations.

As introduced in Chapter 4, we must be careful to calculate the rms value from a noise density graph. We have to first go into the power plot domain and then back into the noise density plot. We have two typical cases. If the spectral density is *uniform*, we can calculate the rms value as

$$W(f) = k_W^2; \quad \overline{v_N^2} = k_W^2 \cdot B$$
$$\rightarrow v_{Nrms} = k_W \cdot \sqrt{f_H - f_L}; \quad k_W \equiv [\text{V}/\sqrt{\text{Hz}}], \tag{7.2}$$

where $W(f)$ is the white noise PSD, k_W^2 is the white noise constant, and $B = f_H - f_L$ is the bandwidth under which we want to calculate the rms value of the random process.

If the power spectrum has a typical "pink" noise spectrum, we can calculate the rms value as

$$P(f) = k_P^2/f; \quad \overline{v_N^2} = k_P^2 \int_{f_L}^{f_H} \frac{1}{f}\, df = k_P^2 \ln\left(\frac{f_H}{f_L}\right) \tag{7.3}$$
$$\rightarrow v_{Nrms} = k_P \cdot \sqrt{\ln\frac{f_H}{f_L}}; \quad k_P \equiv [\text{V}/\sqrt{\text{Hz}}],$$

where k_P^2 is the power value of the pink noise at 1 Hz. Note that the amount of power noise is the same for any decade.

Example Using the data from Fig. 7.2, we have $k_W = 2.2\text{nV}/\sqrt{\text{Hz}}$ so in the bandwidth between 100 Hz and 1 kHz the total rms noise value is $v_{Nrms} = 2.2 \times 10^{-9} \cdot \sqrt{1 \times 10^3 - 100} = 66$ nV and between 1 kHz and 10 kHz the total rms noise value is $v_{Nrms} = 2.2 \times 10^{-9} \cdot \sqrt{10 \times 10^3 - 1 \times 10^3} = 208$ nV.

Example Using the data from Fig. 7.2 again, we have $k_P = 8\text{nV}/\sqrt{\text{Hz}}$. If we want to calculate the noise power in the bandwidth between 0.1 Hz and 1 Hz (pink zone), we have: $v_{Nrms} = 8.0 \times 10^{-9} \cdot \sqrt{\ln(1/0.1)} = 12$ nV. The same rms noise is between 0.01 Hz and 0.1 Hz.

> **Hint** The log-log plot could be misleading for a graphical estimation of the total noise in a bandwidth. Even if the noise value on the pink side is higher than that of the white part, the total noise per decade could be lower than that calculated on some decade on the white side. This is because the true area, which is representable only on the lin–lin plot, should be exponentially expanded as we move on the right side of the abscissa.

If both pink and white noises are present, we have a total PSD noise given by

$$N(f) = W(f) + P(f) = k_W^2 + \frac{k_P^2}{f} = k_W^2\left(1 + \frac{f_C}{f}\right) \text{ with } f_C = \frac{k_P^2}{k_W^2}, \quad (7.4)$$

where f_C is called *corner frequency*. Note how the corner frequency is the frequency where the PSD value is twice (3 dB) above the asymptotic value of the white noise. The big difference between the linear plot of Fig. 7.2A with respect to the log-log of Fig. 7.2B should also be pointed out. The graphical representation of the power of noise characterized as the area below the PSD curve could be done only in the linear plot. Conversely, the logarithmic plot easily shows the corner frequency as the intercept between the pink and white noise asymptotes barely visible in the linear one.

To calculate the total power, we have to cope with two problems: first, we cannot integrate power for $f \to 0$ because the integral does not converge, and second, we have to deal with a relationship between f_C and f_0. For the first problem, we have already discussed the issue in Chapter 6 and the implications of assuming a limited low-bound frequency f_L. For the second, we will assume $f_0 \gg f_C$. Therefore, as shown in (7.5), we integrate the pink noise from f_L to f_0 while the white noise over the whole spectrum

$$\overline{v^2} = \int_{f_L}^{\infty} k_W^2\left(1 + \frac{f_C}{f}\right)|H|^2 df \cong \int_0^{\infty} k_W^2|H|^2 df + \int_{f_L}^{\infty} k_W^2 \frac{f_C}{f} df$$

$$\cong \int_0^{\infty} k_W^2|H|^2 df + \int_{f_L}^{f_0} k_W^2 \frac{f_C}{f} df = k_W^2\left(\frac{\pi}{2}f_0 + f_C\ln\frac{f_0}{f_L}\right). \quad (7.5)$$

In other terms, the effect of pink noise is such that we have an increase of the noise bandwidth from the typical $(\pi/2)f_0$ value for a first-order system to

$$\text{NBW} \approx \frac{\pi}{2} f_0 + f_C \ln \frac{f_0}{f_L} \tag{7.6}$$

in the presence of pink noise.

Therefore, the pink noise percentage over total noise is given by

$$\frac{P_P}{P_{W+P}} = \frac{\beta}{1+\beta} \cdot 100 \equiv [\%] \ \ where \ \beta = \frac{f_C}{\pi/2 \cdot f_0} \ln \frac{f_0}{f_L}. \tag{7.7}$$

This is a good point to show the limited effect of the overall computation choice. As seen in the discussion about the physical derivation of the flicker noise, f_L it could be linked to the reciprocal of the system observation or lifetime, avoiding the integral divergence problem. However, a simple inspection of (7.7) shows that the power increment for low frequencies is a weak function of f_L.

Example Let assume that the cutoff frequency of the system is 100 kHz, f_C equal to 500 Hz, and f_L given by the maximum observation time of our sensor. The estimated pink noise weight over the total noise increases only from 7.7% to 9.6%, considering an observation time incremented from one month to one century.

As a final note, we can observe how the noise power could be considered in log-log plots, as shown in Fig. 7.3. Again, the noise power is *not* proportional to the area subtended by the log–log curve.

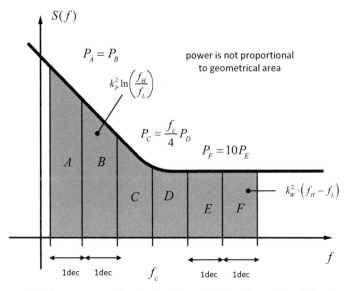

Figure 7.3 Noise power per decade in pink and white noise regions in log-log plots.

Therefore, we can see that the power per decade in the pink region is the same, while the power per decade in the white region is 10 times higher for adjacent increasing decades. Furthermore, it is easy to see that the ratio between white and pink noise power per decade across corner frequency could be higher or lower than 1 depending on the corner frequency value.

7.3   Calculation of Total Noise in Linear Circuits

The noise models discussed in Chapter 5 at the device level allow calculation of the total noise in linear circuits by means of conventional Kickhoff's laws (KL) with power superposition techniques following the steps:

1. Substitute noisy devices with noise models for each component.
2. Derive total circuit noise assuming *uncorrelated* noise sources by applying *noise power superposition*.

The second proposition is supported by the fact that noises are made by completely independent physical processes associated with different components. At the base of this approach is that noise models are ergodic. This means that *the model of a device is the same for any other physical device having the same characteristics*.

To better understand the procedure, let us use two simple examples. The first one is illustrated in Fig. 7.4, where two noisy resistors are series-connected.

To calculate the open-circuit noise of the two noisy resistors in Fig. 7.4A, we substitute them with their noise models as illustrated in Fig. 7.4B. For simplicity, we will assume that the noise is observed in a given portion of the spectrum equal to Δf:

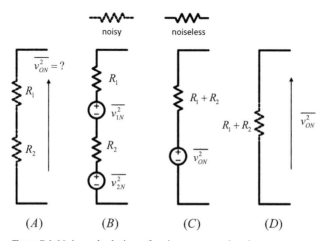

Figure 7.4 Noise calculation of series-connected resistors.

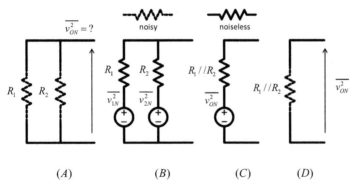

$v_{ON}^2 = ?$ noisy noiseless

R_1 R_2 $R_1 // R_2$

$\overline{v_{1N}^2}$ $\overline{v_{2N}^2}$ $\overline{v_{ON}^2}$ $R_1 // R_2$ $\overline{v_{ON}^2}$

R_1 R_2

(A) (B) (C) (D)

Figure 7.5 Noise calculation of parallel-connected resistors.

$$\overline{v_{ON}^2}(f)\Delta f = \overline{v_{1N}^2}(f)\Delta f + \overline{v_{1N}^2}(f)\Delta f$$
$$= 4kTR_1\Delta f + 4kTR_2\Delta f = 4kT(R_1 + R_2)\Delta f, \tag{7.8}$$

where $\overline{v_{1N}^2}(f)$, $\overline{v_{2N}^2}(f)$, and $\overline{v_{ON}^2}(f)$ are the spectral power densities of the two resistors and of the output, respectively. Note how the result is that the output noise of the two resistors is equivalent to the noise of a unique resistor given by the sum of the two resistance values, according to conventional linear circuit calculation rules as shown in Fig. 7.4C and Fig. 7.4D.

To further illustrate this technique, let us take a second example given by a parallel of two resistors as illustrated in Fig. 7.5, where, to be more explicit, we assume only white noise.

Thus, the PSD of each resistor is given by $\overline{v_N^2}(f) = 4kTR$ and the noise power in Δf by $\overline{v_N^2} = 4kTR_1\Delta f$. Therefore, using the superposition of noise power for linear circuits, we have that

$$\overline{v_{ON}^2} = \overline{v_{1N}^2}(f)\left(\frac{R_2}{R_1 + R_2}\right)^2 \Delta f + \overline{v_{2N}^2}(f)\left(\frac{R_1}{R_1 + R_2}\right)^2 \Delta f$$
$$= 4kTR_1\left(\frac{R_2}{R_1 + R_2}\right)^2 \Delta f + 4kTR_2\left(\frac{R_1}{R_1 + R_2}\right)^2 \Delta f \tag{7.9}$$
$$= 4kT\left(\frac{R_1 R_2}{R_1 + R_2}\right)\Delta f = 4kT(R_1 // R_2)\Delta f,$$

showing that the total noise power of resistors connected in parallel is equivalent to that generated by a single resistor whose value is given by the equivalent one. The same rules could be applied to any *complex linear* circuit once the noise model of any device has been identified.

With the generalization of Nyquist's formula shown in Chapter 5, we have seen that capacitors and inductors do not contribute to noise. For this reason, for a generic impedance Z, we have that the open circuit and short circuit PSDs are

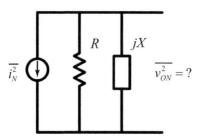

Figure 7.6 Complex impedances noise example.

$$\overline{v_N^2}(f) = 4kT \, \text{Re}\{Z\}$$

$$\overline{i_N^2}(f) = 4kT \, \text{Re}\{Y\}. \tag{7.10}$$

As an example, let us take the scheme illustrated in Fig. 7.6, where a noisy resistance is placed in parallel to a pure imaginary impedance jX (such as a capacitor or an inductor).

Therefore, the impedance seen by the noise generator is

$$Z = \frac{jRX}{R + jX} = \frac{RX^2 + jR^2X}{R^2 + X^2}. \tag{7.11}$$

Since the current power spectral density due to the resistance is $\overline{i^2}_N(f) = 4kT/R$, the output voltage noise is given by the current power dropped onto the parallel of the two impedances:

$$\overline{v_{ON}^2}(f) = \frac{4kT}{R}|Z|^2 = \frac{4kT}{R}\frac{R^2X^2}{R^2 + X^2} = 4kT\frac{RX^2}{R^2 + X^2} = 4kT \, \text{Re}[Z], \tag{7.12}$$

which is a confirmation of (7.10).

7.4   Input-Referred Noise in Circuits

Whenever we deal with a noisy electronic system, it would be convenient (i.e., for comparing noise with signals) to reduce the degree of complexity by moving the causes of noise outside the system, either at the input and/or at the output. In practice, following generalized Thevenin's and Norton's theorems, we can *model* any linear and time-invariant two-port system containing internal generators with a new one where internal generators are replaced by external sources inserted at the input and/or at the output of the system.

Specifically, the concept mentioned above could be applied to linear two-port systems such as those described in Fig. 7.7. A first example is shown in Fig. 7.7B using Thevenin's generalized theorem, where all internal sources are replaced by two voltage generators v_{N1}, v_{N2} placed in series to the input and output. Following

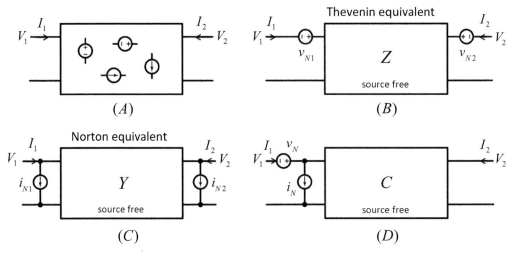

Figure 7.7 Impedance representation of a two-port with internal sources.

Thevenin's approach, the two generators implement the noise seen at the two ports in open circuit conditions (infinite impedance boundary conditions: $I_1 = I_2 = 0$).

Therefore, by using the impedance representation (Z-matrix), the two-port input-output relationship becomes

$$\begin{bmatrix} V_1 \\ V_2 \end{bmatrix} = \begin{bmatrix} Z_{11} & Z_{12} \\ Z_{21} & Z_{22} \end{bmatrix} \begin{bmatrix} I_1 \\ I_2 \end{bmatrix} + \begin{bmatrix} v_{N1} \\ v_{N2} \end{bmatrix}, \tag{7.13}$$

where v_{N1} and v_{N2} are also called *open-circuit equivalent generators*.

In a similar fashion and using Norton's equivalent circuit approach, the same system containing generators could be represented by the one illustrated in Fig. 7.7C, where i_{N1} and i_{N2} are called *short-circuit equivalent current generators*. As before, those sources could be determined as the currents measured by short-circuiting the input and output ports $V_1 = V_2 = 0$.

By using the admittance representation (Y-matrix), the relationship in this case becomes

$$\begin{bmatrix} I_1 \\ I_2 \end{bmatrix} = \begin{bmatrix} Y_{11} & Y_{12} \\ Y_{21} & Y_{22} \end{bmatrix} \begin{bmatrix} V_1 \\ V_2 \end{bmatrix} + \begin{bmatrix} i_{1N} \\ i_{2N} \end{bmatrix}. \tag{7.14}$$

Finally, another important representation could be obtained by using the chain matrix (C-matrix)

$$\begin{bmatrix} V_1 \\ I_1 \end{bmatrix} = \begin{bmatrix} A & B \\ C & D \end{bmatrix} \begin{bmatrix} I_2 \\ V_2 \end{bmatrix} + \begin{bmatrix} v_N \\ i_N \end{bmatrix}, \tag{7.15}$$

where v_N and i_N are *input equivalent open-circuit and short-circuit generators*.

These representations have common characteristics derived from the *linear* properties of the system with two degrees of freedom. More specifically, it could be shown that (except pathological cases) to fully model a noisy two-port linear system with a noiseless system:

- No *fewer than two* external generators are needed.
- For a given configuration, the two generators are *uniquely* determined by the characteristics and the disposition of individual internal generators.
- The models of Fig. 7.7B–D are *exactly equivalent* to the original one shown in Fig. 7.7A. Thus, the values of the two generators are *independent* of any input or output loads that constitute boundary conditions.

Furthermore, the input and output noise sources of Fig. 7.7B–D can be mapped from one model to another. For example, it is easy to show that

$$v_N = -\frac{i_{2N}}{Y_{21}} = v_{1N} - v_{2N}\frac{Z_{11}}{Z_{21}}$$
$$i_N = i_{1N} - i_{2N}\frac{Y_{11}}{Y_{21}} = -v_{2N}Z_{21}. \tag{7.16}$$

Unfortunately, the models of Fig. 7.7B, C do not hold if the input/output loads have asymptotic values. For example, v_{N1} and i_{N1} becomes ineffective in the models of Fig. 7.7B and Fig. 7.7C if the impedance of the source that will be connected has infinite or zero values (i.e., in open circuit or short circuit boundary conditions) respectively. Since these conditions are widespread in electronic design, it is safe to refer to a model such that of Fig. 7.7D that is valid for *any boundary input impedance conditions*.

Since the external sources are independent of input/output impedance, we can use a simple approach to calculate voltage and current sources independently by imposing opposite and asymptotic impedance conditions ($Z_i = \infty$ and $Z_i = 0$) as illustrated in Fig. 7.8. Note that we have used the open-circuit output boundary condition in this case. The procedure will be more apparent when applied to some examples in the following sections.

The relationships discussed so far are related to mapping generic current or voltage generators in linear systems. The results could be applied to noise sources, but we have to keep in mind that in this case, the generators are represented by *random variables*, and therefore they could have some *degrees of correlation*. The following steps can be used when dealing with noise sources to be referred to the boundaries of the system:

- We assume all internal noise sources uncorrelated to each other (for example, assuming they are deriving from different devices or independent physical processes).
- Any internal kth noise power source is linked to the input/output referred sources by means of $|H_k|^2$, where H_k is the single source mapping transfer function.
- The superposition of noise power will be used to calculate the total input/output referred noises.

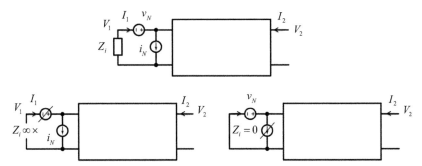

Figure 7.8 Procedure to calculate input-referred voltage and current noises.

- We check the degree of correlation between input/output referred noise source sources.

More specifically, we could determine any external input noise source by linking any internal source by means of the reciprocal of its mapping transfer function $|H_k|^2$ and eventually summing all of them. From the mathematical point of view, if we refer to a bandwidth Δf, the noise power contributions to input sources are

$$\overline{v_{iN}^2} = \sum_k \frac{S_{kN}(f)\Delta f}{|H_k(f)|^2}\bigg|_{Z_i=0} \quad \text{and} \quad \overline{i^2_{iN}} = \sum_k \frac{S_{kN}(f)\Delta f}{|H_k(f)|^2}\bigg|_{Z_i=\infty}, \tag{7.17}$$

where $S_{kN}(f)$ represents the power spectral densities either in the square of voltage or in the square of current: $\overline{v^2_{iN}}(f)$ and $\overline{i^2_{iN}}(f)$. Similarly, the output sources are mapped from each internal source by using its mapping transfer function $|H_k|^2$:

$$\overline{v^2_{oN}} = \sum_k S_{kN}(f)|H_k(f)|^2\Delta f. \tag{7.18}$$

Note that we have not included any output current source because usually, the output is taken in open circuit conditions, such as by high-impedance A/D converters.

An important point is how external sources are linked together. More specifically, even if all internal noise sources are independent, the same source could be mapped onto both of them. Thus, the two generic external sources could be partially correlated. If the two equivalent noises are those described in Fig. 7.7B (as an example), the correlation can be written by means of the correlation coefficient

$$\gamma = \frac{\overline{v_{1N}v_{2N}}}{\sqrt{\overline{v^2_{1N}}\,\overline{v^2_{2N}}}} = \frac{S_{12}(f)\Delta f}{\sqrt{S_1(f)\Delta f \cdot S_2(f)\Delta f}}, \tag{7.19}$$

where $S_{12}(f)$ is the cross-spectrum between the two voltage noises. As already discussed in Chapter 4, even if the two noise generators are real functions, their cross-correlation could be complex, where $Re\{\gamma\}$ takes into account in-phase correlation while $Im\{\gamma\}$ takes into account in-quadrature correlation between noises.

To summarize for stochastic power sources such as noise:

- Any two-port linear system is completely defined in terms of noise by four real numbers: two defining external noise generators and the other two (one complex number) taking into account the complex correlation between them.

When are the two generators completely uncorrelated? Assume that an internal noise source is mapped to both external generators, whatever the model of Fig. 7.7B–D is. In this case, there is a correlation between external noise sources because any change in the internal source is mapped into both external generators. Conversely, if a source is mapped only into one external generator, its contribution will not be correlated between the two sources. For instance, if an internal resistance is connected in series to the input or to the output of the model of Fig. 7.7B, its noise will be mapped only to one of the external sources with no contribution to the other. The next example on the bipolar junction transistor (BJT) noise model will enlighten this issue.

Now, once we have found the external noise generators, we can further simplify the noise model as illustrated in the steps of Fig. 7.9 due to the choice of the "boundary conditions" (polarization of input/output ports) of the linear system. In Fig. 7.9A, are represented the sensor equivalent impedance and conductance Z_S and Y_S, the input-referred voltage and current noise power of the interface $\overline{v_N^2}$, $\overline{i_N^2}$ and finally, the noise of the source (not generated by the interface) $\overline{v_{sN}^2}$. If we choose to use the C-matrix model and Thevenin's input source model, we can derive the structure of Fig. 7.9C where $\overline{v_{oN}^2}$ is called *output-referred voltage noise*. Alternatively, we can easily derive an identical model of Fig. 7.9D where $\overline{v_{iN}^2}$ is called *input-referred noise* voltage. Note how the three models are completely equivalent to each other.

We can now turn back to a behavioral representation as in Fig. 7.10. This is important to generate a link between the circuital model and other forms of representation. In Fig. 7.10A is shown, together with the signal V_i, the input-referred noise $(\overline{v_{iN}^2})$ and the source noise $(\overline{v_{sN}^2})$ as in Fig. 7.9D. Equivalent behavioral representations are shown in Fig. 7.10B and Fig. 7.10C, where the input-referred and the output-referred noises are represented, respectively. Of course, a very similar model could be derived referring to currents instead of voltages in Norton's theorem fashion.

As a final remark, note that the input/output referred noise generally depends on the gain (which affects the mapping transfer function of noise sources). However, we will see in next Section 7.5 that the Friis formula shows that the most effective noise sources are those physically placed at the beginning of the amplifying chain, and thus, the input-referred noise is weakly dependent on the gain. This is why the equivalent SNR of the system (that refers only to the noise of the interface and not to the noise associated with the signal) is almost constant with respect to gain changes.

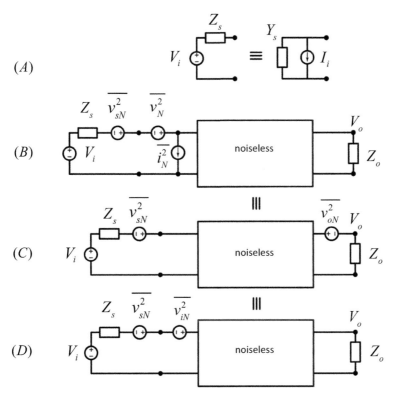

Figure 7.9 Equivalent representations for fixed input and output impedance.

7.5 Noise Factor and Optimal Noise Performance

The situations of Fig. 7.9 show an input signal V_i coming from the sensor and the representation of input-referred and output-referred noise. However, due to the presence of Z_S or Y_S, the sensor itself is noisy, and it is important to understand the relationship between the sensor noise and the amplifier noise at zero input conditions. A parameter to account for sensor and amplifier noise is the *noise factor F* defined as

$$F \triangleq \frac{\text{output noise considering internal system noise sources}}{\text{output noise considering a noiseless system model}} \in [1, \infty], \qquad (7.20)$$

which is the ratio of the output noise of the system comprising the sensor and the interface contribution with the output noise of the same system assuming a noiseless interface. Since the output noise with a noiseless interface is equal to or smaller than the output noise with a real amplifier, the values of F range from 1 to infinity with increasing values due to increasing noise conditions. The noise factor expressed in dB is called *noise figure*.

The definition of noise factor expressed by (7.20) could also be referred to the input

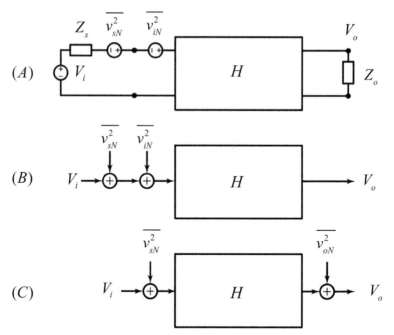

Figure 7.10 Circuital representation Fig. 7.9B where the noise of the source is evidenced (A). Behavioral representation input-referred (B) and output-referred (C) noise.

$$F \triangleq \frac{\overline{v_{oN}^2} + \overline{v_{sN}^2}|H|^2}{\overline{v_{sN}^2}|H|^2} = \frac{\overline{v_{iN}^2} + \overline{v_{sN}^2}}{\overline{v_{sN}^2}} = \frac{\text{total effective input noise}}{\text{input noise due to sensor only}}. \quad (7.21)$$

Furthermore, referring to Fig. 7.10C and defining P_i and P_o as the input and output signal power, and P_{oN} the total output noise power, we have:

$$F \triangleq \frac{\overline{v_{oN}^2} + \overline{v^2}_{sN}|H|^2}{\overline{v_{sN}^2}|H|^2} = \frac{P_{oN}}{\overline{v_{sN}^2}|H|^2} \frac{P_i}{P_i} = \frac{P_{oN}}{P_o} \frac{P_i}{\overline{v_{sN}^2}} = \frac{\text{SNR}_i}{\text{SNR}_o}. \quad (7.22)$$

Furthermore, recalling the relationships between SNRs of Chapter 2, we have

$$\text{SNR}_{i(EQ)} = \text{SNR}_o = \frac{P_o}{P_{oN}} = \frac{\text{SNR}_i}{F}. \quad (7.23)$$

We can now merge the noise of the source with the noise of the interface, as shown in Fig. 7.11, to understand the optimal noise performance of the overall system versus the source impedance. Now, assuming uncorrelated noises, zero-input conditions, and Thevenin's representation of the sensor, the noise of the sensor system is represented as in Fig. 7.11, were v_N and i_N are the voltage and current input-referred interface noise generators. Since the configurations of Fig. 7.11A and Fig. 7.11B are equivalent, we can derive the total input-referred noise as

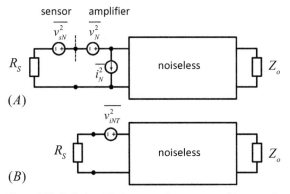

Figure 7.11 Relationship between the sensor and input-referred amplifier noises.

$$\overline{v_{iNT}^2} = \left(\overline{v_N^2} + \overline{i_N^2}R_s^2\right) + \overline{v_{sN}^2} = \overline{v_N^2} + \overline{i_N^2}R_s^2 + 4kTR_s\Delta f. \tag{7.24}$$

The source impedance effect on noise is shown in Fig. 7.12, where the noise components are shown in rms values.

If we calculate the noise factor, we get from (7.21)

$$F = 1 + \frac{1}{4kT\Delta f}\left(\frac{\overline{v_N^2}}{R_s} + \overline{i_N^2}R_s\right). \tag{7.25}$$

Differentiating (7.25) with respect R_S, we get the value of the source resistance that minimizes noise factor as

$$R_{opt} = \sqrt{\frac{\overline{v_N^2}}{\overline{i_N^2}}}. \tag{7.26}$$

On the other hand, if we calculate the signal-to-noise ratio, we have

$$\text{SNR}_i = \frac{V_i^2}{\overline{v_N^2} + \overline{i_N^2}R_s^2 + 4kTR_s\Delta f}. \tag{7.27}$$

Therefore, we have that the maximum SNR is given by $R_S \to 0$ while the minimum noise factor is given for R_{opt}. This could be easily shown in Fig. 7.12. Note how for low R_S the noise is mostly due to voltage interface noise. Then the prevalent contribution is given by source noise, and finally, for high, R_S the main contribution is that from the current.

From the preceding observation, why is R_{opt} called "optimal source resistance" if the minimum SNR is achieved minimizing R_S? Which is thus the best procedure for designing the sensor interface? For the first question, we have to observe that for a given R_S, the total rms noise has a lower bound given by $\sqrt{4kTR_S}$ thus for "optimal source resistance" it is intended the minimum increment of noise due to the interface contributions additional to $\sqrt{4kTR_S}$.

For the second answer, we should point out that we usually cannot change source impedance in electronic sensor design because the sensor characteristics fix it. Therefore, the following steps should be taken in the design:

- Among sensors with the same performance, it is convenient to choose that one with smaller source resistance R_S because this minimizes the SNR in any condition of the network.
- Once the source resistance is set, we can optimize the total noise performance by adapting the interface to the source resistance so that to move R_{opt} toward R_S. Therefore, we should design the interface so as to get noise components related by (7.26).

What if v_N and i_N are instead fixed? We could, in principle, increase the performance by *impedance matching*, and of course, it could be realized only in AC operation (such as in RF low-noise amplifiers) because even if noise is a stochastic signal, it has spectrum components as deterministic ones. Just as a matter of example, we can use, in principle, a coil transformer to change the relationship between v_N and i_N. As well known, the transformer relationships could be mapped in our case as $v_{N2}/v_{N1} = i_{N1}/i_{N2} = N_2/N_1$ where subscript 1 is for the primary transformer circuit, subscript 2 for the secondary one, and N_x are the turns in a winding. Therefore, the balance between the current and voltage noise could be changed by keeping the same product constant: we can increase the amount of voltage noise by decreasing that of the current one and vice versa. Thus, we can adapt the interface to the sensor if we could do that.

If the source resistance is initially lower than the optimal one, we can decrease the original v_N and increase i_N to move R_{opt}(characterized by the crossing point of the two asymptotic lines) on the left toward R_S. On the other hand, if source resistance is initially higher than the optimal one, we can increase the original v_N and decrease i_N to move the corner on the right. If you play the approach on the graph, you can easily find that the final result achieves a better noise performance than the original in both cases.

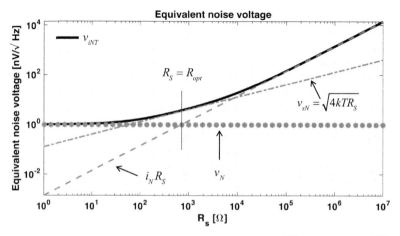

Figure 7.12 Input-referred noise components ($v_N = 1.0\text{nV}/\sqrt{\text{Hz}}$, $i_N = 1.4\text{pA}/\sqrt{\text{Hz}}$).

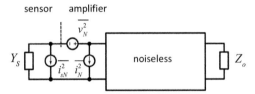

Figure 7.13 Model for calculating the impedance matching considering reactances.

To better analyze the effect of impedance matching in AC, we should consider reactances to use Norton's model, as shown in Fig. 7.13.

The noise factor (in terms of currents) could be expressed as

$$F = \frac{\overline{i_{sN}^2} + \overline{|i_N + Y_s v_N|^2}}{\overline{i_s^2}}. \tag{7.28}$$

Note that we assume superposition of noise power between i_{sN} and the rest of the sources, but not between i_N and v_N, since they could be partially correlated as discussed before. Therefore, we can express i_N as composed of two parts, one, i_u, uncorrelated with respect to v_N and another, i_C, which is correlated:

$$i_N = i_u + i_c = i_u + Y_c v_N, \tag{7.29}$$

where Y_c is called *correlation admittance*, and it is another way to express the input-referred noise current and voltage relationship. Since it is $\overline{i_u v_N} = 0$ by definition, the correlation factor between i_N and v_N γ could be expressed as

$$\gamma = \frac{\overline{i_N v_N}}{\sqrt{\overline{i_N^2 v_N^2}}} = Y_C \sqrt{\frac{\overline{v_N^2}}{\overline{i_N^2}}}. \tag{7.30}$$

Now, if we assume white noise conditions

$$\overline{v_N^2} = 4kTR_N \Delta f; \quad \overline{i_U^2} = 4kTG_u \Delta f; \quad \overline{i_{sN}^2} = 4kTG_s \Delta f; \tag{7.31}$$

where G_s is the real conductance of the sensor while R_N and G_u are model resistance and conductance of the input-referred voltage noise and uncorrelated input-referred current noise, respectively. Thus, using (7.29) and (7.31) into (7.28), we get

$$F = \frac{\overline{i_{sN}^2} + \overline{|i_N + Y_s v_N|^2}}{\overline{i_{sN}^2}} = \frac{\overline{i_{sN}^2} + \overline{|i_u + (Y_c + Y_s)v_N|^2}}{\overline{i_{sN}^2}} = 1 + \frac{\overline{i_u^2} + |Y_c + Y_s|^2 \overline{v_N^2}}{\overline{i_{sN}^2}}$$

$$= 1 + \frac{G_u + |Y_c + Y_s|^2 R_N}{G_s} = 1 + \frac{G_u}{G_s} + \frac{R_n}{G_s}\left[(G_c + G_s)^2 + (B_c + B_s)^2\right]$$

where: $Y_c = G_c + jB_c$ and $Y_s = G_s + jB_s$,

$$\tag{7.32}$$

where G and B symbols refer to conductances and susceptances. Note how (7.32) is equal to (7.25) for uncorrelated noises and negligible susceptances.

It can be easily demonstrated that the optimal values that minimize noise factor are

$$B_{\text{opt}} = B_s = -B_c$$

$$G_{\text{opt}} = G_s = \sqrt{\frac{G_u}{R_N} + G_C^2} = \sqrt{\frac{\overline{i_u^2}}{\overline{v_N^2}} + G_c^2}, \tag{7.33}$$

meaning that to optimize noise, the source susceptance should be made equal to the negative value of the correlation susceptance, and the source conductance should be made equal to the expression in (7.33).

If we turn back, assuming that the frequency is low enough to neglect susceptances. It is interesting now to show that the conclusions that we can achieve assuming totally correlated and totally uncorrelated noise sources are equivalent.

In the first case, let us assume that the two sources are completely uncorrelated. Thus

$$i_c = 0; \ Y_c = 0; \ G_c = 0; \ i_u = i_N$$

$$\rightarrow G_{\text{opt}} = \sqrt{\frac{\overline{i_N^2}}{\overline{v_N^2}}}. \tag{7.34}$$

In the second case, where the two sources are completely correlated, we have

$$i_u = 0; \ Y_c = G_c = i_N/v_N$$

$$\rightarrow G_{\text{opt}} = \sqrt{G_c^2} = \sqrt{\frac{\overline{i_N^2}}{\overline{v_N^2}}} = \frac{1}{R_{\text{opt}}}, \tag{7.35}$$

as before, where $R_{\text{opt}} = \sqrt{\overline{v_N^2}/\overline{i_N^2}}$.

It is worth looking at how noise factors are composed in a chain of amplifiers. The system shown in Fig. 7.14A is equivalent to that in Fig. 7.14B, where all the noise sources are moved at the system's input.

We can set the definition of noise factor of the kth stage following (7.21) as

$$F_k \triangleq \frac{\overline{v_{sN}^2} + \overline{v_{kN}^2}}{\overline{v_{sN}^2}} = \frac{\text{source noise} + \text{input-referred noise of}k\text{th stage}}{\text{source noise}}, \tag{7.36}$$

from which

$$(F_k - 1) \triangleq \frac{\overline{v_{kN}^2}}{\overline{v_N^2}} = \frac{\text{input-referred noise of } k\text{-th stage}}{\text{source noise}} \tag{7.37}$$

Therefore, we have that

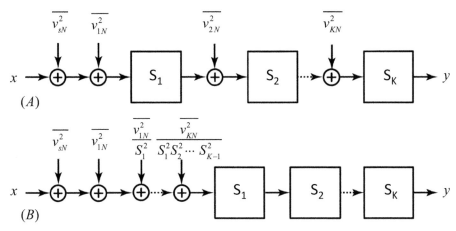

Figure 7.14 Determination of Friis formula.

$$F_T = \frac{\overline{v^2_{sN}} + \overline{v^2_{1N}} + \dfrac{\overline{v^2_{2N}}}{S_1^2} + \dfrac{\overline{v^2_{3N}}}{S_1^2 S_2^2} + \cdots \dfrac{\overline{v^2_{kN}}}{S_1^2 S_2^2 \ldots S_{k-1}^2}}{\overline{v^2_N}}$$
$$= F_1 + \frac{F_2 - 1}{S_1^2} + \frac{F_3 - 1}{S_1^2 S_2^2} + \cdots \frac{F_k - 1}{S_1^2 S_2^2 \ldots S_{k-1}^2}, \tag{7.38}$$

which is referred to as the *Friis formula*.

The formula points out two important aspects: (1) The noise figure of the first stage dominates the total noise figure; (2) high gain in the first stage reduces the contribution of the noise factor (NF) of the following stages. The above aspects show that (in most cases) *input front-end stages are crucial in determining the overall noise* in acquisition chains. The fact that most of the contribution of noise comes from the first stages of amplification implies that the input-referred noise of most real systems is weakly dependent on the gain (for equal bandwidth), and thus the equivalent SNR is almost stable with respect to gain variations.

7.6   Example: Noise in Junction Transistors

To better understand the input-referred noise calculation presented in Section 7.5, we will take the chance to calculate it for the bipolar junction transistor (BJT). This will be further useful for more complex circuits that will be introduced in this book. The idea is to derive a noise model as illustrated in Fig. 7.15A, where all the internal noise sources are compacted at the input of a noiseless device. This technique easily derives noise characteristics in more complex circuits employing single BJT devices.

To do so, we will assume that the transistor will work in the *normal* mode, which is one of the most frequently used operating points of the device, whose small (noiseless) signal model is illustrated in Fig. 7.15B. All the notations are indicated in Fig. 7.15 where B, C, E are the base, collector, and emitter of the device, respectively; β_0 is the

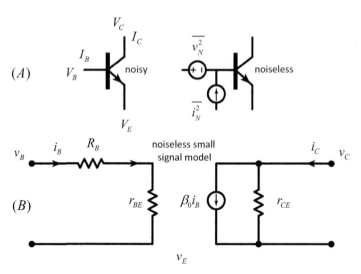

Figure 7.15 Bipolar junction transistor noisy model.

current gain, r_{BE} is the BE model resistance; r_{CE} is the CE model resistance, and R_B is the base resistance. Incremental small-signal quantities are expressed in all lowercase (i_B, i_C, v_B, v_C, etc.), while quiescent large-signal quantities (bias point) are expressed in all uppercase (I_B, I_C, V_B, V_C, etc.). The transconductance of the device is given by $g_m = \beta_0/r_{BE}$.

For the target, the first step is to find the input-referred noises $\overline{i_{iN}^2}$ and $\overline{v_{iN}^2}$ by using the equivalent model as illustrated in Fig. 7.16A. Note how the r_{BE} and r_{CE} do *not* contribute to noise since they are derivative models of current and voltages. Instead, R_B does contribute to noise because it is the physical resistance of the base. Other contributions are $\overline{i^2}_{BN}$ and $\overline{i^2}_{CN}$, the shot noises of base and collector currents due to the BE and CB junction barrier, respectively. Considering *only thermal* and *shot noise*, the power contributions in a bandwidth Δf are $\overline{i_{BN}^2} = 2qI_B\Delta f \overline{i^2}_{CN} = 2qI_C\Delta f$, and $\overline{v_{BN}^2} = 4kTR_B\Delta f$. Of course, flicker noise is always present; however, it will be neglected for simplicity in this example because it is much lower than in other devices (such as metal–oxide–semiconductor). In any case, the procedure is very general and could be easily applied to flicker noise contributions.

The second step is to derive the input-referred current and voltage noises $\overline{i_N^2}$, and $\overline{v_N^2}$ by imposing that the behavior of the circuit of Fig. 7.16A should be equal to that of Fig. 7.16B that is, it should always be $v_i = v_i'$, $v_o = v_o'$, $i_i = i_i'$, $i_o = i_o'$ where the symbols with a prime are related to the bottom model.

To facilitate the derivation of the input-referred noises, we will use the trick introduced in Section 7.5 by setting the input impedance to asymptotic values. Thus, to derive $\overline{v_N^2}$ we will first impose a short circuit to the two models as illustrated in Fig. 7.17. In doing so, the contribution of $\overline{i_N^2}$ becomes ineffective, as shown in Fig. 7.17B. By using the assumption $R_B << 1$ and the usual equivalence $r_{BE} = V_T/i_B = kT/qi_B$, we have:

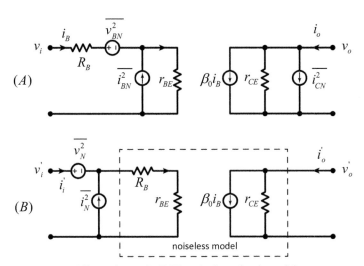

Figure 7.16 Adding noise to the BJT equivalent circuit model.

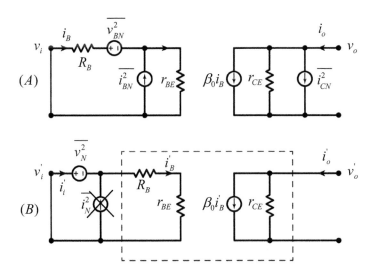

Figure 7.17 Input-referred voltage noise derivation using zero input impedance (short-circuit).

$$v_o = r_{CE}(\beta_0 i_B + i_{CN}) = v'_o = r_{CE}\beta_0 i'_B = r_{CE}\beta_0 \frac{v_N}{r_{BE}}$$

$$\rightarrow v_N = r_{BE}i_B + \frac{r_{BE}}{\beta_0}i_{CN} = v_{BN} + \frac{r_{BE}}{\beta_0}i_{CN}$$

$$\rightarrow \overline{v_N^2} = \overline{v_{BN}^2} + \left(\frac{r_{BE}}{\beta_0}\right)^2 \overline{i_{CN}^2} = 4kTR_B\Delta f + \left(\frac{r_{BE}}{\beta_0}\right)\frac{V_T}{I_C} \cdot 2qI_C\Delta f \qquad (7.39)$$

$$= 4kT\left(R_B + \frac{r_{BE}}{2\beta_0}\right)\Delta f \equiv [\text{V}^2].$$

Instead, to derive $\overline{i_{iN}^2}$ we will use the opposite input impedance conditions of open-circuit as illustrated in Fig. 7.18.

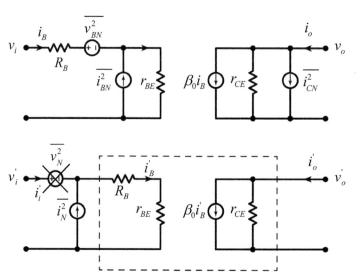

Figure 7.18 Input-referred current noise derivation using infinite input impedance (open-circuit).

Again, by imposing the same equivalence between the two models, we have

$$v_o = r_{CE}(\beta_0 i_{BN} + i_{CN}) = v'_o = r_{CE}\beta_0 i'_N$$

$$\rightarrow \overline{i_N^2} = \overline{i_{BN}^2} + \frac{\overline{i_{CN}^2}}{\beta_0^2} = 2qI_B\Delta f + \frac{1}{\beta_0^2}2qI_C\Delta f = 2qI_B\left(1 + \frac{1}{\beta_0^2}\right)\Delta f \equiv [\text{A}^2]. \qquad (7.40)$$

To summarize, the input-referred noise voltage and current are

$$\begin{cases} \overline{v_N^2} = 4kT\left(R_B + \dfrac{r_{BE}}{2\beta_0}\right)\Delta f \equiv [\text{V}^2] \\[3mm] \overline{i_N^2} = 2qI_B\left(1 + \dfrac{1}{\beta_0}\right)\Delta f \equiv [\text{A}^2]. \end{cases} \qquad (7.41)$$

It is interesting to note how internal noise generators are mapped back to the input. More specifically, rearranging Eq. (7.41) as in Fig. 7.19, we can see that $\overline{i_{CN}^2}$ is mapped in the two generators, and the two contributions are correlated through the input port resistance r_{BE}. However, the input resistance and the base current noises are mapped back to only one source. This behavior makes the two contributions uncorrelated, as illustrated in Fig. 7.19.

This is due especially if some noise generator is placed in particular and "pathological" connection (i.e., a voltage noise generator in series to the input or a current noise generator in parallel to the input) so that no current or voltage input-referred generators could induce the same effect to the system. In conclusion, we can assume for BJT devices for $\beta_0 \gg 1$ the two input-referred generators *uncorrelated*.

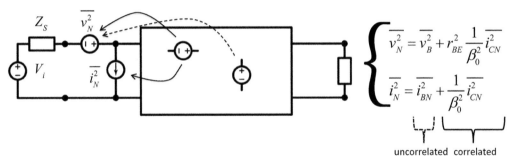

$$\begin{cases} \overline{v_N^2} = \overline{v_B^2} + r_{BE}^2 \dfrac{1}{\beta_0^2} \overline{i_{CN}^2} \\[4mm] \overline{i_N^2} = \overline{i_{BN}^2} + \dfrac{1}{\beta_0^2} \overline{i_{CN}^2} \end{cases}$$

$$\underbrace{}_{\text{uncorrelated}} \quad \underbrace{}_{\text{correlated}}$$

Figure 7.19 Correlation between input-referred noise sources.

7.7   Example: Noise in Metal–Oxide–Semiconductor Transistors

Metal–oxide–semiconductor (MOS) transistors have physical behavior that is quite different from that of BJT transistors; therefore, noise source contributions are different from those of bipolar transistors. MOS noise is dominated by two contributions: flicker and thermal. Other noises are also present, such as shot and burst noise, but they will be neglected in this derivation.

Again, the idea is to derive a noise model as illustrated in Fig. 7.20A, where all the internal noise sources are compacted at the input of a noiseless device. This technique allows to easily derive noise characteristics in more complex circuits using single MOS transistors. The starting point will be using the linearized small-signal model as illustrated in Fig. 7.20B, where C_i is the input capacitance and g_m is the transconductance.

The main MOS noise source is flicker noise. It is still unclear whether such noise is due to fluctuations in the number of carriers due to charge trapping in surface states (MacWhorter model) or to bulk mobility fluctuations that are changing the channel resistance (Hooge model). Recently, several efforts have been aimed at combining both models in a unified one where the idea is that channel charges when trapped, do induce correlated surface mobility fluctuations. In this example, we will simply use an empirical expression showing that input-referred noise power spectrum density (PSD) at the voltage gate is given by

$$\overline{v_{NF}^2}(f) = \frac{K_f}{C_{OX} WL} \frac{1}{f}, \tag{7.42}$$

where C_{OX} is the oxide capacitance, and W and L are the width and length of the channel and $K_f \approx 3 \times 10^{-24} \equiv [V^2 \cdot F]$ is a constant for MOS transistors. Note how the relationship (7.42) is physically consistent since the time-dependent components of the surface charge are mapped back to the threshold voltage that is inversely proportional to C_{OX}.

On the other hand, it could be shown that thermal noise is due to the conductance of the channel that in the saturation region is equal to $2/3g_m$. Therefore, all noise sources

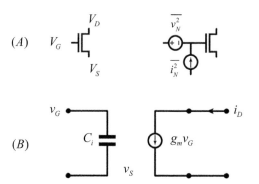

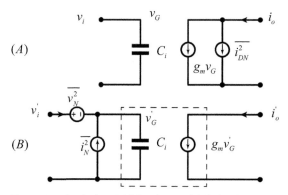

Figure 7.20 MOS transistor noisy model.

Figure 7.21 Equivalent MOS noise model derivation.

could be seen as a drain current noise source $\overline{i_{DN}^2}$, which sums up the contributions of both thermal $\overline{i_{DNT}^2}$ and flicker noise $\overline{i_{DNF}^2}$

$$\overline{i_{DN}^2} = \overline{i_{DNT}^2} + \overline{i_{DNF}^2}$$

$$\overline{i_{DNT}^2} = 4kT\frac{2}{3}g_m\Delta f = \frac{8}{3}kTg_m\Delta f \tag{7.43}$$

$$\overline{i_{DNF}^2} = g_m^2\frac{K_f}{C_{OX}WL}\frac{1}{f}\Delta f.$$

Note how in this representation, the flicker noise is mapped forward by means of g_m^2.

The target input-referred noises $\overline{i_N^2}$ and $\overline{v_N^2}$ can be given by substituting the noisy components of the equivalent model, as illustrated in Fig. 7.21, where the internal sources are mapped back to the input.

The two sources could be easily found by first making a short-circuit of the input, as shown in Fig. 7.22.

The input-referred voltage noise could be found by setting $v_o = v_o'$, thus

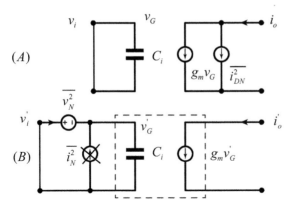

Figure 7.22 Derivation of the input-referred voltage noise.

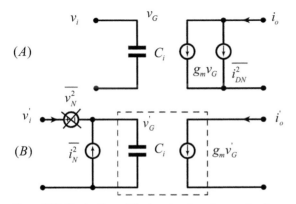

Figure 7.23 Derivation of the input-referred current noise source.

$$v_i = v'_i = 0$$
$$i_o = i_{DN} + g_m v_G = i'_o = g_m v'_G = g_m v_{in}$$
$$\overline{i_{DN}^2} = g_m^2 \overline{v_{in}^2} \rightarrow \tag{7.44}$$
$$\overline{v_N^2} = \frac{1}{g_m^2} \overline{i_{DN}^2} = \frac{1}{g_m} \frac{8}{3} kT\Delta f + \frac{K_f}{C_{OX} WL f} \frac{1}{f} \Delta f.$$

Similarly, the input-referred noise source could be derived by setting $i_o = i'_o$ and using the first-port impedance (Fig. 7.23)

$$i_i = i'_i = 0$$
$$\overline{i_N^2} = \frac{\overline{v_N^2}}{Z_i^2} = (\omega^2 C_i^2) \overline{v_N^2} = (\omega^2 C_i^2) \left[\frac{1}{g_m} \frac{8}{3} kT\Delta f + \frac{K_f}{C_{OX} WL f} \frac{1}{f} \Delta f \right]. \tag{7.45}$$

Note how the two sources are strictly *correlated* by means of the correlation admittance $Y_c = j\omega C_i$, in this case. This is why the noise of a MOS transistor with an open gate and short-circuit gate is the same as illustrated in Fig. 7.24.

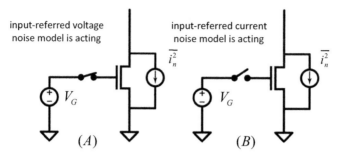

Figure 7.24 Noise of a MOS transistor with shorted gate (A) Noise of a MOS transistor after the gate is left open (B).

When the input switch is closed, as illustrated in Fig. 7.24A, the gate voltage is put at a fixed value so that the bias point of the device gives the output noise. Now, if we open the switch, as illustrated in Fig. 7.24B, the gate charge remains trapped so that the transistor bias is kept constant with an equivalent output noise. From the input-referred noise sources perspective, the two situations are treated with the voltage and current noise power models alternatively but with identical effects because they are strictly correlated.

7.8   Input-Referred Noise Representation in the Spectrum Domain

We have already discussed the relationship between the total input-referred and the total output-referred noise power in linear circuits in a generic bandwidth $\Delta f = f - f_0$:

$$\overline{v_{oN}^2}(f_0)\Delta f = |H(f_0)|^2 \overline{v_{iN}^2}(f_0)\Delta f. \tag{7.46}$$

Now, suppose that *we do not inject any signal/noise into the input*, and we observe only output PSD due to interface noise. We would like to define a generic input-referred noise spectral density as

$$\overline{v_{iN}^2}(f) \triangleq \frac{1}{|H(f)|^2}\overline{v_{oN}^2}(f). \tag{7.47}$$

In other terms, while output PSD $\overline{v_{oN}^2}(f)$ could be experimentally measured, input-referred noise PSD is only a mathematical model. The problem is that, in most cases, the output PSD is bandwidth self-limited, so that the integral on the whole spectrum converges, while the input-referred PSD could be uniform; thus, *we cannot integrate it on the whole spectrum*. To frame this aspect into a consistent model, we will make some assumptions and few examples.

Let *assume that the system is bandwidth self-limited*, so as we can define for it the noise bandwidth NBW as discussed [(4.100) of Chapter 4]: Therefore

$$P_o = \overline{v_{oN}^2} = \int_0^\infty \overline{v_{oN}^2}(f)df = \int_0^{NBW} \overline{v_{oN}^2}(f)df = \alpha^2 NBW. \qquad (7.48)$$

At the same time, from (7.48), assuming that the in-bandwidth gain is constant, we have

$$P_o = \int_0^{NBW} \overline{v_{oN}^2}(f)df = \int_0^{NBW} \overline{v_{iN}^2}(f)|H(f)|^2 df \simeq |H_K|^2 \int_0^{NBW} \overline{v_{iN}^2}(f)df = |H_K|^2 P_i,$$

$$(7.49)$$

where $|H_K|^2 = |H_{max}(f)|^2$ is the *in-bandwidth gain* (e.g., $|H_K|^2 = |H(0)|^2$ for low-pass behavior). In practice, we calculate the total input-referred noise power as the one that induces for the given noiseless system the total measured output power (generated by the interface).

Therefore, we can use the input-referred PSD to calculate the input-referred power as

$$\overline{v_{iN}^2} \triangleq \int_0^{NBW} \overline{v_{iN}^2}(f)df = \frac{\alpha^2}{|H_K|^2} \cdot NBW. \qquad (7.50)$$

Equation (7.50) shows that the (uniform) input-referred noise spectrum *should be integrated into the noise bandwidth* to be consistent with the total output noise power. This opens a new viewpoint of noise bandwidth.

Hint The noise bandwidth is the portion of the spectrum on which we should integrate the input-referred noise PSD to calculate the total input-referred noise of the system.

To make an example, if we refer to Fig. 7.25, we have

$$\overline{v_{oN}^2}(f) = \frac{1}{1 + (f/f_0)^2} \qquad \overline{v_{iN}^2}(f) = \frac{4kTR}{1 + (f/f_0)^2}, \qquad (7.51)$$

where $f_0 = 1/2\pi RC$. Therefore, the output-referred noise is the noise of the resistance shaped out by a Lorentzian form (note carefully that in the figure, we used the rms representation of noise for simplicity).

We can determine the input-referred noise spectrum model by dividing the output noise by the module of the transfer function, as shown in the figure. Note that this could be graphically determined by *subtracting* (because we are in a log-log representation) the graph of the output noise spectrum with that of the transfer function obtaining a uniform distribution

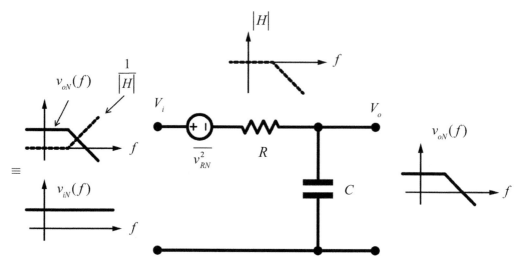

Figure 7.25 Concept of the input-referred noise spectrum. Note that in this case, we are referring to the rms values and not to the power ones since they are equivalent.

$$\overline{v_{iN}^2}(f) = \frac{1}{|H_K|^2}\,\overline{v_{iN}^2}(f) = v_{RN}^2 = 4kTR. \tag{7.52}$$

This is the result that we are expecting since the source noise is the uniform noise of the resistance; however, this step is important to understand the procedure to be applied in more complex cases. To calculate the total output noise

$$\overline{v_{oN}^2} = \int_0^\infty \overline{v_{iN}^2}(f)\frac{1}{1+(f/f_0)^2}\,df = 4kTR \cdot \frac{\pi}{2}f_0 = \frac{kT}{C}. \tag{7.53}$$

To get the total output noise, we integrate the output-referred noise in the whole spectrum; however, we cannot do that on the input-referred noise spectrum since an infinite noise power would result. The answer is in the previous analysis: we should integrate into the NBW.

$$\overline{v_{iN}^2} = \int_0^{NBW} \overline{v_{iN}^2}(f)df = 4kTR \cdot NBW = \frac{kT}{C}. \tag{7.54}$$

Note that the total input and output noise are equal since the in-bandwidth gain is unitary.

Therefore, for any system that is *self-band-limited*, the signal-to-noise ratio could be calculated as

$$\text{SNR}_O = \frac{P_o}{P_{oN}} = \frac{\displaystyle\int_0^\infty V_o^2(f)\,df}{\displaystyle\int_0^\infty v_{oN}^2(f)\,df}. \tag{7.55}$$

Conversely, following what we have found, as far as we refer to the input, the signal-to-noise ratio is found as

$$\text{SNR}_{I(EQ)} = \frac{P_I}{P_{iN}} = \frac{\displaystyle\int_{BW} V_i^2(f)\,df}{\displaystyle\int_{NBW} v_{iN}^2(f)\,df} = \text{SNR}_O, \tag{7.56}$$

where BW is the bandwidth of the signal transfer function (actually, for bandwidth-limited signals, the integral could be extended to infinity) while NBW is the noise bandwidth of the noise gain. In other words, at the input, the signal PSD could be integrated into the bandwidth (BW), while the noise PSD should be integrated into the noise bandwidth (NBW).

In other cases, where the system is not bandwidth self-limited, we must be careful, as we will see in the next example.

Figure 7.26 is very similar to Fig. 7.25, but with the addition of a new resistance (at the moment, do not consider R_3). R_3 acts as the output resistance in Thevenin's model approach. The first step is to consider the output-referred noise as the combination (since they are assumed uncorrelated) of the noises of the two resistances:

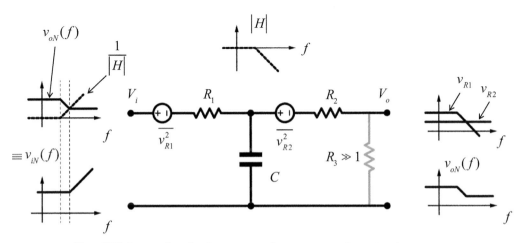

Figure 7.26 Input-referred noise spectrum for a more complex network.

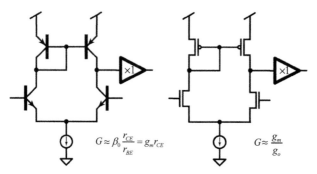

Figure 7.27 Basic scheme of a single-ended operational amplifier.

$$\overline{v^2_{oN}}(f) = v^2_{R2}(f) + v^2_{R1}(f)\frac{1}{1+(f/f_0)^2} \; ; \tag{7.57}$$
$$v^2_{R1}(f) = 4kTR_1; \quad v^2_{R2}(f) = 4kTR_2.$$

Therefore, the total output noise appears as shown in the figure on the graph. Note that the presence of a constant value due to the R_2. This means that the output noise is not self-limited by the system as before. Now, we can deduce the input-referred noise spectrum by dividing the output by the transfer function as already done (by adding in the log-log plot the 0 dB-line-rotated transfer function). As you can see from the figure, the input-referred noise spectrum is characterized by an increasing behavior.

This is not intuitive and physically not consistent because *it is a model* and not a real noise applied to the input. The input-referred spectrum is increasing because the roll-off effect of the filter will result in constant output noise. In this case, the concept of noise bandwidth is more difficult to define because the system is not bandwidth self-limited. An exact calculation of the noise behavior should be performed by closing the second loop. However, using a very high value of R_3 it is easy to show that the result is exactly that illustrated in Fig. 7.27.

If the bandwidth of a measurement instrument does the noise limitation, we can apply the concept of NBW for the calculation of the total input-referred equivalent noise.

7.9   Noise in Operational Amplifier Configurations

One of the most used devices for electronic design is the *operational amplifier* (OPAMP). The use of operational amplifiers makes the design easier thanks to the use of *structured design* where a complex analog system is partitioned in hierarchical sub-blocks interacting with each other according to terminal-defined functions.

The most common architecture of the OPAMP is based on the differential pair circuit that could be implemented by means of both BJT and MOS transistors, as illustrated in Fig. 7.27. In most cases, the larger part of the gain is due to the differential input devices; thus, following the Friis formula, most of the input-referred noise is due to those devices, as shown in Fig. 7.28A. Therefore, using the conventional circuital

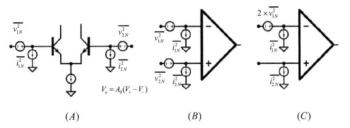

Figure 7.28 Derivation of the input-referred noise sources of a generic operational amplifier.

symbol of the operational amplifier, we can summarize the input-referred noises of the OP as illustrated in Fig. 7.28B. However, since the operational amplifier amplifies the input *voltages difference* $V_o = A_0(V_+ - V_-)$, and since the differential pair noises are uncorrelated, we can summarize the voltage contributions as unique ones, as shown in Fig. 7.28C. Note that this could not be done for current sources because the input impedances should mediate their contribution.

In general, the input-referred voltage and current noise power spectral density are given following the form of the relationship (7.4)

$$
\begin{cases}
\overline{v_N^2}(f) = \overline{v_N^2}\left(1 + \dfrac{f_{CE}}{f}\right) \equiv \left[\dfrac{\mathrm{V}^2}{\mathrm{Hz}}\right] \\[2ex]
\overline{i_N^2}(f) = \overline{i_N^2}\left(1 + \dfrac{f_{CI}}{f}\right) \equiv \left[\dfrac{\mathrm{A}^2}{\mathrm{Hz}}\right]
\end{cases},
\tag{7.58}
$$

where f_{CE} and f_{CI} are the corner frequency of voltage noise and current noise, respectively.

Now, we will take the example of one of the most generic schemes of an amplifier, as shown in Fig. 7.29. Most of the simpler schematics, such as inverting and non-inverting amplifiers, can be easily derived by this scheme. The input–output relationship is given by the transfer function

$$
V_o = -\frac{Z_2}{Z_1}V_X + \frac{1 + (Z_2/Z_1)}{1 + (Z_3/Z_4)}V_Y,
\tag{7.59}
$$

where Z_k are generic impedances as shown in Fig. 7.29A. If we set $Z_4 = 0$ and $Z_3 \to \infty$ and tie V_Y to the ground, we can get the configuration of the inverting amplifier. Conversely, if we connect V_X to the ground, we get a noninverting amplifier.

As already done, noise calculations are performed by substituting each noise impedance with a noiseless impedance with a series-connected voltage noise power generator as indicated by (7.10) such that $v_k^2 = 4kT\,\mathrm{Re}[Z_k]\Delta f$. On the other hand, for the OPAMP, we use the input-referred noise power generators as illustrated in Fig. 7.28A.

> **Note** In the following we will assume for simplicity all impedances either as real (resistances) or imaginary pure (capacitances or inductances). In the latter case, they will be noiseless.

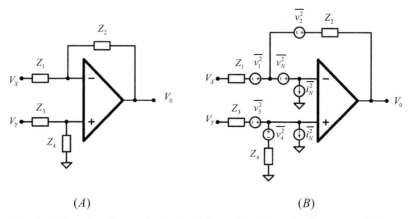

(A) (B)

Figure 7.29 Generic scheme of a differential amplifier (A) and noise sources (B).

The output noise could be calculated by using the power superposition principle, where the output noise is given by the sum of all the single source contributions reflected into the output. Here, an important distinction should be made. While the signals follow their path toward the output by means of a signal transfer function, called *signal gains*, noise perturbations follow a different transfer function, referred to as *noise gains*. For the scheme's specific case, the relationship (7.59) represents its signal gain.

Among all, we will just calculate the input OPAMP voltage noise (Fig. 7.30) as an example, leaving the derivations of other noise contributions to the reader.

> **Note** For simplicity, we will assume in gain calculations for Z all pure resistances (generating noise) or all pure reactances (not generating noise) such as capacitances.

The derivation could be performed assuming that the operational amplifier has a dominant pole behavior so that the open-loop gain is expressed as

$$A(f) = \frac{A_0}{1 + j(f/f_0)},\qquad (7.60)$$

where f_0 is the pole (cutoff frequency) of the open-loop gain of the operational amplifier. Now, the voltage of the inverting input of the OPAMP could be written as

$$v_- = v_N + v_o \frac{Z_1}{Z_1 + Z_2} = -\frac{v_o}{A(f)},\qquad (7.61)$$

from which we can derive the relationship between the output and the noise perturbation as

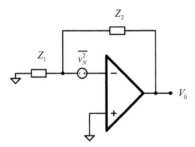

Figure 7.30 Noise gain calculation for the input-referred voltage noise of the OPAMP.

$$H_N(f) = \frac{v_o}{v_N} = -\frac{1}{\dfrac{Z_1}{Z_1 + Z_2} + \dfrac{1}{A(f)}} = -\frac{1}{\dfrac{Z_1}{Z_1 + Z_2} + \dfrac{1}{A_0}} \cdot H'(f)$$

$$\approx -\left(1 + \frac{Z_2}{Z_1}\right) \cdot H'(f) \tag{7.62}$$

where:

$$H'(f) = \frac{1}{1 + j(f/f'_0)} \quad \text{and} \quad f'_0 = f_0\left(1 + A_0 \frac{Z_1}{Z_1 + Z_2}\right),$$

where $H_N(f)$ is the *noise gain* of the input-referred voltage noise of the OPAMP and f'_0 is the new pole (cutoff frequency) of the closed-loop system. Note how the relationship is similar to that of the noninverting amplifier (except the sign). This is because if we model input-referred voltage noise on the noninverting input – and this is equivalent – we have the same result.

If we calculate other noise gains, we always get a relationship implying the function $H'(f)$. For instance, if we refer to the perturbation related to Fig. 7.31, we get the relationship

$$H_1(f) = \frac{v_o}{v_1} = -\frac{Z_2}{Z_1 + \dfrac{Z_1 + Z_2}{A_0}} \cdot \frac{1}{1 + j(f/f'_0)} \approx -\left(\frac{Z_2}{Z_1}\right) \cdot H'(f). \tag{7.63}$$

This presence of $H'(f)$ is because, since the system is linear, any perturbation is reflected onto the output with the same spectral characteristic.[1]

Now, applying the superposition of noise powers, the overall output noise calculation could be expressed as

$$\overline{v_{oN}^2}(f) = \left[\begin{array}{l} \left|1 + \dfrac{Z_2}{Z_1}\right|^2 \left(\overline{v_N^2}(f) + \overline{v_P^2}(f) + Z^2{}_P \overline{i_N^2}(f)\right) \\[3mm] + \left|\dfrac{Z_2}{Z_1}\right|^2 \overline{v_1^2}(f) + Z^2{}_2 \overline{i_N^2}(f) + \overline{v_2^2}(f) \end{array} \right] \cdot |H'|^2 \equiv \left[\dfrac{V^2}{\text{Hz}}\right], \tag{7.64}$$

where $\overline{v_P^2}(f)$ is the noise spectral density of the parallel connection of Z_3 and Z_4.

---

[1] The poles of the transfer functions of linear systems are always the roots of the *characteristic function* of the linear circuit itself.

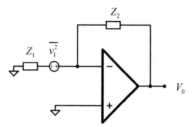

Figure 7.31 Noise gain calculation for the input impedance noise Z_1.

Now, using the concept of *noise bandwidth* (the system is bandwidth self-limited), we have

$$
\overline{v_{oN}^2} = \left[\frac{\left| 1 + \frac{Z_2}{Z_1} \right|^2 \left(\overline{v_N^2}(f) + \overline{v_P^2}(f) + Z^2 \, _P \overline{i_N^2}(f) \right)}{+ \left| \frac{Z_2}{Z_1} \right|^2 \overline{v_1^2}(f) + Z_2^2 \overline{i_N^2}(f) + \overline{v_2^2}(f)} \right] \cdot \text{NBW} \equiv [V^2] \text{ where}
$$

$$
\text{NBW} = \frac{\pi}{2} f_0' = \frac{\pi}{2} f_0 \left(1 + A_0 \frac{Z_1}{Z_1 + Z_2} \right),
$$

(7.65)

where to be more precise, the NBW could contain the correction of the pink noise (7.6) on both current and voltage noise cases

$$
\text{NBW}_E = \frac{\pi}{2} f_0' + f_{CE} \ln\frac{f_0'}{f_L}; \quad \text{NBW}_I = \frac{\pi}{2} f_0' + f_{CI} \ln\frac{f_0'}{f_L}.
$$

(7.66)

7.9.1   Signal and Noise Gain Paths

Signal and noise follow different paths. For example, as shown in Fig. 7.32A, where it is considered only one source of noise, the signal gain is given by the usual expression of the inverting amplifier

$$
H_S(f) = \frac{v_o}{v_i} \approx - \left(\frac{Z_2}{Z_1} \right) \cdot H'(f),
$$

(7.67)

where $H'(f)$ is the expression already found.
On the other hand, the noise gain is

$$
H_N(f) = \frac{v_o}{v_N} \approx - \left(1 + \frac{Z_2}{Z_1} \right) \cdot H'(f).
$$

(7.68)

Note in Fig. 7.32A that the two signals travel along different paths, referred to as *signal path* and *noise path*, respectively.

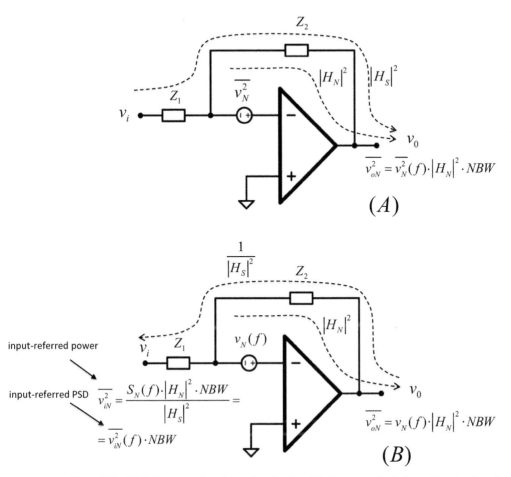

Figure 7.32 (A) Different paths of signal and noise. (B) Reverse calculation of input-referred noise.

Now, once we have derived the output equivalent noise, assuming uniform PSD in NBW, we have

$$\overline{v_{oN}^2} = \overline{v_N^2}(f) \cdot |H_N|^2 \cdot \text{NBW}, \tag{7.69}$$

where $|H_N|$ is the in-bandwidth noise gain.

However, since we want to compare noise with the signal at the input level, we have to return back along the signal path

$$\overline{v_{iN}^2} = \frac{S_N(f) \cdot |H_N|^2 \cdot \text{NBW}}{|H_S|^2} = \overline{v_{iN}^2}(f) \cdot \text{NBW}, \tag{7.70}$$

where $|H_S|$ is the in-bandwidth signal gain.

7.9.2   Example: Noise Calculation for an Operational Amplifier

We will use to evaluate previous results the scheme illustrated in Fig. 7.33 as previously discussed.

In this case, we will use information from a classical low-noise BJT OPAMP OP27/OP37, summarized in Table 7.1.

Furthermore, we assume $f_L = 1/365$ days. To assess the confidence of all expressions of Section 7.9, we made simulations using LTSpice®. The results are summarized in Table 7.2.

The following points should be noted:

- The output noise has been calculated on a large spectrum (due to the self-limited bandwidth) while input-referred noise on NBW [considering pink noise according to Eq. (7.66)] following the arguments and relationships of Sections 7.8 and 7.9.
- The equations are in good agreement with the simulations. However, this is because OP27 has a sharp and clear $1/f$ noise behavior. In other devices, the noise model could be complex (like $1/f^\alpha$ with noises), and the agreement with analytical expressions of Section 9.1 could no longer match.
- The decrease of the input-referred noise is mostly due to the bandwidth decrease because the architecture shows a constant gain-bandwidth product.
- The current noise NBW (not shown) could be different from the voltage noise NBW.

Table 7.1 Data from the OP27 datasheet used in the example.

DATA	Value	Units
Open loop gain	125	[dB]
GBW	8.0	[MHz]
v_N	3.0	[nV/$\sqrt{\text{Hz}}$]
i_N	40.0	[pA/$\sqrt{\text{Hz}}$]
f_{CE}	2.7	[Hz]
f_{CI}	140	[Hz]
Power supply	±5	[V]

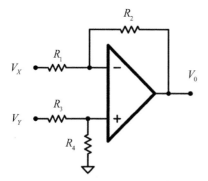

Figure 7.33 Example of calculation of noise in a differential amplifier configuration.

Table 7.2 Comparison between Eq. (7.65) and simulations with LTSpice.

Gain	Output-rereferred noise (7.65)	Output-referred noise (LTSpice)	Abs. Error [%]	Incidence pink noise [%]	Input-referred noise (7.50)	Input-referred noise (LTSpice)	Abs. Error [%]	Voltage NBW	ENOB
1	25.3 μV	22.1 μV	15	0	25.3 μV	23.1 μV	10	7.2 MHz	18.3
10	73.6 μV	63.8 μV	4	0	7.36 μV	6.6 μV	12	1.1 MHz	16.7
100	231 μV	260.3 μV	12	0.1	2.31 μV	2.2 μV	5	124 kHz	15.1
1000	731 μV	844 μV	13	0.8	73.1 μV	70.3 μV	4	13 kHz	13.4
10 k	2.34 mV	2.59 mV	13	3.2	234 nV	227 nV	3	1.3 kHz	11.7
100 k	8.0 mV	8.6 mV	10	20	80 nV	83 nV	4	156 Hz	10.0

- Note that even if for high gain, the input-referred noise decreases for bandwidth reduction, the equivalent resolution levels quantified by ENOB decreases for the reduction of the input full-scale.

7.9.3 Noise Efficiency Factor and Power Efficiency Factor

An efficiency factor has been introduced to compare different amplifying architectures in terms of noise. The idea is to compare all architectures with an asymptotical and ideal case. For this purpose, if we take the expression for the input-referred noise of a BJT where we assume negligible base resistance $R_B \to 0$, negligible effect of noise due to input current $I_B \to 0$, we have from (7.41)

$$\overline{v_N^2} = 4kT \frac{r_{BE}}{2\beta_0} \Delta f = \frac{2kT}{g_m} \Delta f \equiv [V^2].$$ (7.71)

This is considered an "ideal" case since it considers the shot noise across a single ideal electric potential barrier set directly by the input voltage. Therefore, the BJT implements an idealistic situation of shot noise against an electric potential barrier. In other device examples, such as in subthreshold MOS, the electric potential barrier is set by gate voltage through a capacitance divider defining a *gain slope factor*, thus not in an ideal valve-effect condition. Thus, in the noise bandwidth $NBW = BW \cdot \pi/2$, the total noise of this ideal case for a differential couple is

$$v_N = \sqrt{BW \frac{\pi}{2} \frac{4kT}{g_m}} = \sqrt{BW \frac{\pi}{2} \frac{4kTV_T}{I_{tot}}} \equiv [V],$$ (7.72)

where $V_T = kT/q$ is the thermal voltage and I_{tot} is the current supplying the amplifier.

Now, the noise of an amplifying architecture is compared with this ideal case using the *noise efficiency factor* (NEF)

$$\text{NEF} = v_{iN} \cdot \sqrt{\frac{2I_{tot}}{BW\pi \cdot 4kTV_T}} \equiv [\cdot], \qquad (7.73)$$

where v_{iN} is the input-referred noise of the architecture to characterize and I_{tot} is the total current supplied to the amplifier (e.g., that of the differential pair). For a differential pair of "ideal" transconductors, NEF = 1. NEF has been conceived for classical continuous-time architectures. It should be pointed out that for advanced techniques such as *current reusing* and *signal chopping*, NEFs could be lower than 1.

NEF has been defined by looking at a classical analog quality factor: the ratio of the current supplied and the transconductance I/g_m. However, in more advanced architectures, it is better to take into account the total power supplied, thus defining the *power efficiency factor* (PEF) as

$$\text{PEF} = \overline{v_{iN}^2} \cdot \frac{2P_{tot}}{BW\pi \cdot 4kTV_T} = \text{NEF}^2 V_{DD} \equiv [\text{V}], \qquad (7.74)$$

which may be interpreted as the ratio of the amplifier's power consumption to the power required by an "ideal" transconductor pair operating at a 1-V supply to get the same input-referred noise.

7.10   Capacitively Coupling Amplifier Techniques

Capacitively coupling techniques have two advantages: first, a capacitor is intrinsically noiseless, and we could get better overall performances with respect to resistor-biased architectures. Second, thanks to CMOS technology's very high impedance, we could use the discrete-time approach, such as in switched capacitor techniques.

7.10.1   Continuous-Time Techniques for Voltage Sensing

The very high input impedance of CMOS OPAMPs easily allows configurations based on negative feedback as a capacitively coupling inverting amplifier, as shown in Fig. 7.34.

Assuming for the OPAMP the dominant pole approximation where f_0 is the open-loop first-order cutoff frequency and using the same expressions of Section 6.9 where $Z_1 \leftarrow C_i // C_L$ and $Z_2 \leftarrow C_F$ we can easily find as for (7.63) the *signal gain* as

$$\frac{V_o}{V_i} = -\frac{C_I}{C_F + C_T/A_0} \cdot \frac{1}{1 + j(f/f'_0)} = -\frac{C_I}{C_F + C_T/A_0} \cdot H'(f) \approx -\frac{C_I}{C_F} \cdot H'(f)$$

$$\text{where } f'_0 = f_0\left(1 + A_0\frac{C_F}{C_T}\right). \qquad (7.75)$$

where, again, $H'(f)$ is the feedback transfer function and f'_0 is the new cutoff frequency of the system given by the feedback effect and where $C_T = C_L + C_F + C_I$.

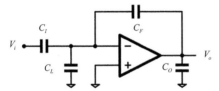

Figure 7.34 CMOS OPAMP inverting amplifier architecture.

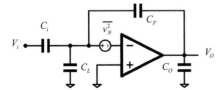

Figure 7.35 Noise model of a capacitively coupling inverting amplifier.

To calculate the noise in the circuit of Fig. 7.35, we can use the expression already discussed in Section 7.7 for CMOS OPAMPs

$$\overline{v_N^2}(f) = 2 \cdot \frac{8}{3} \cdot \frac{kT}{g_m} + \frac{2K_f}{C_{OX}WL} \cdot \frac{1}{f} \tag{7.76}$$

Therefore, the *noise gain* could be calculated as

$$\overline{v_{oN}^2}(f) = \left|1 + \frac{Z_2}{Z_1}\right|^2 |H'(f)|^2 \overline{v_N^2}(f); \text{ where} \tag{7.77}$$
$$Z_1 \leftarrow C_i // C_L \text{ and } Z_2 \leftarrow C_F.$$

Therefore, it follows that

$$\overline{v_{oN}^2}(f) = \left(\frac{C_T}{C_F + C_T/A_0}\right)^2 |H'(f)|^2 \overline{v_N^2}(f). \tag{7.78}$$

Now, we assume that the OPAMP is implemented in a transimpedance amplifier configuration (TIA), which means the amplifier could be modeled by a voltage-controlled current-generator, terminated on an output conductance as the circuit shown in Fig. 7.36A. Using the *dominant-pole* approximation, where we assume that the output capacitance gives the main time constant, we can easily calculate the generic OPAMP noise bandwidth as shown in Fig. 7.36.

As shown in Fig. 7.36B, the small-signal gain is given by g_m/g_o and the dominant pole by g_m/C_o so that the gain-bandwidth product is

$$f_0 A_0 = \frac{g_m}{g_o} \frac{g_o}{2\pi C_O} = \frac{g_m}{2\pi C_O}. \tag{7.79}$$

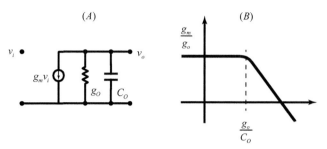

Figure 7.36 Trans-impedance amplifier model using the dominant pole approximation to calculate the NBW.

Using the above expression, we can get the noise bandwidth from the feedback cutoff frequency f_0' as

$$f_0' = f_0 \left(1 + A_0 \frac{C_F}{C_T}\right) \approx f_0 A_0 \frac{C_F}{C_T} = \frac{g_m}{2\pi C_O} \frac{C_F}{C_T} \rightarrow$$
$$\mathrm{NBW} = \frac{\pi}{2} f_0' = \frac{\pi}{2} f_0 \left(1 + A_0 \frac{C_F}{C_T}\right) \approx \frac{g_m}{4 C_O} \frac{C_F}{C_T}. \tag{7.80}$$

Now, the output-referred noise (for thermal noise only) could be easily calculated as the noise power density multiplied by the in-bandwidth noise gain and by the noise bandwidth

$$\overline{v_{oN}^2} = \left(\frac{C_T}{C_F + C_T/A_0}\right)^2 \cdot \mathrm{NBW} \cdot \overline{v_N^2}(f) =$$
$$= \left(\frac{C_T}{C_F + C_T/A_0}\right)^2 \cdot \frac{g_m}{4 C_O} \frac{C_F}{C_T} \cdot \frac{16\, kT}{3\, g_m} \approx \frac{4\, kT}{3} \frac{C_T}{C_O\, C_F}, \tag{7.81}$$

where $C_T = C_F + C_i + C_L$. Note how this calculation was possible because the bandwidth is fixed by the pole of the OPAMP itself. The final result of Eq. (7.81) is the typical expression of the output thermal noise power of a capacitance OPAMP amplifier.

Furthermore, the input-referred *minimum detectable signal* $\overline{v_{iN}^2}$ could be easily calculated by dividing the output noise power $\overline{v_{oN}^2}$ by the square of the in-bandwidth signal gain

$$\overline{v_{iN}^2} = \overline{v_{oN}^2} / \left(\frac{C_I}{C_F}\right)^2 = \frac{4\, C_F}{3\, C_I^2} \frac{C_T}{C_O}. \tag{7.82}$$

Note that Eqs. (7.81) and (7.82) do not hold for $C_F \rightarrow 0$ because of approximations. We can calculate input- and output-referred noise in this case, recalling that for $C_F \rightarrow 0$ we have

$$C_F \to 0$$
$$\text{NBW} \to \frac{g_o}{4C_O} \quad \text{and}$$
$$\overline{v_{oN}^2} \to \frac{4\,kT}{3\,C_O}A_0; \quad \overline{v_{iN}^2} \to \frac{4\,kT}{3\,C_O}\frac{1}{A_0}\left(\frac{C_I + C_L}{C_I}\right)^2, \tag{7.83}$$

which is consistent with the output- and input-referred noise of an open-loop OPAMP configuration.

7.10.2 Continuous-Time Techniques for Current Sensing

Current sensing is a well-known technique widely used in electronic sensors such as radiation detectors, impedance spectroscopy interfaces, mechanical sensors, as well as biosensors. The simplest technique to sense a current is based on a transimpedance amplifier (TIA), as illustrated in Fig. 7.37.

In the TIA scheme, the input current I_i is transformed into an output voltage V_o by using an OPAMP in negative feedback. As before, we will focus on using CMOS OPAMPs with high impedance inputs. Following the same procedure explained in Section 7.10.1, we get the signal gain as

$$\frac{V_o}{I_i} = -Z_2 \cdot H'(f) = -\frac{R_F}{1 + j2\pi f R_F C_F} \cdot H'(f)$$
$$\text{where}: \ H'(f) = \frac{1}{1 + j(f/f'_0)} \quad \text{and}: \tag{7.84}$$
$$f'_0 \approx f_0 A_0 \frac{Z_1}{Z_1 + Z_2} = f_0 A_0 \frac{1 + j2\pi f R_F C_F}{1 + j2\pi f R_F C_T} \approx f_0 A_0,$$

where $Z_1 \leftarrow C_L$ and $Z_2 \leftarrow R_F // C_F$ and $C_T = (C_F + C_L)$. Note that for $C_F \to 0$ we have the basic *current-to-voltage transimpedance amplifier*, where for low frequency, we have

$$\frac{V_o}{I_i} = -R_F. \tag{7.85}$$

The presence of C_F is motivated by the fact that it could easily fix the system's bandwidth. In this case, we will always assume that the pole introduced by C_F is lower

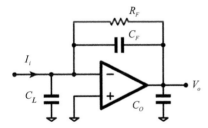

Figure 7.37 Current sensing by a transimpedance amplifier (TIA).

than that given by the negative feedback of the OPAMP: $1/(2\pi C_F R_F) << f_0'$. Conversely, we can use a large value of R_F and use C_F as an integrating capacitor of the current. In this case, the TIA implements an integrator to be used in a discrete-time approach, as illustrated in Section 7.10.3. Using (7.84), we can easily find that the noise bandwidth is given by

$$\text{NBW} = \frac{1}{4R_F C_F}. \tag{7.86}$$

Similarly, using the same procedures, we can define the input-referred noise PSD as

$$\overline{i_{iN}^2}(f) = \frac{4kT}{R_F} + \overline{v_N^2}(f)\left[\frac{1}{R_F} + (2\pi f)^2 C_T\right], \tag{7.87}$$

where $\overline{v_N^2}(f)$ is the OPAMP input noise power density given by $(7.76)^2$ and again $C_T = (C_F + C_L)$.

The input-referred noise power (for thermal noise only) can thus be calculated as

$$\overline{i_{iN}^2} = \overline{i_{iN}^2}(f) \cdot \text{NBW} = \frac{kT}{R_F}\left[\frac{1}{R_F C_F} + \frac{4}{3}\frac{kT}{g_m R_F}\right], \tag{7.88}$$

and the output-referred noise power as

$$\overline{v_{oN}^2} \approx \overline{i_{iN}^2} \cdot R_F^2 = \frac{kT}{C_F} + \frac{4}{3}\frac{kT}{g_m}. \tag{7.89}$$

Note how $\overline{i^2}_{iN}$ is strongly dependent on R_F and one may be induced to increase its value to reduce the noise. However, the result is obtained at the bandwidth reduction cost, as apparent from (7.86). Figure 7.38 shows an input-referred noise power spectrum for standard TIA with finite feedback resistance and infinite feedback resistance.

Equation (7.87) shows how the input-referred noise PSD shape depends on the value of R_F or C_F. For low values of R_F there is a low-frequency noise floor set by $4kT/R_F$ that dominates at low frequencies. For higher frequencies, the input-referred noise increases asymptotically as f^2, by means of the total capacitance connected to the input node C_T. Clearly, the total noise power will decrease for higher R_F, converging to the case of a typical charge integrator. Thus, replacing R_F with a noise-free capacitor represents the optimum choice from the noise standpoint, but the mode of operation should change, as discussed.

To summarize, we have two opposite cases:

1. For $R_F << 1/\omega C_F$ the current interface is acting like a typical TIA.
2. For $R_F >> 1/\omega C_F$ the interface is behaving as an integrator. However, the integrator must be followed by a differentiator to restore the original transfer function.

---

[2] The expression (7.87) is also found in the literature with an additional term $2qI_{in}$ to model the shot noise of input devices current I_{in} whenever it applies such as in BJT or JFET cases.

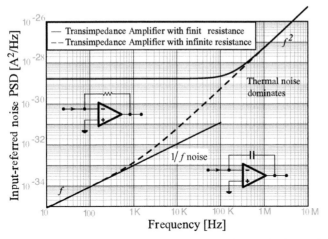

Figure 7.38 Input-referred noise power spectrum for standard TIA with finite feedback resistance RF=1 GΩ (solid line) and infinite feedback resistance. The graph refers to an OPAMP described by a thermal noise voltage of approximately 3 nV/√Hz and C_T= 1.2 pF. [From Crescentini, 2015]

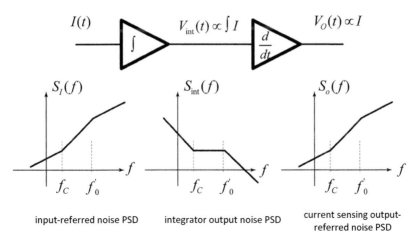

Figure 7.39 PSD noise shapes at different stages of the integrator-differentiator current sensing scheme.

A charge integrator senses the charge that is the integration function of the current. Therefore, the charge integrator has to either be followed by a differentiator block or work in discrete time mode, which approximates the differentiation function, as we will see.

It is interesting to see the evolution of the noise spectral density in case 2, as shown in Fig. 7.39.

The output of the first stage is the typical output of an integrator considering flicker noise and the cutoff frequency of the feedback configuration. Therefore, if we step back to the input using the integration function, we can find again the behavior of input noise shown in Fig. 7.38. However, we have to put a differentiator at the output of the first stage to get the original signal.

As shown in Figs. 7.38 and 7.39, the current sensing is weakly dependent on flicker noise due to the effect of either the integrating function or of the noise dominance of R_F.

7.10.3   Capacitively Coupling Amplifiers in Discrete-Time Techniques

The circuits that will be analyzed in this section are related to a circuit technique where instead of working in an analog domain, the signal is identified with amounts of charge shared between capacitors in a fixed amount of time, called *sampling time*. This discrete-time analysis approach is referred to as *switched capacitor circuits* technique.

Referring to Fig. 7.40, the charge variation ΔQ into a capacitor C in a fixed amount of time T_S is $\Delta Q = C(\Delta V_A - \Delta V_B)$ where ΔV_A and ΔV_B are the voltage variations of the two terminals between two adjacent sampling periods.

Note how the linear relationship between *variations* of charges and voltages is similar to what happens in Ohm's law between current and voltage. Consequently, we can apply Kirchhoff laws on nets and nodes on charge and voltage variations.

A basic sensor interface is referred to as a *charge amplifier*. The charge amplifier is an electronic system that measures an amount of charge during a sampling time by providing an output voltage variation. Figure 7.41 shows a typical implementation of a charge amplifier by using an OPAMP. The amplifier works as follows: An input charge induces an increase of the voltage of the inverting input of the OPAMP V_-. In a negative feedback scheme, the amplifier reacts by restoring the original voltage value

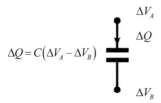

$$\Delta Q = C\left(\Delta V_A - \Delta V_B\right)$$

Figure 7.40 Switched capacitor calculation technique.

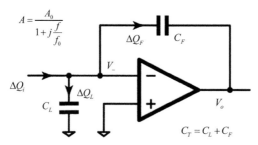

$$A = \frac{A_0}{1 + j\dfrac{f}{f_0}}$$

$$C_T = C_L + C_F$$

Figure 7.41 Charge integration amplifier.

by sinking from the input an equal amount of charge ΔQ_F. This is possible only by decreasing output voltage V_o. Therefore, the output voltage change ΔV_o is a measure of the input charge ΔQ_i.

To calculate the input–output relationship, we can apply the Kirchhoff law on the input node as

$$\Delta Q_i = \Delta Q_L + \Delta Q_F = C_L \Delta V_- + C_F (\Delta V_- - \Delta V_o) = -\frac{C_L + C_F}{A} \Delta V_o - C_F \Delta V_o$$
$$\rightarrow \frac{\Delta V_o}{\Delta Q_i} = -\frac{1}{C_F + \dfrac{C_T}{A}} \approx -\frac{1}{C_F},$$

$$(7.90)$$

where A is the open-loop gain of the OPAMP. Therefore, the transfer function of the charge amplifier is simply given by the feedback capacitor. Note that in this case, the amplifier gain is not infinite $\Delta Q_F \neq \Delta Q_i$ and V_- is not exactly zero. This is a common effect of negative feedback systems when they do not have infinite open-loop gain.

If we want to understand the time response of the charge amplifier, we have to introduce a dominant pole approximation of the OPAMP as usual

$$\frac{\Delta V_o}{\Delta Q_i} = -\frac{1}{C_F + \dfrac{C_T}{A_0}\left(1 + j\dfrac{f}{f_0}\right)} = -\frac{1}{\left(C_F + \dfrac{C_T}{A_0}\right) + j\dfrac{C_T f}{A_0 f_0}}$$

$$= -\frac{1}{\left(C_F + \dfrac{C_T}{A_0}\right)} \cdot \frac{1}{\left(1 + j\dfrac{f}{f'_0}\right)} = -\frac{1}{C_F + \dfrac{C_T}{A_0}} \cdot H'(f) \qquad (7.91)$$

where $f'_0 = f_0 \left(1 + A_0 \dfrac{C_F}{C_T}\right),$

where $H'(f)$ is the feedback transfer function of the system as already used in Section 7.9. To summarize, we can draw the following conclusions:

- The charge amplifier input–output relationship for an infinite gain OPAMP is given by C_F.
- If the gain is finite, the relationship is given by (7.90).
- The larger C_F, the lower sensitivity.
- The larger C_F, the faster response.
- The larger C_L, the slower the response.

7.10.4   Reset Techniques and Related Problems

Since the circuit of Fig. 7.41 should work in discrete time, it needs to be reset. There are several ways to do that; however, one of the most simple architectures is shown in Fig. 7.42, where a switch is inserted between the output and the OPAMP input.

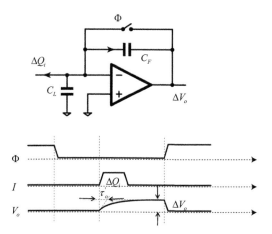

Figure 7.42 Charge amplifier operation technique.

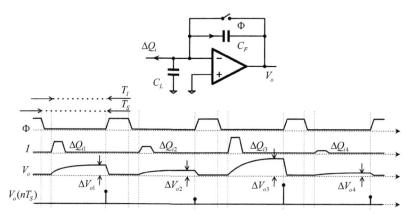

Figure 7.43 Operating principle of a discrete-time charge amplifier.

When the switch has closed, the charge accumulated in C_F is reset, and the OPAMP is placed as in buffer configuration. As shown in Fig. 7.42, after the amplifier is reset, the switch is opened so as it is ready to receive the input charge to quantify. The output settling time $\tau_o' = 1/f_o'$ of the output depends on the feedback cutoff frequency as indicated by (7.91).

A repeated operation of reset and readout of a charge amplifier is shown in Fig. 7.43, where different quantities of charge $(\Delta Q_{i1}, \Delta Q_{i2} \ldots \Delta Q_{i4})$ are injected in the amplifier during the sampling times. The amount of the output voltage swings $(\Delta V_{o1}, \Delta V_{o2} \ldots \Delta V_{o4})$ depends on the input charge. Following the discrete-time architecture, once the output is settled in a stable point, we can use an A/D to convert the output into a digital value every sample time.

The discrete-time operation of a charge amplifier is an example of a generic *integrating sampling* technique, as shown in Fig. 7.44. A generic input is integrated for an amount of time T_I called *integration time*. Then, the amplifier is reset, and the

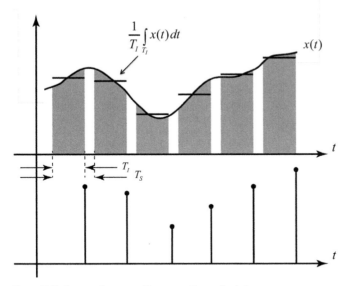

Figure 7.44 Integrating sampling operation principle.

output is read out at a period T_S. Therefore, T_I is a fraction of T_S. This procedure could be useful for using a charge integrator to sense the current, as illustrated in Fig. 7.39, where, following the scheme, the current could be first integrated by a charge amplifier while the sampling process itself performs the differentiation. Therefore, using a charge amplifier as the current amplifier in discrete-time mode implies that its transfer function is the same as that of a transimpedance amplifier (7.85), where the feedback resistance is given by

$$\frac{V_o}{I_i} = -R_{EQ}; \text{ where}$$
$$R_{EQ} = \frac{T_I}{C_F} \approx \frac{T_S}{C_F} = \frac{1}{f_S C_F}. \tag{7.92}$$

The discrete-time approach could be used even for voltage amplification instead of charge amplification, as illustrated in Fig. 7.45. This could be accomplished by simply using a capacitor C_I as a voltage-to-charge converter (Fig. 7.45A).

Thus, following the switched capacitor analysis technique, we get the transfer function of the voltage amplifier:

$$\frac{\Delta V_o}{\Delta V_i} = -\frac{C_I}{C_F + \dfrac{C_T}{A_0}} \cdot \frac{1}{\left(1 + j\dfrac{f}{f'_0}\right)} = -\frac{C_I}{C_F + \dfrac{C_T}{A_0}} \cdot H'(f) \approx -\frac{C_I}{C_F} \cdot H'(f)$$
$$\text{where } f'_0 = f_0 \left(1 + A_0 \frac{C_F}{C_T}\right), \tag{7.93}$$

where again $H'(f)$ is the feedback transfer function.

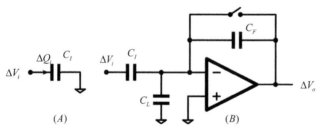

Figure 7.45 Discrete-time voltage amplifier.

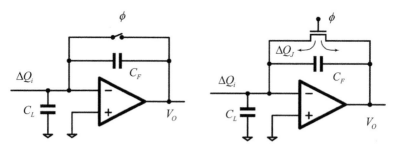

Figure 7.46 The charge injection problem due to the reset transistor.

The reset process has a cost in terms of noise and offset since the switch is usually implemented with a MOS transistor, as shown in Fig. 7.46. When it is switched off, part of the channel charge Q_J could be injected into the input as if it were a signal generating an output offset. This is called MOS channel *charge injection*. It could be demonstrated that the charge injected in the input is inversely proportional to the product of the switch-off time and the gain-bandwidth product of the OPAMP. Thus, the larger the GBW and the slower the switch off, the lower the charge injection into the input.

Another problem is because the switch-off transistor has a resistance generating noise. As illustrated in Fig. 7.47, we can calculate the output-referred noise by determining the noise gain with respect to the noise generator of the reset switch $\overline{v_N^2}$. Thus, the noise gain is

$$
v_- = v_o \frac{\dfrac{1}{sC_L}}{\dfrac{1}{sC_L} + \dfrac{R}{1 + sRC_F}} + v_N \frac{\dfrac{1}{sC_T}}{R + \dfrac{1}{sC_T}}
$$

$$
= v_o \frac{1 + sRC_F}{1 + sRC_T} + v_N \frac{1}{1 + sRC_T} = -\frac{v_o}{A} \tag{7.94}
$$

$$
\text{For } A \gg 1 \quad \frac{v_o}{v_N} \simeq \frac{1}{1 + sRC_F}.
$$

Therefore, by integrating over the total spectrum, we get

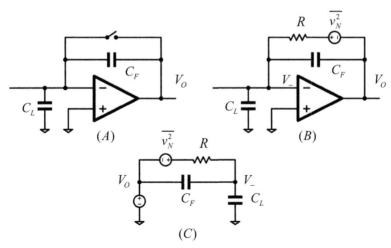

Figure 7.47 Reset transistor noise calculation on a charge amplifier (A), equivalent noise model (B), and equivalent linear model (C).

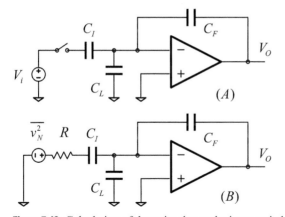

Figure 7.48 Calculation of the noise due to the input switch.

$$\overline{v_{oN}^2} = \int_0^\infty 4kTR \cdot \left|\frac{v_o}{v_N}\right|^2 df = \frac{kT}{C_F}; \quad \text{For } A < \infty \ \overline{v_{oN}^2} = \frac{kT}{C_F + \dfrac{C_T}{A_0}}. \tag{7.95}$$

Another important noise source is the switch used to connect the input to the amplifier. As shown in Fig. 7.48, the real switch has a resistance that generates thermal noise. Therefore, we can easily see that for $A_0 \gg 1$ the noise gain is

$$\frac{v_o}{v_N} = \frac{C_I}{C_F} \frac{1}{1 + sRC_I}. \tag{7.96}$$

Therefore, the noise bandwidth is $1/(4RC_I)$ and the output-referred thermal noise is given by

$$\overline{v_o^2} = \overline{v_N^2}\left(\frac{C_I}{C_F}\right)^2 \cdot \mathrm{NBW} = \left(\frac{C_I}{C_F}\right)^2 \frac{\overline{v_N^2}}{4RC_I} = \left(\frac{C_I}{C_F}\right)^2 \frac{kT}{C_I}. \tag{7.97}$$

The noise analysis done so far is related to the switches in continuous mode (i.e., before they are opened, such as in the kTC noise analysis). However, since the charge amplifier is operating in discrete-time cyclic mode, we have to introduce new concepts related to discrete-time noise analysis.

7.10.5 Summary of Interface Techniques Using a Capacitively Coupling Trans-Impedance Amplifier

In Fig. 7.49 are shown the most common sensing interfaces using operational amplifiers already discussed in previous sections, along with their transfer functions.

For capacitive sensing, we can sense either the input capacitor or the feedback capacitor:

- In the case we use the input capacitance, we can apply repetitive input voltage steps of equal level and read the variation of the variation of the output ($\Delta\Delta V_o$) or fix the input voltage to a reference (V_R) and look at the variation of the output (ΔV_o).
- If we use to sense the variations of the feedback capacitor, we might apply the same two techniques as before even if the transfer function relationship is reverted. For example, if we sense the input capacitance by keeping the same input voltage tied to a reference, we have

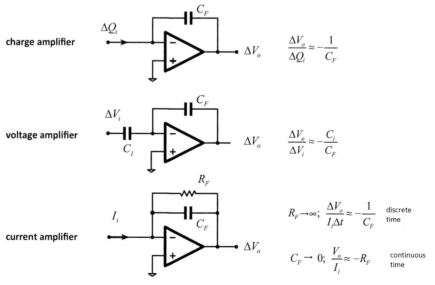

charge amplifier

$$\frac{\Delta V_o}{\Delta Q_i} \approx -\frac{1}{C_F}$$

voltage amplifier

$$\frac{\Delta V_o}{\Delta V_i} \approx -\frac{C_i}{C_F}$$

current amplifier

$$R_F \to \infty; \quad \frac{\Delta V_o}{I_i \Delta t} \approx -\frac{1}{C_F} \quad \text{discrete time}$$

$$C_F \to 0; \quad \frac{V_o}{I_i} \approx -R_F \quad \text{continuous time}$$

Figure 7.49 Various sensor interfaces for a charge, voltage, and current sensing.

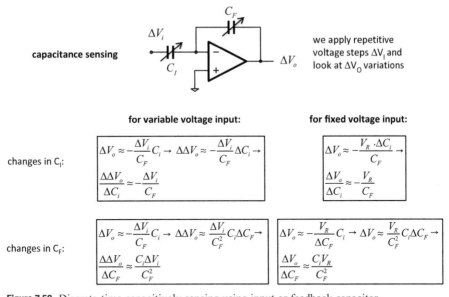

Figure 7.50 Discrete-time capacitively sensing using input or feedback capacitor.

$$\frac{\Delta V_o}{\Delta C_I} = -\frac{V_R}{C_F}. \tag{7.98}$$

For other combinations, refer to Fig. 7.50.

Of course, if we use discrete-time techniques, we have to reset the amplifier, as already discussed.

7.11 Noise Folding in Discrete-Time Techniques

Since the signal is sampled in the discrete-time approach, the Nyquist–Shannon sampling theorem should be observed. In Fig. 7.51A, an ideal sampler using a Dirac function is shown. However, it is better to treat the argument using PSD than Fourier transforms from the noise perspective. It could be shown that for Gaussian white noise, thanks to uncorrelation between samples, the power spectrum of an ideally sampled random signal is given by

$$S_S(f) = \sum_{k=-\infty}^{k=\infty} S\left(f - \frac{k}{T_S}\right), \tag{7.99}$$

where $S(f)$ is the double-sided power spectrum of the random signal and $T_S = 1/f_S$ is the sampling period. If the noise is bandwidth limited, as illustrated in Fig. 7.52 and we undersample noise with respect to the Nyquist–Shannon theorem, overlapping of noise spectra called *noise folding* occurs as discussed in Chapter 3.

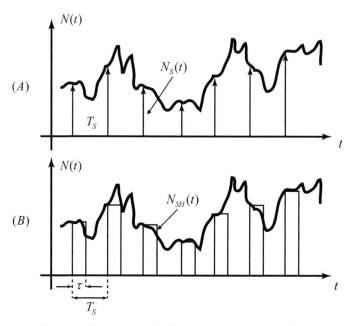

Figure 7.51 Sampling noise with a Dirac function (A) and using a sample and hold function (B).

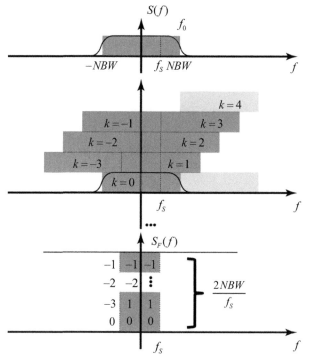

Figure 7.52 Noise folding principle. In the picture, it is illustrated the case where $NBW = 2f_S$

Assuming that the sampling frequency f_S is smaller than NBW, we can define a parameter referred to as *undersampling ratio (USR)*. If a single-pole function limits the bandwidth, we have

$$\text{USR} = \frac{2\text{NBW}}{f_S} = \pi \frac{f_0}{f_S} = \pi f_0 T_S. \tag{7.100}$$

Now, the (double-sided) folded power spectrum defined by Eq. (7.99) becomes

$$S_S(f) = \sum_{k=-\infty}^{k=\infty} S\left(f - \frac{k}{T_S}\right) \approx \text{USR} \cdot S(0) = \pi f_0 T_S \cdot S(0), \tag{7.101}$$

where $S(0)$ is the in-bandwidth (double-sided) spectral density. For example, in the case of an ideal sampling of thermal noise on a capacitor by means of a switch of resistance R, we have that for $f_0 = 1/2\pi RC$, and $S_0 = 2kTR$ the *two-sided* PSD is

$$S_S(f) \approx \text{USR} \cdot S(0) = \pi \frac{1}{2\pi RC f_S} \cdot 2kTR = \frac{1}{f_S} \frac{kT}{C}. \tag{7.102}$$

This means that the sampled power spectral density is the kTC noise divided by the sampling frequency. It should be noted that if we integrate that noise in the Nyquist baseband so that $B = f_S/2$ we have

$$P_N = \int_{-B}^{B} S_S(f) df \approx \frac{kT}{C}. \tag{7.103}$$

Therefore, we get the same result of continuous-time.

Another way for sampling noise is shown in Fig. 7.51B, where the random signal is sampled every T_S and hold for a period of time τ. The procedure is referred to as *sample and hold*.

In this case, it can be shown that the power spectrum is

$$S_{SH}(f) = \left(\frac{\tau}{T_S}\right)^2 \text{sinc}^2\left(\left(\frac{\tau}{T_S}\right)f T_S\right) \sum_{k=-\infty}^{k=\infty} S\left(f - \frac{k}{T_S}\right), \tag{7.104}$$

where $\text{sinc}(x) = \sin(\pi x)/(\pi x)$, as usual.

Now, let us assume that thermal noise generated by a resistor R is periodically sampled at f_S on a capacitor C (Fig. 7.53), where $\overline{v_N^2}(f) = 2kTR$ is the (double-sided) noise power density of the resistor R. The goal is to understand the power spectral density of the output $\overline{v_{oN}^2}(f)$. As discussed previously, energy storage elements cause the system to exhibit a shaped PSD response.

As illustrated in Fig. 7.54A, when the switch is closed, the noise is filtered by an RC low-pass filter whose spectral density is shaped by a Lorentzian

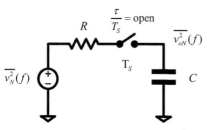

Figure 7.53 Periodic sampling of thermal noise.

$$S(f) = \overline{v_{oN}^2}(f) = \frac{2kTR}{1 + (f/f_0)^2}.$$ (7.105)

However, when the switch is opened, the capacitor voltage is frozen to the value that was immediately before the opening, as discussed in the kTC noise. Therefore the output voltage $v_{oN}(t)$ is composed of two independent contributions: a filtered white noise that acts only in a windowed amount of time (Fig. 7.54B) and a sample and hold noise (Fig. 7.54C). Since the effects of each sampling are independent of all the others (the autocorrelation function decays in times smaller than T_S because $f_0 > f_S$), we can write the total two-sided PSD at the output as the sum of two contributions

$$S_o(f) = S_R(f) + S_{SH}(f),$$ (7.106)

where $S_R(f)$ is the windowed noise of the system and $S_{SH}(f)$ is the sample and hold noise on the capacitor.

For $S_{SH}(f)$, using the relationships (7.101) and (7.104) with the expression of (7.105), we get the expression

$$S_{SH}(f) = \left(\frac{\tau}{T_S}\right)^2 \sin c^2 \left(\frac{\tau}{T_S}fT_S\right) \cdot \frac{1}{f_S}\frac{kT}{C} \approx \left(\frac{\tau}{T_S}\right)^2 \cdot \frac{1}{f_S}\frac{kT}{C}$$ (7.107)

if $f \ll f_S$.

For $S_R(f)$, we simply take into account that the noise is acting on a window of time; thus

$$S_R(f) = \left(1 - \frac{\tau}{T_S}\right)\frac{2kTR}{1 + (f/f_0)^2}.$$ (7.108)

Therefore, the total single-sided power spectral density is given by

$$\hat{S}(f) = \left(1 - \frac{\tau}{T_S}\right)\frac{4kTR}{1 + (f/f_0)^2} + \left(\frac{\tau}{T_S}\right)^2 \sin c^2 \left(\frac{\tau}{T_S}fT_S\right) \cdot \frac{1}{f_S}\frac{2kT}{C}.$$ (7.109)

It is easy to show that by integrating from zero to infinity the total PSD we have

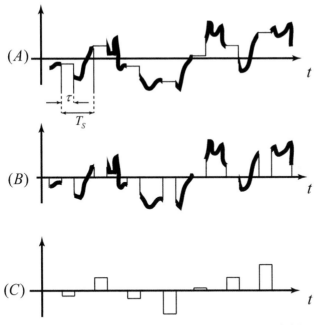

Figure 7.54 Partition of switched-capacitor noise. (A) Sampled thermal noise. (B) Modulated thermal noise. (C) Sample/hold noise.

$$\int_0^\infty \hat{S}(f)df = \left(1 - \frac{\tau}{T_S}\right)\frac{kT}{C} + \left(\frac{\tau}{T_S}\right)^2 \cdot \frac{1}{f_S}\frac{2kT}{C}\frac{1}{2\tau} = \frac{kT}{C}. \tag{7.110}$$

It is easy to show that if $\tau/T_S > 60\%$ SH noise is 8 times larger than windowed thermal noise and for $\tau/T_S > 75\%$ it is 30 times larger.

To summarize, the following rules could be outlined:

- A capacitor that is periodically charged by a resistor shows in the Nyquist baseband an aliasing effect called *noise folding*;
- Noise folding is the dominant noise effect if (1) the charge time constant (that fixes NBW) is lower than the sampling period, following (7.100); and (2) the sampling and hold period is the larger part of the sampling period.

7.11.1 Noise in Discrete-Time Capacitively Coupling Amplifiers

The output noise of a charge amplifier expressed by Eq. (7.80) is valid in continuous time. When a charge amplifier is reset in discrete-time mode, the noise is folded. We can evaluate this considering (7.76), (7.78), (7.79), and (7.101) so that we have

$$\overline{v_{oN}^2} \simeq \left(\frac{C_T}{C_F}\right)^2 \frac{16\,kT}{3\,g_m} \cdot 2\text{USR} = \left(\frac{C_T}{C_F}\right)^2 \frac{16\,kT}{3\,g_m} \cdot \frac{4\text{NBW}}{f_S} = \frac{C_T}{C_F}\frac{16\,kT}{3}\frac{1}{C_o f_S}, \tag{7.111}$$

Table 7.3 Noise sources of a CMOS amplifier operating in discrete time.

Noise source	Output-referred (single-sided) noise PSD $\overline{v_{on}^2}(f) \equiv [V^2/\text{Hz}]$
OPAMP noise	$\dfrac{1}{f_S} \dfrac{16}{3} \dfrac{kT}{C_O} \dfrac{C_T}{C_F}$
Reset switch	$\dfrac{1}{f_S} \dfrac{2kT}{C_F}$
Input switch	$\dfrac{1}{f_S} \dfrac{2kT}{C_I} \left(\dfrac{C_I}{C_F}\right)^2$

These values could be applied to both charge amplifiers, voltage amplifiers, and current amplifiers.[3]

where factor 2 has been used since the USR factor has been used in (7.101), which is a double-sided PSD, while in Eq. (7.111), we are referring to a single-sided PSD.

We can calculate input-referred in the various configurations with capacitance coupled amplifiers. For instance, for the voltage amplifier operating in discrete time, we can summarize noise sources in Table 7.3.

The plots in Fig. 7.55 show the comparison of the expressions of Table 7.3 with simulations using Spectre noise simulator based on harmonic balance computation.

7.11.2 Summary of Input-Referred Noises in Common Discrete-Time Interfaces

It is interesting to note that discrete-time operation of CMOS amplifiers, as evidenced in Table 7.3, have a common background for *charge amplifiers*, *voltage amplifiers*, and *current amplifiers*. If we refer, for simplicity, only to the thermal noise of OPAMP (first line of Table 7.3), we could derive the following conclusion.

Now, we can also calculate the input-referred *minimum detectable signal* of the three CMOS interfaces according to the relationship

$$\overline{\xi_{iN}^2} \sim \frac{\overline{v_{oN}^2}(f) \cdot f_S/2}{|H(0)|^2}, \tag{7.112}$$

where $\overline{\xi_{iN}^2}$ could be either charge, voltage, or current and $H(0)$ is the in-bandwidth transfer function of the interface. We expect to integrate output noise up to the maximum frequency, which is half the sampling one.

Therefore, we could derive Table 7.4 for input noise power

Of course, if we want to consider all the other noise sources summarized in Table 7.4, we have to repeat the procedure of (7.112) for each source and sum them all quadratically.

---

[3] Important note: the expressions could vary according to the structure and clock strategy of the amplifier; therefore, the reader should carefully apply the expressions to the specific context.

Table 7.4 Input-referred noise of discrete-time CMOS interfaces related to only the thermal noise of the OPAMP.

Interface	Transfer function $H(0)$	Ref.	Input-referred noise $\overline{\xi_{iN}^2} \equiv [\xi^2]$	Equation
Charge sensing	$\dfrac{\Delta V_o}{\Delta Q_i} = -\dfrac{1}{C_F}$	(7.90)	$\overline{q_{iN}^2} \approx \dfrac{8}{3}\dfrac{kT}{C_O}C_T C_F$	(7.113)
Voltage sensing	$\dfrac{\Delta V_o}{\Delta V_i} = -\dfrac{C_I}{C_F}$	(7.93)	$\overline{v_{iN}^2} \approx \dfrac{8}{3}\dfrac{kT}{C_O}\dfrac{C_T C_F}{C_I^2}$	(7.114)
Current sensing	$\dfrac{\Delta V_o}{\Delta I_i} = -R_{EQ}$ $= -\dfrac{1}{C_F f_S}$	(7.92)	$\overline{i_{iN}^2} \approx \dfrac{8}{3}C_T C_F \dfrac{kT}{C_O}\dfrac{1}{R_{EQ}^2}$ $\approx \dfrac{8}{3}\dfrac{kT}{C_O}C_T C_F f_S^2$	(7.115)
Input capacitance sensing (constant voltage reference)	$\dfrac{\Delta V_o}{\Delta C_I} = -\dfrac{V_R}{C_F}$	(7.98)	$\overline{c_{iN}^2} \approx \dfrac{8}{3}\dfrac{kT}{C_O}\dfrac{1}{V_R^2}C_T C_F$	(7.116)

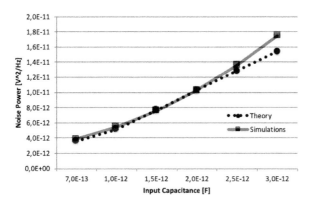

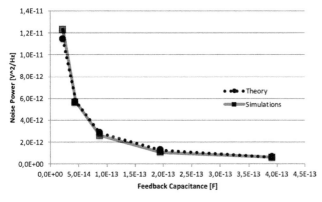

Figure 7.55 Output noise power comparison between Eq. (7.111) and Spectre® simulations versus input and feedback capacitance. CMOS technology 0.35 um $T_S = 100$ μs $C_O = 10$ pF [Courtesy of M. Crescentini]

A final comment on current sensing. While in discrete-time, the input-referred rms noise increases with the power of one of the bandwidth, as in (7.115), for continuous-time current amplifiers, it increases as usual with the power of ½. Therefore, continuous-time current interfaces have better noise performances in large bandwidth with respect to switched capacitor-based interfaces.

7.11.3 Resolution Optimization of a Cascade of Amplifiers

Section 3.7.7 of Chapter 3 discussed the degradation of resolution in two-stage generic amplifiers. Now that we have the instruments to calculate the input-referred noise, we can analyze the tradeoffs on a real amplifier.

To show the behavior, we make calculations on a commercial OPAMP, the LT1028, with the characteristics shown in Table 7.5. Figure 7.56 shows the effect of the gain of a two-stage interface on dynamic range and the number of resolution levels.

In Table 7.6 are shown with N the cases where we decrease the gain of the first amplifier, G1, and increase the gain of the second amplifier, G2, so that the total gain remains the same. For each stage, the dynamic range has been calculated using the minimum detectable signals of previous sections. Then we apply the resolution rule for acquisition chains to calculate the total levels of resolution. The data of the total NL are shown in Fig. 7.57. Therefore, it is apparent that *the more we shift the gain from the first stage to the second one, the lesser the NL.*

As far as the bandwidth is concerned, we can see that the maximum bandwidth of the chain is achieved where the gain is equally split among the stages, as shown in Fig. 7.57. These results confirm the analysis of Section 3.7.7 in Chapter 3 for constant gain-bandwidth product amplifiers configurations. Finally, to characterize an energy figure of merit, we plotted the product of the bandwidth times the resolution levels. This

Table 7.5 Data for noise calculation.

OPAMP	LT1028
GBW	7.50E+07
A0 (open loop gain) [dB] =	146
A0 (open loop gain) [V/V] =	2.00E+07
v_{in} [V/√Hz] =	8.00E-10
i_{in} [A/√Hz] =	1.00E-12
v_{in}^2 [V²/Hz] =	6.40E-19
i_{in}^2 [A²/Hz] =	1.00E-24
f_{ce} (corner frequency voltage) [Hz] =	3.5
f_{ci} (corner frequency current) [Hz] =	250
$R_{1,3}$ [ohm] =	1 k
V_{DD} [V] =	5
rms factor	√2
Coverage factor [sigmas]	6
Total gain	10,000

Table 7.6 Data of the acquisition chain.

N	G1	G2	DR1	DR2	DRT	NL	BW
1	1.000	1	2.54E+06	2.54E+09	2.54E+06	375.83	1.18E+05
2	500	2	1.28E+06	1.28E+09	1.27E+06	266.02	2.35E+05
3	200	5	5.11E+05	5.12E+08	5.10E+05	168.35	5.86E+05
4	100	10	2.55E+05	2.56E+08	2.55E+05	119.07	1.17E+06
5	50	20	1.28E+05	1.28E+08	1.28E+05	84.21	2.31E+06
6	20	50	5.11E+04	5.11E+07	5.11E+04	53.28	2.31E+06
7	10	100	2.56E+04	2.55E+07	2.56E+04	37.69	1.17E+06
8	5	200	1.28E+04	1.28E+07	1.28E+04	26.66	5.86E+05
9	2	500	5.11E+03	5.10E+06	5.11E+03	16.85	2.35E+05
10	1	1.000	2.54E+03	2.54E+06	2.53E+03	11.86	1.18E+05

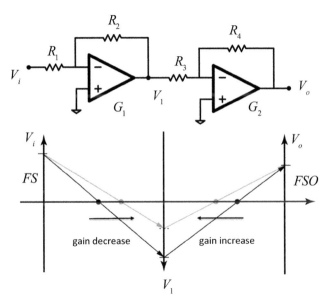

Figure 7.56 Effect of the gain of a two-stage interface on dynamic range (DR) and the number of resolution levels (NL).

shows that from an energetic point of view, we have a maximum until the gain is equally shared between the two stages; then, it decreases as the gain of the second one is greater than that of the first one.

To conclude, for constant GBW two-stage amplifiers:

- The larger the gain in the first stage, the higher the total NL.
- Maximum bandwidth is achieved when the gains of the two amplifiers are equal.
- Maximum energetic performance is achieved when the first stage gain is greater than the second one.

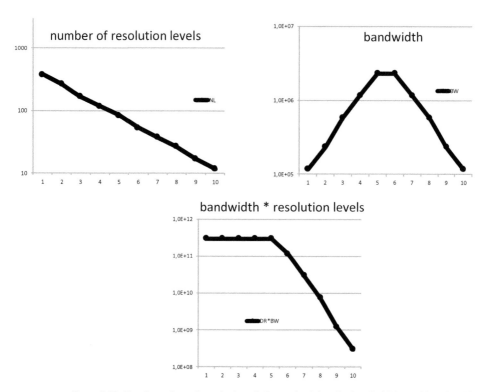

Figure 7.57 Total number of resolution (information) levels, bandwidth, and bandwidth times resolution levels.

Further Reading

Crescentini M., Bennati M., Carminati M., and Tartagni M., "Noise limits of CMOS current interfaces for biosensors: A review," *IEEE Trans. Biomed. Circuits Syst.*, vol. 8, no. 2, pp. 278–92, April 2014.

Enz, C. C., and Temes, G. C., "Circuit techniques for reducing the effects of opamp imperfections: Autozeroing, correlated double sampling, and chopper stabilization," *Proc. IEEE*, vol. 84, no. 11, pp. 1584–1614, 1997.

Gray, P. R., Hurst, P. J., Lewis, S. H., and Meyer, R. G., *Analysis and Design of Analog Integrated Circuits*, 4th ed. New York: John Wiley & Sons, 2001.

Gregorian, R., and Temes, G. C. *Analog MOS Integrated Circuits*. New York: John Wiley & Sons, 1987. Kester, W. Ed., *The Data Conversion Handbook*. Philadelphia; Elsevier, 2004.

Kulah H., Chae J., Yazdi N., and Najafi, K. "Noise analysis and characterization of a sigma-delta capacitive microaccelerometer," *IEEE J. Solid-State Circuits*, vol. 41, no. 2, pp. 352–361, February 2007.

Lee, T. H., *The Design of CMOS Radio-Frequency Integrated Circuits*, 2nd ed. Cambridge: Cambridge University Press, 2004.

Mondal S. and Hall D. A., "An ECG chopper amplifier achieving 0.92 NEF and 0.85 PEF with AC-coupled inverter-stacking for noise efficiency enhancement," *Proc. IEEE Int. Symp. Circuits Syst.*, no. c, pp. 2–5, 2017.

Muller, R. Gambini, S.and Rabaey, J. M. "A 0.013 mm2, 5 μ W, DC-coupled neural signal acquisition ic with 0.5 v supply," *IEEE J. Solid-State Circuits*, vol. 47, no. 1, pp. 232–243, 2012.

Rothe H., and Dahlke, W., "Theory of noisy fourpoles," *Proc. IRE*, vol. 44, no. 6, pp. 811–818, June 1957.

Steyaert, M. S. J., Sansen, W. M. C., and Zhongyuan C., "A micropower low-noise monolithic instrumentation amplifier for medical purposes," *IEEE J. Solid-State Circuits*, vol. 22, no. 6, pp. 1163–1168, 1987.

Tartagni, M., and Guerrieri, R. "A fingerprint sensor based on the feedback capacitive sensing scheme," *IEEE J. Solid State Circuits*, vol. 33, no. 1, pp. 133–142, 1998.

8 Detection Techniques

The noise performance and the main characteristics of electronics devices and elementary building blocks have been discussed in earlier chapters. Here, more complex techniques for sensing interfaces will be presented. We will analyze architectures tailored for specific cases such as resistive and capacitive sensing. Furthermore, modulation, feedback, and time-to-digital techniques for signal detection will be shown.

8.1 From Single-Ended to Differential Architectures

Analog signals could be encoded in several arrangements. In most of the examples examined, we have assumed that all analog signals are referred to a common reference called "ground" (GND) as illustrated in Fig. 8.1A. The electronic circuits designed on this convention are referred to as *single-ended* architectures, meaning that the input/output analog signals are defined as potentials referred to a common ground.

Another technique is to encode the information in the difference between two voltages, namely V_A and V_B (referred to GND) as shown in Fig. 8.1B. Therefore, since we have two degrees of freedom in this new situation, we can identify a *differential signal* $V_D(t) = (V_A - V_B)$ and a *common mode signal* as $V_C(t) = (V_A + V_B)/2$. It is widespread to associate the information with the differential signal and fix the common mode to a reference, for example, half power supply rails: $V_C(t) = V_R = V_{DD}/2$. The differential signal is antisymmetrical with respect to the reference by fixing the common mode. The electronic circuits implemented using this convention are usually referred to as *fully differential* architectures. One important advantage of the differential mode is that we can implement negative signals even if we use single power supply rails. Of course, the fully differential approach could also be used for current/charge signals where, usually, the common-mode current/charge reference signal is the current/charge zero value. When the signal is in the sinusoidal form, it is illustrative to represent the fully differential representation in phasor form, as shown in Fig. 8.1C. This could also be applied as a generic sinusoidal component of a spectrum. Furthermore, a fully differential equivalence from the circuital representation point of view is shown in Fig. 8.1D.

8.1.1 Noise or Interference? The Advantage of the Fully Differential Approach

In Chapter 4, we have seen that the variance of the weighted difference $z = a \cdot x - b \cdot y$ of two generic random signals, x and y is

$$P_{ZAC} = a^2\sigma_X^2 + b^2\sigma_Y^2 - 2ab\rho_{XY} \cdot \sigma_X\sigma_Y, \tag{8.1}$$

$$V_D(t) = (V_A - V_B) \qquad V_C(t) = (V_A + V_B)/2$$

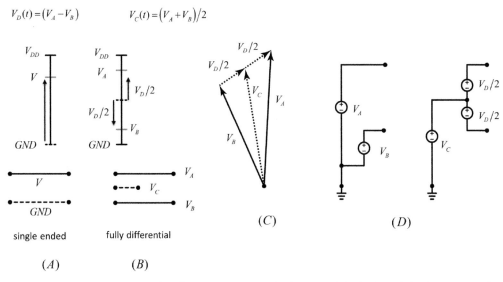

single ended   fully differential

(A)   (B)

(C)   (D)

Figure 8.1 Analog signaling conventions and references. (A) Single-ended scheme. (B) Fully differential. (C) Fully differential phasor representation. (D) Circuital representation of the fully differential signal representation.

where ρ_{XY} is the *correlation coefficient* of the random processes. In general, for low correlation coefficient, the perturbation acting on the differential line is referred to as "noise," while for high values, it is ascribed to the term "interference."

Now, in Eq. (8.1) we assume that a differential signal is composed of the difference of two perturbing signals, A and B, generated by the same random process with equal power $\sigma_{AN}^2 = \sigma_{BN}^2 = \sigma_N^2$. Therefore, the total noise power of the differential signal is

$$P_{DN} = \sigma_{AN}^2 + \sigma_{BN}^2 - 2\rho_{AB} \cdot \sigma_{AN}\sigma_{BN}. \tag{8.2}$$

If the random processes are identical, but realizations totally uncorrelated, $\rho_{AB} = 0$ and $P_{DN} = 2\sigma_N^2$; therefore, there is no improvement in reducing the noise by differential signal architectures. On the other hand, if there is a correlation (i.e., in the case of *interferences* such as 50/60 Hz domestic AC power supply or magnetic couplings), we have a clear reduction of the problem. In the extreme case of totally correlated interference, we have $\rho_{AB} = 1$ and $P_{DN} = 0$ means that the differential architecture totally cancels the perturbation out.

8.1.2   Example: Fully Differential Charge Amplifiers

The fully differential approach could be extended to operational amplifiers (OPAMPs) simply by keeping the original symmetry of the differential pair as illustrated in Fig. 8.2A. The OPAMP is thus called a *fully differential OPAMP* whose symbol is shown in Fig. 8.2B and whose input/output relationship is

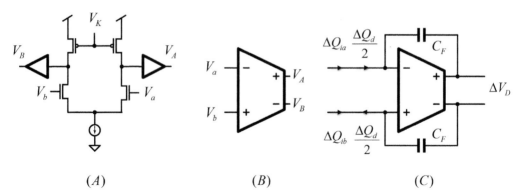

Figure 8.2 Fully differential approach for (A) operational amplifier, (B) symbol, and (C) charge amplifier implementation.

$$V_D = (V_A - V_B) = A \cdot (V_b - V_a). \tag{8.3}$$

The fully differential OPAMP could be used for the same purposes conceived for the single-ended OPAMPs. In general, the common mode of the fully differential amplifiers is set by internal circuits to $(V_A + V_B)/2$. It can be easily shown that a fully differential amplifier biased in feedback configuration has the same gain as the single-ended configuration if differential signals are used in place of the conventional signals. For example, in a fully differential feedback configuration, where $R_2 = R_4$ are the feedback resistances and $R_1 = R_3$ are the input ones, the input/output relationship for the open-loop OPAMP gain $A \gg 1$ is $V_D = (V_{O+} - V_{O-}) = R_2/R_1 \cdot (V_{i+} - V_{i-})$ (where V_{i+}, V_{i-} are the inputs of the feedback amplifier), which is the differential version of the single-ended "inverting" amplifier transfer function, where $V_O = -R_2/R_1 \cdot V_i$. In practice, a symmetrical fully differential configuration acts as the composition of two identical single-ended configurations so that the overall gain is the same as the single-ended one with respect to differential signals. However, we must be careful since the single output values have values placed across common mode in an antisymmetric fashion. As another example, the charge amplifier illustrated in Fig. 8.2C has the following transfer function

$$\Delta V_D = \frac{-\Delta Q_{ia} + \Delta Q_{ib}}{2C_F} = -\frac{\Delta Q_d}{C_F}, \tag{8.4}$$

where $(\Delta Q_{ib} - \Delta Q_{ia})/2$ is the differential signal in terms of charges.

8.2   Resistance Sensing

We will refer to resistive sensing whenever a sensor is implemented by a resistor whose resistance value R_0 is changed by a physical stimulus x by an incremental ratio

$\gamma = \Delta R / R_0 \in [0, 1]$ called a *relative variation* of the resistance. Following the definition of sensitivity given in Chapter 2 and using a first-order approximation, we have

$$R = R_0 + \Delta R = R_0\left[1 + \frac{\Delta R}{R_0}\right] = R_0(1 + \gamma)$$

$$R(x) \cong R(x_0) + \left.\frac{dR}{dx}\right|_0 \Delta x = R_0 + \Delta R = R_0\left[1 + \frac{\Delta R}{R_0}\right] = R_0[1 + S'\Delta x]$$

$$\text{where } \gamma = S'\Delta x = \frac{\Delta R}{R_0} \to S' = \left.\frac{1}{R_0}\frac{dR}{dx}\right|_0 \equiv \left[\frac{1}{\xi}\right] \text{ and } S = \left.\frac{dR}{dx}\right|_0 \equiv \left[\frac{\Omega}{\xi}\right]. \tag{8.5}$$

Above S and S' are the sensitivity and the relative sensitivity of the resistance sensor, respectively. Therefore, following the measuring unit in the above examples, a resistance temperature sensor will have a sensitivity expressed in $[\Omega/^{\circ}C]$ while a resistance strain sensor is expressed in $[\Omega/N]$.

Therefore, the resistance sensor interface is essentially an ohmmeter. A resistance measurement brings errors due to the resistors of the device's connections under test and the interface. As illustrated in Fig. 8.3A, we can estimate the resistance of the resistor R_X by reading the voltage across its terminals under a known current I flowing through it, according to Ohm's law. However, the estimation is affected by an error due to the connection resistances R_{W1} and R_{W2}, especially using long connecting wires.

A smart technique to measure resistance avoiding systematic errors due to the connection wires, is referred to as *Kelvin sensing* or *four-wire sensing*. Using the scheme of Fig. 8.3B, we read the voltage drop across R_X using a high impedance voltmeter so that we have no voltage drops on R_{W3} and R_{W4} while R_{W1} and R_{W2} does not change the reference current, so they do not influence the voltage across the resistor under test. This technique could be implemented in several ways in electronic sensor interfaces.

One of the oldest and most used interfaces for resistive sensors is the *Wheatstone bridge* illustrated in Fig. 8.4. The concept is to measure the voltage unbalance between two resistance (impedance) dividers. Following common notation, the variable resistance is the lower-right one $R_3 = R_0(1 + \gamma)$.

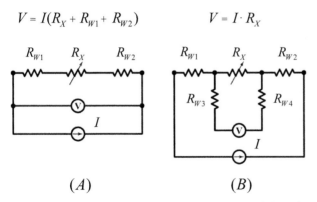

(A)   (B)

Figure 8.3 (A) Typical resistance measurement. (B) Kelvin or four-wire resistance measurement.

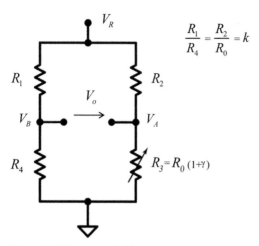

$$\frac{R_1}{R_4} = \frac{R_2}{R_0} = k$$

Figure 8.4 Wheatstone bridge.

For simplicity, we assume that the ratio of the resistances of the two dividers are equal: $R_1/R_4 = R_2/R_0 = k$. Therefore, the output voltage is given by

$$
\begin{aligned}
V_0 = V_A - V_B &= V_R\left(\frac{R_3}{R_2 + R_3} - \frac{R_4}{R_1 + R_4}\right) = V_R\left(\frac{R_0(1+\gamma)}{R_2 + R_0(1+\gamma)} - \frac{1}{k+1}\right) \\
&= V_R\left(\frac{1+\gamma}{k+1+\gamma} - \frac{1}{k+1}\right) = V_R\frac{k\gamma}{(k+1)(k+1+\gamma)},
\end{aligned}
\tag{8.6}
$$

where V_R is the reference voltage of the bridge. Therefore, the bridge's output is directly proportional to the reference voltage. This shows the pros and cons. On the one hand, the sensitivity could be increased by means of the reference voltage. On the other hand, the bridge is prone to interferences (especially from the power supply from which the reference is derived).

We can introduce the *bridge sensitivity* S_B as the variation of the output with respect to the relative resistance variation γ

$$S_B = \frac{1}{V_R}\frac{dV_o}{d\gamma} = \frac{1}{V_R}\frac{dV_o}{dx}\frac{1}{S_R'} \equiv [\cdot]. \tag{8.7}$$

Note that the bridge sensitivity is normalized to the value of V_R to make it independent of the reference value. We can find the best value of k that maximize S_B by derivation of the sensitivity expression with respect to γ

$$
\begin{aligned}
S_B &= \frac{1}{V_R}\frac{dV_o}{d\gamma}\bigg|_{x=0} = \frac{d}{d\gamma}\left(\frac{k\gamma}{(k+1)(k+1+\gamma)}\right) \\
&= \frac{k(k+1)(k+1+\gamma) - k\gamma(k+1)}{[(k+1)(k+1+\gamma)]^2}\bigg|_{x=0} = \frac{k}{(k+1)^2}.
\end{aligned}
\tag{8.8}
$$

To find the best value, we should now derive the sensitivity with respect to k.

$$\frac{dS_B}{dk} = \frac{d}{dk}\frac{k}{(1+k)^2} = \frac{(1+k)^2 - 2k(1+k)}{(1+k)^4} = \frac{1-k^2}{(1+k)^4} = 0 \rightarrow k = 1. \qquad (8.9)$$

Using this value, we could rewrite (8.6) as

$$V_0 = V_0(\gamma) = V_R\frac{\gamma}{2(2+\gamma)} \approx V_R\frac{\gamma}{4}. \qquad (8.10)$$

Note how the relationship is nonlinear and proportional to γ only under the first-order approximation. Therefore, the Wheatstone bridge brings nonlinearity systematic errors that should be quantified to understand the accuracy of the measurement.

As shown in Fig. 8.5, we can see that if the relative resistance variation is within $\pm 25\%$, the real characteristic seems a good approximation of the ideal one. However, it is easy to show that we still have about 10% of integral nonlinearity (INL). Therefore, this kind of bridge should be used only for very small input variations.

We can calculate the total sensitivity of the bridge, that is, the transfer function of the resistor sensor coupled with the bridge

$$S_T = S_B \cdot S_R' = \frac{1}{V_R}\frac{dV_o}{dx} \equiv \left[\frac{V}{V}\frac{1}{\xi}\right], \qquad (8.11)$$

whose unit of measurement is the reciprocal of the sensed quantity.

To reduce the nonlinearity distortion of (8.6), we can apply an antisymmetrical variation on the resistances of the same divider of the bridge, as shown in Fig. 8.6A. This arrangement is called the *half-Wheatstone bridge*.

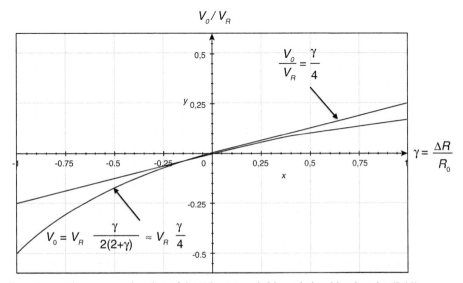

Figure 8.5   Linear approximation of the Wheatstone bridge relationship given by (8.10).

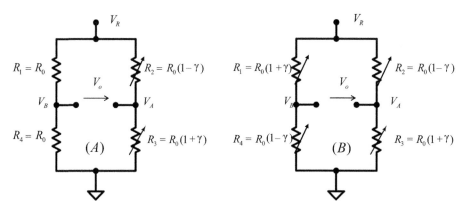

Figure 8.6 Half-bridge (A) and full-bridge (B) configurations.

Therefore, the response of the bridge is linear with respect to the relative variation according to the relationship

$$V_0 = V_A - V_B = V_R\left(\frac{R_0(1+\gamma)}{R_0(1-\gamma)+R_0(1+\gamma)} - \frac{1}{2}\right) = V_R\left(\frac{1+\gamma}{2} - \frac{1}{2}\right) = V_R \cdot \frac{\gamma}{2}.$$

$$(8.12)$$

Further evolution of the half-bridge is achieved using antisymmetrical variations on both legs of the bridge as shown in Fig. 8.6(B), which is referred to as the *full Wheatstone bridge* scheme, where the output relationship is

$$V_0 = V_A - V_B = V_R\left(\frac{R_0(1+\gamma)}{R_0(1-\gamma)+R_0(1+\gamma)} - \frac{R_0(1-\gamma)}{R_0(1-\gamma)+R_0(1+\gamma)}\right)$$

$$= V_R\left(\frac{1-\gamma}{2} - \frac{1+\gamma}{2}\right) = V_R \cdot \gamma$$

$$(8.13)$$

Typical examples of half and full bridges will be examined for strain sensors in Chapter 11.

Alternative techniques for resistive sensing are shown in Fig. 8.7, where the variable resistance is inserted in the feedback loop of an inverting OPAMP amplifier. Since it is, $I_o = V_R/R_0$ we have that the output voltage is given by

$$V_o = V_R + I_o R_0(1+\gamma) = V_R(2+\gamma),$$

$$(8.14)$$

so it is linearly dependent on the relative variation γ. Another technique is based on the difference between two resistive paths, where the currents are set equal by a current mirror, as shown in Fig. 8.7B, where the output relationship is

$$V_o = V_A - V_B = I R_0(1+\gamma) - I R_0 = I R_0 \gamma,$$

$$(8.15)$$

where again, the output is linearly proportional to the relative resistance variation.

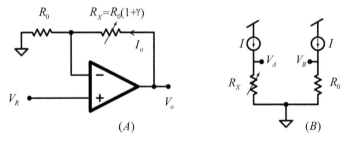

Figure 8.7 Alternative techniques for resistive sensing.

The main advantages of the techniques illustrated in Fig. 8.7 stem from their moderate immunity with respect to V_R. In fact, one of the problems related to the Whetstone bridge is due to the direct proportionality of the output versus the reference value. Therefore, any perturbation that is acting on the power supply could be easily conveyed into the output. In the first case of Fig. 8.7, the immunity with respect to the disturbances is related to the power supply rejection ratio (PSSRR) of the OPAMP, while in the second case to the ideality of the current sinks.

8.2.1 Ratiometric Readout

Another technique to counteract disturbances arising from the power supply is called *ratiometric readout* and is illustrated in Fig. 8.8 with an example. The technique is based on the linear output dependence of both the A/D converter and the converted output of the analog interface (e.g., Wheatstone bridge or impedance divider). Thus, referring to the example, we have that the absolute readout value is

$$V_{O(\text{bin})}[n] \equiv \text{Round}\left(\frac{V_I}{V_R} 2^N\right)$$

$$= \text{Round}\left(\frac{1}{V_R} \cdot V_R \frac{R_0(1 - \gamma)}{R_0(1 - \gamma) + R_0(1 + \gamma)} 2^N\right) \tag{8.16}$$

$$= \text{Round}\left(\frac{1}{2} - \frac{\gamma}{2}\right) 2^N \equiv [\text{LSB}],$$

where the subscript notation is for binary representation in N bits.

Following the same idea, we can read out a Wheatstone bridge using an instrumentation amplifier (IA) that keeps the output of the bridge well balanced and symmetric, as shown in Fig. 8.9, followed by an A/D converter *whose reference is the same as that of the bridge.*

The relationship of the IA is given by a fully differential frontend made of two OPAMPs followed by a differential to the single-ended stage so that the transfer function is

$$V_1 = V_X\left(1 + \frac{R_B}{R_A}\right) - V_Y \frac{R_B}{R_A}; \quad V_2 = V_Y\left(1 + \frac{R_B}{R_A}\right) - V_X \frac{R_B}{R_A}$$

$$\rightarrow V_o = V_1 - V_2 = \left(1 + \frac{2R_B}{R_A}\right)(V_X - V_Y) = (1 + G)(V_X - V_Y), \tag{8.17}$$

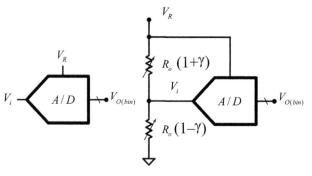

generic A/D   implementation in ratiometric scheme

Figure 8.8 Ratiometric sensing technique.

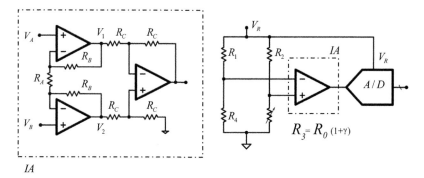

IA

Figure 8.9 Wheatstone bridge readout by instrumentation amplifier (IA).

where G is the gain of the instrumentation amplifier.

Alternatively, we can use a high-impedance differential A/D converter (biased with the same reference as the bridge) to achieve the same result.

8.3   Capacitive Sensing

Capacitive sensing is required whenever a sensor is implemented by means of a capacitor whose capacitance value is variable with respect to the physical stimulus. Capacitance is the effect of the electric field on charges between two conductors. If the electric field is uniform and parallel between two opposite plates, the value of the capacitance C_0 would be

$$C_0 = \frac{\varepsilon A}{h}, \tag{8.18}$$

where ε is the dielectric constant of the material between the plates, A is the area of the plate, and h is their distance. Unfortunately, this neat relationship does not hold in

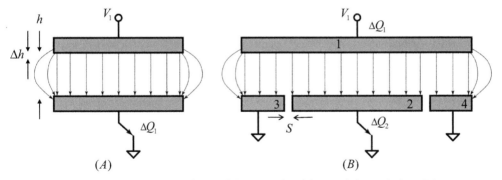

Figure 8.10 Capacitance fringing field (A) and Kelvin guard ring technique (B).

reality because the shape of the electric field follows the dipolar behavior of symmetrical charges at boundaries.

As shown in Fig. 8.10A, the real shape of the electric field between two plates is characterized by the presence of border effects named "fringing field." This effect acts, so that (8.18) does not hold and in general $C_0 > \varepsilon A / h_0$.

A smart technique is shown in Fig. 8.10B to avoid fringing field effects referred to as the *Kelvin guard ring*. The bottom plate is segmented into three parts, where each is tied to the same reference potential, such as ground. If the space S between the electrodes is small enough, the capacitance between electrodes 1 and 2 is not affected by the fringing field, and thus

$$C_{12} = \frac{\Delta Q_2}{\Delta V_1} \simeq \frac{\varepsilon A_2}{h};$$

$$C_{13} > \frac{\varepsilon A_3}{h}; \; C_{14} > \frac{\varepsilon A_4}{h},$$

(8.19)

where ΔQ_2 is the charge variation on the second electrode with respect to a voltage variation ΔV_1 applied to the top electrode, A_2 is the area of the second plate, and C_{xy} are the interelectrode capacitances. In conclusion, using the Kelvin guard ring, we can have a good approximation on Eq. (8.18).

Now, referring to the relative distance variation $\delta = \Delta h / h_0$ of the resting distance h_0, we have that, at first-order approximation, the relative capacitance variation $\gamma = \Delta C / C_0$ is given by

$$C = \frac{\varepsilon A}{h_0(1 + \delta)} = \frac{\varepsilon A}{h_0}[1 - \delta + \delta^2 - \delta^3 + \ldots] \approx C_0(1 - \delta)$$

$$C(\gamma) = C_0(1 + \frac{\Delta C}{C_0}) = C_0(1 + \gamma) = C_0(1 - \delta)$$

(8.20)

$$\rightarrow \gamma = \frac{\Delta C}{C_0} = -\frac{\Delta h}{h_0} = -\delta.$$

The Kelvin guard ring might be placed in a charge amplifier configuration either in place of the input capacitor (Fig. 8.11A) or in place of the feedback one (Fig. 8.11B).

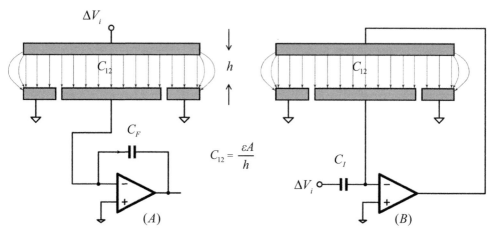

Figure 8.11 Use of the Kelvin guard ring in the capacitive sensing. Sensing of the capacitance (A) and sensing of the distance (B).

Note how the virtual short-circuit of the negative feedback ensures the condition that all the bottom plates of the Kelvin guard ring are at the same potential. Therefore, plate 2 needs to be connected to the input of the inverting input of the amplifier.

In the first case, the output is proportional to the distance of the plates

$$\frac{\Delta V_o}{\Delta V_i} \approx -\frac{C_I}{C_{12}} \to \Delta V_o = -C_I \Delta V_i \cdot \frac{h}{\varepsilon A}. \tag{8.21}$$

In the other case, the output is linearly related to the area of the plate and/or the dielectric constant ε

$$\frac{\Delta V_o}{\Delta V_i} \approx -\frac{C_{12}}{C_F} \to \Delta V_o = -\frac{\Delta V_i}{C_F} \cdot \frac{\varepsilon A}{h}. \tag{8.22}$$

As an example, the configuration of Fig. 8.11A could be used to implement a distance sensor, while the configuration of Fig. 8.11B to realize an area or a dielectric sensor.

A technique using a differential capacitance and a single-ended OPAMP is employed to reduce the nonlinearity of (8.20) is shown in Fig. 8.12.

In the scheme in Fig. 8.12, symmetrical voltage steps are applied to both plates of a capacitor, and the charge is collected from the intermediate plate having a degree of freedom of displacement with respect to the others.

Therefore, the relationship becomes

$$\begin{cases} C_1 \approx C_0(1 - \gamma + \gamma^2 + \ldots) \\ C_2 \approx C_0(1 + \gamma + \gamma^2 + \ldots) \end{cases}$$

$$\to \Delta Q_i = (\Delta V_R C_1 - \Delta V_R C_2) = \Delta V_R C_0 (1 - 1 - \gamma - \gamma + \gamma^2 - \gamma^2 + \ldots)$$

$$\approx -2C_0 \Delta V_R \gamma = 2C_0 \Delta V_R \frac{\Delta h}{h} \to \quad \Delta V_o = -\frac{\Delta Q_i}{C_F} = -2\Delta V_R \frac{C_0}{C_F} \frac{\Delta h}{h}. \tag{8.23}$$

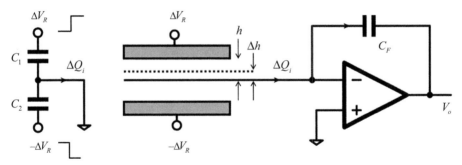

Figure 8.12 Differential capacitor sensing technique.

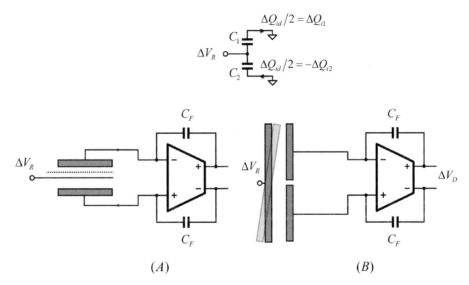

Figure 8.13 Capacitive sensing by means of fully differential OPAMPs. (A) Stack plates architecture. (B) Planar plate architecture.

Note how the differential approach eliminates all the even terms of the power expansion, greatly reducing the effect of nonlinear behavior of the capacitance with respect to the plates' distance.

The same result could be obtained using a fully differential OPAMP with a different architecture, as shown in Fig. 8.13. Instead of applying symmetrical voltage steps, a unique step is applied to the mid-plate, and the charges of single capacitors are sensed through a differential charge amplifier.

Again, the differential output charge is related to the distance variation as

$$\Delta Q_{iD} = -(\Delta Q_{i1} + \Delta Q_{i2}) \rightarrow \quad \Delta V_{OD} = -\frac{\Delta Q_{id}}{C_F} = -2\Delta V_R \frac{C_0}{C_F} \frac{\Delta h}{h}, \qquad (8.24)$$

which is the same as (8.23).

8.3.1   Example: Capacitive Accelerometer

The ideas mentioned above could be used to implement a capacitive accelerometer, as shown in Fig. 8.14. The moving inner plate is anchored to a proof mass M whose displacement is related to an elastic constant α. Under a uniform acceleration exerted along the degree of freedom, a displacement is achieved at mechanical equilibrium

$$F = Ma = -\alpha\Delta h \rightarrow \Delta h = \frac{M}{\alpha}a. \tag{8.25}$$

Therefore, we have

$$\Delta V_O = -2\Delta V_R \frac{C_0}{C_F}\frac{\Delta h}{h} = -2\Delta V_R \frac{C_0}{C_F}\frac{M}{h\alpha}\cdot a, \tag{8.26}$$

showing that (at the first order of approximation) the output is linearly dependent with respect to the acceleration.

The treatment mentioned above is only conceptual, and real implementations are, of course, more complex, as discussed in Chapter 3 and later in this one.

8.3.2   AC Capacitive Sensing

Another way to implement capacitive sensing is based on a harmonic regime, as shown in Fig. 8.15, where a fixed amplitude sinusoidal signal is applied to the input and the peak-to-peak or the rms value of the readout is measured.

Using the first-order approximation of the OPAMP, we can know the cutoff pole given by the feedback f_0', as discussed in Chapter 7. Furthermore, the effect of the feedback resistance acts as a high-pass filter according to the relationship

$$\left|\frac{V_O}{V_I}\right| = -\frac{Z_F}{Z_i} \text{where: } Z_F = \frac{R/sC_F}{R + 1/sC_F}, Z_i = \frac{1}{sC_i}$$

$$\rightarrow \left|\frac{V_O}{V_I}\right| = -\frac{sRC_i}{1 + sRC_F}; f_1 = \frac{1}{2\pi RC_F}. \tag{8.27}$$

Therefore, if poles and zero are well spaced apart, and we use a frequency between f_0' and f_1 the relationship becomes

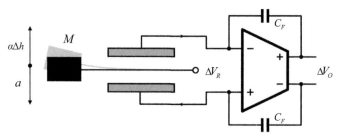

Figure 8.14 Conceptual representation of a capacitive sensing accelerometer.

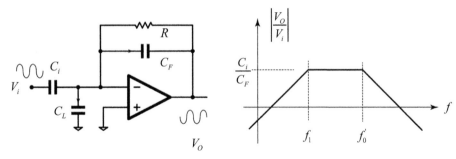

Figure 8.15 AC capacitive sensing.

$$\left|\frac{V_O}{V_I}\right| \simeq \frac{C_i}{C_F},\tag{8.28}$$

so that by comparing the amplitudes of the output signal with respect to the input one, we can use this circuit to read out a capacitive sensor placed either in C_i or C_F.

8.4 Resistance and Capacitive Readout by Transient Techniques

In general, a resistor capacitor (RC) circuit transient could be used to estimate the value of either the resistance or of the capacitance of the components once the other element is known.

Therefore, using the relationship of an RC transient where V_∞ and V_0 are the asymptotic and the initial value of the input potential, we can get the time to reach a generic threshold V_T (see Fig. 8.16)

$$V(t) = V_o - (V_o - V_\infty)\left(1 - e^{-\frac{t}{RC}}\right)$$

$$\rightarrow \Delta t = -RC\ln\left(1 - \frac{V_T}{V_\infty}\right).\tag{8.29}$$

Now, the idea is to compare two transients, one using a reference resistance R_R and another one using an unknown resistance R_X, so that we have the following relationship

$$R_X = R_R \frac{\Delta t_X}{\Delta t_R}.\tag{8.30}$$

Note how the same technique could be applied even to unknown capacitances, thus acting also as a capacitive sensing interface with small variations of the circuitry (i.e., using a reference capacitor instead of a reference resistor).

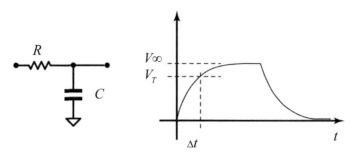

Figure 8.16 RC transient.

We can calculate the resolution achievable with this technique. From Chapter 5, we have that signal and noise of an RC transient are

$$V_C(t) = V_\infty(1 - e^{-t/RC}) \quad \text{for signal;}$$
$$\overline{V_{CN}^2(t)} = \frac{kT}{C}(1 - e^{-2t/RC}) \quad \text{for noise.}$$

(8.31)

Therefore, the SNR could be calculated as

$$\text{SNR} = \frac{V_\infty^2(1 - e^{-t/RC})^2}{(kT/C)(1 - e^{-2t/RC})}$$

(8.32)

which is achieving the asymptotic case of the signal-to-noise ratio (SNR) for the *kTC* noise discussed in Chapter 5, and it is about half of that value for one time constant.

Referring to Fig. 8.17, we can get two values of minimum and maximum of input resistances that we would like to estimate with this technique. Therefore, following the discussion of Section 2.11.2 of Chapter 2, we can estimate the achievable resolution number of levels (i.e., dynamic range) as

$$DR_{dB} = \text{SNR}_{dB} + 20\log\frac{R_{\max}}{R_{\min}}; \quad NL_b = \frac{1}{2}\log_2(1 + DR).$$

(8.33)

Conversely, once the required resolution level is set, we can determine the minimum and maximum resistances and reference capacitance using (8.32).

A possible technique could be implemented through a microcontroller (MCU), as illustrated in Fig. 8.18. Suppose that we would like to measure the resistance of a sensor $R_X = R_0(1 + \gamma)$ with respect to a reference resistor R_R. Those resistances are connected to two input/output (I/O) pins of the MCU as shown in Fig. 8.18A and the transient monitored by means of another I/O pin. In general, I/O pins could be either used to force a voltage or to monitor a voltage, as illustrated in Fig. 8.18B. The operation of each I/O pin is governed by the firmware embedded into the MCU.

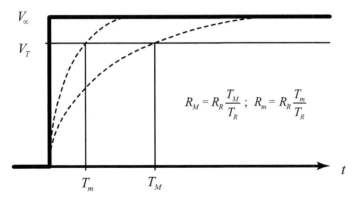

Figure 8.17 Transient technique to estimate two values of resistances.

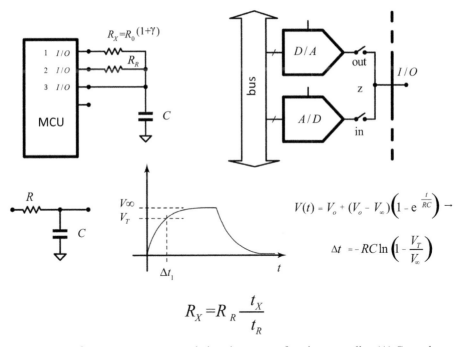

Figure 8.18 Resistance measurement technique by means of a microcontroller. (A) General scheme. (B) Detailed diagram of an input/output pin.

8.5   Integration of a Sensing System Using a Sigma–Delta Modulator Feedback

8.5.1   The Sigma–Delta Converter Concept

The sigma–delta (SD) (or also referred to as delta-sigma (DS)) converter is an over-sampled converter based on a feedback loop.

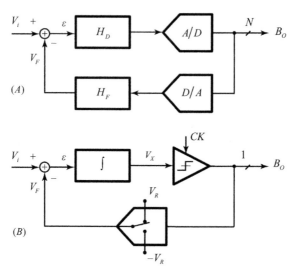

Figure 8.19 Concept of sigma–delta converter. (A) A system with a quantized feedback loop. (B) The synchronous sigma–delta concept with a 1b feedback loop.

In Fig. 8.19A is illustrated a generic feedback system where the feedback loop is made by first converting the analog signal into digital and then converting it back into analog again. In general, we should expect a very similar behavior to that of an analog feedback system, except for quantization effects. The quantization levels should be infinite to match the analog case. However, in the presence of a reduced number of bits, the quantization noise becomes relevant, and it acts as it were summed up in the loop (in place of the A/D converter). We will see that the quantization noise is well suppressed and controlled by the feedback loop, so we should not be concerned about reducing the number of bits.

In view of this, we reduce the minimum number of bits to the lower limit of 1, as shown in Fig. 8.19B. In this case, the A/D converter is simply a comparator, and the D/A converter is a two-way demultiplexer. The direct block is an integrator in order to ensure stability to the system for a required range of inputs, as known from control theory.

From now on, we assume that input voltage is stationary and $-V_R < V_i < V_R$. If the integrator voltage is greater than zero, the feedback loop subtracts the maximum value of the input, i.e., the reference so that the error voltage is negative, and the output of the integrator should have a negative slope. On the other hand, if the integrator's output is equal to or lower than zero, the feedback reacts so that a maximum possible value (a reference voltage) is added to the input and the integrator has a positive slope. In other terms, the feedback acts always to reduce the error signal as in the "negative" feedback approach. However, in this case, the reduction is not immediate, but it is effective on average in several clock cycles.

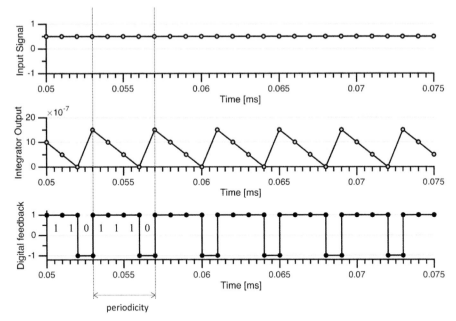

Figure 8.20 Time-domain analysis of a sigma–delta converter where $V_R = 1$ V and $V_i - 0.5$ V (simulation with SimuLink®).

This behavior is shown in Fig. 8.20, where a fixed input $V_i = 0.5$ V $= 1/2 \cdot V_R$ is applied. Every time the integrator output is > 0, the digital output is "high," and a negative reference is added to the input. Conversely, when the integrator output is ≤ 0 the digital output is "low."

Since, *on average*, the feedback minimizes the error voltage,

$$V_\varepsilon \to 0$$

$$V_i - \frac{1}{N}\sum_{i=1}^{N} V_F(i) \to 0. \tag{8.34}$$

Therefore,

$$V_F(i) = \begin{cases} +V_R \\ -V_R \end{cases}; \ V_i \cong V_R \underbrace{\frac{n_1 - n_0}{n_1 + n_0}}_{N}, \tag{8.35}$$

where n_1 is the number of "high" or "ones" bits and n_0 is the number of "low" or "zeros" bits in the output stream considered. Therefore, the input could be estimated from a moving average filter (low-pass filter) of the bitstream. N is the suitable number of bits of the stream and could be set to the number of bits of periodicity.

Now turning back to Fig. 8.20, we can see that the periodicity is of 4 bits and the pattern is "1110." We have chosen to fix the input at a value that is exactly a multiple of 1/4 of the full scale. Therefore, applying (8.35) we have $V_i \approx (3-1)/(4)V_R = 1/2V_R$, which is consistent with our setup. The (subset of) combinations with a 4-bit pattern are

$$\begin{cases} 1111 \to 4/4 = 1V_R \\ 1110 \to 2/4 = 1/2V_R \\ 1010 \to 0/4 = 0 \\ 1000 \to -2/4 = -1/2V_R \\ 0000 \to -4/4 = -1V_R. \end{cases} \tag{8.36}$$

Note that not all the possible combinations of 0's and 1's patterns are considered, but only those consistent with the SD behavior.

Let us take another example, in Fig. 8.21, where $V_i = 0.1$ V $= 1/10 \cdot V_R$. In this case, the pattern is recursive in 20b, and more specifically, we have the pattern "01010101011010101011" so that $V_i \approx (11-9)/(20)V_R = 1/10V_R = 0.1V_R$. Of course, the pattern is more complex in this case, and it is necessary now to collect 20b to cover 20 + 1 levels as it was before using 4b to cover 4 + 1 discrete levels. To summarize, the number of bits of the recursive pattern is proportional to the number of discretization levels required. From the mathematical perspective, to detect an input whose value is an irrational number, we need an infinite number of bits since it has no periodicity.

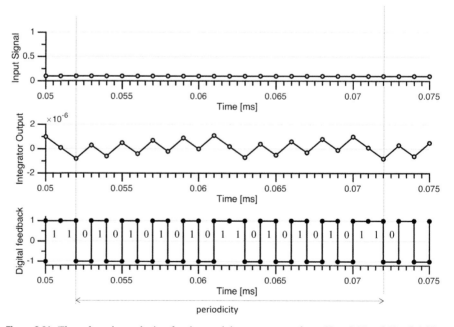

Figure 8.21 Time-domain analysis of a sigma–delta converter where $V_R = 1$ V and $V_i = 0.1$ V (simulation with SimuLink®).

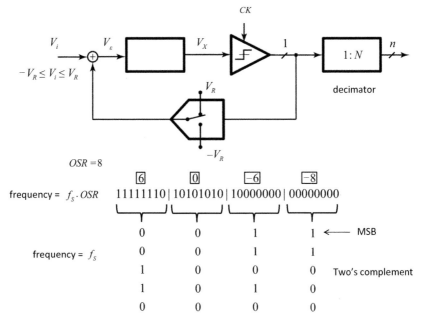

Figure 8.22 Concept of the decimator. OSR is the oversampling ratio of the converter.

However, from the implementation point of view, we have to fix the discrete conversion levels and the window of the bitstream to compute the average. In this case, if the input level belongs to some intermediate value between the required discrete levels, the average is fluctuating from window to window. The only way to reduce this variation is to take a larger window. This has a drawback in the bandwidth of the system: once again, we find a resolution-bandwidth tradeoff.

To compute the average on a bitstream window, we can use a digital filter referred to as a *decimator*. A concept example is shown in Fig. 8.22, where the pattern is composed of 8b, and it is encoded into a 5b two's complement notation of the difference between the 1's and 0's. From the implementation point of view, to compute a digital low-pass filter, the best choice is to use a finite impulse response (FIR) filter on the bitstream. Usually, $sinc^3$ and/or *Kaiser* filters are used as decimators, whose performance is similar to the low-pass, such as a moving average but with better selectivity.

If we want an estimation of the input at Nyquist's frequency f_S, the feedback loop must be running faster, OSR times f_S, where OSR is the *oversampling ratio*. This is why the converter belongs to the class of the oversampling converters. As discussed in Section 3.5 of Chapter 3, if we oversample at $OSR \cdot f_S$ the same power of quantizing noise P_N is uniformly distributed between $-OSR \cdot f_M$ and $OSR \cdot f_M$. However, by filtering back to the baseband, the quantizing noise is reduced to $P_N' = P_N/OSR$. Therefore, oversampling implies a reduction of quantizing noise. A sigma–delta scheme can do more than that. To see this, refer to Fig. 8.23. As already said, the quantization noise could be represented as an additive node in the loop.

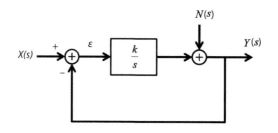

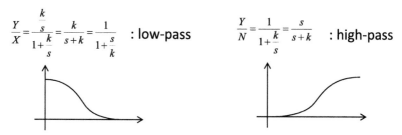

Figure 8.23 Signal gain and noise gain of the sigma–delta converter.

Therefore, we can define two distinct transfer functions: the signal gain and the noise gain:

$$\frac{Y(s)}{X(s)} = \text{signal gain} = \frac{1}{1 + \dfrac{s}{k}}$$

$$\frac{Y(s)}{N(s)} = \text{noise gain} = \frac{s}{s + k},$$

(8.37)

where k is a constant of the integrator. Note that for the definition of signal and noise gain, we apply the superposition principle of linear systems so that when one input is active, the other is set to zero. The result is that signal gain is low-pass bandwidth limited, while noise gain is a high-pass shaped function. Therefore, as shown in Fig. 8.24, the high-pass filtering of noise further reduces the quantization noise. This kind of process is referred to as *noise shaping*.

The noise shaping of Fig. 8.23 is referred to as a first-order sigma–delta architecture. It can be shown that by using higher-order architectures (with more than one feedback loop), we can further decrease the quantization noise.

8.5.2   Example: The Electrostatic Feedback Accelerator

The sigma–delta architecture could be very useful whenever we take advantage of feedback sensing, as discussed in Section 3.2.1 of Chapter 3. As an example, in Fig. 8.25, an electrostatically activated accelerometer is shown. The concept is shown in Fig. 8.25A: as previously discussed, in the open-loop condition, the acceleration induces a displacement Δx_a of the proof mass given by equilibrium of elastic and acceleration

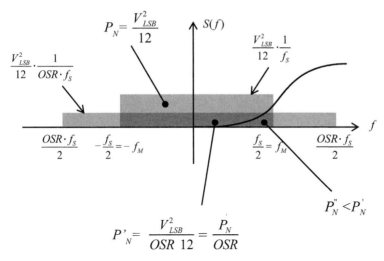

Figure 8.24 Effect of the "noise shaping" in sigma–delta converters in the reduction of the quantization noise.

forces. This is prone to nonlinearities and other error effects. The idea is to insert the scheme in a feedback loop, inducing an electrostatic force that is opposite to the acceleration one. Therefore, the negative feedback is acting to reduce the displacement error Δx_ε as much as greater the loop gain is. Thus, the acceleration is indirectly readout by means of the voltage that is used to counteract the acceleration displacement. This is equivalent to make the proof mass as rigid as possible, reducing the displacements around the resting place. This has benefits in confining the effects of the nonlinearities.

Now, the idea is to use a sigma–delta scheme to implement the feedback, as shown in Fig. 8.25B. The charge amplifier implements the integration function in discrete-time mode. The advantage of this scheme is to implement in the same architecture (and thus on the same integrated circuit (IC)) an oversampling A/D conversion with reduced quantization noise due to noise shaping.

8.6 The Correlated Double Sampling Technique

We have shown in Chapter 4 that in the case of pure white noise (uniform power spectral density (PSD)), we do not correlate with samples at any time difference. This means that in the presence of white noise, any attempt to reduce noise effects by subtracting outcomes (such as in consecutive samples) results in doubling the total noise from the statistical point of view, which is different in other situations, such as filtered white noise and pink noise. In such cases, since PSD is not uniform, necessarily the autocorrelation function is not a Dirac pulse, and thus there is the possibility to reduce the effect of noise by subtracting two consecutive samples as discussed in Section 4.7 of Chapter 4. Several circuital techniques could be used to perform this

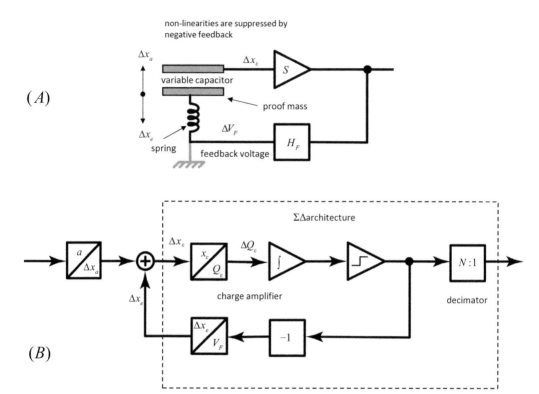

Figure 8.25 (A) Concept scheme of an electrostatic feedback accelerometer. (B) Implementation of the concept in a sigma–delta scheme.

approach that is also valid to reduce fixed systematic errors given by input amplifier(B) offsets.

An example that is called *autozeroing* is illustrated in Fig. 8.26. The concept is based on sampling the unwanted quantity such as offset and noise during the reset phase and then subtracting it in the readout value. As shown in Fig. 8.26A, a differential amplifier is represented with a voltage generator that models both the offset and the input-referred noise. The discrete-time operation is organized in two phases. During the first one (t_1), the input of the input amplifier is shorted to the ground so as the offset and the noise at the output are sampled onto an analog memory implemented with a capacitor placed between the output and ground. During the second phase (t_2), the switches phases are inverted. So, the system's output is the amplified input to which the offset and the noise sampled are subtracted from the previous phase.

Following the notation of Fig. 8.26, the autozeroing could be described as

$$
\begin{aligned}
V_{\mathrm{OA}}(t_1) &= -G\left(v_{\mathrm{off}} + v_{\mathrm{N}}(t_1)\right) \\
V_{\mathrm{OB}}(t_1) &= 0 \\
V_{\mathrm{OA}}(t_2) &= GV_I - G\left(v_{\mathrm{off}} + v_{\mathrm{N}}(t_2)\right) \\
V_{\mathrm{OB}}(t_2) &= V_{\mathrm{OA}}(t_2) - V_{\mathrm{OA}}(t_1) = GV_I + G\left(v_{\mathrm{N}}(t_1) - v_{\mathrm{N}}(t_2)\right),
\end{aligned}
\tag{8.38}
$$

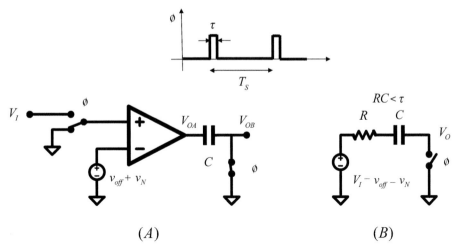

Figure 8.26 Autozero technique principle (A) and equivalent circuit (B).

where G is the gain of the amplifier. Therefore, since the offset is fixed and it is a deterministic error, it is eliminated. The problem now is how effective the difference of noise samples from the power spectral density standpoint is. This autozeroing technique for reducing noise power (other than offsets) is referred to as *correlated double sampling* (CDS).

The entire process could be analyzed to calculate the effect on noise PSD using a simplified model as that of Fig. 8.26B, where we assume that the time constant of the circuit is smaller than the switching time $RC \ll \tau$. In the presence of thermal and $1/f$ noise contributions already described in Chapter 4 the PSD spectrum is

$$S_N(f) = \frac{k_W^2}{1 + (f/f_0)^2}\left(1 + \frac{f_C}{f}\right),\tag{8.39}$$

where f_0 is the cutoff frequency of the system and f_C the pink-noise corner frequency. Considering that the circuit is similar to that discussed in Chapter 7 (even if now the output is not the voltage across the capacitor and the signal is passing through a high-pass filter and not a low-pass one), it could be shown[1] that the PSD is composed of two components, one related to direct noise $S_D(f)$ and another $S_{SH}(f)$ related to the folding process due to sample and hold. The PSD of the autozeroing process, for: $\tau \ll T_S$, $\pi f_C T_S \gg 1$ and $\pi f T_S \ll 1$ is

$S_{AZ}(f) = S_D(f) + S_{SH}(f)$ where

$$S_D(f) \cong (\pi f T_S)^2 k_W^2(f)\tag{8.40}$$
$$S_{SH}(f) \cong \left((\pi f_0 T_S - 1) + 2f_C T_S\left(1 + \ln\left(\frac{2}{3}f_0 T_S\right)\right)\right)k_W^2\,\mathrm{sinc}^2(fT_S),$$

[1] See C. C. Enz and G. C. Temes (1996) for details.

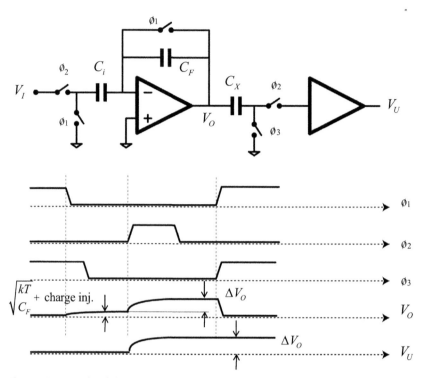

Figure 8.27 Correlated double sampling implementation for capacitive/charge sensing.

where $f_S = 1/T_S$ is the sampling frequency and $USR = \pi f_0 T_S$ is the undersampling ratio (see Chapter 7). Note that the first direct term $S_D(f)$ acts as a high-pass filter with respect to the input noise, thus greatly reducing the flicker noise. For that reason, within the Nyquist bandwidth, $|fT_S| < 1/2$ the direct noise $S_D(f)$ is in general smaller than $S_{SH}(f)$. However, we must be aware that the autozeroing implies a noise folding due to the sample-and-hold included in $S_{SH}(f)$.

A practical implementation of the correlated double sampling is shown in Fig. 8.27, where a capacitive amplifier is coupled with a storage capacitor C_X acting as analog memory. The working behavior is as follows. Before connecting the input to the amplifier, C_X, C_F and C_I are shorted by means of ϕ_1 and ϕ_3. Then ϕ_1 is opened so as C_I and C_F are set free. Thus the kTC noise and charge injection are stored into C_X as an offset. Then ϕ_3 is opened and ϕ_2 is closed so as the input is applied to the amplifier from which the offset and the noise are subtracted by means of C_X.

8.7   The Lock-In Technique

In some sensing technology applications (e.g., in chemical and temperature sensors and biosensing), signal bandwidths are typically very small and overlapping the spectrum where $1/f$ noise becomes dominant over thermal noise, on the left side of the corner frequency, as illustrated in Fig. 8.28. This is especially emphasized when the architectures are implemented in complementary metal–oxide–semiconductor (CMOS) technologies where flicker noise is dominant at low frequencies due to charge traps in the MOS channel.

As shown in Fig. 8.28A, once the signal and the interface noise are summed together at the input, it is difficult to improve the SNR. A possible solution could be to act on the signal *before* entering the amplification stage by making a signal modulation, as shown in Fig. 8.29. The signal $x(t)$ is multiplied by a reference sinusoidal waveform $V_R(t) = V_R \sin(\omega_0 t + \theta_R)$ before being amplified by performing an amplitude modulation (AM). Then, the signal and the noise are amplified by a factor G; it is multiplied again with a sinusoidal signal $V_L(t) = V_L \sin(\omega_0 t + \theta_L)$ in a demodulation technique fashion.

The overall effect can be mathematically shown applying the prosthaphaeresis relationships

$$
\begin{aligned}
V_B(t) &= x(t) \cdot G \cdot V_R \sin(\omega_0 t + \theta_R) \\
\rightarrow V_P(t) &= x(t) \cdot G \cdot V_R \cdot V_L \cdot \sin(\omega_0 t + \theta_R) \cdot \sin(\omega_0 t + \theta_L) \\
&= x(t) \cdot G \cdot V_R \cdot V_L \cdot \frac{1}{2} \left[\cos(\theta_R - \theta_L) - \cos(2\omega_0 t + \theta_R + \theta_L) \right],
\end{aligned}
\tag{8.41}
$$

where the last term is a sinusoidal signal at the double frequency with respect to the modulating ones. Therefore, by filtering it out, we have

$$
V_O(t) = x(t) \cdot \frac{G \cdot V_R \cdot V_L}{2} \cdot \cos(\theta_R - \theta_L).
\tag{8.42}
$$

It should be noted that the output signal is the input one multiplied by the gain G and a constant value given by a $\cos(\theta_R - \theta_L)$ factor. This demands strict control of the

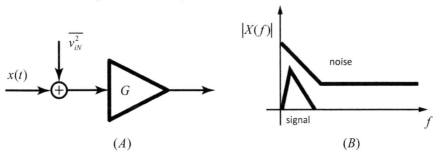

(A)   (B)

Figure 8.28 Block-diagram (A) and spectrum (B) representation of the input-referred noise in the sensing technology.

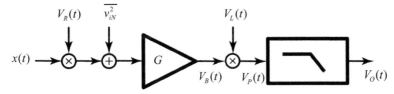

Figure 8.29 Lock-in amplifier principle.

From the spectrum standpoint, we should note that multiplying a signal for another one implies an AM modulation, as illustrated in Fig. 8.30. In an AM modulation, the signal baseband is shifted to higher frequency into symmetrical sidebands replica around the carrier ($f_0 = \omega_0/2\pi$). Since the operation is performed before the noise is added, noise and signal spectrum occupy disjointed bandwidth before the amplification process. Conversely, both signal and noise are shifted into the baseband during the demodulation operated by $V_L \sin(\omega_0 t + \theta_L)$. Therefore, the noise $1/f$ band is replicated around f_0 while the signal bandwidth is folded back to the baseband.

To avoid the systematic error given by the $\cos(\theta_R - \theta_L)$ factor, a slightly more complex system could be used, as shown in Fig. 8.31 using a *quadrature demodulator* architecture, employing a reference signal shifted by $\pi/2$

$$V'_L(t) = V_L \sin\left(\omega_0 t + \theta_L + \frac{\pi}{2}\right) = V_L \cos(\omega_0 t + \theta_L). \tag{8.43}$$

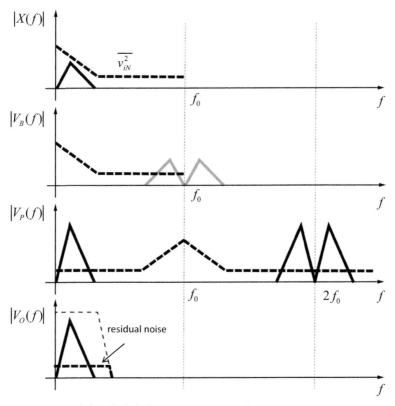

Figure 8.30 Lock-in principle from the spectrum point of view.

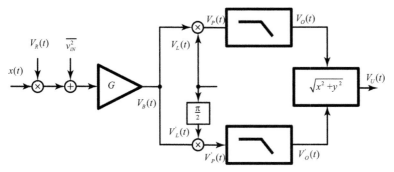

Figure 8.31 Quadrature demodulation lock-in approach.

The signals on the two paths are

$$V_P(t) = V_B(t) \cdot V_L(t)$$
$$= x(t) \cdot G \cdot V_R \cdot V_L \cdot \frac{1}{2} \left[\cos(\theta_R - \theta_L) - \cos(2\omega_0 t + \theta_R + \theta_L) \right]$$
$$V'_P(t) = V_B(t) \cdot V'_L(t)$$
$$= x(t) \cdot G \cdot V_R \cdot V_L \cdot \frac{1}{2} \left[\sin(2\omega_0 t + \theta_R + \theta_L) + \sin(\theta_R - \theta_L) \right] \quad (8.44)$$

$$\left. \begin{array}{l} V_O(t) = x(t) \cdot G \cdot V_R \cdot V_L \cdot \dfrac{1}{2} \cos(\theta_R - \theta_L) \\[2mm] V'_O(t) = x(t) \cdot G \cdot V_R \cdot V_L \cdot \dfrac{1}{2} \sin(\theta_R - \theta_L) \end{array} \right\} \rightarrow V_U(t) = \dfrac{G \cdot V_R \cdot V_L}{2} x(t) \quad .$$

Finally, the output signal $V_U(t) = \sqrt{V_O^2 + V_O'^2}$ is recovered using the Euler formula. In this case, the output is no longer dependent on the reference phase difference.

As an application example, the quadrature scheme could be employed to determine complex impedance according to the scheme shown in Fig. 8.32. A sinusoidal voltage is applied to a complex impedance to be measured

$$\overline{Z} = R + jX = |\overline{Z}| e^{j \sphericalangle (\overline{Z})} \quad (8.45)$$

and the corresponding current is readout and amplified by means of a transimpedance amplifier. Therefore

$$V_P(t) = \frac{A^2 G}{|\overline{Z}|} \cdot \frac{1}{2} \left[\cos(\theta) - \cos(2\omega_0 t + \theta) \right]$$

$$\rightarrow V_O(t) = \frac{1}{2} \frac{A^2 G}{|\overline{Z}|} \cos(\theta) = x$$

$$V_P(t) = \frac{A^2 G}{|\overline{Z}|} \cdot \frac{1}{2} \left[\sin(\theta) + \sin(2\omega_0 t + \theta) \right]$$

$$\rightarrow V'_O(t) = \frac{1}{2} \frac{A^2 G}{|\overline{Z}|} \sin(\theta) = y.$$

$$(8.46)$$

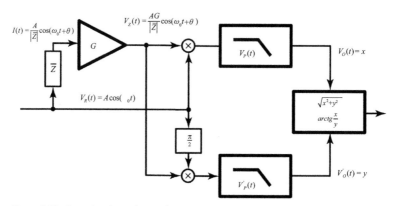

Figure 8.32 Complex impedance detection using a lock-in quadrature scheme.

From the x and y signals, we can recover the impedance according to the following relationships

$$\frac{1}{|\overline{Z}|} = \frac{2\sqrt{x^2 + y^2}}{A^2 G}$$

$$\theta = \sphericalangle \overline{Z} = \arg(\overline{Z}) = \text{arctg}\frac{y}{x}. \tag{8.47}$$

Returning back to the single-branch lock-in scheme, if we use in place of sinusoidal references a square-wave (digital) signal, we could get a similar technique to reduce noise and offsets called *chopping technique* as illustrated in Fig. 8.33.

As in the lock-in scheme, the input is multiplied by a +1, −1 square-wave before being applied to the amplifier. The output signal shows a modulated signal and an amplified offset. Then, the signal is multiplied again by a +1, −1 square reference. Thus, the signal is reconstructed as a DC value again while the offset is "chopped." Finally, a low-pass filter separates the amplified DC signal with respect to the modulated amplified offset. Note how the square-wave multiplication is equivalent to an AM modulation where the carrier is composed of (odd) harmonics. Therefore, the spectrum diagrams are as illustrated in Fig. 8.34.

The chopping technique is based on a "digital multiplier," as shown in Fig. 8.35. The schematic symbol currently used is illustrated in Fig. 8.35A, and a possible realization using switches is shown in Fig. 8.35B while another one using inverting/non-inverting buffers is illustrated in Fig. 8.35C.

8.8   Oscillator-Based Sensing

Oscillator-based sensing techniques refer to systems where the signal is linked to the evolution of a harmonic resonator, such as frequency or phase.

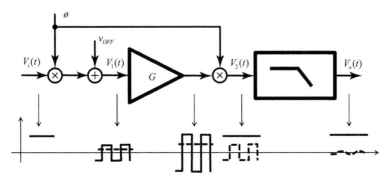

Figure 8.33 Chopping technique for reducing amplifier offset influence over a DC signal.

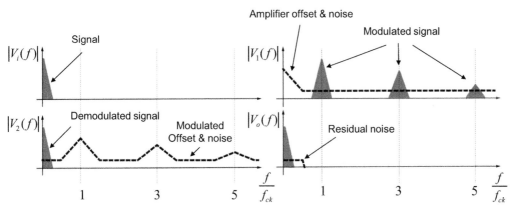

Figure 8.34 Noise reduction by chopping technique.

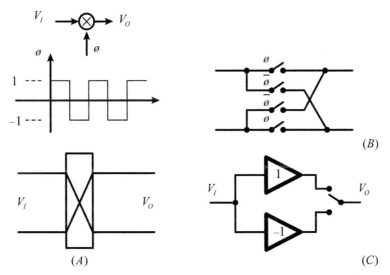

Figure 8.35 Implementation of a "digital multiplier." (A) Schematic symbol. (B) Implementation by switches. (C) Implementation by buffers.

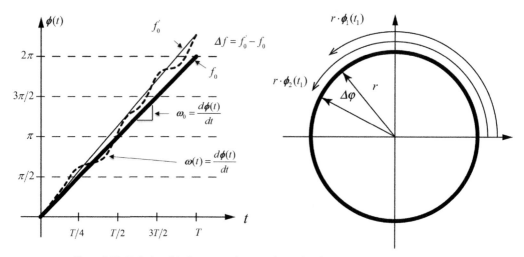

Figure 8.36 Relationship between phase and angular frequency.

Some important relationships related to frequency and phase are graphically recollected in Fig. 8.36. *Phase* is proportional to the circular *length* traveled by a phasor in its harmonic motion, and *frequency* is the *velocity* under which the length is covered. If the movement is uniform, then $d\phi(t)/dt = k$ and we can easily calculate the constant *angular frequency* as the ratio between one phase cycle and the time T taken by the phasor to travel it $\omega_0 = d\phi(t)/dt = 2\pi/T$, where $f_0 = \omega_0/2\pi = 1/T$ is the constant *ordinary frequency*. Therefore, in this case, the harmonic signal could be described as

$$y(t) = A_0 \cos[\phi(t)] = A_0 \cos[\omega_0 t] = A_0 \cos[2\pi f_0 t], \qquad (8.48)$$

where A_0 is the amplitude of the harmonic waveform.

However, if the speed is not uniform, we have to take into account the travel covered by integrating the *instantaneous angular frequency* (or velocity) $\omega(t) = d\phi(t)/dt \equiv [\text{rad/s}]$ or *instantaneous ordinary frequency* $f(t) = \omega(t)/2\pi \equiv [\text{Hz}]$

$$y(t) = A_0 \cos[\phi(t)] = A_0 \cos\left[2\pi \int_0^t f(t)dt\right], \qquad (8.49)$$

where $\phi(t)$ is the *instantaneous phase*. The phase's evolution with a variable frequency is shown in Fig. 8.36A. Note that if at the end of the period, following changes of frequency (speed), the phase (path traveled) is greater than 2π, the harmonic evolution reaches the final condition *as if* the ordinary frequency were constant and equal to $f_0' = f_0 + \Delta f$, graphically illustrated by the thin straight line. In other words, the average frequency in T is different from f_0.

Now assume that a generic signal $x(t)$ is used to make a phase modulation by changing the "natural frequency" f_0 of an oscillator according to the linear function

$f(t) = f_0 + K_X x_i(t)$ where K_X is a constant. Therefore, the previous relationship becomes

$$y(t) = A_0 \cos\left[\underbrace{2\pi \int_0^t \left(f_0 + K_X x(t)\right) dt}_{\phi(t)}\right] = A_0 \cos\left(2\pi f_0 t + \varphi(t)\right), \qquad (8.50)$$

where $\varphi(t) = \Delta\phi(t)$. From the dimensional point of view, the constant is expressed in frequency over the stimulus unit $K_X = \Delta f/\Delta x \equiv [\text{Hz}/\xi]$. Therefore, after a time, ΔT we have the following evolution of the function:

$$y(t) = A_0 \cos\left[\underbrace{2\pi f_0 t}_{\phi_0} + \underbrace{2\pi \underbrace{K_X \Delta x}_{\Delta f} t}_{\Delta\phi}\right] \quad \text{where} \quad \Delta x = \frac{1}{t}\int_0^t x(t) dt, \qquad (8.51)$$

where Δx is the average value of the signal in t. Therefore, for a given natural reference frequency, f_0 we can estimate the input variation in two ways:

1. By fixing a reference phase ϕ_0 and reading the amount of time $\Delta t = t_2 - t_1$ necessary to achieve that phase

$$\Delta x = \frac{\phi_0 \Delta t}{2\pi K_X T(T + \Delta t)}; \qquad (8.52)$$

2. By fixing the time (e.g., to n periods $\Delta t = nT$) and reading the change of phase $\Delta\phi$

$$\Delta x = \frac{\Delta\phi}{2\pi K_X n T}; \qquad (8.53)$$

Note that in both cases $\Delta x = \Delta f/K_x$. It is important to point out that the variation of signal in the integration time should not carry information; for a bandwidth-limited signal, it should be $1/T > 2BW$ where BW is the bandwidth of the signal x. It is interesting to note that time-domain sensing performs a "natural" integration of signal as discussed in Section 7.10.4 of Chapter 7 with the term *integrated sampling* where instead, in that case, a charge amplifier was employed.

If we use a voltage as an input signal, the system acts as a voltage-controlled oscillator (VCO) whose functional block diagram is shown in Fig. 8.37, where $V_i = x$ and $K_V = K_X$.

Depending on the application, we can use either the frequency or the phase as a VCO output variable. In the second case, we need an integrator and, as shown in the same figure, it can be modeled in the transform domain by

$$\varphi(s) = \frac{2\pi K_V}{s} V_i(s), \qquad (8.54)$$

as shown in Fig. 8.37B.

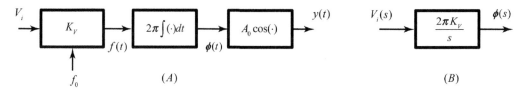

Figure 8.37 VCO interface representation in the time-domain (A) and in the transform domain (B). Depending on the case, the output of the VCO could be either the frequency or the phase. In the second case, we need a further integration transform.

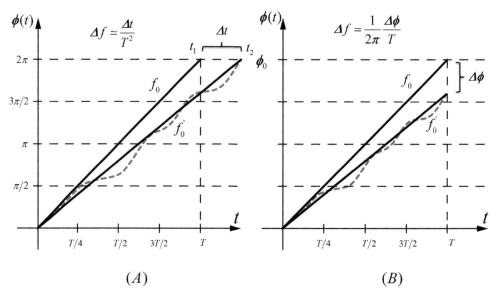

$$(A) \qquad\qquad (B)$$

Figure 8.38 Frequency change detection by (A) time difference variation and (B) phase difference variation. It is assumed that frequency changes very slowly between two measurements in the two cases; thus, the plot is represented by lines.

As discussed, the calculation of the oscillator frequency change (that is, the related to the measuring signal) could be based on either *time difference variation* (Fig. 8.38A) or *phase difference variation* (Fig. 8.38B). In the first case, the frequency could be estimated by the relationship

$$\Delta f = f_0 - f_0' = \frac{\Delta t}{T(T + \Delta t)} \simeq \frac{\Delta t}{T^2}. \tag{8.55}$$

On the other hand, phase detection relies on assessed techniques such as phase-locked-loop architectures. It is based on fixing a reference time and by comparing the phase difference as shown in Fig. 8.38B. In this case, the frequency change could be found as (see Fig. 8.38B)

$$\Delta f = f_0 - f_0' = \frac{1}{2\pi}\frac{\phi_0}{T} - \frac{1}{2\pi}\frac{\phi_0'}{T} = \frac{1}{2\pi}\frac{\Delta\phi}{T}, \tag{8.56}$$

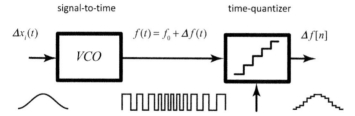

Figure 8.39 A signal quantizer based on time conversion. In here, "time-to-digital" refers to several conversion techniques.

where T is the reference period time. Again, in a VCO presence, the phase difference is the easiest way to recover information from the input signal.

In the analog domain for a sinusoidal waveform, phase difference could be computed at any point of the harmonic evolution. On the other hand, phase differences could be computed only across edges for digital signals. However, we will see effective alternative solutions of phase-detection even in digital systems.

It should be noted that the two estimations are related at the first order by the following:

$$\frac{\Delta \phi}{\Delta t} \simeq \frac{\phi_0}{T + \Delta t},\tag{8.57}$$

where the Eq. (8.57) approximation is valid if the differences are small.

The general scheme for the VCO sensing technique is shown in Fig. 8.39. A signal-controlled oscillator generates a time-dependent signal whose frequency is related to the signal amplitude. Then, a time-quantizer converts the intermediate signal into an output that is usually affected by quantization error.

Therefore, the overall transfer function of the chain is

$$\Delta f = F(\Delta x_i)\tag{8.58}$$

where F is the characteristic function of the interface.

Now, the time mentioned above and phase-detection techniques could be implemented by using a time-to-digital (TDC) converter or a frequency-to-digital (FDC) converter, respectively, as shown in Fig. 8.40A and in Fig. 8.40B, respectively.

8.8.1 Time-to-Digital Conversion Sensing

Time-to-digital conversion is a technique aimed at estimating the difference of time events and could be used in different contexts, especially for measuring distances in optics and ultrasonics. Thus, the results and the techniques of this section could be used independently from the use together with a VCO. However, we will also take the chance to link the use of TDC for frequency estimation based systems, as illustrated in Figs 8.38 and 8.40.

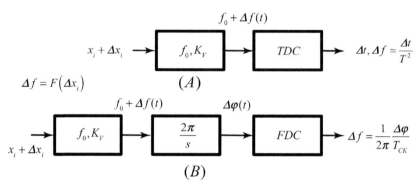

Figure 8.40 Approaches for VCO sensing. (A) A case where the phase is fixed (i.e., during the clock transitions) and frequency estimated by a time-to-digital converter (TDC). (B) A case in which the time is fixed, and the frequency is calculated from the phase difference using a frequency-to-digital converter (FDC).

A possible time-to-digital conversion estimation of the frequency change is shown in Fig. 8.40A using digital signals as in Fig. 8.41, where we refer to a single or multiple periods, where $\Delta T(N)$ is the time difference after N periods. If the frequency is changing from the reference f_0 to $f_0 + \Delta f$ we can measure a single period change as

$$\Delta T = \frac{1}{f_0} - \frac{1}{f_0 + \Delta f} = \frac{\Delta f}{f_0^2 + f_0 \Delta f} \approx \frac{\Delta f}{f_0^2}, \tag{8.59}$$

where $f_0 = 1/T$ and we assumed $\Delta f \ll f_0$. For example, if we have $f_0 = 300$ kHz and $\Delta f = 20$ kHz we have a single period change of about 222 ns. Note that Eq. (8.59) is equivalent to Eq. (8.55). If the measurement of this amount of time is difficult for our system, we can consider N periods to have a more dynamic range, as shown in Fig. 8.41B. In this case

$$\Delta T(N) = N \cdot \Delta T \approx \frac{N \Delta f}{f_0^2}. \tag{8.60}$$

However, the increase on N should be traded off with the bandwidth since we have to assume that the signal is almost stable, at least for NT. In other terms, the signal Nyquist bandwidth should be $BW < 1/2(NT)$; thus

$$N < \frac{f_0}{2BW}. \tag{8.61}$$

Eq. (8.61) states that N could not be greater than a certain value for a required bandwidth.

Two approaches to evaluating a time difference are shown in Fig. 8.42. One of the simplest ways to implement a TDC converter is to use a counter, as shown in Fig. 8.42A, that sums up the clock's number of transitions between a start time and a stop time. Note that this scheme could detect the change of frequency directly by inverting

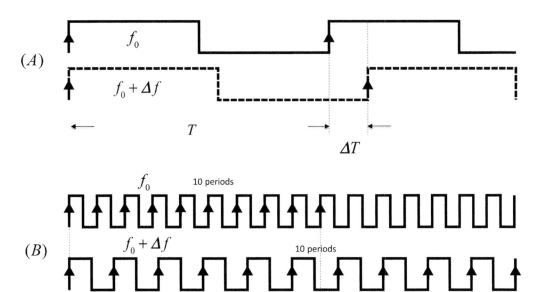

Figure 8.41 Frequency detection of oscillator-based sensing systems using digital signals in a single period (A) and multiple periods (B). Note that in this case $\Delta f < 0$.

the two inputs. This implementation is simple but could be difficult for very short time difference estimations. An alternative implementation of TDC uses a delay chain, as shown in Fig. 8.42B. In this case, the start edge travels through the chain in quantized steps, and at the stop time, the situation is frozen in a thermometer code in a register whose bits are summed by an adder.

In both approaches, the time-difference estimation is affected by quantization noise as in A/D signal conversion. The situation could be seen in Fig. 8.43, where the input-output function of the quantizer is composed of K levels. K could be either the number of states of the counter or the number of delay stages of the delay line. If we assume that the TDC has a maximum measurement time ΔT_{max} and that the discretized time T_{LSB} is given by the delay time of each stage, we have that the digitizer output $\Delta T[n]$ is given by

$$\Delta T[n] = \text{Round}\left(\frac{\Delta T}{\Delta T_{max}} \cdot K\right); \quad \Delta T[n] \in [0, 1, \ldots, K], \quad (8.62)$$

where $\Delta T = t_{stop} - t_{start}$ and $K = \Delta T_{max}/T_{LSB}$ is the number of the delay buffers.

The influence of noise sources in a VCO-based sensing technique implementing a TDC could be modeled as shown in Fig. 8.44, where it is evidenced by both random and quantization noise. Therefore, the acquisition chain scheme of Fig. 8.44A should include noise contributions as illustrated in Fig. 8.44B. More specifically, since the frequency is derived from time differences, it is better to use the absolute jitter by using the random variable ΔT_N instead of the phase noise to which it is anyway related.

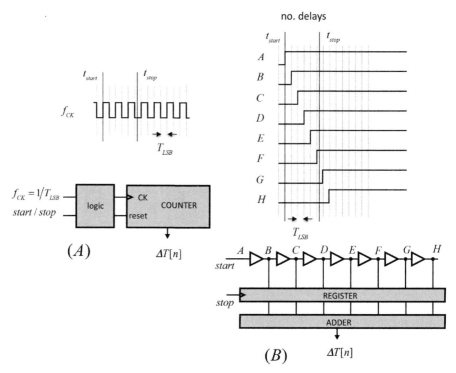

Figure 8.42 Time-to-digital converter based on a counter (A) and a single-delay chain (B).

Therefore, following Fig. 8.44B, we can evaluate the dynamic range (i.e., resolution levels) due to each noise contribution apart and then compose together using the rule of resolutions in acquisition chains (RRC) as discussed in Chapter 3.

If we had an ideal time difference detector, we could derive the dynamic range of a TDC on the basis of this data. Since we referred to time differences, it is easier to calculate the DR with respect to the jitter, so using (8.60)

$$DR_O = \frac{\left(\Delta T(N)\right)^2}{\sigma_{abs}^2(NT)} = \frac{N^2(\Delta f)^2}{f_0^4} \cdot \frac{1}{\alpha^2(NT)} = \frac{N(\Delta f)^2}{\alpha^2 f_0^3}, \quad (8.63)$$

where $\sigma_{abs}^2(NT) = \alpha^2 NT$ is the variance of the absolute jitter in the time calculation (see Chapter 5) and $\Delta f = K_V \Delta x$.

It is apparent that the higher N, the greater the dynamic range (DR); however, this inevitably reduces the bandwidth according to (8.61)

$$DR_O = \frac{N(\Delta f)^2}{\alpha^2 f_0^3} < \frac{(\Delta f)^2}{2BW \cdot \alpha^2 f_0^2}. \quad (8.64)$$

As far as the quantization error is concerned, we have seen (Chapter 2) that the dynamic range of a K-level quantizer is

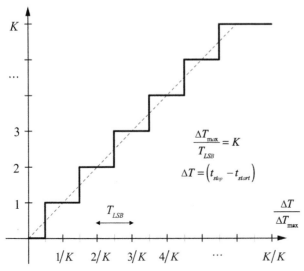

Figure 8.43 Quantized function of the time-to-digital converter.

Within the figure:

$$\frac{\Delta T_{max}}{T_{LSB}} = K$$

$$\Delta T = \left(t_{stop} - t_{start}\right)$$

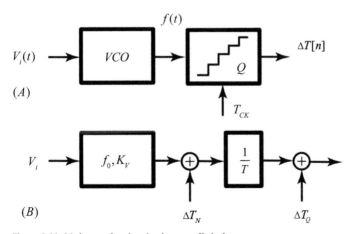

Figure 8.44 Noise evaluation in time-to-digital converters.

$$DR_Q = K^2 \frac{3}{2}, \tag{8.65}$$

where in our case, $K = \Delta T_{max}/T_{LSB}$. Therefore, using the resolution rule of chain acquisition (RRC), the overall dynamic range of the TDC is

$$\frac{1}{DR_T} = \frac{1}{DR_Q} + \frac{1}{DR_O} \tag{8.66}$$

that could also be mapped into the total resolution (information) levels

$$NL_b = \frac{1}{2}\log_2(1 + DR_T). \tag{8.67}$$

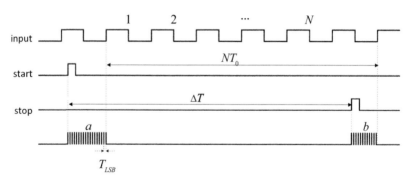

Figure 8.45 Nutt time-of-flight interpolator.

An alternative method for detecting time changes is by using a *Nutt interpolator*, as shown in Fig. 8.45. Coarse and fine measurements determine the time difference between the start and the stop signals.

An oscillator whose period (i.e., resolution) is $T_{\text{LSB}} \ll T$ is activated as a reference from the *start* rising edge. Then it is stopped by the next rising edge of the input signal and the number of counts, a, stored (fine measurement). Then, the number of periods T is counted until the rising edge of the *stop* signal (coarse measurement). Finally, the reference oscillator is activated by the rising edge of the stop signal, and the number of transitions, b, are counted until the next rising edge of the input signal (fine measurement).

Therefore, the difference of time measurement is given by

$$\Delta T = NT + aT_{\text{LSB}} - bT_{\text{LSB}}, \tag{8.68}$$

where the resolution of the measurement is given by T_{LSB}. The trick consists of using a coarse counter to calculate the time difference (NT) and a finer counter to calculate the fraction of the period. Therefore, from the quantization noise point of view, we can consider the situation as if *there were* a number of discrete levels equivalent to

$$K \simeq \frac{NT}{T_{\text{LSB}}} \tag{8.69}$$

where, again, $1/NT > 2BW$ and the dynamic range is

$$DR_{QNC} = K^2 \frac{3}{2}, \tag{8.70}$$

which is much greater than DR without an interpolator. Therefore, the resolution (denominator of the DR) is kept constant, and the range (the numerator of the DR) is arbitrarily increased by several periods, where of course, the drawback is the reduction of bandwidth.

In Table 8.1 is shown a numerical example where a Nutt counter with 10 ns of resolution is used for different signal bandwidths.

Table 8.1 Dynamic range calculation for a Nutt interpolator in different bandwidths.

Inputs					
	Signal BW [Hz]	10.00	100.00	10.000.00	20.000.00
	T_{LSB} [s]	1.00E-08	1.00E-08	1.00E-08	1.00E-08
	f0 [Hz]	3.00E+05	3.00E+05	3.00E+05	3.00E+05
	Δf [Hz]	2.00E+04	2.00E+04	2.00E+04	2.00E+04
	alpha^2 [s]	3.80E-16	3.80E-16	3.80E-16	3.80E-16
	$T = 1/f0$ [s]	3.33E-06	3.33E-06	3.33E-06	3.33E-06
	ΔT [s]	2.22E-07	2.22E-07	2.22E-07	2.22E-07
	$N = f_0/2BW$ (8.61)	15,000.00	1,500.00	15.00	7.50
	$N \cdot T = 1/2BW$ [s]	5.00E-02	5.00E-03	5.00E-05	2.50E-05
	$\Delta T(N)$ [s]	3.33E-03	3.33E-04	3.33E-06	1.67E-06
VCO					
	DR_O (8.64)	5.85E+11	5.85E+10	5.85E+08	2.92E+08
	NL_O [b]	19.54	17.88	14.56	14.06
Quantizer					
	$K = N \cdot T/T_{LSB}$ (8.69)	5.00E+06	5.00E+05	5.00E+03	2.50E+03
	$DR_Q = K^2 \cdot 3/2$ (8.70)	3.75E+13	3.75E+11	3.75E+07	9.38E+06
	NL_C [b]	22.55	19.22	12.58	11.58
TOTAL					
	DR_{TOT}	5.76E+11	5.06E+10	3.52E+07	9.08E+06
	NL_{TOT}	758.825.24	224.922.70	5.936.35	3.013.93
	ENOB [b]	19.53	17.78	12.54	11.56

We have a VCO whose central frequency is 30 MHz with a VCO constant of $K_x = 20$ kHz/V, in practice. Now we have several inputs characterized by different signal bandwidths. The idea is to characterize the dynamic range of the VCO on the jitter data first and then compose it with the DR of the Nutt's counter to evaluate the overall chain resolution.

In the example, as long as the bandwidth is quite low, it is worthwhile to increase the number of periods to calculate the time (phase difference). For example, in a 10 Hz bandwidth, the signal is almost stable for 15,000 periods, and the total time variation is about 3 ms. This increases the DR of the VCO since jitter increases as the square root of the time difference while the signal is proportional to the time difference itself. The increase of the time difference further increases the DR of the quantizer giving a total effective number of bits of about 19b.

If we increase the bandwidth, we reduce the number of periods on which calculate the time difference with the decrease of the dynamic ranges. Interestingly, the further bandwidth increase shows how the DR of the VCO surpasses the DR of the quantizer. This is because the quantization noise is linked to T_{LSB}, which is fixed. This means that if we need measurements on a small number (or just one) period, we have to use T_{LSB} on the order of picoseconds.

8.8.2   Frequency-to-Digital Conversion Sensing

Other than using digital counters, there are more sophisticated ways to implement the frequency to digital sensing based on the difference of phase. An interesting solution is the use of a VCO-based ring oscillator quantizer, as shown in Fig. 8.46. The waveform of each node of the ring oscillator is shifted by an inverter delay time with respect to the adjacent ones. As the polarization of the ring oscillator varies, the inverter delays increase/decrease to change the ring oscillator frequency. As already discussed, the frequency could be estimated by making a difference between phase statuses. For this task, the statuses of the oscillators are stored into two registers during the start and stop times. Then the two registers are thus XORed to get the quantized frequency by phase status difference $f[n] = \varphi[n] - \varphi[n-1]$. More specifically, as illustrated in Fig. 8.46, the XOR operation shows the number of transitions between the start and stop times, thus the number of inverter delays in the lapse time, which is related to the ring

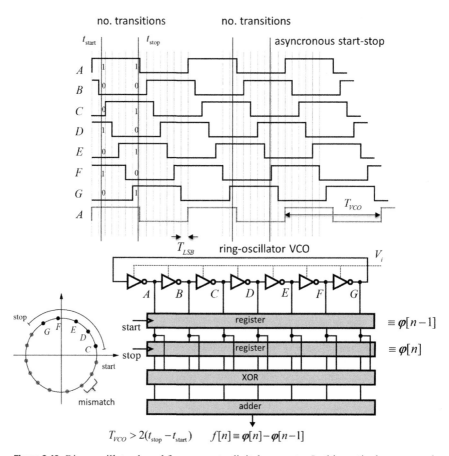

Figure 8.46 Ring-oscillator-based frequency-to-digital converter. In this particular case, we have the average of the frequency of five transitions between the start and stop times. The phasor circle is only showing the concept, and it is not exactly mapped on the ring oscillator of the figure.

oscillator frequency. To better understand this approach, it is fruitful to map the ring oscillator transitions into the harmonic circle of Fig. 8.36. In doing so, in the phasor circle, we can map the ring-oscillator transitions of each stage. If we increase the frequency, the fixed start-stop time interval is graphically enlarged in the figure enclosing more transition edges. To make the thing work, we must ensure that $T_{\text{VCO}} > 2T_{CK}$ where $T_{CK} = (t_{\text{stop}} - t_{\text{start}})$. In ideal conditions, transitions are equally distributed in the circle. Thus, frequency determination is given by counting the number of points (transitions) in each time difference $(t_{\text{stop}} - t_{\text{start}})$. In real cases, the mismatch between single cases places the transition points not equally spaced between them. One of the main advantages of this approach is that the quantized result is an average of the sampling times of the ring oscillator status and thus reducing by averaging multiple results the mismatch distribution of inverters in the ring oscillator. The noise calculation could be modeled as in Fig. 8.47B, where the random noise is added as phase noise power spectral density. The use of the differentiator has the advantage of shaping the quantization noise as in sigma–delta converters. For the DR calculation, we can refer the phase noise to the input. The input-referred phase noise is thus

$$S_I(\Delta f) = S_\varphi(\Delta f) \left(\frac{\omega}{2\pi K_V f_0} \right)^2 = S_V(\Delta f) \frac{2\Delta f^2}{K_V^2 f_0^2}, \tag{8.71}$$

where phase noise $S_V(\Delta f) = S_\varphi(\Delta f)/2$ is defined for normalized input power.

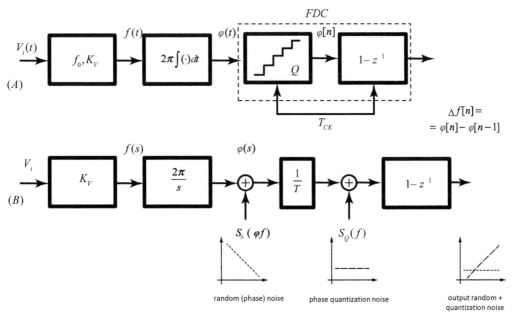

Figure 8.47 Noise evaluation in frequency-to-digital converters. Note how the $(1 - z^{-1})$ is the Z-transform function of the difference of two adjacent samples.

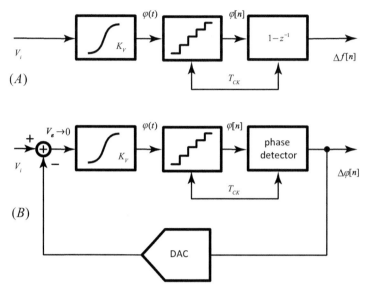

Figure 8.48 (A) Open-loop frequency-to-digital converter oscillator-based sensing. (B) Closed-loop oscillator-based sensing. In this case, the phase difference is not between adjacent samples but between the clock reference and the quantizer output.

A final further improvement of an oscillator-based sensing is, once again, based on the feedback. In fact, the VCO has intrinsic nonlinearities (e.g., think about ring-oscillator VCOs) that seriously compromise the accuracy of the overall system, as shown in Fig. 8.48A. In other terms, the function F of (8.58) is not linear or equivalently $K_V = K_V(x_i)$.

The idea is to keep the VCO in the proximity of the function origin by using a feedback system as discussed in Chapter 3, as shown in Fig. 8.48B. Thus, the phase difference between the output and a fixed reference is fed back in a negative loop. Therefore, the "error" estimate is reduced as much as possible ($V_\varepsilon \to 0$) by the open-loop gain. Therefore, the VCO input is kept around the origin so that the nonlinearities of F are greatly reduced.

8.9   Time-Based Techniques for Resistance and Capacitive Sensing

8.9.1   Relaxation Oscillator Technique

One of the most common techniques to estimate either the value of resistance or capacitance is to use a relaxation oscillator, as shown in Fig. 8.49. In Fig. 8.49A is illustrated the concept where a hysteresis comparator, characterized by two upper and lower thresholds V_{TH} and V_{TL}, is used to reset the capacitor when the RC transient has reached the high threshold level V_{TH}. In this case, the value of the period T is given exactly by (8.29), where $V_T = V_{TH}$.

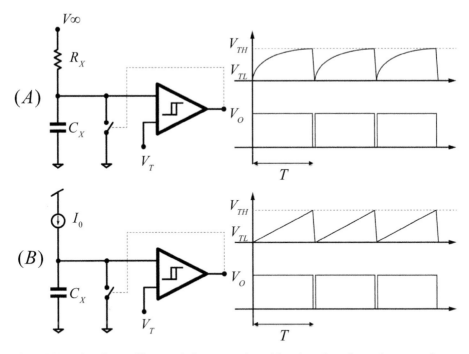

Figure 8.49 Relaxation oscillator technique to readout either the value of capacitance or of resistance. (A) RC transient technique. (B) Constant current technique.

In Fig. 8.49B is shown another technique where the unknown capacitor C_X could be estimated by the period as

$$I_0 = C_X \frac{dV_O}{dt} \rightarrow \frac{V_{TH} - V_{TL}}{T} = \frac{I_0}{C_X}$$
$$\rightarrow T = C_X \frac{V_{TH} - V_{TL}}{I_0}.$$

(8.72)

Therefore, for known thresholds and current, the measurement of the period allows the estimation of the capacitance value.

8.9.2   Bang–Bang Phase Locked Loop Sensing Technique

A very smart technique to sense in the time domain is to employ a binary phase detector, also referred to as *bang–bang phase-locked loop* (BBPLL), originally used for clock recovery. The idea is illustrated in Fig. 8.50 in the case of capacitance sensing using ring oscillators.

The capacitance to be sensed is tied to a ring oscillator node whose frequency is determined by its value, thus acting as a sensor-controlled oscillator whose sensitivity is given by K_V. Note that even if the capacitance affects only one node, its effect determines a uniform (in the ideal case) frequency variation of the ring oscillator on all

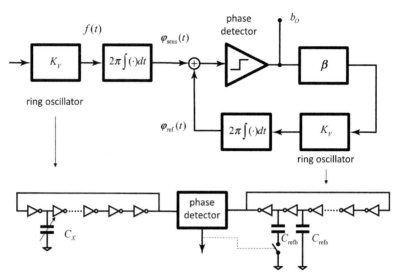

Figure 8.50 Bang-bang phase-locked loop capacitance sensing technique using ring oscillators. (Adapted from Danneels et al., 2011)

the other nodes. Instead of measuring the frequency directly, its phase is compared with that of a reference ring oscillator having binary frequency values through a switched capacitance, C_{refb} as shown in Fig. 8.50. Since we are dealing with binary signals, a D flip-flop could easily implement the phase detector. Now, the output of the phase detector determines the frequency switch of the reference oscillator in a negative feedback fashion by governing the switch of C_{refb}. Thus, the reference oscillator is steered between two frequency values, and the comparator detects if the sensing oscillator leads or lags the reference oscillator. Given f_0 the quiescent frequency of the oscillators, it is

$$\begin{aligned}
f_{sens} &= f_0 + \Delta f = f_0 + K_V \Delta x; \\
f_{dig} &= f_0 + \varepsilon \cdot f_{bb} = f_0 + \varepsilon \cdot \beta K_V; \qquad \varepsilon \in \{+1, -1\},
\end{aligned} \tag{8.73}$$

and the result of the comparison is a binary error $\varepsilon = \text{sign}[\varphi_{sens} - \varphi_{ref}]$ that is fed back to the reference oscillator, determining a frequency shift of $\varepsilon \cdot \beta K_V$. At operating speed, the output of the phase detector reveals if the frequency of the reference oscillator is greater or lower than that of the sensing oscillator. To help to understand, refer to Fig. 8.38, where the sensing frequency evolution is a straight line of the slope f_0. Conversely, the reference oscillator curve is changing its slope to binary values (higher and lower than f_0) so that it evolves in a zig-zag fashion around the sensing oscillator line, similar to what happens in conventional sigma–delta converters around zero (Fig. 8.21) even if in this case it could be represented in the time-phase plot. Thus, if D is the average fraction of time when the phase detector output is on one binary level, we have that, on average, the frequency of the reference will be D times f_{bb} and $(1 - D)$ times $-f_{bb}$. Therefore, the average frequency shift $\overline{\Delta f}$ and D are related to the signal as

$$\overline{\Delta f} = D \cdot f_{bb} + (1 - D)(-f_{bb}) \rightarrow D = \frac{1}{2} + \frac{\overline{\Delta f}}{2 f_{bb}} = \frac{1}{2} + \frac{\Delta f}{2 K_V \beta} = \frac{1}{2} + \frac{\Delta x}{2 \beta}. \quad (8.74)$$

In conclusion, the duty cycle at the output of the PLL is, on average proportional to the stimulus $\Delta x = \Delta C_X$.

8.9.3 Frequency Locked Loop Sensing Technique

Another detection technique for resistances and capacitances is referred to as *frequency locked loop* (FLL) and is shown in Fig. 8.51.

The key idea is to use a switched capacitor circuit, as illustrated in Fig. 8.51A, where the capacitor is first charged to V_I by means of a resistor and then reset to V_R. The difference between the resistance current and the switched capacitor current is fed into an integrator steering a voltage-controlled oscillator that generates the two-phase switch signals. Therefore, the system implements negative feedback, and if the loop gain is very high, the integrated error current I_ε tends to be as small as possible. Therefore, the average current $\overline{I_C}$ flowing through R should be equal to the switched capacitor one

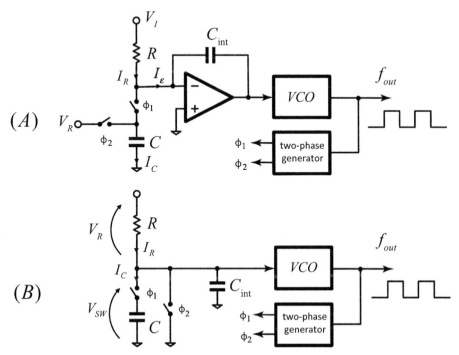

Figure 8.51 (A) Basic scheme of a frequency locked loop. (B) Simplified version.

$$\overline{I_C} = C \cdot V_R \cdot f_{\text{out}} = \frac{V_I}{R}$$
$$\rightarrow f_{\text{out}} = \frac{V_I}{V_R} \frac{1}{RC}. \tag{8.75}$$

Therefore, the output frequency is inversely proportional to the values of the resistance and of the capacitance, namely the time constant.

A further simplification of the implementation is shown in Fig. 8.51B, where the integrator is simply a capacitor C_{int} connected to the ground. Again, the average switched current is

$$\overline{I_C} = C \cdot V_{SW} \cdot f_{\text{out}} = \frac{V_R}{R}$$
$$\rightarrow f_{\text{out}} = \frac{V_R}{V_{SW}} \frac{1}{RC}, \tag{8.76}$$

which is inversely proportional to the time constant, where V_{SW} and V_R are the voltages across the switched capacitor and the resistance, respectively, where the higher the loop gain, the more stable their values. Therefore, the output frequency is an oversampled value of the inverse of R and C values. The main advantage of the FLL technique is that the negative feedback neutralizes the nonlinearities of the VCO, and thus the approach could be implemented without any regard to the implementation of the oscillator linearity as long as the open-loop gain is high.

Further Reading

Abidi, A. A. and Meyer, R. G., Noise in relaxation oscillators, *IEEE J. Solid-State Circuits*, vol. 18, pp. 794–802, 1983.

Barbe, D. F., Imaging devices using the charge-coupled concept, *Proc. IEEE*, vol. 63, no. 1, pp. 38–67, 1975.

Danneels, H., Coddens, K., and Gielen, G. A fully-digital, 0.3 V, 270 nW capacitive sensor interface without external references, in 2011 Proceedings of the ESSCIRC (ESSCIRC), 2011, pp. 287–290.

Enz, C. C. and Temes, G. C., Circuit techniques for reducing the effects of opamp imperfections: Autozeroing, correlated double sampling, and chopper stabilization, *Proc. IEEE*, vol. 84, no. 11, pp. 1584–1614, 1996.

Henzler, S., *Time-to-Digital Converters*, vol. 29. Dordrecht: Springer, 2010.

Jang, T., Jeong, S., Jeon, D., Choo, K. D., Sylvester, D., and Blaauw, D., A noise reconfigurable all-digital phase-locked loop using a switched capacitor-based frequency-locked loop and a noise detector, *IEEE J. Solid-State Circuits*, vol. 53, no. 1, pp. 50–65, 2018.

Kester, W. Ed., *The Data Conversion Handbook*. Philadelphia: Elsevier, 2004.

Navid, R. Lee, T. H., and Dutton, R. W., Minimum achievable phase noise of RC oscillators, *IEEE J. Solid-State Circuits*, vol. 40, no. 3, pp. 630–637, March 2005.

Nutt, R., Digital time intervalometer, *Rev. Sci. Instrum.*, vol. 39, no. 9, pp. 1342–1345, Sept. 1968.

Pallas-Areny, R. and Webster, J., *Sensors and Signal Conditioning*. Hoboken, NJ: John Wiley & Sons, 2001.

Park, M. and Perrott, M. H., A 78 dB SNDR 87 mW 20 MHz bandwidth continuous-time delta-sigma ADC with VCO-based integrator and quantizer implemented in 0.13 μm CMOS, *IEEE J. Solid-State Circuits*, vol. 44, no. 12, pp. 3344–3358, Dec. 2009.

Schreier, R. and Temes, G. C., *Understanding Delta-Sigma Data Converters*. New York: IEEE Press, 2005.

Straayer, M. Z. and Perrott, M. H., A 12-bit, 10-MHz bandwidth, continuous-time sigma-delta ADC with a 5-bit, 950-MS/s VCO-based quantizer, *IEEE J. Solid-State Circuits*, vol. 43, no. 4, pp. 805–814, 2008.

Part III

Selected Topics on Physics of Transduction

9   Selected Topics on Photon Transduction

Photon transduction is a fundamental process of any optical detector or image sensor where the primary task is to estimate an average quantity of photons versus time and/or space. We will start from basic physical phenomena of optical transduction, considering photon flux as an average quantity, disregarding the quantum mechanics characteristics of a single photon. Then, we will investigate the role of noise in the transduction process better to assess design rules in the electronic design of interfaces. As in Chapters 10 and 11, we will treat only a very small part of existing optical sensor implementations to serve as examples of applying the transduction principle.

9.1   Overview of Basic Concepts

9.1.1   Electromagnetic and Visible Spectra

The object of this section is related to optical transduction, that is, the transformation of a luminous (or more in general electromagnetic) signal into an electrical one.

Electromagnetic propagation could be described by either waveform or particle models, referred to as the *wave-particle duality*. We will model electromagnetic radiation as either a wave or a flux of photons, depending on the convenience.

In any sinusoidal propagating wave, the relationship between wavelength λ, its velocity u, and frequency f is given by the relationship

$$\lambda \cdot f = u. \tag{9.1}$$

This expression is valid for any sinusoidal propagating waveform, either mechanical or electromagnetic and thus holds also for wave pressure propagation such as the sound. Referring to electromagnetic waves, we can classify particular electromagnetic radiation in both *wavelength* and *frequency* for a given speed of light in a vacuum $u = c$. Figure 9.1 shows a broadband electromagnetic spectrum ordered in both wavelength and frequency variables. The electromagnetic radiation that is perceivable to our eyes is a small part of the spectrum, typically from 400 nm (violet) to 800 nm (red).

As in other contexts, the total electromagnetic power could be composed of the sum of sinusoidal contributions at different wavelengths (or frequencies) in the signal

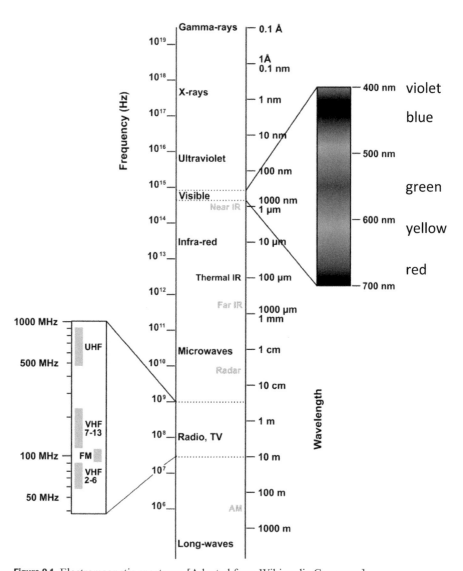

Figure 9.1 Electromagnetic spectrum. [Adapted from Wikimedia Commons]

spectrum. Thus, the electromagnetic signal could be measured in terms of total spectrum power and, more specifically, in *radiant flux*

$$\Phi = \int_0^\infty \Phi(\lambda) d\lambda \equiv [\text{W}], \tag{9.2}$$

where

$$\Phi(\lambda) \triangleq \frac{\text{EM power}}{d\lambda} = \frac{d\Phi}{d\lambda} \equiv \left[\frac{\text{W}}{\text{m}}\right] \tag{9.3}$$

is the *spectral flux.*[1]

When we refer to *luminous radiation*, we are relating to the electromagnetic radiation spectrum that is visible to the human eye, whose sensitivity has been standardized since 1931 by the International Commission on Illumination.

Therefore, the total spectral power related to the human eye sensitivity spectrum is referred to as *luminous flux*

$$\Phi_O = \int_0^\infty \Phi(\lambda)V(\lambda)d\lambda = K\int_0^\infty \Phi(\lambda)V'(\lambda)d\lambda \equiv [\text{lm}], \qquad (9.4)$$

where $V(\lambda)$ is the "bandpass" filter of the eye sensitivity called *luminosity function*. The human eye is characterized by two kinds of sensitivity: one for day vision, called *photopic*, and another for twilight vision, called *scotopic*, as shown in Fig. 9.2. The common derived unit of measurement of luminous power is the *lumen* (lm), and therefore, the luminous function has units of [lm/W]. Alternatively, we can use a *normalized luminosity function* $V'(\lambda) = V(\lambda)/K \in [0, 1]$ expressed in dimensionless units. Therefore, the K scaling factor in (9.4) is given in [lm/W] units. As shown in Fig. 9.2, we can say that we get 683 lm for 1 W of monochromatic light at 555 nm for

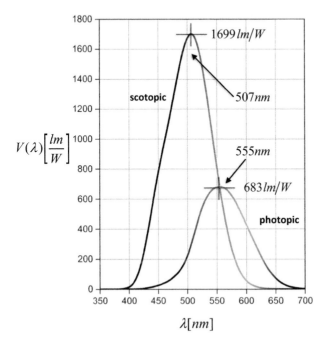

Figure 9.2 Spectral sensitivity of the human eye. [Adapted from Wikimedia Commons]

[1] We can also use the frequency variable instead of the wavelengths so that $\Phi(f) \equiv [\text{W/Hz}]$.

photopic sensitivity (thus, $K = 683$ lm/W), and we have 1699 lm for 1 W of monochromatic light at 507 nm at scotopic vision sensitivity (thus, $K = 1699$ lm/W).

In the case of polychromatic emission of a light source, we can define how much its power is falling in the sensitivity of the human eye by defining the *luminous efficiency* as

$$\text{luminous efficiency} = \frac{\int_0^\infty \Phi(\lambda) \cdot V'(\lambda) d\lambda}{\int_0^\infty \Phi(\lambda) d\lambda} \in [0, 1], \tag{9.5}$$

which is the area subtended by the spectral density and weighted by the photopic response over the total power. We can also define *the luminous efficacy,* where in (9.5) is substituted $V(\lambda)$ in place of $V'(\lambda)$ so that it is measured in [lm/W] units. There are two cases for having maximum luminous efficacy/efficiency. The first is where the emitting spectrum has the same shape of human eye sensitivity ($\Phi(\lambda)V'(\lambda) = \Phi(\lambda)$) so that all the electromagnetic power is visible. The second is a monocromatic emitting power (Dirac's function) in the maximum of sensitivity.

Figure 9.3 shows the graphical interpretation of luminous efficacy/efficiency in four different emission: monochromatic emission at 555 nm, red–green–blue light-emitting diodes (RGB LEDs), phosphor LEDs, and incandescent light. Clearly, in the case of monochromatic emission at 555 nm we have luminous efficacy = 1 (or 100%) according to (9.5) or luminous efficacy equal to 683 lm/W.

9.1.2   Photometry/Radiometry Quantities

Since the human eye is sensitive to a particular bandwidth of the electromagnetic spectrum, from now on we will therefore use the term *luminous* or *optical* for those electromagnetic signals belonging to this spectrum. In general, we will refer to *photometry* as the science considering the relationships of EM quantities in the visible spectrum, while we will use the term *radiometry* as the science ascribed to relationships related to the whole electromagnetic spectrum.

We will shortly introduce the concept of solid angle Ω, which is a measure of the field of view subtended from a certain viewpoint as shown in Fig. 9.4 and is expressed in *steradian*, which is a dimensionless derived SI unit. Following Fig. 9.4, one steradian is defined as the solid angle subtended at the center of a unit sphere by a unit area on its surface. Therefore, in general, for the case of a sphere, the measure of a solid angle in steradians is

$$\Omega \triangleq \frac{\text{area subtended}}{\text{total area sphere}} \cdot 4\pi = \frac{S}{4\pi r^2} \cdot 4\pi = \frac{S}{r^2} \equiv [\cdot] \equiv [\text{sr}]. \tag{9.6}$$

Note that the steradian is defined as a dimensionless ratio similar to the *radian* defined as the length of the arc divided by the radius of the circle.

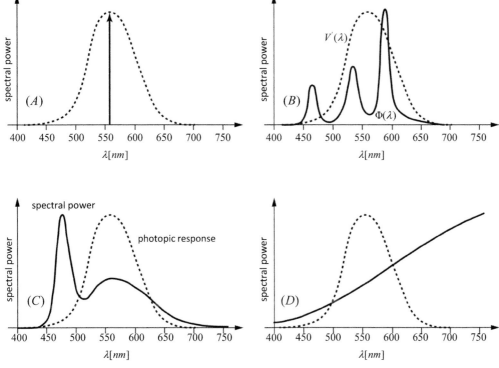

Figure 9.3 Luminous efficacy graphical representation of different light source. (A) Monochromatic 555 nm power with luminous efficiency 100%. (B) RGB LED. (C) Phosphor LED. (D) Incandescent lamp. LED sources could have an efficiency as high as 40%, while tungsten incandescent light sources have an efficiency of around 2%. [Adapted from National Research Council and *Assessment of Advanced Solid State Lighting*, The National Academies Press]

For small solid angles,[2] the increase of the subtended area due to its tilting could be taken into account by using the *cosine law* for a constant solid angle, as shown in Fig. 9.5A:

$$d\Omega = \frac{dS}{r^2} = \frac{dS' \cos\theta}{r^2}, \tag{9.7}$$

where dS is the area of the surface orthogonal to the longitudinal axis while $dS' > dS$ is the tilted one. This effect is valid also for constant areas as shown in Fig. 9.5B

$$d\Omega' = \frac{dS \cos\theta}{r^2} = d\Omega \cdot \cos\theta \le d\Omega, \tag{9.8}$$

where $d\Omega$ is the solid zenith angle whilst $d\Omega'$ is the bent one.

---

[2] From now on, we will refer to *differential or infinitesimal* quantities because they will be used in differential ratios. Therefore, the quantities at the numerator (power, energy, flux) are proportional to the quantities at the denominator; however, their ratio converges to a fixed value for infinitesimal values according to differential calculus rules.

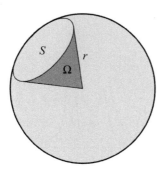

Figure 9.4 Concept solid angle and steradian.

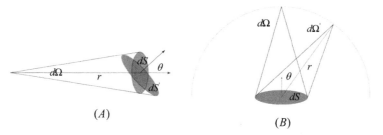

(A)

(B)

Figure 9.5 Effect of the tilting of the area subtended by a solid angle. (A) Constant solid angle. (B) Constant area.

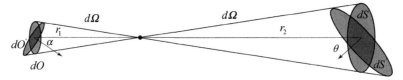

Figure 9.6 Effect of the equivalence between two solid angles.

The solid angle could be used for determining the ratio between areas of surfaces subtended by equal solid angles, as shown in Fig. 9.6, where

$$d\Omega = \frac{dO}{r_1^2} = \frac{dO' \cos \alpha}{r_1^2} = \frac{dS}{r_2^2} = \frac{dS' \cos \theta}{r_2^2}$$

$$\rightarrow \frac{dO}{dS} = \frac{r_1^2}{r_2^2}; \quad \frac{dO'}{dS'} = \frac{r_1^2 \cos \theta}{r_2^2 \cos \alpha},$$

(9.9)

showing how the subtended areas are related to the square of the ratio of the radii with the correction of the cosine law in the case of tilted surfaces.

A first definition in photometry is the *illuminance* (*irradiance* in radiometry) which is the luminous flux power incident on a surface

$$E \triangleq \frac{\text{optical power}}{\text{surface area}} = \frac{d\Phi_O}{dS} \equiv \left[\frac{W}{m^2}\right] \equiv \left[\frac{lm}{m^2} \triangleq lux\right],$$

(9.10)

where the lux, a derived unit, has been defined as a lumen per square meter. Therefore, using the constant value for the photopic vision of (9.4), we have the relationship between the electromagnetic power versus luminous power given by

$$1\left[\frac{W}{m^2}\right] = 683[\text{lux}]. \tag{9.11}$$

In the case the luminous power is emitted from a surface, it is called *emittance* (*radiance* or *radiosity* or *radiant exitance* in radiometry).

If we refer to the luminous power distributed on the spectrum, we have the *spectral illuminance*

$$E(\lambda) \triangleq \frac{\text{optical power}}{\text{surface area} \cdot \text{wavelength}} = \frac{d^2\Phi_O}{dS \cdot d\lambda} \equiv \left[\frac{\text{lm}}{m^2} \cdot \frac{1}{m}\right];$$

$$E = \int_0^\infty E(\lambda)\, d\lambda. \tag{9.12}$$

Referring to the wavelength–particle duality, we can express the concept of EM power in terms of a flux of photons instead of radiant wavelength energy. Thus, another way to define *spectral illuminance* is by taking into account the *photon flux,* where the energy of each photon is related to its wavelength

$$E(\lambda) \triangleq \frac{\text{no. photons} \cdot \text{photon energy}}{dS \cdot dt \cdot d\lambda} = \Phi_P(\lambda) \cdot \frac{hc}{\lambda} \equiv \left[\frac{\text{lm}}{m^2}\frac{1}{m}\right] \equiv \left[\frac{W}{m^2}\frac{1}{m}\right]$$

$$\Phi_P(\lambda) \triangleq \frac{\text{no. photons}}{dS \cdot dt \cdot d\lambda}, \tag{9.13}$$

where $\Phi_P(\lambda)$ is the *photon flux* and hc/λ is the energy of a single photon of wavelength λ.

Notation Since we will primarily refer to optical sensors, for simplicity of notation, we will use from now on the symbol Φ to denote luminous flux power Φ_O.

Note that the photon flux is *not* proportional to the luminous power because we can have the same power using a reduced number of photons with a shorter wavelength. In the case the power is emitted by a surface, the irradiance is called *radiosity*.

Another important quantity of photometry is the *luminous intensity* (*radiant intensity* for radiometry), which is the luminous power per unit solid angle

$$I \triangleq \frac{\text{optical power}}{\text{solid angle}} = \frac{d\Phi}{d\Omega} \equiv \left[\frac{W}{\text{sr}}\right] = \left[\frac{\text{lm}}{\text{sr}} \triangleq \text{cd}\right], \tag{9.14}$$

where Ω is the solid angle. The intensity is usually measured in *candle* units, a lumen per *steradian*.[3] Therefore, the total luminous intensity of a source radiating into a solid angle remains constant irrespective of the distance from the source.

---

[3] Actually, the reference SI unit is the *candle*; therefore it is more correct to define the *lumen* as a candle times a steradian.

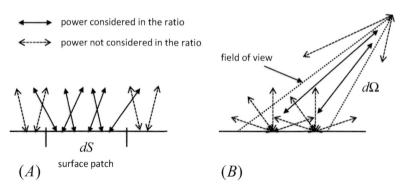

Figure 9.7 Concept of illuminance (A) and intensity (B) using photon ray tracing. The ray trajectories could be considered in both directions.

The *illuminance* and *intensity* quantities could be visualized with the use of the ray-tracing abstraction (ray optics), where *rays* could be considered extremely small pencils of light traveling in the space along straight-lines. To apply this technique in already discussed definitions, we have drawn the ray trajectories as shown in Fig. 9.7, where the intersects of ray traces with a surface could represent the impact of photons with that surface[4]. In the case of *illuminance*, we first consider the ratio of the power given by photons that are hitting the surface (solid line), and we do not consider the photons impinging outside (dotted line), as shown in Fig. 9.7A. The more decreasing the area, the more decreasing the luminous power considered, so that the limit of the ratio converges to the illuminance. If we consider now the ratio limit of the luminous power related to all rays whose trajectory is contained in the field of view subtended by the solid angle as shown in Fig. 9.7B with the solid angle, we get the intensity. Therefore, according to this view, we can redefine the quantities mentioned above as

$$\text{illuminance} = \text{photon flux received (emitted) per unit surface S}$$
$$\text{intensity} = \text{photon flux coveyed (emitted) per unit solid angle } \Omega. \tag{9.15}$$

A significant relationship should be pointed out between the intensity and the illuminance

$$I = \frac{d\Phi_O}{d\Omega} = \frac{d\Phi}{dS} r^2 = E \cdot r^2, \tag{9.16}$$

---

[4] The relationship between the ray tracing abstraction with photon flux could be very misleading. Rays could be considered as statistical averages of photon flux (as in lasers). We can count photons as discrete effects in contributing to energy/power once they are received on an object, but we cannot consider them traveling space as a macroscopic particle, following straight lines because they are subject to quantum dynamics laws. In other terms, we can assume where photons are emitted and where they are impinging, but we will disregard their true trajectories even if we can model them as ray traces as a valuable abstraction for our needs.

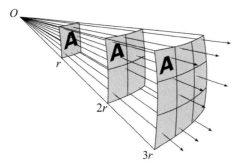

Figure 9.8 The inverse-square law applied between illuminance and intensity. [Adapted from Wikimedia Commons]

showing how the illuminance is inversely proportional to the square of the distance from the source of that physical quantity of a given intensity, following the *inverse-square law* (frequently used in EM propagation) as shown in Fig. 9.8.

A third important quantity is the luminous power per unit solid angle per unit projected source area, called *luminance* (in radiometry is called *radiance*), usually measured in *nit* units

$$L = \frac{d^2\Phi}{dS \cdot d\Omega} \equiv \left[\frac{\text{lm}}{\text{m}^2 \cdot \text{sr}} \right] \equiv [\text{nit}]. \tag{9.17}$$

The concept of luminance could be tricky, and it deserves more explanation. The point with the luminance is that it has been conceived to calculate the amount of power going from a surface to another. As shown in Fig. 9.9A we will start to calculate the power transfer from a surface S_B to another one S_A placed in front of the other in parallel fashion. Note that, as displayed in Fig. 9.9A, some ray traces (solid line) go from one side to the other contributing to the power transfer while some others (dotted lines) do not contribute to the power transfer.

Notation: In the following, we will refer to Ω_X as the solid angle *with a vertex on the surface S_X* and subtending something *different* from S_X.

To calculate the amount of energy going from S_B to S_A, we can imagine that S_B is subdivided into smaller subsections s_{iB} emitting photons in all directions as shown in Fig. 9.9B. However, only those rays that are fired in the field of view of the corresponding solid angle Ω_{iB} hit the target. Therefore, the power that is transmitted is proportional (by a factor L) to the size of those spots and to the size of the solid angle

$$\text{optical power } B{\rightarrow}A = L \cdot s_{1B} \cdot \Omega_{1B} + L \cdot s_{2B} \cdot \Omega_{2B} + \ldots, \tag{9.18}$$

where $s_{1B} + s_{2B} + \ldots = S_B$. Now, if we decrease the size of the areas to infinitesimals $S_A{\rightarrow}dS_A$ and $S_B{\rightarrow}dS_B$ the solid angles become equal $\Omega_{1B} = \Omega_{2B} = \ldots = d\Omega_B$ and therefore

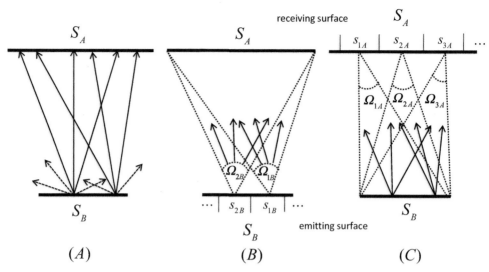

Figure 9.9 Luminous energy transfer between two parallel surfaces, with the emitting shown at the bottom and the receiving at the top. (A) Power transfer by ray traces. Dotted ray traces are those with no power transfer contribution. (B) Luminance calculated with respect S_A field of view. (C) Luminance calculated from S_B field of view.

$$\text{differential optical power } dB{\to}dA = d^2\Phi = L \cdot \underbrace{(s_{1B} + s_{2B} + \dots)}_{dS_B} \cdot d\Omega_B, \quad (9.19)$$

from which we find that the proportionality factor is the *luminance* of the system as

$$L = \frac{d^2\Phi}{dS_B \cdot d\Omega_B}. \quad (9.20)$$

Conversely, we can reverse the calculation using the field of view seen from S_A; therefore

$$\text{optical power } B{\to}A = L' \cdot s_{1A} \cdot \Omega_{1A} + L' \cdot s_{2A} \cdot \Omega_{2A} + \dots$$
$$\text{differential optical power } dB{\to}dA = d^2\Phi = L' \cdot \underbrace{(s_{1A} + s_{2A} + \dots)}_{dS_A} \cdot d\Omega_A. \quad (9.21)$$

Since the power transferred is the same, then the two proportionality constants are equal

$$L' = \frac{d^2\Phi}{dS_A \cdot d\Omega_A} = \frac{d^2\Phi}{dS_A \cdot dS_B} \cdot r^2 = \frac{d^2\Phi}{dS_B \cdot d\Omega_B} = L, \quad (9.22)$$

where r is the distance between the two areas. In other words, the luminance is the optical power emitted by the surface dS_B in the solid angle $d\Omega_B$ (defined by dS_A) or the optical power received by dS_A in the field of view $d\Omega_A$ (defined by dS_B).

Now, we will cope with tilted surfaces. Starting from (9.22), we have to take into account the cosine law so that, using the graphical convention of Fig. 9.11A in Fig. 9.10, we have

$$L = \underbrace{\frac{d^2\Phi}{dS_B \, \cos\theta_B \cdot d\Omega_B}}_{\text{tilted surface}} = \frac{d^2\Phi \cdot r^2}{dS_B \, \cos\theta_B \cdot dS_A \, \cos\theta_A}$$

$$= \underbrace{\frac{d^2\Phi}{dS_A \, \cos\theta_A \cdot d\Omega_A}}_{\text{tilted surface}}, \tag{9.23}$$

which justifies the equivalence of the luminance determination construction of Fig. 9.10B. It should be noted that, even if we have used the opposite solid angles to get the same result if the two surfaces have different emission characteristics, there is no reciprocity in power transfer; that is, the *luminance definition is different if we exchange the emitting and receiving surface role.*

The derivation we did so far clarifies some graphical interpretations since luminance is usually drawn as in Fig. 9.11A, implying the amount of energy conveyed by a patch dS_B into the solid angle $d\Omega_B$. However, it should be understood that we are summing up all the rays (integral sum) stemming from solid angles in each point of the emitting patch dS_B in the field of view of dS_A as shown in Fig. 9.11B. Those solid angles are as much similar to each other as the smaller the patch. Following (9.22), the same result could be achieved summing up all the rays emitted by dS_B and received by each angle of view of the points in dS_A, as illustrated in Fig. 9.11C.

Hint The luminance could be calculated using the solid angle defined by the field of view of receiving surface or by the solid angle defined by the field of view of the emitting surface.

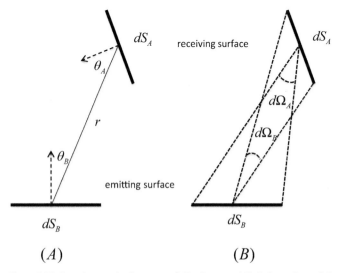

Figure 9.10 Luminance in the case of tilted areas. (A) Orientation of the areas. (B) Definition of the solid angles.

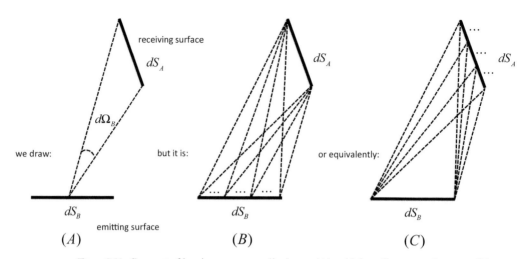

Figure 9.11 Concept of luminance as usually drawn (A), which really means the sum of the rays from all emitting points in the field of view of the receiving surface (B), which is, in turn, equivalent to the sum in any point of the receiving surface of all the rays coming from the field of view of the transmitting surface.

Another important relationship is that one between the *luminance* with the *illuminance*

$$L = \frac{d^2\Phi}{dS_B \cdot d\Omega_B} = \frac{dE_B}{d\Omega_B} = \frac{dE_A}{d\Omega_A}, \tag{9.24}$$

where E_A is the *illuminance* of the receiving surface and E_B is the *emittance* of the emitting surface.

Furthermore, the relationship with respect to the intensity is

$$L = \frac{d^2\Phi}{dS_B \cdot d\Omega_B} = \frac{dI_B}{dS_B} = \frac{\text{luminous intensity of the source}}{\text{source area}}, \tag{9.25}$$

where I_B is the intensity of the emitting surface.

We can now refer to a relationship between the luminous power impinging onto a surface and the one reflected in a particular direction, as shown in Fig. 9.12. A photon flux coming from the direction of polar coordinates (θ_I, φ_I) and falling onto a surface dS_P is scattered in any direction. Specifically, the figure is considered the reflected direction (θ_R, φ_R). We will assume that the surface characteristics are symmetrical with respect to the normal axis; thus, there is no dependence versus φ that is $L_R(\theta_R, \varphi_R) = L_R(\theta_R)$.

We can define the *bidirectional reflectance distribution function* map (BDRF) as the ratio of the reflected *luminance* of a patch in the θ_R direction over the *illuminance* of the same patch from the θ_I direction

$$\text{BDRF}(\theta_I, \theta_R) = \frac{L_R(\theta_R)}{E_I(\theta_I)}, \tag{9.26}$$

where $L_R(\theta_R)$ is the luminance reflected on the patch dS_O of the hemisphere and $E_I(\theta_I)$ is the part of the impinging illumination coming from the patch dS_I of the hemisphere.

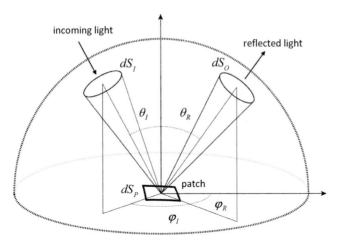

Figure 9.12 Geometrical construction of the bidirectional reflectance distribution function map (BDRF).

Now, for a power balance of the patch, the total reflected energy (*emittance*) is part of the total *illuminance* on the patch

$$E_O = k \cdot E_I; \text{ with } k \in [0, 1] \tag{9.27}$$

because part of the energy could be absorbed by the material, where k is also referred to as *albedo*.

Now, if the *reflected power is constant over any direction*, we have $L_R(\theta_R) = L_R$. In this case, the surface is called *Lambertian*. Therefore, in this case

$$E_O = \int_\Omega L_R(\theta_R)d\Omega_R = L_R \int_\Omega d\Omega_R = L_R \int_0^{2\pi} \int_0^{\pi/2} \cos\theta_I \, \sin\theta_I d\theta_I d\varphi_I = L_R \cdot \pi = k \cdot E_I,$$

$$\tag{9.28}$$

where Ω is the hemisphere space and the infinitesimal patch subtended by solid angle has sides of $\sin\theta_I d\theta_I$ and $\sin\varphi_I d\theta_I$, respectively. Therefore, for a Lambertian surface

$$L_R = \frac{k}{\pi} E_I, \tag{9.29}$$

Consequently, a Lambertian surface, under constant illumination directions, has a constant BDRF. Furthermore, looking at Fig. 9.13, we can see that the solid angles defined between dS_O and dS_P are

$$d\Omega_P = \frac{dS_O}{r^2}; \quad d\Omega_O = \frac{dS_P \cos\theta_R}{r^2}. \tag{9.30}$$

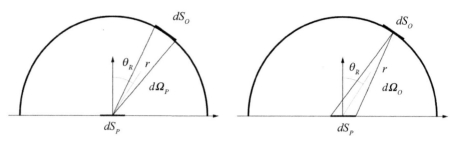

Figure 9.13 Geometrical constructions for properties of a Lambertian surface.

Therefore,

$$L_R = \frac{k}{\pi} E_I = \frac{d^2\Phi}{d\Omega_P dS_P} = \frac{dI_P}{dS_P} \quad \to \quad dI_P = L_R \cdot dS_P;$$

$$L_R = \frac{d^2\Phi}{d\Omega_O dS_O} = \frac{dI_O}{dS_O} \quad \to \quad dI_O = L_R \cdot dS_O; \tag{9.31}$$

$$L_R = \frac{d^2\Phi}{d\Omega_O dS_O} = \frac{dE_O}{d\Omega_O} \quad \to \quad dE_O = L_R \cdot d\Omega_O.$$

This means that in a Lambertian surface, the intensity dI_P emitted by the patch dS_P and the intensity dI_O seen on the field of view $d\Omega_O$ are constant in any direction. However, the amount of power impinging on dS_O depends on θ_R because combining Eqs. (9.30) and (9.31)

$$dE_0 = L_R \frac{dS_P}{r^2} \cos\theta_R \propto \cos\theta_R. \tag{9.32}$$

9.1.3   Power Transfer in Image Projection Systems

Every time an image sensor interacts with an environment, it is by means of an image projection system. Our task is to understand the luminous power ratio between the source (object) and destination (sensor). We will use the definitions and relationships so far derived from doing so. As a first step, we will consider two simple cases of projection systems.

As shown in Fig. 9.14, we will use a pinhole projection system where an emitting surface dS is projected onto a plane surface dO by using a pinhole made on a separation sheet. The distance of the hole from the projection plane is also referred to as *focal length* f. If the pinhole area is small, each point in the object is mapped into a unique point in the projection plane. The smaller the pinhole size, the more precise the correspondence is.

However, making the hole very small, together with the problem of light diffraction, will lead to the problem of luminous power transfer. The relationships between the luminous fluxes could be better understood by using the construction of Fig. 9.14A and that of Fig. 9.14B by equating the luminance of the hole

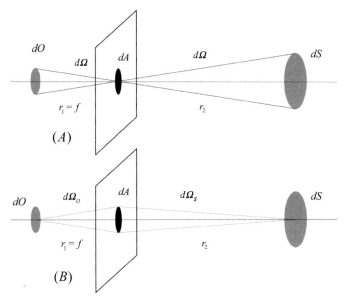

Figure 9.14 Pinhole projection system.

$$L = \frac{d^2\Phi}{d\Omega \cdot dA} = \frac{d^2\Phi}{d\Omega_S \cdot dS} = \frac{d^2\Phi}{d\Omega_O \cdot dO}. \tag{9.33}$$

What is important is to calculate the amount of luminous power passing through the system

$$d^2\Phi = L \cdot dA \cdot d\Omega$$
$$\rightarrow \frac{d\Phi}{dO} = E_O = \frac{L \cdot dA}{f^2}, \tag{9.34}$$

which means that the illuminance of the projected image is proportional to the size of the hole times the luminance of the system, divided by the square of the focal length. Therefore, the amount of the luminous power or of the luminance clearly becomes negligible as $dA \rightarrow 0$. This is one of the main problems of pinhole camera: the smaller the hole the smaller the absolute luminous power received by the sensor.

A well-known approach to overcome this problem is to use a lens projection system, as shown in Fig. 9.15. It could be shown that the lens projection system is exactly behaving like in the pinhole: each point of the object is mapped into a unique point of the projection plane. However, in this case, the number of photons that are transmitted from side to side depends on the lens aperture.

More precisely, in this case, we have

$$d^2\Phi = L \cdot S_L \cdot d\Omega$$
$$\rightarrow \frac{d\Phi}{dO} = E_O = \frac{L \cdot S_L}{f^2}, \tag{9.35}$$

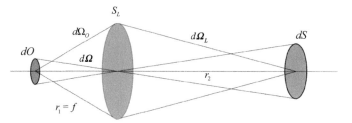

Figure 9.15 Lens projection system.

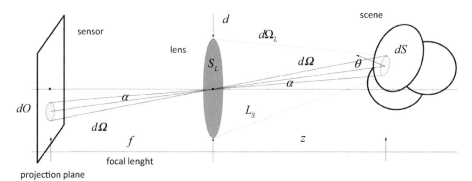

Figure 9.16 Lens image projection system. The distance of the lens from the projection plane is called *focal length, f*.

where S_L is the diameter of the lens. The area of the lens plays a fundamental role in the energy transfer, ensuring the same mapping of the pinhole system but providing a greater luminous power transfer that is proportional to the diameter of the lens. We can modulate the total energy by reducing the size of S_L using a *diaphragm* placed in front of the lens, as commonly used in cameras.

We will now complete the previous discussion by analyzing a more realistic lens projection system considering the deviation of projection from the central axis. The main question is: how much energy (power) is impinging the sensor surface for a given illumination of the scene? To get an answer to this, we draw a simple projection scheme based on a lens, such as those illustrated in Fig. 9.16. In the picture, a patch dS of an illuminated object is mapped into a patch dO of the projection plane where the sensor resides. The answer to the previous question is to calculate the ratio of the object's illuminance with the illuminance of the sensor to understand the power ratio between the two sides of the system.

We have seen that according to (9.35), the luminous power passing through the system is

$$d^2\Phi = L_S \cdot S_L \cdot d\Omega, \tag{9.36}$$

where L_S is the luminance of the scene patch in the field of view of the lens of area S_L. Therefore, the illumination of the patch dO is

$$E_O = \frac{d\Phi}{dO} \cos \alpha = \frac{L_S \cdot S_L \cdot d\Omega}{dO} \cos \alpha, \tag{9.37}$$

where α is the angle between the projection and central axes, and the $\cos\alpha$ correction is necessary because the area is enlarged with respect to a projection of the same patch on the central axis. Then, since

$$d\Omega = \frac{dS \cos\theta}{(z/\cos\alpha)^2} \tag{9.38}$$

and using (9.9)

$$\frac{dS}{dO} = \frac{\cos\alpha}{\cos\theta} \left(\frac{z}{f}\right)^2, \tag{9.39}$$

we have

$$E_O = \frac{dS \cos\theta}{(z/\cos\alpha)^2} \frac{L_O \cdot S_L}{dO} \cos\alpha = \frac{1}{f^2} \cdot S_L \cdot L_S \cdot \cos^4\alpha, \tag{9.40}$$

which is a generalized form of (9.35) considering the deviation of the projection from the central axis. The $\cos^4\alpha$ factor determines the effect of *vignetting* in the sensor border (as in old photography where no lens correction system were employed).

Now considering the diameter d of the circular lens and assuming the surface of the object scene Lambertian, we can use (9.29) so that

$$E_O = L_S \left(\frac{d}{f}\right)^2 \cdot \frac{\pi}{4} \cdot \cos^4\alpha = L_S \frac{\pi}{4} \left(\frac{d}{f}\right)^2 \cos^4\alpha \approx \frac{1}{4} \left(\frac{d}{f}\right)^2 k \cdot E_S, \tag{9.41}$$

where f/d; that is, the ratio of the focal length with the lens diameter (or aperture) is called the *f-number* or *focal ratio*. The power transmitted could be reduced by inserting a diaphragm with variable aperture in front of the lens.

Using typical albedo (k) values for mixed scenes and *f*-numbers of high-performance objectives (≈ 1.4), we get that the luminous power impinging the sensor is about an order of magnitude smaller than the luminous power impinging the object. This is an important point is sensor design because in the best case we have to know that the number of photons falling in the sensor surface is about ten times smaller than the number of photons falling on the scene.

> **Hint** The luminous power impinging the sensor surface could be about one order of magnitude smaller than that impinging the scene even in the better power transfer condition of the projection system (higher diaphragm aperture).

9.2 Blackbody Emission

A *blackbody* is by definition an idealized model of a body that absorbs all incident electromagnetic radiation. Therefore, it does not transmit (as in transparent objects) or reflects (as in specular objects) incident radiation regardless wavelength and angle of incidence, acting as a perfect absorber for all incident radiation. Real objects could just approximate the definition of blackbody because it represents a physical abstraction.

Since the body absorbs energy, it will eventually reach thermal equilibrium, thus emitting electromagnetic radiation. One of the earliest and seminal achievements of quantum mechanics, called *Planck's law for blackbody emission*, states that the emission spectrum of the blackbody depends only on the temperature of the object, irrespective of the composition of the material of the body.

Planck's law is usually referred to in terms of *spectral radiance*

$$L(\lambda,\ T) = \frac{2hc^2}{\lambda^5} \frac{1}{e^{\frac{hc}{\lambda kT}} - 1} \equiv \left[\frac{W}{m^2} \frac{1}{sr \cdot m}\right], \tag{9.42}$$

where h and k are Planck's and Boltzmann's constants, respectively; c is the light speed in vacuum[5]; and T is the temperature in Kelvin degrees. The same law could be seen in terms of *spectral irradiance*, by multiplying the radiance of a factor π for Lambertian surfaces

$$E(\lambda,\ T) = \frac{2\pi hc^2}{\lambda^5} \frac{1}{e^{\frac{hc}{\lambda kT}} - 1} \equiv \left[\frac{W}{m^2} \frac{1}{m}\right]. \tag{9.43}$$

By integrating the blackbody emission over the entire spectrum, we find the Stefan–Boltzmann law

$$E = \frac{2\pi^5 k^4}{15h^3 c^2} T^4 = \sigma T^4 \equiv [W/m^2], \tag{9.44}$$

where σ is the Stefan–Boltzmann constant. Therefore, the total power emitted by a blackbody per surface area is proportional to the fourth power of the temperature.

Furthermore, taking the first derivative of Planck's expression, we get how the maximum peak of the radiation depends on the temperature according to *Wien's law*

$$\lambda_{max} = \frac{b}{T}; \quad b = 2.89 \times 10^6 \ nm \cdot K. \tag{9.45}$$

The blackbody irradiance spectrum is shown in Fig. 9.17, where is evidenced the luminous spectrum bandwidth of eye sensitivity. Note how the higher the temperature the shorter the wavelength, that is, the "cooler" is the color of the emitting source. More specifically, the *color temperature* of an emitting source is defined as the temperature in Kelvin degrees of an ideal blackbody spectrum that matches (as much as possible) that one of the source. Typical 60 W incandescent, "warm" and "cold" light fluorescent lamps have the color temperature of about 2700 K, 3500 K, and

---

[5] $h = 6.62 \times 10^{-34} J \cdot s$; $k = 1.38 \times 10^{-23} J/K$; $c = 2.99 \times 10^8 m/s$

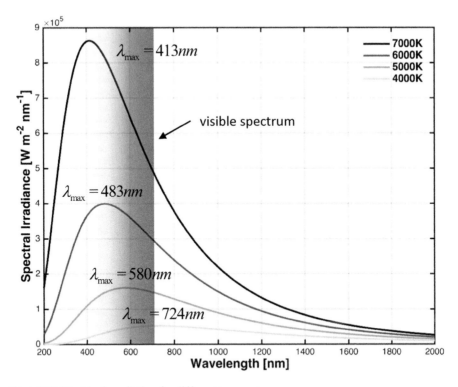

Figure 9.17 Blackbody radiation for different temperatures.

5500 K, respectively. Among them, the source that most resembles to the blackbody emission is the incandescent lamp. Tuning back to Fig. 9.3D referring to luminous efficacy of an incandescent light we can see that the eye sensitivity curve is overlapped to a "tail" of the blackbody spectrum at 2700 K. Since the estimated temperature of the Sun is 5778 K, using Wien's law, one finds a peak emission at a wavelength of about 500 nm in the green part of the spectrum near the peak sensitivity of the human eye.

Figure 9.18 shows the comparison of the Sun irradiance (above Earth's atmosphere) against the spectral emission of a blackbody with a temperature of 5777 K. It should be noted that the values are in absolute value much different from the values of the spectral irradiance given by (9.44) and shown in Fig. 9.17 since they are taken at the ideal surface of the Sun while the last ones are taken at Earth's surface; therefore we should take into account the distance between the Sun and Earth. The areas subtended by the two functions are very similar and equal to

$$\int_0^\infty E(\lambda)d\lambda = 1362 \text{ W/m}^2, \tag{9.46}$$

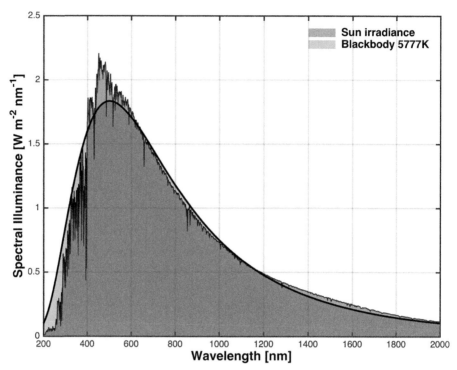

Figure 9.18 Spectral illuminance of the Sun compared with the spectral emission of a blackbody at 5777 K.

which is called *solar constant*, which is the maximum power per aera received by the Sun above Earth's atmosphere. As a conclusion, the sun is a very good approximation of the blackbody radiator, even if its emission is not in general Lambertian.

In photon transduction, we have to consider that the energy of each photon changes according to the wavelength. Therefore, for equal power, we could have different photon fluxes. The relationship is given by rearranging (9.13)

$$\Phi_P(\lambda) = \frac{\text{no. photons}}{dS \cdot dt \cdot d\lambda} = E(\lambda)\frac{\lambda}{hc} \equiv \left[\frac{1}{\text{m}^2\text{s}}\frac{1}{\text{m}}\right]. \tag{9.47}$$

Usually, it is more convenient to consider the spectral photon flux instead of the spectral irradiance, as shown in Fig. 9.19 when considering the photon-to-charge conversion. Note how the shape of the spectrum is more pronounced than irradiance for higher wavelengths, according to (9.47). Therefore, the number of photons coming from a source having a spectral irradiance $E(\lambda)$ and hitting a unit of surface in a unit of time is

$$\int_0^\infty \Phi_P(\lambda)d\lambda = \int_0^\infty E(\lambda)\frac{\lambda}{hc}d\lambda. \tag{9.48}$$

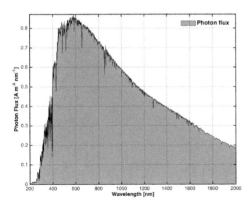
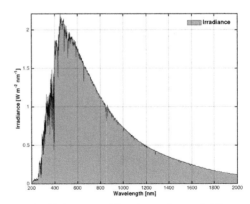

Figure 9.19 Photon flux of the sun (top) compared with its spectral irradiance (bottom). The difference is due to the fact that the energy per photon is different according to the wavelength.

9.3   Photon Interaction with Semiconductors

According to the physics of semiconductor devices, when a photon is able to penetrate a semiconductor, it has some probability to convert its energy into the generation of a charge carrier couple (i.e., one electron and one hole) or into the increment the kinetic energy of the lattice, as shown in Fig. 9.20A.

The transduction of a photon into charge carriers is called *photogeneration*, whilst the conversion into kinetic energy could be modeled as the generation of a virtual "particle" called a *phonon*. Here we will not take into consideration phonon generation.

The photogeneration process produces many charges that could be collected by an electronic device to quantify the intensity (energy) of the incoming light. Most electronic image sensors are based on this principle of conversion. The amount of photogenerated charges in a specific area and in a unit of time could be quantified by a current density called photogenerated current density

$$J_P(\lambda) = q \cdot \text{flux of charges} = q \cdot \frac{\text{no. photogenerated charges}}{dS \cdot dt \cdot d\lambda} \equiv \left[\frac{A}{m^2}\frac{1}{m}\right]. \quad (9.49)$$

The effectiveness of the material into photogeneration with respect to other conversions is called *photon (or quantum) efficiency* and could be calculated as the ratio between the flux of charges generated versus the flux of photons

$$\eta(\lambda) = \frac{\text{flux of charges}}{\text{flux of photons}} = \frac{1}{q}\frac{hc}{\lambda}\frac{J_P(\lambda)}{E_P(\lambda)} = \frac{hc}{q\lambda}R(\lambda) \equiv [\cdot], \quad (9.50)$$

where $R(\lambda)$ is called *responsivity*

$$R(\lambda) = \eta(\lambda) \cdot \frac{q\lambda}{hc} = \frac{J_P(\lambda)}{E_P(\lambda)} \equiv \left[\frac{A}{W}\right]. \quad (9.51)$$

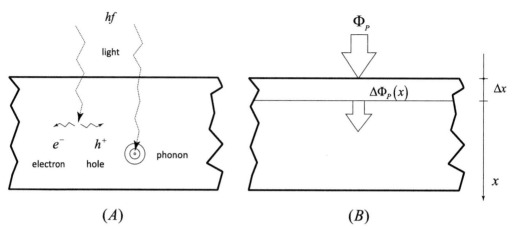

Figure 9.20 Photon interaction with matter. (A) Transduction of a photon into a charge carrier pair or into a phonon. (B) Variation of the photon flux versus depth.

Quantum efficiency is related to several factors under which the photon could be ineffective in the charge transduction process, such as photon energy match with the semiconductor bandgap, phonon conversion, or electron-hole recombination.

It is important to note that in the case we have the conversion of each photon into one electron, that is, $\eta = 1$,

$$R(\lambda) = \frac{q\lambda}{hc}.$$

(9.52)

Equation (9.52) could be read in the following way: we have the maximum efficiency if every quantum of energy given by a photon in a unit of time and in a unit of area (W/m^2) yields a charge per unit of time per unit of area (A/m^2). Therefore, this is referred to as an *asymptotic limit* of the responsivity that could never be exceeded due to the quantum nature of light.

We want to focus on the efficiency (and thus on the responsivity) due to geometrical considerations only. In other words, we assume that for a generic technology we can collect photocharges in a defined region and all the others generated outside are lost (for example suddenly recombined), thus decreasing the efficiency. To understand this phenomenon, we have to take into account the relationship of the photogeneration with the depth of penetration of photon flux into the material.

If we refer to the photon flux behavior on the depth, x we expect it to be decremented by a factor proportional to either the flux amount and the length of the path. Therefore, if we refer to a small discrete element Δx

$$-\Delta\Phi_P(x) = a \cdot \Phi_P(x) \cdot \Delta x = \Phi_{P0} - \Phi_P(x),$$

(9.53)

where Φ_{P0} is the photon flux impinging the surface and a is the constant of proportionality.

Therefore, using the differential calculus with the boundary condition $\Phi_P(0) = \Phi_{P0}$

$$-\frac{d\Phi_P(x)}{dx} = \alpha \cdot \Phi_P(x) \ \rightarrow \ \Phi_P(x) = \Phi_{P0}e^{-\alpha x}$$

$$\alpha = \text{ absorption coefficient} \equiv \left[\frac{1}{m}\right].$$

(9.54)

where α is called *absorption coefficient*, which is in general dependent on the wave-length $\alpha = \alpha(\lambda)$. In conclusion, the photon flux is decaying along its path into the material following an exponentially decreasing function.

Note also that the quantity $\alpha \cdot \Phi_P(x)$

$$-\Delta\Phi_P(x) = \frac{\text{no. photon lost}}{\Delta S \cdot \Delta t} = \frac{\text{no. photon lost}}{\Delta V \cdot \Delta t} \cdot \Delta x$$

$$\rightarrow \alpha \cdot \Phi_P(x) = \frac{\text{no. photon lost}}{\Delta V \cdot \Delta t} = G(x)$$

(9.55)

is the amount of photons converted into charges in a unit of volume per unit of time. This is called the *photogeneration function*.

The quantity $1/\alpha$ also expresses the mean penetration depth of the flux into the matter and, for this reason, is also called *penetration depth*. Owing to a complex interaction of photon with the matter, for most semiconductors, the higher wavelength (i.e., the lower the photon energy) the higher the penetration depth as shown in Fig. 9.21.

As far as the silicon is concerned, the plot is shown in Fig. 9.22, where first the absorption coefficient is shown and then the absorption depth. Note how the penetration depth ranges from about 0.1 μm for ultraviolet radiation to about 10 μm for infrared in the optical window. Therefore, infrared radiation is penetrating deeper in the bulk with respect to other radiations. In any case these ranges are fully compatible with modern microelectronic technology.

So far, we have referred to a uniform silicon crystal bulk. Unfortunately, the situation in microelectronic devices is quite different. The objective of an electronic image sensor is to grab as many photogenerated charges as possible, however this could be possible only in specific regions of the silicon technology structure. We will

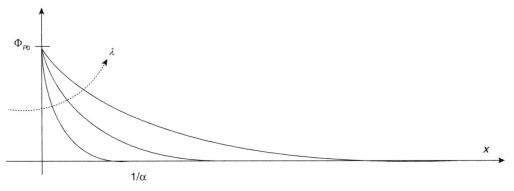

Figure 9.21 Behavior of the photon flux with respect to the wavelength.

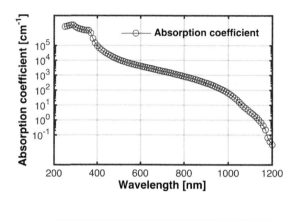

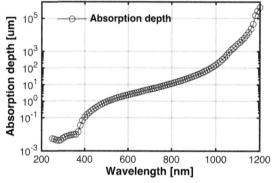

Figure 9.22 Absorption coefficient of silicon and related absorption depth.

refer to the area where photocharges are collected to as a *photosite*. Photosites could be single or arranged in arrays where they are also called *pixels*.

Therefore, we have to assume that we will *not* be able to grab photocharges at any depth, but only in fixed boundaries, as shown in Fig. 9.23, depending on the technology. This will happen for either charge-coupled devices (CCDs), complementary metal–oxide–semiconductor (CMOS) or other technologies.

Now the density of photocharges G_l collected per unit of the area within the boundaries x_1 and x_2 is given by the integral of the photogeneration function in that interval

$$
\begin{aligned}
G_l &= \frac{\text{no. carriers}}{\Delta V \cdot \Delta t} \cdot \Delta x = \int_{x_1}^{x_2} G(x) dx = \int_{x_1}^{x_2} \alpha \cdot \Phi \cdot e^{-\alpha x} dx \\
&= \alpha \cdot \Phi \cdot \int_{x_1}^{x_2} e^{-\alpha x} dx = \Phi \cdot e^{-\alpha x_1} [1 - e^{-\alpha l}] = \frac{\text{no. carriers}}{\Delta S \cdot \Delta t} \equiv \left[\frac{1}{\text{m}^2 \text{s}} \right],
\end{aligned}
\tag{9.56}
$$

where $l = x_2 - x_1$ is the thickness of the collecting zone.

Therefore, the flux of the photogenerated charges will be

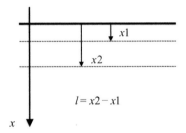

Figure 9.23 Area in which the photogeneration is collected in an electronic device technology.

$$J_P(\lambda) = q \cdot G_l = \frac{\text{no. charges}}{\Delta S \cdot \Delta t} = q \cdot \Phi_P \cdot e^{-\alpha x_1} [1 - e^{-\alpha l}]$$

$$\approx q \cdot \Phi_P \cdot [1 - e^{-\alpha l}] \equiv \left[\frac{A}{m^2}\right],$$

(9.57)

which is the *photogeneration current*.

Therefore, we can calculate again the responsivity $R(\lambda)$ and efficiency $\eta(\lambda)$ of a given material in a collecting zone as

$$R(\lambda) = \frac{J(\lambda)}{E(\lambda)} = \frac{\text{photocurrent}}{\text{illumination}} = \frac{q \cdot \Phi_P \cdot e^{-\alpha x_1} [1 - e^{-\alpha l}]}{\dfrac{hc}{\lambda} \cdot \Phi_P}$$

$$= \frac{q\lambda}{hc} \cdot e^{-\alpha x_1} [1 - e^{-\alpha l}] \equiv \left[\frac{A}{W}\right]; \ \alpha = \alpha(\lambda)$$

(9.58)

and

$$\eta(\lambda) = e^{-\alpha x_1} [1 - e^{-\alpha l}] \equiv [\cdot].$$

(9.59)

Note that in this case, the calculated responsivity and efficiency depends only on the efficiency of the technology to collect charges in the selected regions. Again, $\eta = 1$ is an asymptotic value called the *maximum response line* of phototransduction that we can achieve in this case when $x_1 = 0$ and $x_2 = \infty$.

In other terms, the line determines the maximum possible sensitivity assuming that any photon is converted into a charge at any depth and for a given efficiency.

In Fig. 9.24 is shown the response of a hypothetical photosite response based on silicon determined using (9.58) with data as illustrated in Fig. 9.22. Note how the sensitivity is, in general, more shifted on the infrared values than scotopic or photopic human sensitivity: this is the reason why silicon image sensors are usually susceptible to near-infrared radiations where the human eye is not.

As shown in Fig. 9.24A, the silicon responsivity increases at lower wavelengths for lower values, even if it is below the maximum response line (unitary efficiency) given by (9.59). However, it is in practice very difficult to grab photocharges in proximity of the surface due to the increasing number of superficial traps.

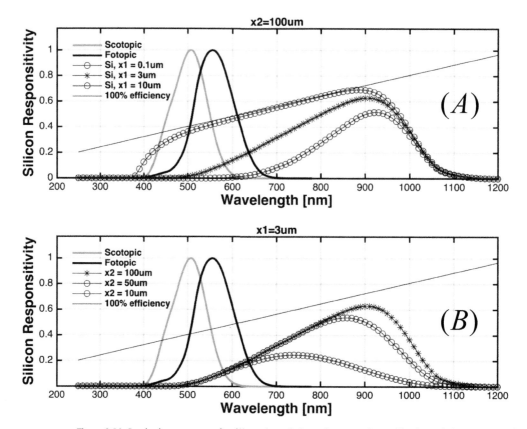

Figure 9.24 Intrinsic response of a silicon-based photosite assuming collection of photogenerated charges between two quotas. (A) Variation of the top quota. (B) Variation of bottom quota. These curves are ideal responses and do not take into account situations such as nonuniform drug concentration and complex electric field patterns as in real devices.

On the other hand, as illustrated in Fig. 9.24B, a reduced bottom quota shifts the peak on the right even if it reduces the overall responsivity. The behavior of the responsivity could also be graphically seen in terms of efficiency, following (9.59) as shown in Fig. 9.25: the efficiency at each wavelength could be found by using the intersect of the corresponding maximum response line with the responsivity plot.

In the case we know the spectral emissivity of the source (i.e., blackbody or other), we can determine the *total responsivity with respect to a source* for a photosite in the following way:

$$R_T = \frac{\displaystyle\int_0^\infty R(\lambda) \cdot E(\lambda)\, d\lambda}{\displaystyle\int_0^\infty E(\lambda)\, d\lambda} \equiv \left[\frac{A}{W}\right]. \tag{9.60}$$

Typical values are between 0.3 and 0.8 A/W.

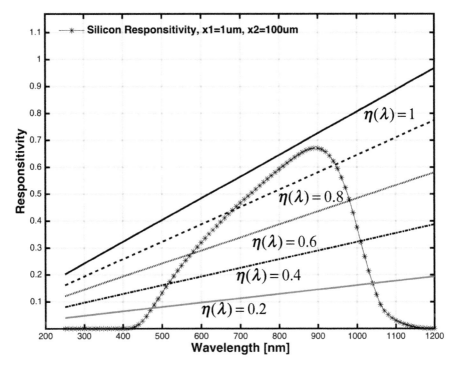

Figure 9.25 Silicon responsivity versus efficiency.

9.4 Image Sensor Devices and Systems

This part aims to apply the concepts of previous sections to common basic implementations of image sensors. Therefore, the structures illustrated are intentionally highly simplified and do not cover all the large classes of possible implementations. Basic knowledge of electronic devices is required.

9.4.1 Photosite Examples: CCDs and Photodiodes

One of the earliest solid-state optical image sensor implementation is the charge coupled devices (CCD), whose physical cross-section and basic operating principle is illustrated in Fig. 9.26. A semiconductor's surface is covered by a periodic array of electrodes partly transparent to light. By using a voltage pattern of electrodes as illustrated in the top of the figure, an electrical potential well (or "bucket") is created to trap the photogenerated carriers. Therefore, the previously referred region where the charges are collected is located within the volume depth of the potential well. After a fixed amount of time, referred to as *integration time*, where the photogenerated charges are collected, the information is moved to a readout section (not shown in the figure) using a periodic pattern of the voltage's phases applied to the electrodes.

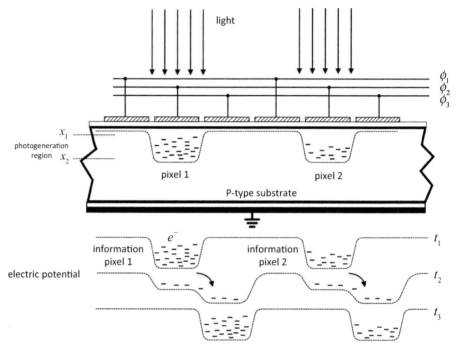

Figure 9.26 Simplified structure of a charge-coupled device.

The procedure consists of enlarging the potential well to the adjacent right electrode and then restricting the well under this new position. If the technique is executed correctly, the photogenerated charges (whose number represents the information) are not lost and shifted alongside the array raw, together with all those of the other pixels, in analog shift register fashion.

In the last decades, CCD devices encountered difficulties of integration with the CMOS technology's scaling trend, especially for the moderately high electrical potentials required to generate the wells to ensure high photon efficiency. For this reason, the consumer electronics market moved to a different technology based on photodiodes implemented in CMOS technology. After decades of development and improvements, CMOS sensor technology offers today the capability to integrate on the same silicon chip both sensor devices with analog and digital signal processing with excellent overall quality.

Photodiodes are nothing other than semiconductor junctions exposed to the effect of light. A possible cross-sectional implementation of the device is shown in Fig. 9.27. A p–n junction can be fabricated by diffusing a p-type impurity (anode) into an n-well created as a p-type substrate. The area exposed to the light is called active area and usually it is covered with a coating to reduce the reflection of the light or to implement color filters. Contacts are created in the top surface for electrical connections.

The photosensitive junction concept is illustrated in Fig. 9.28 using the *abrupt junction* approximation.[6] This model assumes that the depletion zone across the

---

[6] For more details about the model, please refer to electronic device physics textbooks.

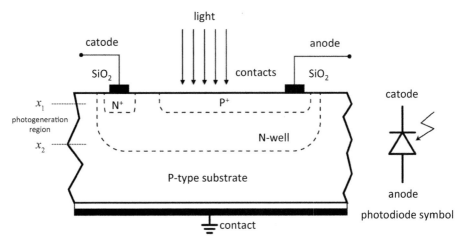

Figure 9.27 Photodiode physical structure.

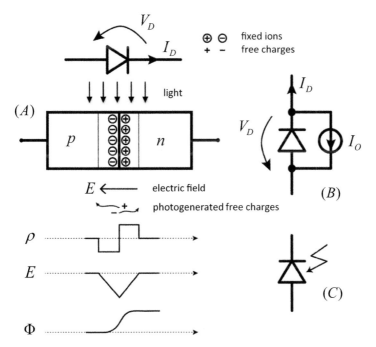

Figure 9.28 (A) Physical structure of the photodiode using abrupt junction model. (B) Electrical symbol of photodiode with evidenced photocurrent. (C) Compact electrical symbol of the photodiode.

junction has a constant distribution of space charge, ρ as shown in Fig. 9.28A, forming the *depleted region*. Away from the depleted region there are neutral regions where net charge concentration and electric field are almost zero. Using Poisson's equation, we find that the electric field profile is linear in the depleted region and that electrical potential has a parabolic function characteristic, thus creating a potential barrier for the free carriers. When the junction is exposed to light, the energy of a photon might create an electron-hole charge couple according to the physics of semiconductor devices.

The most important zone (but not the only one) where the photogenerated charges contribute to the photocurrent is the *depletion region*. In that area, the electron and hole carriers generated by the photon are separated by the high electric field in the opposite directions. Therefore a photogenerated current I_O is generated in the opposite direction with respect to the direct current of the diode, as shown Fig. 9.28B. The electrical symbol of the photodiode is illustrated in Fig. 9.28C.

Let us take into account the physical cross-section of Fig. 9.27 under the assumptions mentioned above. We can set up an approximate physical model considering the photogeneration quotas x_1 and x_2 as those that are delimiting the depletion region, also illustrated in Fig. 9.27. Therefore, following the reference of voltages and currents of Fig. 9.28B, we can easily derive the static *I–V* characteristics of the photodiodes from the known one of the semiconductor junction diode

$$I_D = I_S \left(e^{\frac{V}{V_T}} - 1 \right) - I_O, \tag{9.61}$$

where I_O is the *photocurrent* or *photogenerated current*.

The photocurrent effect in the *I–V* characteristic is shown in Fig. 9.29 for three different photocurrent values. As shown, the *I–V* characteristic behavior could be

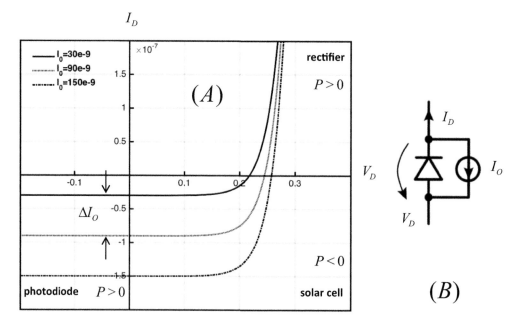

Figure 9.29 Characteristic of the photodiode for different photocurrents. $I_S = 5\text{E-}12$, $T = 300$ K.

characterized according to the quadrants: in the first, the photodiode is acting as a rectifier, therefore consuming external power $P = V \cdot I > 0$. In the fourth quadrant, the power is negative; thus, it could generate power from the light, as in solar cells. Finally, in the third quadrant it behaves in the typical photodiode mode, where it shows the most effective variation of the curve with respect to the light signal. Therefore, the photodiode is mostly used in the reverse-mode for sensing purposes in the microelectronic context.

9.4.2 Continuous-Time Readout Mode

A first way to readout the photodiode is by directly sensing its current in the *photovoltaic mode,* as shown in Fig. 9.30A, where a current-to-voltage converter senses directly the photocurrent.

Therefore, the output is directly proportional to the photocurrent

$$V_O = V_R + I_O(t) \cdot R. \tag{9.62}$$

The capacitor C_F sets the optimal configuration by setting the feedback gain of the system. The theory of feedback control systems states that the total transfer function poles move according to the amount of the feedback (root locus). Therefore, if the feedback is quite strong the poles could become complex conjugates determining an overall strongly underdamped response (Chapter 6). On the one hand this could be beneficial because large overshoot and ringing. Moreover, the strongly underdamped behavior could harm the signal-to-noise ratio (SNR), since noise power accumulates in the resonance peak (Chapter 6). We can look better at the situation using a graphical approach.

The closed-loop gain of a feedback system is given by

$$G = \frac{A}{1 + A\beta} = \frac{1}{\beta} \frac{1}{1 + \dfrac{1}{A\beta}}, \tag{9.63}$$

where A is the *open-loop gain* of the OPAMP, β is the *feedback gain* of the amplifier, and $A\beta$ is called the *open-loop gain* of the system.

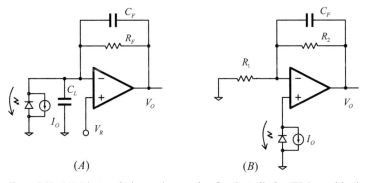

Figure 9.30 (A) Photovoltaic sensing mode of a photodiode. (B) Logarithmic sensing mode.

For high loop gain, the gain G is given by the reciprocal of the feedback gain $1/\beta$, while for moderate loop gain we should use (9.63). The feedback gain (i.e., the ratio of the output that is fed back to the input) is given by the impedance ratio (Chapter 7)

$$\beta = \frac{Z_1}{Z_1 + Z_2},$$ (9.64)

where Z_1 is the parallel of the stray capacitance C_L(comprising the diode capacitance) and the diode equivalent resistance R_S(not shown in the figure) while Z_2 is the feedback impedance given by the parallel of R_F with C_F

$$\frac{1}{\beta} = \frac{Z_1}{Z_1 + Z_2} = \frac{\dfrac{1}{R_F}(1 + sR_FC_F)}{\dfrac{R_F + R_S}{R_FR_S} + s(C_F + C_L)}.$$ (9.65)

Therefore, the transfer function $1/\beta$ has a zero at f_1 (which is a pole for β) given by the feedback impedance. Furthermore, it has a pole at f_2(which is a zero for β) given by the parallel of all resistances with the parallel of all capacitances

$$f_1 = \frac{\dfrac{R_FR_S}{R_F + R_S}}{2\pi(C_F + C_L)}; f_2 = \frac{1}{2\pi R_FC_F}.$$ (9.66)

The overall behavior is shown in Fig. 9.31, where are plotted the A and $1/\beta$ functions. The geometrical distance between the two log-log plots is the loop gain $A\beta$. Therefore, the crossing point of the two functions corresponds to the frequency f_U, where the loop gain $A\beta = 1$ which is important for the stability of the system. The closed-loop gain G follows the $1/\beta$ behavior until $A\beta = 1$; then it rolls off following A.

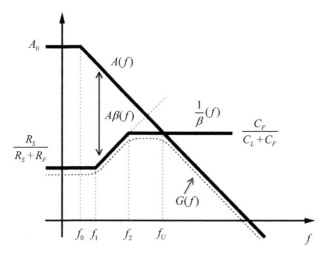

Figure 9.31 Noise peaking effect. The dotted plot is showing the noise gain of the system (i.e., the transfer function of the input OPAMP noise to the output).

The problem of this system is that the phase of the loop gain at f_U has to be lower than 180° to ensure stability (phase margin); therefore, it is important that the the zero for β (at f_2) is lower than f_U to avoid a strongly underdamped behavior that could also have negative effects on the noise of the system, since it tends to be integrated in the resonance peak of the spectrum. This effect is referred to as *noise peaking*. This effect is graphically illustrated in Fig. 9.31.

The problem mentioned above could be avoided by using the following design approach:

- Selecting a C_F so that f_2 is set well before f_U
- Increasing the GBW of the operational amplifier (OPAMP) for the same target.

Another way to read out a photodiode in continuous mode is shown in Fig. 9.30B, where it is connected directly to the noninverting input of the amplifier. Using this approach, we get

$$
\begin{aligned}
I_D &= 0 \\
I_O &= I_S\left(e^{\frac{V}{V_T}} - 1\right) \rightarrow V_D = V_T\ln\left(\frac{I_O}{I_S} + 1\right) \\
V_O &= \left(1 + \frac{R_2}{R_1}\right) \cdot V_T\ln\left(\frac{I_O}{I_S} + 1\right).
\end{aligned}
\tag{9.67}
$$

Therefore, the output of the interface is proportional to the logarithmic function of the intensity of the light. For the aforementioned reason, this is called a *logarithmic photodiode interface*.

9.4.3   The Storage Mode Concept

One of the most common techniques to read the information from a photodiode is to use a discrete-time readout sensing technique called *storage mode*. The basic point of the storage mode is the integration of the photocurrent in the depletion charge across the junction when the photodiode is kept electrically floating. More specifically, the photocurrent acts to reduce the charge of the depletion region due to its backward direction with respect to the junction current. The main steps of the *storage mode* are

1. The photodiode is reset to a fixed reference voltage in reverse mode.
2. The photodiode is kept electrically floating for an amount of time, called *integration time*, during which the photocurrent is integrated across the junction.
3. The integrated charge is read out by either voltage- or charge-sensitive techniques.

It should be pointed out that because the photocurrent reduces the depletion charge, the voltage across the diode decreases versus time as in a capacitor, as shown in the model of Fig. 9.32A. Therefore, for a given photocurrent $I_0(t)$ and a reset voltage V_R, the voltage V_D across the diode at the end of integration time T_i is

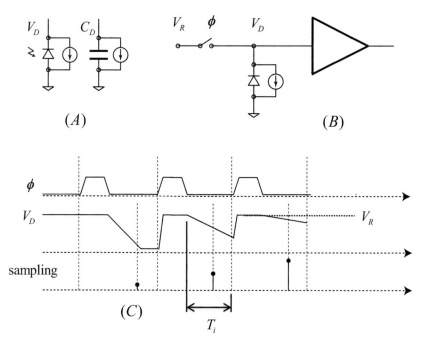

Figure 9.32 Storage mode in voltage readout. (A) Photodiode model for storage mode. (B) Schematic example. (C) Waveforms of the voltage readout.

$$V_D = V_R - \frac{1}{C_D}\int_0^{T_i} I_o(t)dt = V_R - \frac{Q_o}{C_D}, \qquad (9.68)$$

where C_D is the capacitance of the photodiode. The signal could also be represented by the charge generated by the light during integration time Q_O, which is also called *photocharge*.

A schematic example of the storage mode using *voltage readout* is shown in Fig. 9.32, while in Fig. 9.32A, the equivalent charge model of the photodiode is shown in Fig. 9.32B is displayed the technique of the voltage readout. The photodiode is alternatively reset using a discrete-time approach while a high-input impedance voltage buffer monitors its voltage.

In summary, the steps of the voltage-readout storage mode are

- During the high value of the phase, ϕ the photodiode is biased in reverse mode by means of a reference voltage V_R.
- In the lower value of the phase, ϕ the photodiode is kept floating so as the photocurrent discharges the junction capacitance. Assuming linear capacitance and constant photocurrent (i.e., constant illumination), the photodiode voltage decreases almost linearly, where the slope of the discharge depends on the intensity of the light.

- An A/D converter (not shown in the figure) samples the photodiode voltage after a fixed period of time, called *integration time*. The sampled value linearly encodes the photocharge of the integration time according to (9.68).
- The photodiode is reset again following a cyclic period governed by the discrete-time approach.

The scheme mentioned earlier is shown in a waveform diagram in Fig. 9.32C. Note that the illumination is stable during the integration time but lowering from one cycle to the other. Therefore, the response of the system is inversely proportional to the illumination.

How long could the integration time be? The voltage ramp could not decrease indefinitely. In fact, the ramp could decrease as long as the photodiode is kept in reverse mode. Therefore, the voltage could decrease until the forward current equates the photocurrent. Once the equilibrium is achieved, the voltage is given by the intersection of the horizontal axis, corresponding to the $I_D = 0$ condition of the *I–V* plot shown in Fig. 9.29. This limit also defines the *saturation* limit of the photodiode, useful for calculating the dynamic range.

So far, we have assumed a linear junction capacitance for simplicity, but this is not true. However, even if this does not affect the perceived performance of image sensors, due to the limited sensitivity with respect to small nonlinearities in intensity functions of our visual system, the problem could be avoided using a *charge readout* of the storage mode, as shown in Fig. 9.33.

A photodiode is connected to a charge amplifier that senses the charge variation induced by the photocurrent during the integration time.

The charge readout of the storage mode could be summarized in these steps:

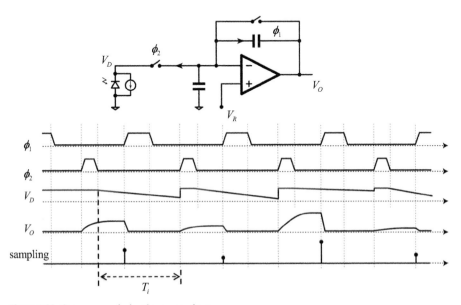

Figure 9.33 Storage mode in charge readout.

- During the high value of the phase, ϕ_2 the photodiode is biased to the value V_R by means of the virtual short circuit of the OPAMP, independent of the status of ϕ_1.
- ϕ_2 is opened to start the integration time, leaving the photodiode floating. Then, ϕ_1 is opened, allowing the charge to be readout.
- The closure of ϕ_2 sets the end of the integration time and the readout of the charge. The photodiode is concurrently reset during the readout.
- An A/D converter (not shown in the figure) samples the output of the OPAMP. The sampled value linearly encodes the photocharge of the integration time.
- The OPAMP is reset again, and the cycle restarts.

Differently from the voltage readout, the output is now directly proportional to the photocharge signal (containing the information), which is now integrated on C_F

$$V_O = V_R + \frac{1}{C_F} \int_0^{T_i} I_o(t)dt = V_R + \frac{Q_o}{C_F}. \tag{9.69}$$

Note that the sign is now changed because the photodiode has to restore the charge removed by the photocurrent; therefore, the current should be drained from the charge amplifier during the reset/readout phase.

9.5   Noise in Photodiodes

The noise in photodiodes could be modeled for both *continuous-time mode* and *discrete-time mode* readouts. We will recall concepts already explained in Chapter 6.

For the continuous mode, we will model a noisy photodiode as a noiseless photodiode with noise current sources placed in parallel to contribute to the current.

In a bandwidth Δf of the system, the main noise sources are

- Shot noise due to the photocurrent $\overline{i_O^2} = 2qI_O\Delta f$, where I_O is the photocurrent
- Shot noise due to the dark current $\overline{i_D^2} = 2qI_D\Delta f$, where I_D is the dark current

- Thermal noise due to the resistance of the photodiode $\overline{i_R^2} = \sqrt{\dfrac{4kT\Delta f}{R_P}}$ where R_P is the series resistance of the photodiode

The *dark current* is essentially the leakage current of the junction when no light is acting on the photodiode, that is, at dark conditions. Note that R_P is *not* the equivalent model resistance of the photodiode (as it is the result of a model it does not create noise), but it is the real series resistance of the photodiode contacts and substrates.

From the expressions mentioned earlier, it is possible to characterize the *intrinsic* photodiode SNR as

$$\text{SNR}_{P(dB)} = 20\log\frac{I_O}{\sqrt{\overline{i_O^2} + \overline{i_D^2} + \overline{i_R^2}}}, \tag{9.70}$$

where again I_O is the signal photocurrent. Note that for high illumination levels, the shot noise of the photocurrent is dominant with respect to the other noise sources, and thus the signal-to-noise ratio

$$\text{SNR}_{\text{P(dB)}} \approx 20\log\frac{I_O}{\sqrt{2qI_O\Delta f}} = 20\log\frac{\sqrt{I_O}}{\sqrt{2q\Delta f}}, \tag{9.71}$$

is proportional to the square root of the photocurrent.

We used the term *intrinsic* to emphasize the fact that this is the SNR due only to the photodiode. In the case of sensing by means of an interface, we may define an *input-referred SNR* that also considers the equivalent input-referred noise (Chapter 7) of the electronic interface $\overline{i_E^2}$

$$\text{SNR}_{\text{I(dB)}} = 20\log\frac{I_O}{\sqrt{\overline{i_O^2} + \overline{i_D^2} + \overline{i_R^2} + \overline{i_E^2}}}. \tag{9.72}$$

For the *discrete-time mode* (such as in the *storage mode*), it is convenient to refer to the characteristic of the process following Poisson distribution as described in Chapter 6, where the signal is considered as the mean collected photocharges during the integration time T_i

$$\overline{n} = \lambda T_i = \frac{1}{K}\sum_{i=1}^{K} n[i] = N \equiv [\cdot], \tag{9.73}$$

where λ is the mean rate of collected charges, T_i is the integration time, and $n[i]$ is the number of the charges collected in the photosite during the ith experiment. Following Poisson's approach, the variance of the process is equal to the mean

$$\sigma^2 = \lambda T_i = \overline{(n-N)^2} = \frac{1}{K}\sum_{i=1}^{K}(n[i] - N)^2 = N \equiv [\cdot^2]$$
$$\sigma = \sqrt{N} \equiv [\cdot]. \tag{9.74}$$

Therefore, similarly to the continuous-time for expression (9.72), we can define the photodetector SNR in terms of shot noise. This SNR definition also considers the noise of the readout referred back to the input in terms of charges

$$\text{SNR}_{\text{P(dB)}} = 20\log\frac{N_O}{\sqrt{N_O + N_D + N_E}}, \tag{9.75}$$

where N_O, N_D, and N_E are the mean collected charges for the photocurrent, dark current, and equivalent charges of the noise of the electronic interface, respectively. N_E also includes several kTC noise effects due to the photodiode reset and the wiring lines to the readout system, as shown in Section 9.6. Note that N_E is not a measurable amount of charges but it is an input-referred "equivalent" amount of charges that would

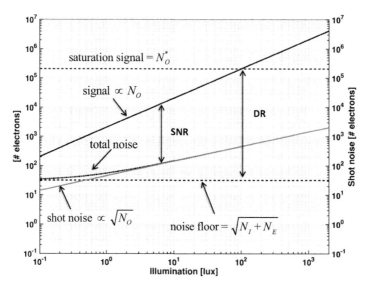

Figure 9.34 Graphical interpretation of the SNR and of the DR of photodetectors operating in discrete-time mode.

induce the same noise at the output of the interface. We neglected the noise of the resistance because usually, storage mode is used for very small photodiodes (such as in the arrays of area image sensors) so that its contribution is small compared to the shot noise.

The SNR calculated with respect to shot noise could be represented in the log-log plot as shown in Fig. 9.34. The SNR is the geometrical distance between the signal plot, which is rising with a 1:1 slope, and the noise plot, which is rising as a 1:2 slope. On the other hand, the dynamic range (*DR*) is the distance between the maximum allowable signal given by the saturation and the plafond of the input-referred noise and dark current noise, which is often called *noise floor*.

In this case, *DR* is *not* the maximum SNR due to the dependence of the noise on the signal intensity.

In the plot, we used the following relationship:

$$N_O = E \cdot \frac{A \cdot K \cdot R_T \cdot T_i}{q};$$
$$N_D = \frac{I_D \cdot A \cdot T_i}{q};$$
$$N_T = \sqrt{N_O + N_D + N_E},$$

(9.76)

where E is the illumination in lux, $T_i = 30$ ms is the integration time, $R_T = 0.3$ A/W is the responsivity, $K = 683$ lux/W is the luminous conversion factor, $I_D = 0.6$ nA/cm^2 is the dark current $A = 5 \times 5$ μm^2 is the area of the junction, and $N_E = 1000$ is the input-referred noise of the interface, including several kTC noise sources. The saturation charges N_O^* could be calculated from the junction capacitance C_D according to

$$N_O^* = \frac{C_D \cdot V_R}{q},$$ (9.77)

where $C_D = 10f\,F$ is the junction capacitance, and $V_R = 3.3$ V is the reset voltage.

9.6 CMOS Area Image Sensor Architectures

The CMOS photodiode structure is particularly suited to be arranged in sensor arrays, also referred to as CMOS optical sensor arrays. One of the earliest implementations of a photodiode array is the *passive pixel sensor* (PPS) architecture, as shown in Fig. 9.35. The approach is very similar to the architecture of random access memories (RAM), reading out analog instead of binary values.

In this arrangement, a set of photodiodes is organized into an $M \times N$ array that is selectively addressed by a vertical or horizontal shift register or decoder. The location of a single photodiode and of a switch is the physical implementation of the *pixel* abstraction. The array scan is performed by addressing only one pixel per time by concurrently enabling a *column switch* and a *pixel switch*. The vertical shift register activates a line of pixel switches, and the horizontal one scans all the columns so that the integrated photocharge is read out by the charge amplifier. Every time one pixel is read out, the charge amplifier is reset. Once the horizontal shift register reaches the end of the line, the vertical shift register is incremented, thus selecting another line to be scanned by horizontal addressing that restarts from the beginning. This approach to address the array is referred to as *raster scanning mode* as shown in Fig. 9.36. The time of scanning of the entire array is called *frame time* and its reciprocal is called the *frame rate*.

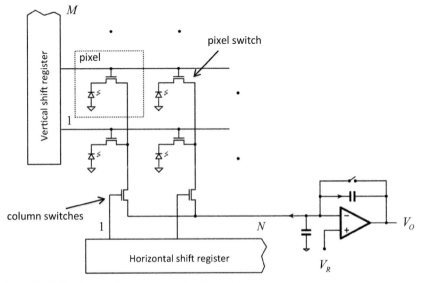

Figure 9.35 Passive pixel array sensor implementation.

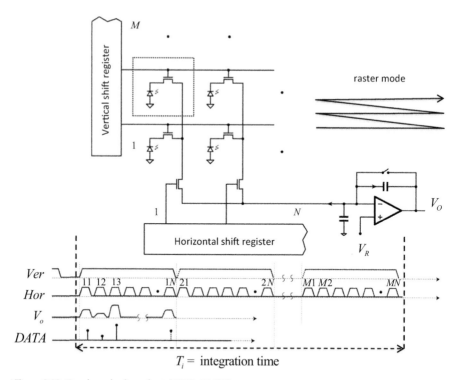

Figure 9.36 Passive pixel readout (PPS) CMOS sensor.

We have to refer to the integration time as the time during which each photodiode is kept floating; thus, it corresponds to the frame time. This is one of the problems of the PPS that could be here summarized:

- The charge of each pixel should travel on long wiring lines before being readout, and it is prone to noise and interferences. The larger the area is, the longer the lines.
- Once the readout is operated, the information is lost.
- The photodiodes have the same integration times, but they are not synchronized with each other. Therefore, we must assume that the image variations (i.e., image signal bandwidth) should be much slower than the frame rate.

A solution to the drawbacks of the PPS at expenses of the complexity of the pixel architecture is the *active pixel sensor* (APS) readout. The key concept of the APS is to operate a storage-mode voltage readout of the photodiode directly in the pixel area as. As shown in Fig. 9.37A, each photodiode voltage is buffered and addressed using a raster mode approach as in the PPS scheme. The buffer is easily implemented by using a "source follower" scheme as shown in Fig. 9.37C, where a constant current I_D is drawn into the channel of a MOS transistor, and therefore the input–output relationship could be found using the classical MOS saturation relationships

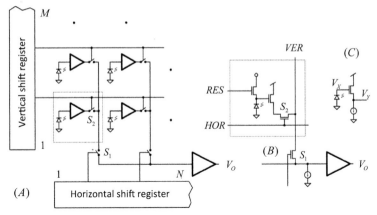

Figure 9.37 Active pixel sensor (APS) CMOS sensors. (A) Array organization. (B) Structure of the pixel. (C) "Source follower" approach for the buffer.

$$I_D = \frac{1}{2}\frac{W}{L}\mu C_{OX}(V_{GS} - V_T)^2$$
$$\rightarrow V_{GS} = V_T + \sqrt{\frac{2I_D}{\frac{W}{L}\mu C_{OX}}} = V_K \rightarrow V_Y = V_X - V_K, \qquad (9.78)$$

where W, L, μ, C_{OX}, V_{GS}, V_T are the width, length, oxide capacitance, gate-source voltage, and the threshold voltage of the MOS transistor, respectively. Therefore, since the current is constant, the source follower configuration acts as a high-impedance voltage level shifter.

The low-impedance output of the buffer could thus be easily read out along the addressing lines. Note how the current generator could be unique and placed outside of the array, as shown in Fig. 9.37B. The arrangement of the active pixel sensor (APS) requires two more MOS transistors into the pixel,[7] but the drawback of a reduced area of the active junction is compensated by a more efficient readout approach. In summary, the main advantages of the APS scheme are:

- Reduced readout noise
- Nondestructive readout mode, thus allowing multiple readouts
- Possibility to parallelize the readout on lines or columns

More specifically, the latter statement could be shown in the arrangement of Fig. 9.38, where a readout A/D conversion is applied to each column of the array. The difference of the capacitance and resistance of the vertical lines induces a typical "pattern noise," which is a systematic area noise. The reset value of the line is stored into a capacitance while readout one into another identical capacitance to reduce pattern noise by differentiation. A differential A/D converter amplifies the difference

---

[7] There are many APS architectures. Here we have shown only the simplest one to illustrate the concept.

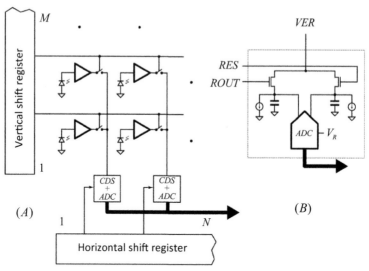

Figure 9.38 (A) Parallel readout of the APS scheme. (B) Correlated double sampling applied to the column readout.

operating a *correlated double sampling* (CDS), as already explained in Chapter 8, that reduces the *kTC* noise and the offsets of the column lines.

Nowadays, APS CMOS image sensors represent the most used architecture in solid-state image sensors implementation.

9.7   Appendix: Summary of Photometry/Radiometry Definitions

Physical definition	Name	Units	Comment
$\Phi = \int\limits_{0}^{\infty} \Phi(\lambda)d\lambda$	Radiant flux (power)	[W]	
$\Phi_O = \int\limits_{0}^{\infty} \Phi(\lambda)V(\lambda)d\lambda$	Luminous flux (power)	[lm]	
$\Phi(\lambda) \triangleq \dfrac{\text{EM power}}{d\lambda} = \dfrac{d\Phi}{d\lambda}$	Spectral radiant flux	$\left[\dfrac{W}{m}\right]$	
$\Phi_O(\lambda) \triangleq \dfrac{\text{optical power}}{d\lambda} = \dfrac{d\Phi_O}{d\lambda}$	Spectral luminous flux	$\left[\dfrac{lm}{m}\right]$	

(cont.)

Physical definition	Name	Units	Comment
$E \triangleq \dfrac{\text{optical power}}{\text{area}} = \dfrac{d\Phi_O}{dS}$	Illuminance	$\left[\dfrac{\text{lm}}{\text{m}^2} \triangleq \text{lux}\right]$	Received
	Luminous exitance		Emitted
$E \triangleq \dfrac{\text{EM power}}{\text{area}} = \dfrac{d\Phi}{dS}$	Irradiance	$\left[\dfrac{\text{W}}{\text{m}^2}\right]$	Received
	Radiosity		Emitted
$I \triangleq \dfrac{\text{optical power}}{\text{solid angle}} = \dfrac{d\Phi_O}{d\Omega}$	Luminous intensity	$\left[\dfrac{\text{lm}}{\text{sr}} \triangleq \text{cd}\right]$	
$I \triangleq \dfrac{\text{EM power}}{\text{solid angle}} = \dfrac{d\Phi}{d\Omega}$	Radiant intensity	$\left[\dfrac{\text{W}}{\text{sr}}\right]$	
$L = \dfrac{d^2\Phi_O}{dS \cdot d\Omega}$	Luminance	$\left[\dfrac{\text{lm}}{\text{m}^2 \cdot \text{sr}}\right] \equiv \left[\dfrac{\text{cd}}{\text{m}^2}\right] \equiv [\text{nit}]$	
$L = \dfrac{d^2\Phi}{dS \cdot d\Omega}$	Radiance	$\left[\dfrac{\text{W}}{\text{m}^2 \cdot \text{sr}}\right]$	

Further Reading

Barbe, D. F., Imaging devices using the charge-coupled concept, *Proc. IEEE*, vol. 63, no. 1, pp. 38–67, 1975.

Fossum, E. R., Active pixel sensors – are CCD dinosaurs?, *Proc. SPIE*, vol. 1900, pp. 2–14, 1993.

Horn, B. K. P., *Robot Vision*. Cambridge: Cambridge University Press, 1986.

Mendis, S. K., Kemeny, S. E., and Fossum, E. R., A 128×128 CMOS active pixel image sensor for highly integrated imaging systems." In *Proceedings of the IEEE International Electron Devices Meeting*, pp. 583–586, 1993.

Smith, L., and Sheinghold, D., "AN-358 analog devices application note," 1969.

Sze, S. M., and Ng, K. K., *Physics of Semiconductor Devices*. Hoboken, NJ: Wiley-Interscience, 2007.

Wurfel, P., *Physics of Solar Cells*. Weinheim, Germany: Wiley-VCH, 2005.

Yadid-Pecht, O., Ginosar, R., and Shacham-Diamand, Y., A random access photodiode array for intelligent image capture, *IEEE Trans. Electron Devices*, vol. 38, no. 8, pp. 1772–1780, 1991.

10 Selected Topics on Ionic–Electronic Transduction

Ion–electron transduction is at the base of biosensing. We start addressing some basic principles by emphasizing the common background between the electronic and ionic behavior based on some classical statistical mechanics concepts. Then, we will focus on more specific examples of application in this area. Of course, the examples covered are only a tiny part of the subject. They are intended as a proof of application to consider the transduction process in the framework of the design of biosensor interfaces' electronic design.

10.1 Statistical Thermodynamics: Background Overview

It is useful to review some basic statistical thermodynamics concepts, especially regarding the interaction between electrons and ions. Even if we will achieve some fundamental physical relationships, what follows is far from a rigorous mathematical treatment even if we get the same formal result. Notwithstanding that these ideas turn out to be oversimplifications, they are useful for creating a mental image for understanding how single processes in biosensing are all linked together.

10.1.1 Maxwell–Boltzmann Statistics

Following what we introduced in Chapter 6, the main task is now to understand a gas of particles' behavior when subject to a conservative force. Figure 8.1 shows a gas of identical particles partially confined in an open container under the gravitational force's effect. The system is subject to a uniform temperature T, so that it is said to be in *thermodynamic equilibrium*. On the one hand, the particles' kinetic agitation tends to diffuse the gas out of the container; on the other hand, the gravitational force tends to drag the particles down to the bottom. For this reason, the particles accumulate in a greater concentration at the bottom of the container, creating a pressure gradient, similar to what happens in Earth's atmosphere. The difference of the pressure between the level h and $h + \Delta h$ shown in Fig. 10.1A is given by the increment of the weight of the slice of thickness Δh:

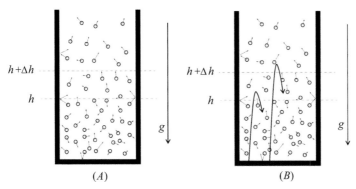

Figure 10.1 Effect of a force on a gas of particle (A) and relationship with kinetic energy (B).

$$P_{h+\Delta h} - P_h = \Delta P = -mgn\Delta h, \tag{10.1}$$

where n is the particle concentration and m is the single particle's mass. However, for an ideal gas, the relationship between pressure and temperature is $P = nkT$; therefore, using the limit of the ratio of increments to get a differential equation with the boundary condition $n = n_0$ at the reference level, we have

$$\Delta nkT = -mgn\Delta h$$
$$\rightarrow \frac{\Delta n}{\Delta h} = -\frac{mg}{kT}n \underset{\Delta h \to 0}{\rightarrow} n = n_0 e^{-\frac{mgh}{kT}}, \tag{10.2}$$

stating that the concentration of particles of the gas is an exponentially decaying function with respect to the height (and thus potential energy), where n_0 is the concentration of molecules at the reference quote.

Looking at Fig. 10.1B, we can see that the particles reaching quote h should have an initial velocity u at ground level that is given by the energy equivalence $1/2mu^2 = mgh$. Furthermore, the fraction of particles reaching the level h with a velocity v greater than zero $n_{v>0}(h)$ is equal to those having the initial velocity $v > u$ (due to a bounce at the zero level) $n_{v>u}(0)$. However, since we are in thermodynamic equilibrium holds

$$n_{v>u}(0) = n_{v>0}(h)$$
$$\rightarrow \frac{n_{v>0}(h)}{n_{v>0}(0)} = e^{-\frac{mgh}{kT}} = \frac{n_{v>u}(0)}{n_{v>0}(0)} = e^{-\frac{mu^2}{2\,kT}} \rightarrow n_{v>u}(0) \propto e^{-\frac{E_C}{kT}}. \tag{10.3}$$

Thus, the fraction of particles passing through a level quote is exponentially related to the barrier's height and to their initial kinetic energy. In Eq. (10.3), we have assumed (1) uniform temperature, thus uniform velocity distribution and (2) no particle and/or border collisions. However, it can be shown from other arguments that the last statement is not necessary to establish (10.3).

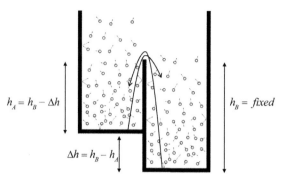

Figure 10.2 Merging of two containers where the particles on the left side have a better probability of jumping on the right than particles in the opposite way.

We can now imagine joining two containers with about the same number of particles at the same temperature in the fashion illustrated in Fig. 10.2. The barrier seen by the particles on the left container is h_A while the barrier seen by the particles on the right one is h_B. For this reason, it is easier for particles on the left to cross the barrier to the right because less kinetic energy is required. Therefore, there is a larger fraction of particles in the population on the left than that on the right with this amount of kinetic energy. We can estimate the *rate* of particles crossing the barrier, considering all those having the sufficient energy:

$$\text{rate} = \frac{\text{no. particles crossing the barrier}}{\Delta t} \equiv \left[\frac{1}{s}\right]$$

$$\text{rate}_{A \to B} \propto e^{-\frac{mgh_A}{kT}} \; ; \quad \text{rate}_{B \to A} \propto e^{-\frac{mgh_B}{kT}} \tag{10.4}$$

Therefore, after merging the containers, we have a transient state with a net rate of particles moving from left to right. This temporary (nonequilibrium) situation gives a net flow given by the differences of the two rates:

$$\text{rate}_{tot} = \text{rate}_{A \to B} - \text{rate}_{B \to A} \propto e^{-\frac{mgh_B}{kT}} \left(e^{\frac{mg\Delta h}{kT}} - 1 \right) = k \left(e^{\frac{mg\Delta h}{kT}} - 1 \right). \tag{10.5}$$

After the transient state, the system reaches a stationary state by reducing the number of particles in the container having the easiest barrier to be crossed, as shown in Fig. 10.3.

In practice, the system achieves equilibrium (rates in opposite directions are equal) by lowering the container concentration where it is easier to cross the barrier. In other words, the difficulty of crossing the barrier, given by its height, is counteracted by the probability of the population crossing it. Any further change of the equilibrium condition, such as the shift of the barrier, determines a "flow" (i.e., a current) given by (10.5).

In general, the relationship (10.2) could be extended in any case where a gas is subject to a *conservative force*, that is, derived from a potential gradient such as the

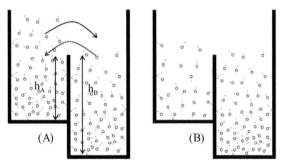

Figure 10.3 From nonequilibrium to equilibrium state between two compartments. In nonequilibrium, A to B's rate is more significant than from B to A (A). In equilibrium, the rates are equal and A's concentration should be lower than in B (B).

electrical one. In this situation, the force on each particle times particle concentration (the *mgh* factor of relationship (10.1)) should be balanced by pressure variation

$$F \cdot n \cdot \Delta x = \Delta P = kT \cdot \Delta n$$

$$\rightarrow n = n_0 e^{-\frac{U}{kT}}, \tag{10.6}$$

where $U = -\int_S F dx \; \forall S \in$ any path joining x_1, x_2 is the potential energy determining the conservative force $F = -\nabla U$.

In the case of the electric field, we have that the force exerted on a charged particle is $F = zqE$, where z is the *ion valency* (i.e., the number and sign of unit charges held by a carrier). In other terms, the force exerted on a carrier by an electric field is z times its unit of charge. Therefore, for the *electric potential energy* U_E, we have that $U_E = zq\Phi$, where Φ is the *electric potential*, so that $E = -d\Phi/dx$. Thus, the previous relationship becomes

$$n = n_0 e^{-\frac{zq\Phi}{kT}}, \tag{10.7}$$

frequently used in semiconductor device physics (with $z = 1$), where n_0 carriers' concentration is at the reference potential. To conclude, the relationships so far achieved could be applied to gases of atoms or molecules and gases of charged particles (such as electrons and holes in semiconductors).

The last relationship in (10.6) is one of the *Boltzmann laws* showing the distribution of free charges under conservative forces determined by potentials. The behavior upon which a gas of particles is subject to Boltzmann's law relates to *Maxwell–Boltzmann statistics*. It is often regarded as the statistics of "distinguishable" classical particles. When electrons and photons are involved, there are no distinguishable particles due to quantum mechanics principles, and new models are required to describe in-depth these kinds of gases, whereas the Maxwell–Boltzmann one is an approximation. However, it is sufficient to describe the processes of interest for our tasks' purpose.

If we look back at the flow relationship (10.5), we can note the similarity with the current in a semiconductor junction:

$$I = I_0 \left(e^{\frac{q\Delta\Phi}{kT}} - 1 \right). \tag{10.8}$$

This is because Boltzmann statistics could model the behavior of a generic "electronic rectifier" (either semiconductor or vacuum diodes) in the presence of a potential barrier. Of course, in semiconductor physics, we have to consider two charge carriers seeing the same barrier on two different sides and contributing at the same current. Why does direct conduction of the junction not reach equilibrium? The fact is that in semiconductor junctions, differently from the example shown (where the amount of gas is limited), the free carriers are continuously generated at the contacts to keep the direct flow constant.

Suppose now we build a device in which another terminal controls the barrier. In that case, we have a typical behavior at the base of any three-terminal "active" electronic device such as thermionic valves, bipolar junction transistors (BJTs), metal oxide semiconductors transistors (MOSTs) working in the subthreshold region, and other devices. Therefore, active electronic devices could be seen as valves controlling potential barriers between electron gases behaving according to Maxwell–Boltzmann statistics.

10.1.2   Some Applications of the Maxwell–Boltzmann Statistics

The kT/q factor frequently used in physics (referred to as *thermal voltage* in semiconductor device physics) could be viewed in terms of moles in (electro)chemistry. More specifically, we have shown in Chapter 6 that the ideal gas law is also written in terms of the number of moles $\tilde{n}$ of particles:

$$PV = NkT = \tilde{n}RT, \tag{10.9}$$

where N is the number of particles and R is the *ideal gas constant*:

$$R = \frac{N}{\tilde{n}} k = N_A \cdot k = 8.314 \equiv \left[\frac{J}{mol \cdot K} \right], \tag{10.10}$$

where $N_A = 6.02 \times 10^{23}$ is the Avogadro number and k the Boltzmann constant. Therefore, dividing both terms by the elementary electric charge, we have

$$\frac{k}{q} = \frac{R}{q \cdot N_A} = \frac{R}{F}$$

$$\rightarrow \frac{q}{kT} = \frac{F}{RT}, \tag{10.11}$$

where F is called *Faraday's constant*, that is, the charge in Coulombs (C) carried by a mole[1] of elementary charges (the electron charge)

$$F = \frac{\text{total charge}}{1\text{mol of elementary charges}} = \text{electron charge} \cdot N_A$$
$$= 1.6 \times 10^{-19} \cdot 6.02 \times 10^{23} = 96,500 \equiv \left[\frac{C}{\text{mol}}\right]. \tag{10.12}$$

This constant is fundamental in electrochemistry, where it is important to refer to processes in moles' units.

Boltzmann statistics could show the behavior in a large number of chemical and physical processes. For instance, we can use it in chemical equilibrium. It is known that a chemical reaction is governed by *Gibbs free energy G* (work done per mole), which is a thermodynamic potential (therefore, it has the same characteristic of potentials so far discussed) expressing the capacity of a system to do reversible *nonmechanical work*, such as chemical bonds. Therefore, a chemical reaction could be represented in the *reaction coordinates system*, as shown in Fig. 10.4, where different plane points identify reactants and products. To transform one point to another in the coordinates, we need to do nonmechanical work on chemical bonds without any possible mechanical work and heat exchange.

We will refer to the simplest reaction in which A and B are the *reactants* and AB is the *product*

$$A + B \underset{k_B}{\overset{k_F}{\rightleftarrows}} AB. \tag{10.13}$$

Following Boltzmann's thermodynamics, we can analyze what happens in the reaction as illustrated in Fig. 10.4: the *forward* and *backward* rates k_f and k_b of the reaction depend on the energetic barriers as

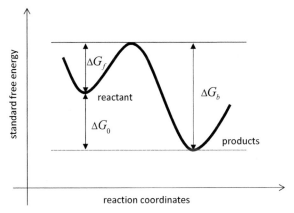

Figure 10.4 Thermodynamics of a (bio)chemical reaction at chemical equilibrium.

---

[1] A mole is an Avogadro number $N_A = 6.02 \times 10^{23}$ of objects. Since the mole is a basic unit of the SI, N_A has the unit of $1/\text{mol}$.

$$k_f \propto e^{-\frac{\Delta G_f}{RT}}$$

$$k_b \propto e^{-\frac{\Delta G_b}{RT}}.$$

$$(10.14)$$

Therefore, if we consider the *rate* of transitions toward products, it should be proportional to the concentration of *both* reactant A *and* reactant B. On the other hand, the backward rate should be proportional to the concentration of the product AB. At chemical equilibrium, we have that the net forward and backward *velocities* (rate times concentration) should be equal

$$\begin{cases} v_f = k_f C_A \cdot C_B \\ v_b = k_b C_{AB} \end{cases} v_{\text{NET}} = k_f C_A \cdot C_B - k_b C_{AB} = 0$$

$$C \equiv [\text{M}] \equiv \left[\frac{\text{mol}}{\text{m}^3}\right]; \quad v \equiv \left[\frac{\text{M}}{\text{s}}\right]; \quad k \equiv \frac{1}{\text{s}},$$

$$(10.15)$$

where C_A, C_B, and C_{AB} are concentrations in *molar concentration* units, M.[2] Note that the rate coefficients have units of the frequency. Therefore, under equilibrium, we have

$$k_f C_A \cdot C_B - k_b C_{AB} = 0$$

$$K = \frac{k_f}{k_b} = \frac{C_{AB}}{C_A \cdot C_B} = e^{-\frac{\Delta G_0}{RT}},$$

$$(10.16)$$

where ΔG_0 is the reaction *standard Gibbs energy change of the reaction*, K is the *reaction constant*, and the concentrations are *referred to those at the chemical equilibrium*.[3] Therefore

$$\Delta G_0 = -RT\ln K \equiv \left[\frac{\text{J}}{\text{mol}}\right].$$

$$(10.17)$$

It is important to point out that K and ΔG_0 are dependent (except temperature) only on the kind of reaction. The reaction will always (if allowed) tend to the concentration ratio of its reaction constant in long or short times depending on the kinetics of the reaction. From thermodynamics, we know that if $\Delta G_0 < 0$ the reaction is spontaneous and if $\Delta G_0 > 0$ it is not. Among the most used reactions in biosensing, there are those in which electrons are exchanged, referred to as *reduction-oxidation* (redox) reactions, where

$$O + ze^- \underset{k_B}{\overset{k_F}{\rightleftarrows}} R,$$

$$(10.18)$$

---

[2] We will skip the concept of "*activity*" involved in chemical reactions for simplicity.

[3] Chemical equilibrium is the state of the system in which the concentrations of reactants and products are steady versus time. It is necessary for thermodynamic equilibrium but not vice versa.

where O is the oxidized species or form, R is the reduced species, and z is the number of electrons involved in a single reaction. We will use this reaction as a reference in faradaic processes, as shown in Section 10.2.5. When we are not at equilibrium, the ratio of the concentrations of products over those of reactants species (taking stoichiometric coefficients as exponents of concentration) is referred to as *reaction quotient* Q_R . In the simple case of the reaction of Eq. (10.14), we have $Q_R = C'_{AB}/(C'_A \cdot C'_B)$, and it tends to the equilibrium constant $Q_R \rightarrow K$ with the progress of the reaction.

Another example of reactions involving charge exchange is in the case of self-ionization of water; the reaction is

$$H_2O \rightleftharpoons OH^- + H^+, \tag{10.19}$$

where H^+ denotes a solvated proton (often written as the *hydronium* ion H_3O^+) and OH^- is referred to as *hydroxide*.

Using the common notation where $[\cdot]$ refers to molar concentrations per liter (or dm^{-3}), we have that the constant of the reaction is

$$\frac{[H^+][OH^-]}{[H_2O]} \approx [H^+][OH^-] = K_W, \tag{10.20}$$

where K_W is referred to as the *self-ionization water constant*.

Usually, the constant K_W is expressed in logarithmic form: $pK_W = -\log(K_W) = 13.99$ at $25°C$. Therefore, the concentrations of hydronium and hydroxides are in the relationship

$$[H^+][OH^-] \approx 1 \times 10^{-14} \frac{mol}{L}. \tag{10.21}$$

Furthermore, the concentration of solvated protons in log form is called pH

$$pH = -\log[H^+]. \tag{10.22}$$

Therefore, pH=7 means that we have an equal concentration of the two species, and the water is considered neutral.

10.1.3 Relationship Between Potentials in a Redox

During the evolution of a chemical reaction, the free energy plot changes in the reaction coordinates, as shown in Fig. 10.5. In the case of the exchange of charges, as in a redox reaction, we can link the free energy change during the reaction progress to the change of an electric potential.

The free energy is given in work done per mole, while the electric potential energy $U_E = zq\Phi$ is the work done by z elementary charges to reach the electric potential Φ from its reference. Therefore, a small change of free energy occurring during the progress of the reaction determines a charge transfer at a given electric potential

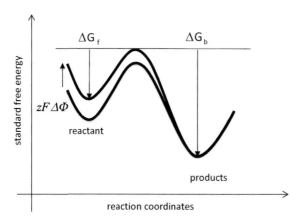

Figure 10.5 Reaction coordinate evolution of a chemical reaction in nonequilibrium. In the case of exchanges of charges as in a redox reaction, the free energy variation is related to a change of electric potential (10.23) (the picture is a simple and schematic representation of the concept and does not cover all specific cases).

$$|\Delta G| \equiv \left[\frac{J}{mol}\right] = \frac{energy}{moles \ (no. \ molecules)}$$

$$= \frac{energy}{charge \ transfered} \cdot \frac{charge \ transfered}{moles} = zF \cdot |\Phi| \qquad (10.23)$$

$$\equiv \left[\frac{J}{C} \cdot \frac{C}{mol}\right] \equiv \left[V \cdot \frac{C}{mol}\right].$$

Now, referring to (10.17), we have

$$\Delta G_0 = -RT\ln K = -zF\Phi_0, \qquad (10.24)$$

where Φ_0 is the *standard emf of the cell* reaction or redox potential, and, again, z is the number of electrons involved in the reaction.

Equation (10.23) shows a fundamental rule of biosensing: by using electrical conductive devices referred to as *electrodes* in strict contact with the chemical reaction, we could link a variation of electric potential to the thermodynamic properties of a (bio)chemical reaction. On the one hand, the electrodes could be used to sense the status of the reaction from their voltage (using the Nernst equation, as will be discussed later); on the other hand, we can force electric variations at the electrodes to sense the reaction characteristics as will be discussed in Section 10.2.9.

Now, if the reaction is away from chemical equilibrium and recalling the reaction quotient, we have

$$\Delta G = \Delta G_0 + RT\ln Q_R, \qquad (10.25)$$

That for a redox reaction becomes

$$\Delta G = \Delta G_0 + RT \ln \frac{C_R}{C_O}, \tag{10.26}$$

where C_R and C_O are the concentrations of oxidized and reduced species, respectively. This relationship also shows that ΔG_0 is the difference of free energy when the species are at unitary concentrations [and is also related to K by means of (10.17)].

However, using (10.23), we have

$$\Phi = \Phi_0 + \frac{kT}{zq} \ln \frac{C_O}{C_R} \quad \text{where}$$

$$\Phi_0 = \frac{kT}{zq} \ln K = \frac{RT}{zF} \ln K, \tag{10.27}$$

where Φ is the potential at the electrodes. Equation (10.27) is referred to as the *Nernst equation*.

It shows how electrode potential is related to the concentration of reactants. Note that Eq. (10.27) gives a hint about the standard potential, which is the potential observed at the electrodes when redox concentrations are in unitary concentrations.[4]

We can apply the Nernst equation whenever the reaction is spontaneous and not yet at the chemical equilibrium. In that case, the Nernst equation gives us the potential at the electrodes for given concentrations at thermodynamic (not chemical) equilibrium, i.e., using open circuit condition

$$\Phi_{OC} = \Phi_0 + \frac{kT}{zq} \ln \frac{C_O}{C_R} . \tag{10.28}$$

In the case of galvanic cells, used for generating electrical energy, such as in batteries, at the beginning, when the redox reaction is away from equilibrium, we have that usually the concentrations of oxidants and reductants are unitary. Therefore, the potential that we see at the electrodes is $\Phi = \Phi_0$ (actually, this is the composition of two semi-reaction potentials). Then, when we discharge the battery (with a load until the current is zero), we have an increase of C_R and a decrease of C_O so that the potential decreases with respect to Φ_0. When the battery is completely discharged, the reaction achieves equilibrium and $\Delta G = 0$ $\Phi = 0$.

In the case the reaction is not spontaneous, the exchange of charges is so weak that open circuit potential could not be defined. This is the typical case of polarizable electrodes, as shown in the plots of Fig. 10.24 (where the current is negligible for a large set of potentials) and discussed in Section 10.3.2. Equation (10.28) also shown that we can force a small variation of external potential and sense the corresponding current to get information of the reaction properties, as will be discussed in Section 10.2.5.

---

[4] Actually, to sense such a potential, we need two electrodes using a standard reference electrode. Refer to standard electrochemistry textbooks.

10.1.4   Drift and Diffusion Effects

In a gas of particles in thermodynamic equilibrium, we can imagine observing a single particle for an amount of time T_0 and count the number of hits with other particles. We can determine two averages in the same experiment:

$$\text{average no. hits} = \frac{1}{\tau} = \lim_{T_0 \to \infty} \frac{\text{no. hits}}{T_0}$$

$$\to \text{average collision time} = \tau.$$

(10.29)

where τ has the dimension of time. If the hit event is associated with an identically independent distributed process, the larger T_0, the better the average according to the law of large numbers.

Now let's consider N particles: the number of hits in a small timeframe Δt is N times what we expect from a single particle; that is, $N \cdot \Delta t / \tau$. Therefore $\Delta t / \tau$ is the fraction of a given number of particles having hits in the amount of time. Therefore, if we call $N(t)$ the number of particles that after a time t have *no* collisions, we have

$$N(t + \Delta t) = N(t) - N(t)\frac{\Delta t}{\tau}$$

$$\to \frac{dN(t)}{N(t)} = -\frac{N(t)}{\tau}$$

$$\to N(t) = N_0 e^{-t/\tau},$$

(10.30)

where $N(t) \cdot \Delta t / \tau$ is the fraction of particles having a collision in a time Δt and N_0 is the number of particles we consider at the observation time $t = 0$. Therefore, the number of free particles from any hits exponentially decays versus time with a time constant τ. Recalling the argument presented earlier, it is not difficult to show that τ is both the time after which 63% of the population has a hit and the average collision time of a single particle.

Following previous results, in a classical mechanics model, we can consider the movement of charged particles (electrons or ions) in a conductor upon the electric field driving force and having collisions with other molecules, as shown in Fig. 10.6. In a vacuum, they would have a uniformly accelerated motion, but in this case, the motion is stochastically changed by collisions. Therefore, even if they are subject to hits that could revert their direction, on average, they are moving in the direction of the electric field. Therefore, since we have defined an average collision time τ, we can determine an average distance between collisions, called *mean free path* $l = \tau \cdot v$, where v the particle's average velocity. In other words, since the collision of particles involves a continuous loss of kinetic energy, *on average*, the particle's movement is uniform with velocity v. Furthermore, the loss of kinetic energy is associated with the dissipation of heat at the base of the electrical resistance concept.

Therefore, similarly to Eq. (10.30), the number of particles free from hits after a path x is

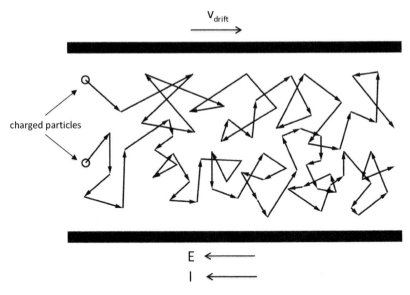

Figure 10.6 Simplified model of conduction of charged particles under the effect of an electric field.

$$N(t) = N_0 e^{-x/l}. \tag{10.31}$$

Therefore, a particle's chance to have a hit (or the fraction of particles having hits) in a very small depth Δx is $\Delta x/l$.

If a particle is subject to a force that is driving it, we can define the aforementioned velocity, referred to as *drift velocity*, as

$$v_{\text{drift}} = \frac{\text{force} \cdot \tau}{\text{equivalent mass}} = \frac{F \cdot \tau}{m^*}, \tag{10.32}$$

where again τ is the average collision time, F is the force applied, and the equivalent mass m^* is not the real mass of the particle, but its value is related to the environmental conditions where the particle is moving.

Following Eq. (10.32), *particle mobility* is defined as

$$\tilde{\mu} = \frac{v_{\text{drift}}}{\text{force}} \equiv \left[\frac{\text{s}}{\text{kg}}\right]. \tag{10.33}$$

However, in the case of electrodynamics, it is preferable to refer to the electric field E instead of the force F using the equivalence $F = zqE$, where z is again the *ion valency* (i.e., the number of equivalent charges with sign carried by a carrier, such as an ion). Therefore, the *electrical mobility* is defined as

$$\mu = \frac{v_{\text{drift}}}{E} = |z|q \cdot \frac{v_{\text{drift}}}{F} = |z|q \cdot \tilde{\mu} \equiv \left[\frac{\text{m}^2}{\text{V} \cdot \text{s}}\right]. \tag{10.34}$$

The force is responsible for an electric current density of particles referred to as *drift current*

$$J_{\text{drift}} = zqn \cdot v_{\text{drift}} = zqn \, \mu E, \tag{10.35}$$

where $n \equiv [1/m^3]$ is the number of particles per unit volume, E is the electric field, and q is the electron charge; therefore, $|z|qn$ is the charge density per unit volume. Drift is also referred to in electrochemistry as *migration*. Since we know that the current density is also proportional to the electrical field by the factor conductivity σ,

$$\begin{aligned} J_{\text{drift}} &= \sigma \cdot E = zqn \, \mu E \\ \rightarrow \sigma &= \frac{1}{\rho} = zqn\mu, \end{aligned} \tag{10.36}$$

where ρ is the material *resistivity*.

A single particle has a larger number of kicks in places where particles are more crowded, so on average, it will tend to move where there is a smaller concentration. This behavior is called *diffusion*. Since a gradient of concentration gives diffusion, we can define a *diffusion coefficient* D such that[5]

$$J_{\text{diff}} = -zqD \cdot \frac{dn}{dx}. \tag{10.37}$$

Hint We will use conventional definition for electric current, where it is made of a flow of positive charges in the electric field direction. This follows the Kirchhoff laws of electrical circuits despite the physical reality where the current in metal conductors is made of electrons.

The diffusion tendency is often balanced by a driving force, as shown in Fig. 10.7. The case is precisely that of Fig. 10.1A, where the force is electric and not gravitational. However, in this case, we want to understand the role of the newly introduced diffusion effect and the relationship with mobility. In the figure, the gas tends to go from left to right by diffusion, but at the same time, it is driven from right to left by the electric field. Thus, we might expect an equilibrium state in which these two tendencies are balanced.

We would like to understand what the relationship is between D and other parameters such as the mobility. For doing so, we apply a force acting on the gas of charged particles such that we keep maintaining gradient differences thanks to the current balancing

---

[5] If not necessary for the discussion, we will use $z = 1$ to simplify and make a link with semiconductor physics.

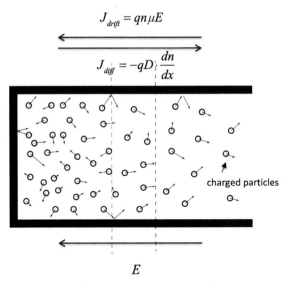

Figure 10.7 Equilibrium between drift and diffusion currents. For simplicity, we assume here $z = 1$.

$$J_{\text{diff}} = -zqD \cdot \frac{dn}{dx} = -J_{\text{drift}} = zqn\mu E \rightarrow \frac{dn}{dx} = \frac{n\mu}{D}E. \tag{10.38}$$

By differentiating the Maxwell–Boltzmann statistics expression (10.25) $n = n_0 \exp (-zq\Phi/kT)$, we have

$$\frac{dn}{dx} = -n_0 e^{-\frac{zq\Phi}{kT}} \frac{zq}{kT}\frac{d\Phi}{dx} = -n\frac{zq}{kT}\frac{d\Phi}{dx} = n\frac{zq}{kT}E. \tag{10.39}$$

Now, substituting in (10.38) the expression $E = -d\Phi/dx$, we get

$$n\frac{zq}{kT}E = \frac{n\mu}{D}E, \tag{10.40}$$

obtaining

$$D = \frac{\mu kT}{|z|q} \equiv \left[\frac{\text{m}^2}{\text{s}}\right], \tag{10.41}$$

which is the relationship between the diffusion coefficient and the mobility as first derived by Einstein. It could be shown that this relationship is also valid in nonequilibrium using Boltzmann statistics concepts. Therefore, if we combine the effect of drift with that of diffusion in the contribution to current, we have

$$J = J_{\text{diff}} + J_{\text{driff}} = -zqD\frac{dn}{dx} + zqn\mu E = -zqD\frac{dn}{dx} - zqn\mu \frac{d\Phi}{dx}, \tag{10.42}$$

which is called the drift (migration)–diffusion equation (or Nernst–Planck equation) for charged particles.

However, it is better to refer to moles when dealing with solutions. For this reason, using (10.12) and (10.41) and integrating (10.42) in the cross-sectional area of the device, A, it is better to refer to another form used in electrochemistry. Specifically, considering that

$$zqn = \text{no. charges per particle} \cdot \frac{\text{no. particles}}{\text{volume}} = \frac{\text{charge}}{\text{volume}}$$

$$= \text{no. charges per particle} \cdot \frac{\text{no. particles}}{1 \text{mole elementary charges}} \cdot \frac{\text{no. moles}}{\text{volume}} = zFC.$$

(10.43)

Therefore, also considering that we have, $q/kT = F/RT$,

$$I = -zFAD\frac{dC}{dx} - \frac{(zF)^2}{RT}ADC\frac{d\Phi}{dx}$$

(10.44)

where again $C \equiv [\text{mol}/\text{m}^3]$, F is the Faraday constant, and z is the ion valency.

Another application of Boltzmann statistics is the permeable membrane, as shown in Fig. 10.8. A selective membrane separates two chambers containing solutions (such as NaCl ionic salt). We have both positive (cations) and negative (anions) ions on the left chamber. At the starting status, the solution is completely neutral. However, the membrane allows cations to pass through while the anions do not due to their size. Therefore, the diffusion pushes cations on the right part, but this creates a potential gradient that counteracts the diffusion until an equilibrium is reached. In other words, there is a rate of cations from left to the right due to diffusion and a rate of cations from right to left due to the electric drift. We can therefore use the model so far introduced, where at the equilibrium drift and diffusion currents are equal:

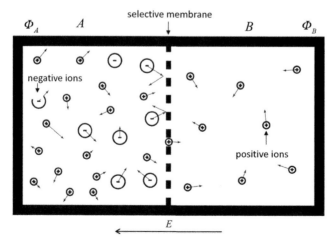

Figure 10.8 Effect of a selective membrane on ions.

$$qD\frac{dn}{dx} = -qn\mu E = qn\mu\frac{d\Phi}{dx}, \tag{10.45}$$

which could be easily solved by separation of variables, giving

$$\Delta\Phi_{AB} = \frac{kT}{q}\ln\frac{n_A}{n_B} = \frac{RT}{zF}\ln\frac{C_A}{C_B}, \tag{10.46}$$

where $n_{A,B}$ and $C_{A,B}$ are concentrations of the two reservoirs in particles and moles per unit of volume, respectively. This is nothing other than the *Nernst equation* applied to porous membranes, equivalent to (10.46).

Note that we could reach the same result applying (10.7). This means that the diffusion current is another way to look at the Boltzmann statistic.

10.2   Electrical Conduction and Polarization of Matter

10.2.1   Conductivity

The electrical resistance of a conductor, as shown in Fig. 10.9, is proportional to its length and inversely proportional to its area; therefore

$$R = \rho\frac{L}{W\cdot H} = \rho\frac{L}{S} \equiv [\Omega];$$

$$\rho = \text{resistivity} = \frac{R\cdot S}{L} \equiv [\Omega\cdot\text{m}] \tag{10.47}$$

$$\sigma = \text{conductivity} = \frac{1}{\rho} = \frac{L}{R\cdot S} \equiv \left[\frac{1}{\Omega\cdot\text{m}}\right] = \left[\frac{S}{\text{m}}\right],$$

where R is the resistance of the elementary structure. For the same elementary structure, we can define the current density (i.e., current per unit of area) as

$$J = \frac{I}{A} = \frac{V}{A\cdot R} = \frac{E\cdot L\cdot S}{A\cdot\rho\cdot L} = \frac{E}{\rho} = \sigma\cdot E \equiv \left[\frac{A}{\text{m}^2}\right], \tag{10.48}$$

where A is the area of the cross section.

Following the definition of current density and (10.44), we can also express the *conductivity* σ of a conductor/solution as

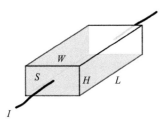

Figure 10.9 Structure definitions of a material crossed by charged particles.

Table 10.1 Mobility and diffusion coefficient of several ions in water solutions at 25°C

Type	z_i	μ_i [m²/V/s]	Diffusion coefficient [m²/s]
H^+	+1	36.3×10^{-8}	9.31×10^{-9}
OH^-	−1	20.4×10^{-8}	5.27×10^{-9}
K^+	+1	7.62×10^{-8}	1.96×10^{-9}
Cl^-	−1	7.91×10^{-8}	2.03×10^{-9}
Ca^{2+}	+2	6.17×10^{-8}	0.79×10^{-9}
Na^+	+1	5.2×10^{-8}	1.33×10^{-9}
Li^+	+1	4.0×10^{-8}	1.03×10^{-9}

$$J = znq\mu E; \quad \text{where} \quad \sigma = znq\mu = zFC\mu = \frac{(zF)^2}{RT}DC, \qquad (10.49)$$

where E is the electric field, n the density of charges, and μ the mobility. Equation (10.49) is also referred to as the Einstein–Smoluchowski equation.

In the case of round particles in motion in a liquid, it has been shown that their mobility is

$$\mu = D\frac{q}{kT} \equiv \left[\frac{m^2}{V \cdot s}\right]; \text{where } D = \frac{kT}{6\pi\eta a} \equiv \left[\frac{m^2}{s}\right]$$
$$\rightarrow \mu = \frac{q}{6\pi\eta a}, \qquad (10.50)$$

where a is the effective radius of the particle and η is called *viscosity* of the medium.

Table 10.1 shows several common ions and the corresponding mobility and diffusion coefficients.

If the conductivity is sustained by several ions, for the ith one we have

$$\sigma_i = z_iF\mu_iC_i, \qquad (10.51)$$

where z_i is the valency of the ion species. It is common to use the concentration in terms of moles per liter; therefore, Eq. (10.51) relationship becomes

$$\sigma_i = z_iF\mu_i \cdot \underbrace{c_i \cdot 1000}_{\text{n.mol} \atop m^3}; \quad \text{where} \quad c_i \equiv \left[\frac{mol}{L}\right]. \qquad (10.52)$$

In the presence of several species of ions, the total conductivity could be

$$\sigma = 1000 \cdot \sum_i c_i \cdot F \cdot z_i \cdot \mu_i = 1000 \cdot \sum_i c_i\lambda_i, \qquad (10.53)$$

where λ_i is called the *equivalent ionic conductivity* of the species. For symmetrical electrolytes, it is

$$\sigma_i = (\lambda_i^+ + \lambda_i^-)c_i = \Lambda_i c_i;$$

$$\sigma = 1000 \cdot \sum_i \Lambda_i c_i, \tag{10.54}$$

where Λ_i is the *molar conductivity* and is frequently tabulated for different ionic species.

Example A solution 10 mM of KCl has a conductivity at 25°C equal to $\sigma = 96500 \cdot 1000 \cdot 0.01 \cdot (7.62 + 7.91) \times 10^{-8} = 1.49 \times 10^{-3}\,\text{S/m}$, close to the experimental one of $1.41 \times 10^{-3}\,\text{S/m}$.

Example A highly deionized water at pH=7 has an ion concentration equal to $[H^+] = [OH^-] = 1 \times 10^{-7}\,\text{mol/L}$. Therefore, the conductivity at 25°C is $\sigma = 96500 \cdot 1000 \cdot 0.01 \cdot (36.4 + 20.4) \times 10^{-8} = 5.5 \times 10^{-6}\,\text{S/m}$, close to the experimental one.

Unfortunately, the experimental evidence shows no more direct proportionality between the conductance and the concentration for high concentrations, and there is a nonlinear law for strong electrolytes following the square root of the concentration.

10.2.2 Polarization of Matter

Electrically neutral materials could be composed of domains where opposite sign electric charges might have a limited degree of freedom of displacement induced by an external electric field. Therefore, under the field's application, charges tend to separate and/or displace themselves in opposite directions, as shown in Fig. 10.10. This effect is referred to as *electric polarization,* and the displacement configuration of opposite sign charges barycenters is referred to as *electric dipole.* This is possible at various levels of space dimensions such as atomic, molecular, and macromolecular. There are two main polarization effects. In the first one, as shown in Fig. 10.10A, the barycenters displacement of positive and negative charges is induced only after applying the electric field. In a second case, as shown in Fig. 10.10B, the material is made of units that already have an electric dipole, but they are randomly oriented. After applying the electric field, elementary dipoles are oriented in the same direction. This is referred to as *orientation polarization,* and it is typical, for example, in water molecules.

Polarization could be modeled by a first-order approximation given by two charges displaced in the distance, δ as shown in Fig. 10.10C. In other terms, the electric potential and the electric field generated by complex polarizations is "as if" it were generated by two opposite sign charges placed at a given distance. We could increase the accuracy of the model by incrementing the order of the polarization by adding to the dipole the effects of higher-order terms (quadrupole, octupole, etc., referred to as multipole expansion).

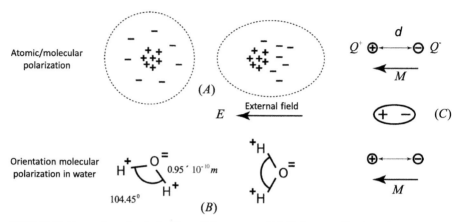

Atomic/molecular polarization

Orientation molecular polarization in water

$0.95 \cdot 10^{-10}\,m$

104.45^0

External field

Figure 10.10 Examples of polarization of the matter. (A) Before application of an external electric field. (B) After application of an external electric field. (C) Equivalent model.

Notation In the following, we will use for vectors and their transforms the notation $\overline{V}(x,t) = \mathrm{Re}\{\overline{V}(x,j\omega)e^{j\omega t}\} = \mathrm{Re}\{|\overline{V}|e^{j\omega t + \varphi}\} = |\overline{V}|\cos(\omega t + \sphericalangle \overline{V})$, where $\overline{V}(x,j\omega)$ is the *phasor* of the sinusoidal (or harmonic) regime and $\sphericalangle \overline{V}$ its argument. However, for simplicity of notation, we will use $\overline{V} \leftarrow \overline{V}(x,t)$ or $\overline{V} \leftarrow \overline{V}(j\omega)$ $\leftarrow \overline{V}(x,j\omega)$, wherever it is implied that we are dealing with time-domain or phasor transformations. Furthermore, sometimes we will use $V \leftarrow |\overline{V}|$ as symbol of the module (or amplitude) of the vector. Note that if two vectors are collinear in the space, it does not imply that they are collinear in the complex phasor space.

We can measure the effect of electric polarization by defining a vector called *electric dipole moment* in coulomb-meter units[6]

$$\overline{M} = Q \cdot \overline{\delta} = \text{dipole moment} \equiv [C \cdot m], \tag{10.55}$$

where it is defined by two opposite sign points of charge $|Q|$ displaced at a distance $\boldsymbol{\delta}$. We can also define the *polarization density* as

$$\overline{P} = \frac{\text{no. of dipole moments}}{\text{volume}} \equiv \left[\frac{C \cdot m}{m^3}\right] = \left[\frac{C}{m^2}\right], \tag{10.56}$$

which is a vector. The polarization density is useful for finding the polarization surface charge density using the scalar product with the versor $\vec{n}$ of a generic surface of a polarized matter

---

[6] In molecular or atomic physics, the unit for dipole moment is still in use, the *debye*: $1\,D = 3.33 \times 10^{-30}\ C \cdot m$.

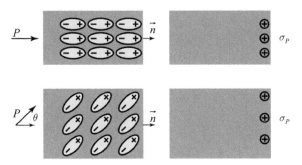

Figure 10.11 Relationship of the polarization density with the charge surface density.

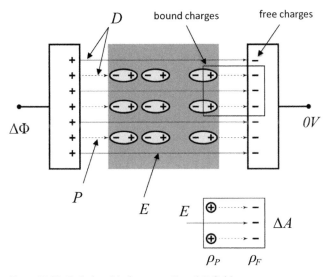

Figure 10.12 Relationship between D and E fields.

$$\overline{P} \cdot \vec{n} = \frac{\text{no. of dipole moments}}{\text{volume}} = \frac{\text{no. of charges}}{\text{volume}} \cdot \overline{\delta} \cdot \vec{n}$$

$$= \frac{\text{no. of charges}}{\text{area}} \cdot \cos\theta = \sigma_P \equiv \left[\frac{\text{C}}{\text{m}^2}\right]. \qquad (10.57)$$

Figure 10.11 shows how the surface density σ_P[7] decreases when the polarization vector is not orthogonal to the surface. Note that the charge density "seen" at the surface comprises bounded and not free charges.

The preceding discussion helps understand the relationship between free charges and polarization charges in capacitor dielectrics, as shown in Fig. 10.12.

Applying Gauss theorem in the area indicated in Fig. 10.12, we have[8]

---

[7] Do not confuse surface charge densities $\sigma_P, \sigma_F \equiv [\text{C/m}^2]$ with conductivity $\sigma \equiv [\text{S/m}]$.
[8] $\varepsilon_0 = 8.85 \times 10^{-12}\,[\text{F/m}]$.

$$\nabla \cdot \overline{E} = \frac{\rho_F + \rho_P}{\varepsilon_0} = \frac{\rho}{\varepsilon_0} \rightarrow \int_S \overline{E} \cdot \vec{n} \, dS = \frac{Q}{\varepsilon_0}, \tag{10.58}$$

where ρ_P and $\rho_F \equiv [C/m^3]$ are the volume density of bound and free charges, respectively.[9] Therefore

$$-\overline{E} \cdot \vec{n} \, \Delta A = \frac{(\sigma_P - \sigma_F)}{\varepsilon_0} \cdot \Delta A$$

$$\rightarrow \overline{E} \cdot \vec{n} = \frac{\sigma_F - \sigma_P}{\varepsilon_0} = \frac{\sigma_F - \overline{P} \cdot \vec{n}}{\varepsilon_0} \equiv \left[\frac{V}{m}\right]$$

$$\rightarrow \varepsilon_0 \overline{E} + \overline{P} = \sigma_F = \overline{D} \equiv \left[\frac{C}{m^2}\right]$$

$$\nabla \cdot \overline{D} = \rho_F \rightarrow \int_A D \cdot \vec{n} \, dA = Q_F, \tag{10.59}$$

where D is called *electric displacement*, which is composed of polarization and electric field lines, as shown in Fig. 10.12. Therefore, the Gauss theorem's application for the electric displacement field, as illustrated in (10.59), shows that D is a measure of the total free charges on the electrode.

However, the polarization density is proportional to the electric field by a factor called *electric susceptibility* χ

$$\overline{P} = \chi \cdot \varepsilon_0 \cdot \overline{E}; \quad \overline{E} \cdot \vec{n} \stackrel{=}{} \frac{\sigma_F - \overline{P} \cdot \vec{n}}{\varepsilon_0} \rightarrow E = \frac{\sigma_F}{(1 + \chi) \cdot \varepsilon_0} = \frac{\sigma_F}{\varepsilon_R \cdot \varepsilon_0};$$

where $\varepsilon_R = (1 + \chi)$

$$\rightarrow \overline{D} = \varepsilon_R \varepsilon_0 \overline{E} = \sigma_F = \varepsilon_0 \overline{E} + \overline{P}$$

$$\rightarrow \overline{P} = \varepsilon_0 (\varepsilon_R - 1) \overline{E}, \tag{10.60}$$

where $\varepsilon_R \equiv [\cdot]$ is referred to as the *relative dielectric constant*.

The increase of the relative permittivity due to polarization is the reason why a capacitor has a greater capacitance if the dielectric is implemented by using a polarizable material, as can be seen by the following relationship

$$C = \frac{Q}{\Phi} = \frac{Q}{E \cdot L} = \frac{\sigma_F \cdot A \cdot \varepsilon_R \cdot \varepsilon_0}{\sigma_F \cdot L} = \frac{A \cdot \varepsilon_R \cdot \varepsilon_0}{L} > \frac{A \cdot \varepsilon_0}{L}, \tag{10.61}$$

where A is the area of the plates of the capacitor and L their distance.

To better understand the aforementioned concept from different viewpoints, in Fig. 10.13 are shown two experiments. In the first one, we apply a constant potential on the plates and then insert a polarizable dielectric between them. The external potential

---

[9] Do not confuse volume charge densities $\rho_P, \rho_F \equiv [C/m^3]$ with conductivity $\rho \equiv [\Omega \cdot m]$.

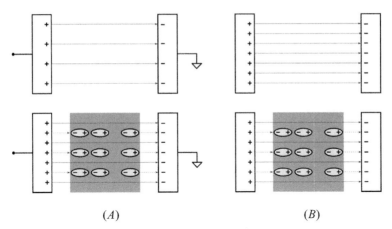

Figure 10.13 Experiments for capacitance variation effects. With constant electric field (A) and constant plate charges (B).

fixes the electric field; however, the polarization dipoles attract a larger amount of free charges, as shown in Fig. 10.13A, the capacitance increases according to (10.61). In the second experiment, we keep the plates electrically floating in an open-circuit configuration so that no charge variation of free charges occurs in the plates. Several electric field lines are interrupted by material dipoles by inserting the polarizable material, as shown in Fig. 10.13B. Therefore, again, the capacitance increases according to (10.61).

10.2.3 Complex Dielectric Constant and Debye Relaxation Model

A complex dielectric constant is a way to model the effects of both conduction and polarization on a medium in a sinusoidal regime. Using a bottom-up approach, we will start with a particular example; then, we will make more general considerations on E and D fields' physical relationships.

Consider some material that is placed between two electrodes, as in Fig. 10.14, showing both electrical conductivity (either ions or electrons) σ and electrical polarizability represented by the dielectric constant ε.

Assuming a constant and uniform electric field within the material, we could represent the conductivity and the polarizability with an electric lumped model based on a resistor and a capacitor as illustrated in Fig. 10.14 so that

$$C = \frac{\varepsilon_0 \varepsilon_R A}{L} = \varepsilon_R C_0; \ C_0 = \frac{\varepsilon_0 A}{L}$$

$$G = \frac{\sigma S}{L} = \sigma \frac{C_0}{\varepsilon_0},$$

(10.62)

where L and A are the distance and the area of the plates and C_0 is the capacitance measured between the plates in a vacuum. Note that the relative permittivity could be defined as a ratio of two capacitors $\varepsilon_R = C/C_0 \equiv [\cdot]$.

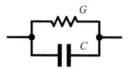

Figure 10.14 Simple example of a lumped circuit model for matter subject conductivity and polarization.

The time constant of the system could be found as

$$\tau = \frac{C}{G} = \frac{\varepsilon_R C_0}{\sigma \dfrac{C_0}{\varepsilon_0}} = \frac{\varepsilon_0 \varepsilon_R}{\sigma} = \frac{\varepsilon}{\sigma}, \tag{10.63}$$

showing the relationship between the lumped model parameters, R and C, with the physical parameters, ε and σ.

Under a sinusoidal regime, with no conductance, we have that the current phasor $\bar{I}(j\omega)$ is related to the voltage one $\bar{V}(j\omega)$ as

$$\bar{I}(j\omega) = j\omega C_0 \bar{V}(j\omega) = j\omega \cdot \varepsilon_0 \cdot \frac{C_0}{\varepsilon_0} \bar{V}(j\omega). \tag{10.64}$$

The idea is to have a similar relationship in the presence of conduction. Therefore, we could define a *complex dielectric constant* $\tilde{\varepsilon}$ so that

$$\bar{I}(j\omega) = j\omega \cdot \tilde{\varepsilon} \cdot \frac{C_0}{\varepsilon_0} \bar{V}(j\omega). \tag{10.65}$$

Therefore, the complex dielectric constant considers the phase and magnitude of the current and voltage relationship in both conduction and polarization.

To find the components of $\tilde{\varepsilon}$ we can rearrange the preceding expressions:

$$\bar{I}(j\omega) = [G + j\omega C]\bar{V}(j\omega) = \left[\sigma \frac{C_0}{\varepsilon_0} + j\omega \varepsilon_R C_0\right]\bar{V}(j\omega)$$

$$= \left[\sigma \frac{C_0}{\varepsilon_0} + j\omega \varepsilon_R \varepsilon_0 \frac{C_0}{\varepsilon_0}\right]\bar{V}(j\omega) = [\sigma + j\omega \varepsilon_R \varepsilon_0] \frac{C_0}{\varepsilon_0} \bar{V}(j\omega), \tag{10.66}$$

from which we can derive the relationship

$$j\omega \cdot \tilde{\varepsilon} \cdot \frac{C_0}{\varepsilon_0} z \triangleq [\sigma + j\omega\varepsilon_0\varepsilon_R] \frac{C_0}{\varepsilon_0} \rightarrow \sigma + j\omega\varepsilon_0\varepsilon_R = j\omega \cdot \tilde{\varepsilon}$$

$$j\omega(\frac{\sigma}{j\omega} + \varepsilon_0\varepsilon_R) = j\omega \cdot \tilde{\varepsilon} \rightarrow \tilde{\varepsilon} = \varepsilon_0\varepsilon_R - j\frac{\sigma}{\omega} = \varepsilon' - j\varepsilon''$$

$$\rightarrow \begin{cases} \varepsilon' = \varepsilon_0\varepsilon_R \equiv \left[\frac{F}{m}\right] \\[2mm] \varepsilon'' = \frac{\sigma}{\omega} \equiv \left[\frac{F}{m}\right]. \end{cases} \tag{10.67}$$

Following the preceding case, we can create a more general model of the complex dielectric constant starting from the nature of the current flowing in an impedance. We can identify two contributions of the total current due to conductance and polarization. The first contribution, due to conduction, is referred to as the *conductance current* $J_C(x) = \sigma E(x)$, as we have already shown. The second contribution is due not to an electric current of moving charges but to the time-varying electric field. It exists only in a dynamic regime (like the AC current in a capacitor $i(t) = C dv(t)/dt$). Therefore, using electric field and current density, we have $J_D(x) = \varepsilon_0\varepsilon_R(dE(x)/dt)$ called *displacement current* that in the sinusoidal regime phasor space is written as

$$\overline{J}_D = j\omega \cdot \varepsilon' \overline{E} = j\omega \cdot \overline{D} \equiv \left[\frac{A}{m^2}\right]. \tag{10.68}$$

The AC current flowing in a capacitor's terminals is due to the time-varying accumulation of free charges in plates. However, we have seen that those free charges are directly proportional to the electric displacement vector D, from which (10.68) follows. In this case, the displacement current is *in quadrature* with the applied voltage and thus does not contribute to heat generation. Conversely, the conductance current is *in phase* with the applied voltage and contributes to heat dissipation.

Therefore, the total current at the terminals of an impedance working in harmonic regime could be considered as the sum of two (in-phase and in-quadrature) contributions

$$\overline{J} = \overline{J}_C + \overline{J}_D = \sigma\overline{E} + j\omega\varepsilon_0\varepsilon_R\overline{E} = (\sigma + j\omega\varepsilon_0\varepsilon_R)\overline{E} = j\omega\tilde{\varepsilon}\overline{E}$$
$$= j\omega(\varepsilon' - j\varepsilon'')\overline{E}, \tag{10.69}$$

from which we find again (10.67) because

$$\overline{J} = j\omega(\varepsilon' - j\varepsilon'')\overline{E} = \sigma\overline{E} + j\omega\varepsilon_0\varepsilon_R\overline{E}$$
$$\rightarrow \begin{cases} \varepsilon' = \varepsilon_0\varepsilon_R \\[2mm] \varepsilon'' = \frac{\sigma}{\omega}. \end{cases} \tag{10.70}$$

This is a general case, and it is not limited to the particular model of Fig. 10.14. Furthermore, we can extend the relationship between E and D shown in (10.60) in the sinusoidal regime so that

$$\overline{D} = \varepsilon \cdot \overline{E} = \varepsilon_0 \varepsilon_R \cdot \overline{E}$$
$$\rightarrow \overline{D} = \widetilde{\varepsilon} \cdot \overline{E} = (\varepsilon' - j\varepsilon'') \cdot \overline{E}. \tag{10.71}$$

This relationship has essential energetic properties hidden in it. We know that, from the theory of electrical circuits in a sinusoidal regime, whenever current and voltage are in-phase, we have electric power dissipated in heat, like a resistor. On the other hand, whenever we have current and voltage in quadrature, the electric power is exchanged between the systems with no dissipation, like in a capacitor or in an inductor. Now, since the total device current is composed of free charges, it is proportional to the derivative of D. Therefore, we note in (10.71) that the second side of the equation, once derived, contains two addends of the current: one in-phase (contributing to heat) holding ε'' and the other in-quadrature (making an only exchange of energy), containing ε'. As a result, ε'' is the part of the complex dielectric constant that contributes to the loss of power in heat.

More specifically, the power imparted per unit volume is composed of two contributions, one supplied to the electric field and the other to the polarization

$$\frac{\text{power delivered}}{\text{unit volume}} = P_D = \frac{d}{dt}\left(\frac{1}{2}\varepsilon_0 \overline{E} \cdot \overline{E}\right) + \overline{E} \cdot \frac{d}{dt}\overline{P}$$
$$= \overline{E} \cdot \frac{d}{dt}\varepsilon_0 \overline{E} + \overline{E} \cdot \frac{d}{dt}\overline{P} = \overline{E} \cdot \frac{d}{dt}(\varepsilon_0 \overline{E} + \overline{P}) = \overline{E} \cdot \frac{d\overline{D}}{dt}. \tag{10.72}$$

It is better to look at (10.72) in the harmonic regime in phasor space

$$P_D = \overline{E} \cdot j\omega\varepsilon_0 \overline{E} + \overline{E} \cdot j\omega\overline{P} = \overline{E} \cdot j\omega\overline{D}. \tag{10.73}$$

Therefore, the first part of the sum is given by a product of the phasor always in quadrature; thus, it is always a reactive power that is exchanged with the electric field. On the other hand, the second term gives an active or reactive power according to the phase lag induced by the polarization. For example, if the polarization is always in phase with the electric field (instantaneous polarization), we have total reactive power. Otherwise, if polarization has some phase lag with respect to the electric field, the second term could contribute to a dissipative contribution, and therefore, we have active power.

Now, in a sinusoidal regime, since the mean energy stored in the field has a net exchange throughout the system, the time average gives the average power dissipated in heat:

$$\langle P_D \rangle = \left\langle \overline{E} \cdot \frac{d\overline{D}}{dt} \right\rangle, \tag{10.74}$$

similar to the current and voltage relationships for the lumped electric circuits model. Therefore,[10] since the time average in a sinusoidal regime could be expressed in phasor notation,

---

[10] Recall that for harmonic phasors it is $\langle A(t)B(t)\rangle = \frac{1}{2}\text{Re}\{\overline{A} \cdot \overline{B}^*\}$, where $\langle \ \rangle$ is the time average and $\cdot$ is the scalar product in the phasor space.

$$\langle P_D(j\omega) \rangle = \frac{1}{2}\text{Re}\{ j\omega \overline{D} \cdot \overline{E}^* \} = \frac{1}{2}\omega|\overline{E}|^2\varepsilon'', \tag{10.75}$$

because $\overline{D} = (\varepsilon' - j\varepsilon'')\overline{E}$. The relationship shows that the sinusoidal regime power dissipation due to polarization is proportional to ε''.

An index of the loss due to conductivity in dielectrics could be found in the "tangent-delta" factor

$$\tan\delta = \frac{|J_C|}{|J_D|} = \frac{\varepsilon''}{\varepsilon'} = \frac{\sigma}{\omega\varepsilon}. \tag{10.76}$$

We can now start from experimental measurements of the complex permittivity to derive a lumped electric model. More specifically, it has been observed that the permittivity starts with a high value at low frequencies and then has a drop after a given frequency. To cope with observed experimental behaviors, Debye introduced a relaxation model based on the variation of the permittivity with respect to frequency, starting with higher values ε_S at low frequency and then decaying at lower values ε_∞ as the frequency goes to infinity. Therefore, using a single-pole approximation, the following is set:

$$\tilde{\varepsilon}_R = \varepsilon_\infty + \frac{\varepsilon_S - \varepsilon_\infty}{1 + j\omega\tau}, \tag{10.77}$$

so that the complex dielectric constant becomes

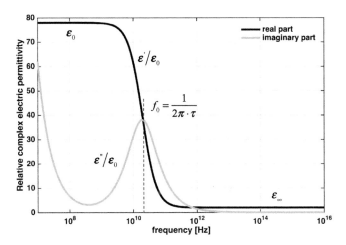

Figure 10.15 Behavior of (10.79) versus frequency for a simplified model of highly deionized water, where $\varepsilon_0 = 78$, $\varepsilon_\infty = 2$, $f_0 = 20$ GHz, $\sigma = 5.5$ μS/m.

$$\tilde{\varepsilon} = \varepsilon_0 \tilde{\varepsilon}_R - j\frac{\sigma}{\omega} = \varepsilon_0 \left[\varepsilon_\infty + \frac{\varepsilon_S - \varepsilon_\infty}{1 + j\omega\tau} \right] - j\frac{\sigma}{\omega}$$

$$= \varepsilon_0 \varepsilon_\infty + \varepsilon_0 \frac{\varepsilon_S - \varepsilon_\infty}{1 + j\omega\tau} - j\frac{\sigma}{\omega} = \varepsilon_0 \varepsilon_\infty + \varepsilon_0 \frac{\varepsilon_S - \varepsilon_\infty}{(1 + j\omega\tau)} \frac{(1 - j\omega\tau)}{(1 - j\omega\tau)} - j\frac{\sigma}{\omega} \qquad (10.78)$$

$$= \varepsilon_0 \varepsilon_\infty + \varepsilon_0 \frac{\varepsilon_S - \varepsilon_\infty}{1 + \omega^2\tau^2} - j\omega\tau\varepsilon_0 \frac{\varepsilon_S - \varepsilon_\infty}{1 + \omega^2\tau^2} - j\frac{\sigma}{\omega},$$

giving these two expressions:

$$\varepsilon' = \varepsilon_0 \varepsilon_\infty + \varepsilon_0 \frac{\varepsilon_S - \varepsilon_\infty}{1 + \omega^2\tau^2}; \qquad \varepsilon'' = \varepsilon_0 \frac{(\varepsilon_S - \varepsilon_\infty)\omega\tau}{1 + \omega^2\tau^2} + \frac{\sigma}{\omega}. \qquad (10.79)$$

Note that the real part of (10.79) is a Lorentzian form.

An example of this behavior is shown in (10.14), where some values are taken from data on highly purified water. This is an oversimplified model since it is based on a single relaxation time.

We could comment on the behavior shown in the following way: for low frequency, the polarization follows the electric field almost instantaneously. As the frequency approaches the inverse of the dipoles' time constant, the polarizable charges' movement becomes slower, incrementing the phase lag. The dipole does not have time to follow the external field's pace for very high frequency, and there is no more polarization. This is the reason for the drop of permittivity at high frequency. The first "bump" of the ε'' variable is due to the static conduction. The second is due to the dipole's work against the surrounding molecule/atoms in its displacement.

In the case that we have multiple relaxation times, we can consider them in the expression

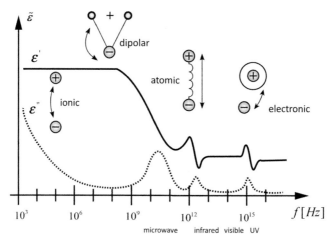

Figure 10.16 Example of multiple polarization of matter. [Adapted from WikiCommons]

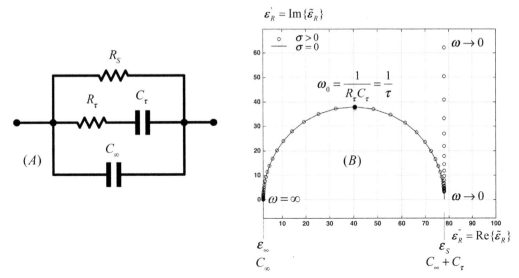

Figure 10.17 Complex permittivity Debye relaxation lumped model (A) for highly deionized water ($\sigma = 5.5 \times 10^{-6}$S/m, $\varepsilon_\infty = 2$, $\varepsilon_S = 78$, $f_t = 20$ GHz) and graphical representation in the complex plane (conductance Cole–Cole plot) (B). This approximate model of the physical system considers only the first relaxation time.

$$\tilde{\varepsilon}_R = \varepsilon_\infty + \frac{\varepsilon_S - \varepsilon_\infty}{(1 + j\omega\tau_1)(1 + j\omega\tau_2)(1 + j\omega\tau_3)\cdots}. \tag{10.80}$$

The effect of this model in the complex dielectric plot is shown in Fig. 10.16. For example, we can have different phenomena of polarization (ionic, atomic, etc.) where we have a different time constants for each of them. The effect is shown in multiple "transitions" and "bumps" in the real and imaginary parts of the plot.

Following Debye's relaxation model, it is interesting to model the material conduction and polarizability with an electric lumped circuit. This is important because the electric model could be integrated into a unified design approach for the electronic interface simulation. Thus, we start putting the relaxation model in a conductance model

$$\tilde{\sigma} = \frac{J(j\omega)}{E(j\omega)} = j\omega\tilde{\varepsilon}$$

$$= j\omega\left[\varepsilon_0\varepsilon_\infty + \varepsilon_0\frac{\varepsilon_S - \varepsilon_\infty}{1 + \omega^2\tau^2}\right] + \left[\varepsilon_0\frac{(\varepsilon_S - \varepsilon_\infty)\omega^2\tau}{1 + \omega^2\tau^2} + \sigma\right] \equiv \left[\frac{1}{\Omega \cdot m}\right]. \tag{10.81}$$

Then, we set up a lumped model as shown in Fig. 10.17A using *normalized* components. From (10.47) and (10.62), we see that resistances are multiplied by electrode distance and divided by their area, while the capacitances are multiplied by electrode area and divided by their distance. Thus, R_S models resistivity of the material, C_∞ the

$$\varepsilon'_R = \mathrm{Im}\{\tilde{\varepsilon}_R\}$$

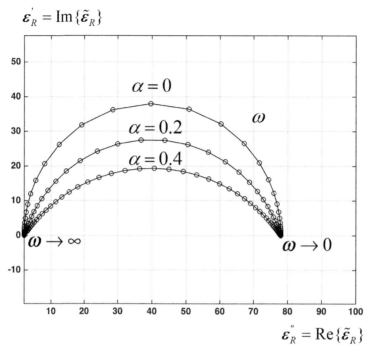

$$\varepsilon''_R = \mathrm{Re}\{\tilde{\varepsilon}_R\}$$

Figure 10.18 Cole–Cole plot in the case of distribution of relaxation times.

permittivity at very high frequency, and $R_\tau C_\tau$ the relaxation time constant. Solving Kirchhoff's equations, we get

$$\tilde{\sigma} = Y \cdot \frac{L}{A} = j\omega \left[C_\infty + \frac{C_\tau}{1 + \omega^2 R_\tau^2 C_\tau^2} \right] \frac{L}{A} + \left[\frac{\omega^2 R_\tau C_\tau}{1 + \omega^2 R_\tau^2 C_\tau^2} + \frac{1}{R_S} \right] \frac{L}{A} = \left[\frac{S}{m} \right]. \quad (10.82)$$

Comparing (10.81) with (10.82), we have

$$\tau = R_\tau C_\tau \equiv [s]$$

$$\sigma = \frac{L}{A} \frac{1}{R_S} \equiv [S]$$

$$C_\infty = \frac{L}{A} \varepsilon_\infty \varepsilon_0 \equiv [F] \qquad (10.83)$$

$$C_\tau = \frac{L}{A} (\varepsilon_s - \varepsilon_\infty) \varepsilon_0 \equiv [F].$$

It is interesting to represent the complex permittivity with an alternative plot with respect to that of Fig. 10.15 with a parametric (the parameter is the frequency) curve in a complex plane of the real and imaginary parts of the permittivity, as shown in Fig. 10.17B. This kind of plot (similar to the Nyquist one) is referred to as an *admittance Cole–Cole plot*.

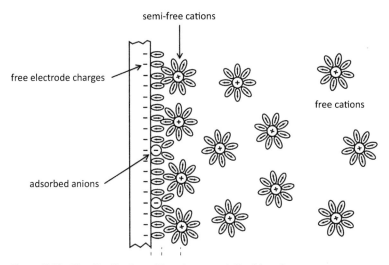

semi-free cations

free electrode charges

free cations

adsorbed anions

Figure 10.19 The distribution of ions in a metal–liquid surface.

> **Caution** The relaxation time of (10.77) is not related to the relaxation time (10.63). The first is the typical time related to the dipole's work against the surrounding molecule/atoms in its displacement. The second is a characteristic of the entire cell with respect to the external signal.

The plot describes a semicircle whose top corresponds to the system's time constant. However, experimental results on various materials and solutions show a more "squeezed" shape of the plot. Several authors suggested that this is due to a *quasi-Gaussian* distribution of time constant around a dominant τ_0. This observation leads to a model described by

$$\widetilde{\varepsilon}_R = \varepsilon_\infty + \frac{\varepsilon_S - \varepsilon_\infty}{1 + (j\omega)^{(1-a)}\tau_0}, \tag{10.84}$$

where α is an exponent that is related to the spread of time constants around τ_0. The larger α, the larger the spread. The effect of the spread is also shown in the complex plot in Fig. 10.18, where the amount of depression is related to the spread of time constant values.

> **Hint** The dielectric constant of a solution is also dependent on the type and concentration of dissolved ions. For example, that of 1 M NaCl is 67 compared with water, which is about 78.

10.2.4   The Double Layer Interface in Ionic Solutions

When an electrode with a given electric potential is immersed in a water solution with no charge exchange between the two interfaces (no redox reactions), we observe

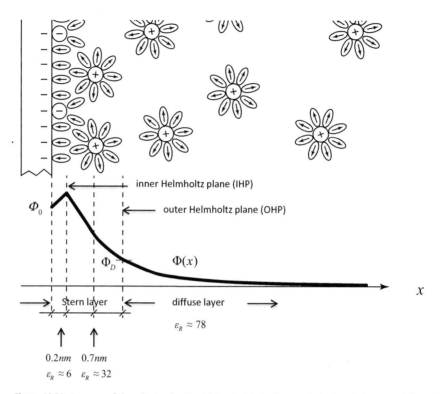

Figure 10.20 Layers of the electrode–liquid (water) interface and electrostatic potential profile. [Adapted from Morgan and Green, 2003]

a charge distribution, as shown in Fig. 10.19. In general, the potential attracts counter-ions and repels co-ions[11] in the proximity of the surface. Therefore, we have an accumulation of counter-ions with a greater density than the bulk. This discrepancy becomes smaller as we go away from the electrode until the bulk concentration is reached. The region where free charges and counter-ions are accumulated is referred to as the *double layer*. More specifically, referring to Fig. 10.19, we have that

- Ions are considered with their *ionic atmosphere (or cloud)*, a region around the ion where we have the adhesion and orientation of nearby water molecules due to the ion-generated electric field.
- Negative free electrons and cations tend to accumulate on the metal surface and on the liquid side, respectively, to set the overall charge neutrality. The shorter the distance to the interface, the greater the concentration of cations. The concentration of the free ions with respect to the electrode's distance depends on the electric potential profile, as we will discuss later.

---

[11] *Co-ions* are ions having the same charge sign for a charge reference (in this case, that of the electrode), and *counter-ions* are the opposite.

- Water molecules, co-ions, or counter-ions without ionic atmosphere could be *adsorbed* (i.e., attached by adhesion to the surface[12]). Even ions of the same charge could be adsorbed to find a local minimum of total energy. This is due to a difference between anions and cations. Anions have a less rigid ionic cloud, and they could become more easily desolvated going closer to the surface while cations, in general, keep their solvation shell.

The above configuration determines an electrostatic profile as that shown in Fig. 10.20. According to the Gouy–Chapman–Stern–Graham model, the following regions with respect to the distance from the electrode could be identified:

- A *diffuse region* of the double layer extends outside the electrode toward the bulk with a decreasing concentration of counter-ions. Here, we have an equilibrium between the diffusion and the electrostatic force, as already discussed.
- A *Stern layer*, where electric charges on the solution side are relatively bound to the electrode. A further refinement of the model identifies two other regions within the Stern layer:
  - The Inner Helmholtz Plane (IHP), characterized by adsorbed ions without an ionic cloud together with polarized water molecules. The electrostatic potential profile could be increasing or decreasing according to the sign of the adsorbed molecules. Fig. 10.20 illustrates the case of adsorbed co-ions. This layer's depth could be about a few angstroms with a small relative dielectric coefficient due to reduced polarization degrees of freedom.
  - The Outer Helmholtz Plane (OHP), where counter-ions are bound. Its depth is about 1 nm, and the dielectric constant is greater than that of the IHP.

In general, to observe charge neutrality of the system the following should hold:

$$\sigma_E = -\sigma_S - \sigma_D, \qquad (10.85)$$

where σ_E is the charge per unit of area of the electrode and $\sigma_S \sigma_D$ the Stern and diffuse layer's charge density, respectively. Often, the Stern layer's charge density is smaller than the others and so is neglected.

In general, the total charge density in the solution is given by the Boltzmann distribution as

$$\rho(x) = q \sum_i z_i n(x) = q \sum_i z_i n_{0i} e^{-\Phi(x) \frac{z_i q}{kT}}, \qquad (10.86)$$

where z_i is the valency and n_{0i} is the volume density of ith ion species, respectively. We can take the particular case of a charged sphere of radius a using spherical coordinates $x \leftarrow r$. At a great distance from the sphere, $r \gg a$ the electric potential is equal to that one generated from a point charge, then using the Poisson equation, we have

---

[12] *Adsorption* is a surface-based electrode process while *absorption* involves the volume of electrode's material.

$$\nabla^2\Phi(r) = -\frac{\rho(r)}{\varepsilon} \rightarrow \nabla^2\Phi(r) = -\frac{q}{\varepsilon}\sum_i n_{0i}z_i e^{-\Phi(r)\frac{z_i q}{kT}}, \qquad (10.87)$$

where $\Phi(r)$ is the unknown electrostatic potential to be found with respect to the distance r from the point of charge. Then, assuming a first-order approximation of the exponential expression[13]

$$\nabla^2\Phi(r) = -\sum_i \frac{qn_{0i}z_i}{\varepsilon} + \sum_i \frac{q^2 n_{0i}z_i^2}{\varepsilon kT}\Phi(r) = \sum_i \frac{q^2 n_{0i}z_i^2}{\varepsilon kT}\Phi(r), \qquad (10.88)$$

where the first addendum is equal to zero due to the neutrality condition of the solution. Therefore, the differential equation becomes a typical second-order expression:

$$\nabla^2\Phi(r) = \alpha^2\Phi(r) \text{ with } \alpha^2 = \sum_i \frac{q^2 n_{0i}z_i^2}{\varepsilon kT}, \qquad (10.89)$$

so that the electrostatic potential decays exponentially as

$$\Phi(r) = \Phi_0 e^{-\alpha r}, \qquad (10.90)$$

where Φ_0 is the potential for $r = 0$. This is the conclusion of the Debye–Hückel model theory.

For the particular case of symmetrical solutions, where $z = z^+ = z^-$, using (10.36) and (10.41), we have the following relationship

$$\alpha^2 = \frac{2q^2 n_0 z^2}{\varepsilon kT} = \rightarrow \frac{1}{\alpha} = \sqrt{\frac{\varepsilon kT}{2z^2 q^2 n_0}} = \sqrt{\frac{D\varepsilon}{\sigma}} = \sqrt{D\tau} = \lambda_D \equiv [\text{m}], \qquad (10.91)$$

where λ_D is called the *Debye length*. This quantity will be used as a reference for further expressions involving electric potential across electrodes. The Debye length sets a dimension beyond which the point charge electric field's effect is negligible. Note that τ is the same as in (10.63), which is thus related to the system's time to build the double layer.

Turning back to the first-order approximation that gave rise to this expression, we can see that this is valid if $r \ll \lambda_D$. This is true when the effective radius of the particle $a \ll \lambda_D$. This condition is called a *thick double layer*. Therefore, any ion is surrounded by an electrostatic exponentially decaying potential whose typical length is λ_D.

For a planar surface such as that illustrated in Fig. 10.20, the Debye–Hückel theory is no longer valid. More specifically, a planar electrode is a limiting case where the particle's radius $a \gg \lambda_D$. This is referred to as a *thin double layer*. For this case, an electric potential model was determined by Gouy and Chapman, where the potential profile is found to be

[13] $e^{-\Phi(r)\frac{z_i q}{kT}} \approx 1 - \frac{z_i q}{kT}\Phi(r).$

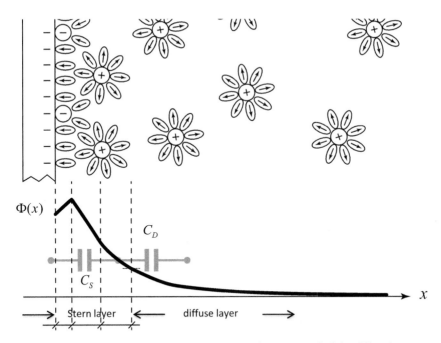

Figure 10.21 Representation of the Stern layer capacitance C_S and of the diffuse layer capacitance C_D. [Adapted from Morgan and Green, 2003]

$$\Phi(x) = \frac{2kT}{q} \ln\left(\frac{1 + \gamma e^{-\frac{x}{\lambda_D}}}{1 - \gamma e^{-\frac{x}{\lambda_D}}}\right) \approx \frac{4kT}{q} \gamma e^{-\frac{x}{\lambda_D}} \tag{10.92}$$

where $\gamma = \tanh(q\Phi_0/4kT)$,

where, similarly to Gouy–Chapman expression, the potential has an exponential decay by the Debye length factor.

For low potential values ($\Phi_0 < 50$ mV) (10.92) becomes an approximation of the Debye–Hückel equation (10.90) with the same exponential decay.

The Gouy–Chapman theory could be used to estimate the charge per unit of area of the diffuse layer for a planar electrode by using Gauss's law in the derivative of (10.92), yielding the Grahame equation

$$\sigma_D = -\frac{\varepsilon}{\lambda_D}\Phi_D \frac{\sinh\left(\frac{z_i q}{2kT}\Phi_D\right)}{\left(\frac{z_i q}{2kT}\Phi_D\right)} \approx -\frac{\varepsilon}{\lambda_D}\Phi_D, \tag{10.93}$$

where Φ_D is the *diffuse layer potential* as illustrated in Fig. 10.20. Often, the effect of the Stern layer is neglected so that the diffuse layer potential is equal to the electrode potential $\Phi_D \approx \Phi_0$.

So far, we have dealt with a time stationary situation. However, if we apply a variation of the electrode potential (e.g., a small-signal harmonic one), we have a corresponding change of the whole potential profile, thus a variation of the double-layer charge. We can estimate the charge variation ratio with that of the potential, which is also referred to as *differential capacitance*. Therefore, using the Grahame equation, we can find the value of the *diffuse layer differential capacitance*

$$C_D = -\frac{d\sigma_E}{d\Phi_0} \approx \frac{d\sigma_D}{d\Phi_D} = \frac{\varepsilon}{\lambda_D} \cosh\left(\frac{z_i q}{2kT}\Phi_D\right) \approx \frac{\varepsilon}{\lambda_D} \equiv \left[\frac{F}{m^2}\right]. \tag{10.94}$$

More easily, we can calculate the capacitances of the Stern layer since the potential variation is linear. Therefore, for the calculation of the total double-layer capacitance, we should consider the series of the capacitances of the different contributions as shown in Fig. 10.21:

$$\frac{1}{C_{DL}} = \frac{1}{C_D} + \frac{1}{C_{S1}} + \frac{1}{C_{S2}} = \frac{\varepsilon_1}{d_1} + \frac{\varepsilon_2}{d_2} + \frac{\varepsilon}{\lambda_D} \approx \frac{\varepsilon}{\lambda_D}, \tag{10.95}$$

where d_1 and d_2 are the widths of the IHP and OHP layers, respectively. The approximation is because usually $C_{S1}, C_{S2} >> C_D$.

Example For 1 mM KCl solutions we have $\lambda_D \approx 10$nm and $C_D \approx 70 \cdot 10^{-3}$ F/m² = 7.0 μF/cm² which is a quite high value, typically, double layer capacitance on the order of microfarads per square centimeter values.

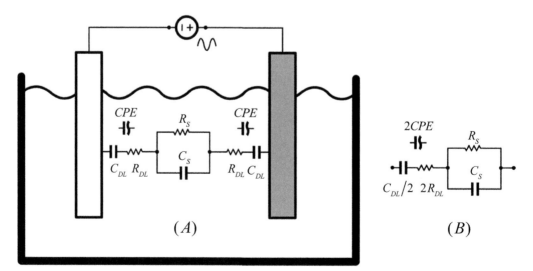

Figure 10.22 (A) Small-signal lumped model of a non-faradaic process. (B) Simplified equivalent lumped model.

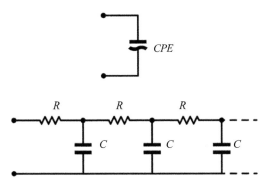

Figure 10.23 Representation of a constant phase element by an infinite network of RC circuits.

Unfortunately, the double-layer capacitance should include other effects. A first simplified model introduces a relaxation time at the interface by using a double-layer resistance, R_{DL} as shown in Fig. 10.1A, that becomes effective at low frequencies. At high frequencies, the electrode potential is acting as if it were applied directly to the solution. Thus, we can estimate the interface impedance by subtracting from the total impedance that one of the solution. In the case of equal electrode characteristics, we can simplify the model with the one of Fig. 10.1B. An important note is that in the solution model, the relaxation model of water molecules is *not* included, as in the case of Fig. 10.17, because it occurs at very high frequencies with respect to those taken into consideration in this model.

From the experimental point of view, the double layer impedance shows more complex behaviors than that one modeled by a simple resistor and capacitor in series. The behavior follows the model of a *constant phase element* (CPE), defined as

$$Z_{DL} = Z_{CPE} = \frac{A}{(j\omega)^\beta} = \frac{A}{\omega^\beta}\left[\cos\left(\frac{\pi}{2}\beta\right) - j\sin\left(\frac{\pi}{2}\beta\right)\right]; \quad \beta \in [0,1]$$
$$\beta = 1 \ Z_{CPE} \rightarrow C$$
$$\beta = 0 \ Z_{CPE} \rightarrow R.$$

(10.96)

At the boundaries of the range, when β has the value 0 or 1, the CPE behaves as the impedance of a resistor or a capacitor.

As known in circuit theory, a lumped linear circuit model is described by polynomial equations in the transform domain. Therefore, we cannot represent a CPE effect with a combination of resistors or capacitors in a lumped model. However, it could be shown that a CPE is described by an infinite ladder network of resistor-capacitor (RC) circuits, as shown in Fig. 10.23.

This means that, from the physical standpoint, we are in the presence of a distribution of relaxation times at the interface. Therefore, turning back to Fig. 10.22, we can substitute the double layer resistance and capacitance composition with a CPE element. The effects of the two interfaces could be simplified in the model by taking into account the sum of the two CPE impedances, as in the previous case.

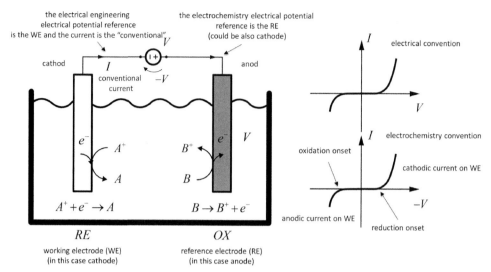

Figure 10.24 The electrolytic cell. A is reduced at the cathode, and B is reduced at the anode. Note that in this case, the cathode is the place where the electrons are supplied, in contrast with the galvanic cell, where instead they are gathered. As a consequence, for this cell, the electrical current is departing from the cathode.

It could be easily shown using (10.96) that[14]

$$\beta = -\frac{2}{\pi}\tan^{-1}\frac{\text{Im}\{Z\}}{\text{Re}\{Z\}} = -\frac{2}{\pi} \sphericalangle Z. \tag{10.97}$$

Note that for $\beta = 0.5 \rightarrow \sphericalangle Z = 45^0$ and $\beta = 1 \rightarrow \sphericalangle Z = 90^0$.

10.2.5   Faradaic Processes in Electrolytic Cells

Whenever charge transfer exists between one electrode and the solution due to a redox reaction, this is referred to as *faradic process*. In general, the interaction of two electrodes with a solution could be described as an *electrochemical cell,* as shown in Fig. 10.24. Electrochemical cells could be subdivided into two classes: (1) *galvanic cells,* where spontaneous redox reactions are used to perform electric work to external loads, and (2) *electrolytic cells,* where a driving electromotive force E is forcing a current into the cell where nonspontaneous reactions occur.

In the next sections of this chapter, we will mostly refer to electrolytic cells as the most frequent biosensing case. A half-cell reaction occurs to exchange electrons between the electrodes and the solution on each electrode. Half-reactions can be

---

[14] The calculation of β by means of (8.1) is valid only for a pure CPE. The relationship is no longer valid if the CPE is inserted in a complex network. In that case, the exponential parameter could be estimated in the range of frequencies where the CPE is dominant for the current/voltage relationship as $\beta = -(2/\pi)\tan^{-1}(\Delta\text{Im}\{Z\}/\Delta\text{Re}\{Z\})$.

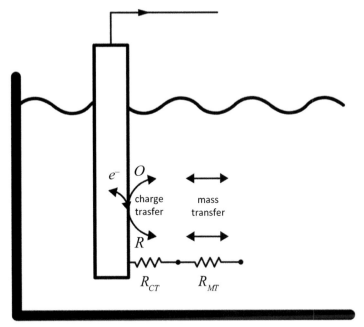

Figure 10.25 Charge transfer in electrochemical interfaces. [Adapted from Bard and Faulkner, 2001]

written to describe both the electrode undergoing oxidation and reduction. Referring to Fig. 10.24, we can summarize the behavior of the electrodes as

- *Cathode*: This is the electrode where electrons supported by the driving current combine with cations present in the solution to yield a neutral product. Therefore *reduction* $A^+ + e^- \rightarrow A$ occurs at the cathode, and species A is said to be "reduced" at the cathode. The cathode is characterized by electrons entering the electrode, referred to as cathodic current.
- *Anode:* This is the electrode where a species B in the solution is *oxidized* $B \rightarrow B^+ + e^-$, thus providing electrons to the electrode and cations to the solution. The species B is said to be "oxidized" at the anode. The anode is characterized by electrons flowing out of the electrode, referred to as anodic current.

Warning Figure 10.24 shows the most used convention in Electrochemistry (EC) and Electrical Engineering (EE). In EC, the electrical potential reference is given by the reference electrode (RE), and a positive current is considered that one (made of electrons) entering the cathode. Since the EE conventional current is made of positive charges, for electric power balance convention, it is better to use the working electrode (WE) as a reference and consider as positive the conventional current exiting from the cathode. This duality is evidenced in the two graphs of Fig. 10.24. Note that the WE could work as either cathode or anode in a full characterization of the $I(V)$ plot.

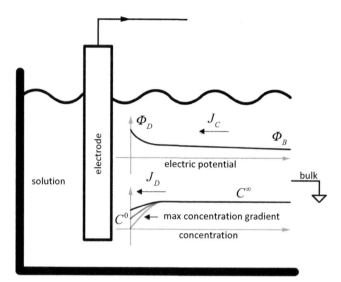

Figure 10.26 Electrode–solution potential and concentration in the presence of a faradaic process.

10.2.6   Charge and Mass Transfer Effects

The pathway of charges in the interactions with the electrode is shown in Fig. 10.25. We can devise two main classes of charged carriers to/from the electrode processes: (1) by means of electrons due to the redox, and in this case it is referred to as the *charge transfer effect* and (2) by means of ions, carrying a charge, referred to as the *mass transfer effect*.

We have seen that any time we have equilibrium between opposite rates, according to Boltzmann statistics, we have an electric potential. Therefore, any double arrow indicated in Fig. 10.25 is an exchange of rate between a difference of electric potential. Therefore, since the net charge rate is constant along the pathway, a lumped element model could describe the process by means of a resistor. We, therefore, have different resistance values for each process, as shown in Fig. 10.25.

Mass transfer in solution occurs in several forms, such as *migration (drift), diffusion, convection,* and *electroosmosis.* Other than migration, already discussed, we will refer solely to the effect of diffusion.

In the case of nonspontaneous redox (as in the electrolytic cell), if we apply an external potential signal to the electrode, we force the reaction to move away from equilibrium so that a current could be seen at the electrode. One task of this section is to model the electrolytic cell's effects with an electric lumped model, representing the complex impedance by making the ratio between the voltage signal and the current one.

When we have a faradaic process, the situation is illustrated in Fig. 10.26 and could be summarized in the following points:

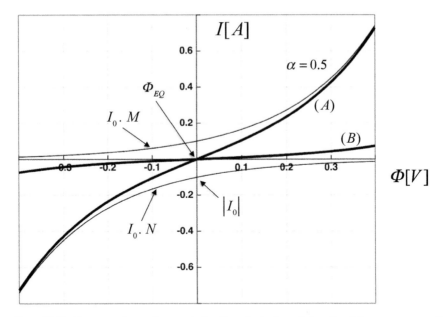

Figure 10.27 Current-voltage characteristic of an electrode–solution faradaic process according to (10.100) not considering the limiting current effect. (A) $I_0 = 1\ \mu A$. (B) $I_0 = 0.1\ \mu A$.

- The total static current is always given by the sum of the conduction and diffusion currents, and it is constant in any different section of the system: $I = A \cdot (J_C + J_D)$.
- Away from the electrode, the current is sustained by the electric field because the concentration of charged species is constant and equal to the bulk concentration C^∞.
- Close to the electrode, ions are reduced by the electrode, and thus their concentration presents a "depression" that creates a diffusion current, only partially compensated by the gradient of the electric potential in the Debye length. Therefore, the total current is mostly given by diffusion current at the interface.
- The gradient of the concentration in the electrode's proximity depends on the amount of the total current. The higher the current, the higher the gradient. However, the interface concentration C^0 could not be negative. Therefore, the condition $C^0 \approx 0$ sets the maximum current that the reaction could obtain and is referred to as *limiting current I_L*.

Away from thermodynamic equilibrium, using (10.13), (10.15), and (10.23) and some simple analysis of the reaction coordinate plot, it can be shown that

$$v_c = C_O k_0 e^{-\frac{q\alpha}{kT}\Delta\Phi'} \quad ; \quad v_a = C_R k_0 e^{\frac{q(1-\alpha)}{kT}\Delta\Phi'}, \tag{10.98}$$

where v_c and v_a are the cathodic and anodic velocities flows, $\alpha \in [0, 1]$ is called *charge transfer coefficient* that could be found from the slope of the chemical reaction plot (see Fig. 10.5), k_0 is a proportionality constant called the *standard rate constant*, and

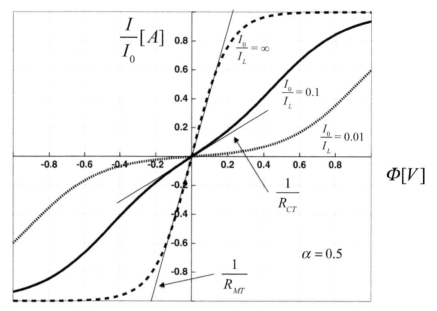

Figure 10.28 Current–voltage characteristic of an electrode–solution faradaic process considering the limiting effect of mass transfer due to diffusion, considering (10.101) for different values of exchange/limiting currents ratio.

$\Delta\Phi' = \Phi - \Phi_0$ is a variation of the electrode potential with respect to the standard potential of the cell. Here, we have used the electrical convention for currents.

Therefore, the total current will be related to the imbalance between the forward and backward velocity rates:

$$I = zFA[v_a - v_c] = zFAk_0 \left[C_Re^{\frac{q(1-\alpha)}{kT}\Delta\Phi'} - C_Oe^{-\frac{q\alpha}{kT}\Delta\Phi'} \right], \qquad (10.99)$$

which is referred to as the *current-overpotential equation* since $\Delta\Phi'$ is a variation with respect to the standard potential of the cell and z is the stoichiometric number of electrons involved in the electrode reaction $O + ze^- \leftrightarrow R$.

It can be shown that using as a reference, not the standard potential but the open-circuit potential $\Delta\Phi = \Phi - \Phi_{OC}$ (see Eq. (10.28) and the discussion of 10.1.3), the current-overpotential could be further simplified as

$$I = zFA[v_a - v_c] = I_0 \left[\frac{C_R^0}{C_R^\infty}e^{-\frac{q\alpha}{kT}\Delta\Phi} - \frac{C_O^0}{C_O^\infty}e^{\frac{q(1-\alpha)}{kT}\Delta\Phi} \right] = I_A + I_C, \qquad (10.100)$$

which is referred to as the *Butler–Volmer equation* where I_0 is the *exchange current* and I_A I_C are the anodic and the cathodic currents, respectively.

Since the concentrations at the interface could not be negative (see Fig. 10.26), there is an upper bound of the concentration gradient that limits the total current. Therefore, we can relate such a condition to the experimental *limiting currents* $I_{LO} I_{LR}$:

$$\frac{C_O^0}{C_O^\infty} = \left(1 - \frac{I}{I_{LO}}\right); \quad \frac{C_R^0}{C_R^\infty} = \left(1 - \frac{I}{I_{LR}}\right), \tag{10.101}$$

where clearly $C^0 = 0$ when $I = I_L$. Now, we can consider the current function of three variables, $I = I(C_O^0, C_R^0, \Phi)$, so using the calculus of variations,

$$\begin{aligned}
\frac{\Delta I}{I_0} &= \frac{dI}{dC_O^0}\bigg|_o \Delta C_O^0 + \frac{dI}{dC_R^0}\bigg|_o \Delta C_R^0 + \frac{dI}{d\Phi}\bigg|_o \Delta\Phi \\
&= \frac{\Delta C_R^0}{C_R^\infty} - \frac{\Delta C_O^0}{C_O^\infty} + \frac{q}{kT}\Delta\Phi
\end{aligned} \tag{10.102}$$

where the derivatives are calculated in the reference point $0 \equiv \{C_O^0 = C_O^\infty, C_R^0 = C_R^\infty, \Delta\Phi = 0\}$. By using the relationships of Eq. (10.101), Eq. (10.102) becomes

$$\frac{\Delta I}{I_0} = -\frac{\Delta I}{I_{LR}} + \frac{\Delta I}{I_{LO}} + \frac{q}{kT}\Delta\Phi. \tag{10.103}$$

For small variations of electrode potential, we have the following:

$$\frac{\Delta\Phi}{\Delta I} = \frac{kT}{q}\left[\frac{1}{I_0} + \frac{2}{I_L}\right] = R_{CT} + R_{MT}, \tag{10.104}$$

where we have used $I_{LR} = -I_{LO} = I_L$ for simplicity. Furthermore, as usual, $kT/q = RT/F$. It should not be surprising that Eq. (10.104) is similar to the small-signal resistance of a semiconductor junction since both are derived from Maxwell–Boltzmann thermodynamics.

It is interesting to plot (10.100) on the basis of the two components, as shown in Fig. 10.27. The exchange current is the intersection of the two components with the y-axis, and the greater it is, the higher the slope of the $I = F(\Phi)$ characteristic.

The plot of (10.100) coupled with (10.101) is shown in Fig. 10.28. Since the exchange current is related to the reaction's kinetics, the lower it is, the slower the charge exchange. Therefore, the higher should be the overpotential to achieve the same current. Note that the charge transfer resistance is the reciprocal of the slope of the characteristic in the origin, and in any case, it is limited by the mass transfer resistance. In general, we will assume charge transfer resistance greater than mass transfer resistance due to limiting current.

In conclusion, the potential–current plot could show important characteristics from which we could evaluate lumped model components such as mass transfer and charge transfer equivalent resistances.

10.2.7 Complex Effects of Diffusion

So far, we have dealt with mass transfer effect due to diffusion in its implication to the limiting current. However, there are other mass transfer effects due to the diffusion under modulation of the concentration profile. The goal is to derive an equivalent lumped electric model considering the diffusion profile effect. We have to start with the governing laws of diffusion and transform them into a sinusoidal regime.

Fick's second law of diffusion shows the evolution in time of a concentration profile $C = C(x,t)$

$$\frac{dC}{dt} = D\frac{d^2C}{dx^2} \text{ where } C = C(x,t). \tag{10.105}$$

Now, using a sinusoidal small-signal regime, the variable could be written as

$$\begin{aligned} C(x,t) &= \langle C(x,t)\rangle + c(x,t) \\ &= \langle C(x,t)\rangle + \mathrm{Re}\{c(x,j\omega)e^{j\omega t}\} \xrightarrow{F} C(x,j\omega), \end{aligned} \tag{10.106}$$

where $c(x,t)$ is the perturbation of the concentration profile and $C(x,j\omega)$ is the transform of the *time variation* of the concentration $c(x,t)$, that is, the amplitude of the oscillation around the mean value.[15]

To characterize an equivalent lumped model, we have to evaluate the ratio of two time-dependant variables: the electric potential and the corresponding current. First, we have that

$$\begin{aligned} \Phi(x,t) &= \frac{kT}{zq}\ln\frac{C(x,t)}{C^\infty} = \frac{kT}{zq}\ln\frac{C^\infty + c(x,t)}{C^\infty} \simeq \frac{kT}{zq}\frac{c(x,t)}{C^\infty} \\ &\to \Phi(x,j\omega) = \frac{kT}{zq}\frac{1}{C^\infty}C(x,j\omega). \end{aligned} \tag{10.107}$$

Second, we use the transform of the diffusion current (since it is the predominant contribution at the interface in the drift-diffusion model) were using (10.44),

$$\begin{aligned} I(x,t) &= -zFAD \cdot \frac{dC(x,t)}{dx} \\ &\to I(x,j\omega) = -zFAD \cdot \frac{dC(x,j\omega)}{dx} = -zFAD \cdot C'(x,j\omega), \end{aligned} \tag{10.108}$$

where $C'(j\omega)$ is the derivative of the concentration transform with respect to space.

Therefore, the impedance seen at the electrode could be found as the ratio of the electric potential with the corresponding current at the interface

$$Z(j\omega) = \frac{\Phi(0,j\omega)}{I(0,j\omega)} = -\frac{kT}{(z)^2qFADC^\infty} \cdot \frac{C(0,j\omega)}{C'(0,j\omega)}. \tag{10.109}$$

Then, applying Fourier's transform (on time) on both members of (10.105), we have

---

[15] In $C(x,j\omega)$ (transform of the variation of concentration with respect to time) we did not use the symbol Δ to avoid confusion with the variation with respect to space.

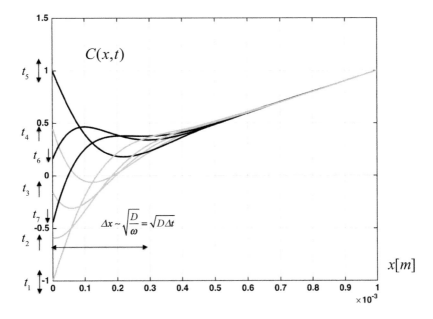

Figure 10.29 Behavior of the diffusion process on the concentration profile under sinusoidal perturbation at different time-lapses. The behavior has been derived from $\mathrm{Re}[C(x,j\omega)exp(j\omega t)]$, where $C(x,j\omega)$ is given by (10.112). The perturbation is superposed to a diffusion gradient, as shown in Fig. 10.26. $D = 7.93 \times 10^{-9}$ m^2/s (Ca++), $\omega = 0.1$ rad/s, M = 1. Note also that the profile of the concentration depends for the same $C(0,t)$ on the direction of motion (see arrows).

$$j\omega \cdot C(x,j\omega) = D\frac{d^2C(x,j\omega)}{dx^2} \rightarrow C(x,j\omega) = \frac{D}{j\omega}\frac{d^2C(x,j\omega)}{dx^2}, \qquad (10.110)$$

whose general solution is of the form

$$C(x,j\omega) = M\sinh\left(\sqrt{\frac{j\omega}{D}}x\right) + N\cosh\left(\sqrt{\frac{j\omega}{D}}x\right), \qquad (10.111)$$

where M and N are constants.

Now, in the case of an infinite bulk length, we have that the perturbation is negligible very far from the electrode; therefore, applying this boundary condition,

$$C(x,j\omega) \overset{x\rightarrow\infty}{\rightarrow} 0, \qquad \Leftrightarrow \qquad M = -N$$

$$C(x,j\omega) = Me^{-\sqrt{\frac{j\omega}{D}}x}. \qquad (10.112)$$

Therefore,

$$\frac{C(0,j\omega)}{C'(0,j\omega)} = -\sqrt{\frac{D}{j\omega}}. \tag{10.113}$$

To better understand the effect of the diffusion, in Fig. 10.29 is shown the variation of the concentration derived by the inverse transform of (10.112), which is the time response of the concentration profile to a sinusoidal perturbation exerted on the electrode. The aforementioned variation is superposed to the concentration gradient shown in Fig. 10.26, only in part shown in Fig. 10.29, which is asymptotically tending to the bulk one.

As shown, the diffusion process should cope with the angular velocity upon which the perturbation is exerted. The slower the perturbation, the deeper the modulation of the concentration. In any case, the balance of the diffusion process and the angular velocity changes the gradient of the profile, which is related to the interface's current. Therefore, understanding the sinusoidal regime relationship between the external potential and the resulting current helps us to model an equivalent circuit lumped element.

Considering the contribution of both oxidants and reductants, we have

$$
\begin{aligned}
Z_W(j\omega) &= -\frac{kT}{(z)^2 qFAD} \cdot \left[\frac{1}{C_O^\infty} \frac{C_O^\infty(0,j\omega)}{C_O^{\infty\prime}(0,j\omega)} + \frac{1}{C_R^\infty} \frac{C_R^\infty(0,j\omega)}{C_R^{\infty\prime}(0,j\omega)} \right] \\
&= \frac{kT}{(z)^2 qFAC_O^\infty} \frac{1}{\sqrt{j\omega D_O}} + \frac{kT}{(z)^2 qFAC_R^\infty} \frac{1}{\sqrt{j\omega D_R}} = \sigma\left(\frac{1}{\sqrt{\omega}} - j\frac{1}{\sqrt{\omega}} \right); \quad \text{where} \\
\sigma &= \frac{kT}{(z)^2 qFA\sqrt{2}(C_O^\infty\sqrt{D_O} + C_R^\infty\sqrt{D_R})},
\end{aligned}
$$

$$\tag{10.114}$$

where C^∞ refers to the bulk concentration of species, we have used the identity $\sqrt{j} = (1/\sqrt{2})(1+j)$ and $\sigma \equiv [\Omega \cdot s^{1/2}]$ is the *Warburg constant*.[16] Note that we have changed the sign of the reductant term because it has an opposite sign in the current direction.

Therefore, considering (10.104) and neglecting the mass-transfer contribution, the overall expression of the resistive path of the faradaic process is given by

$$R_{TOT} = R_{CT} + Z_W = \frac{kT}{qI_0} + \frac{\sigma\sqrt{2}}{\sqrt{j\omega}}. \tag{10.115}$$

The Warburg impedance is a particular case of CPE, where $\beta = 0.5$, however, it has an entirely different physical process.

It should also be observed that the diffusion Eq. (10.105) also describes the propagation of signals in a dissipative wireline as shown in Fig. 10.23:

---

[16] The Greek symbol sigma does *not* have the dimension of the conductance and should not be confused with it.

$$\frac{dV}{dt} = \frac{1}{RC}\frac{d^2V}{dx^2} \quad \text{where } V = V(x,t), \tag{10.116}$$

where $V(x,t)$ is the voltage signal on the wireline. For the aforementioned reason, the Warburg impedance (such as the CPE element) could not be modeled by a finite number of lumped circuit elements, because in that case, the impedance would be based on a polynomial function transform variables. Instead, any CPE elements could be modeled by an infinite number of circuital elements, as shown in Fig. 10.23.

Note that (10.112) is dependent on the square of the angular frequency. In fact, the mean squared displacement of a particle subject to Brownian motion is at the base of the diffusivity $\Delta x \sim \sqrt{D\Delta t} = \sqrt{D/\omega}$. Therefore, as shown in Fig. 10.29, the perturbation depth is related to the average time needed by particles to cover such a path according to Brownian motion.

10.2.8   Diffusion in Finite-Length Conditions

The effect of diffusion could be subject to different boundary conditions than those discussed so far, as shown in Fig. 10.30A, where only the concentration perturbation $c(x,t)$ is illustrated. More specifically, there are two interesting cases: one in which the diffusion perturbation is much greater than the electrode's distance, as shown in Fig. 10.30B. In this case, the concentration perturbation profile could be approximated at first order as a linear one since the counter electrode fixes the concentration on the other side. The latter behavior is typical in fuel cells and is called the finite-length Warburg (FLW) behavior. From the potential profile standpoint, it is apparent that the behavior is very similar to that of a resistor.

It can be shown that the general solution for the diffusion Eq. (10.105) in the case of a finite-length solution is

$$C(x,j\omega) = M\sinh\left(\sqrt{\frac{j\omega}{D}}(x-L)\right) + N\cosh\left(\sqrt{\frac{j\omega}{D}}(x-L)\right). \tag{10.117}$$

Therefore, since the length of the device is shorter than the diffusion length, we have

$$C(L,j\omega) = 0, \quad \Leftrightarrow \quad N = 0$$

$$C(x,j\omega) = M\sinh\left(\sqrt{\frac{j\omega}{D}}(x-L)\right) \rightarrow -\frac{C(0,j\omega)}{C(0,j\omega)} = \sqrt{\frac{D}{j\omega}}\sinh\left(\sqrt{\frac{j\omega}{D}}L\right). \tag{10.118}$$

Then by substituting this expression in (10.114), the FLW impedance expression is

$$Z_W(j\omega) = \frac{\sigma\sqrt{2}}{\sqrt{j\omega}}\tanh\left(\sqrt{\frac{j\omega}{D}}L\right). \tag{10.119}$$

A second interesting case is where the counter-electrode does not fix the concentration, as shown in Fig. 10.30C. This case is referred to as the finite-space Warburg (FSW)

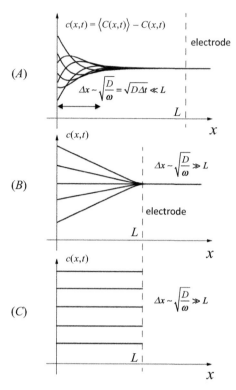

Figure 10.30 Boundary conditions in diffusion. (A) Semi-infinite diffusion. (B) Finite-length Warburg impedance (FLW). (C) Finite-space Warburg impedance behavior (FSW).

behavior and is typical for lithium ion batteries. As shown, there is no constraint on the concentration on the counter-electrode; therefore, there is no redox reaction. For this reason, it is also referred to as a *blocking* or *impermeable* electrode condition.

In this case, we have that the derivative of the concentration at the boundaries is zero:

$$C'(L, j\omega) = 0, \quad \Longleftrightarrow \quad M = 0$$

$$C(x, j\omega) = N \cosh\left(\sqrt{\frac{j\omega}{D}}(x - L)\right) \rightarrow -\frac{C(0, j\omega)}{C(0, j\omega)} = \sqrt{\frac{D}{j\omega}} \coth\left(\sqrt{\frac{j\omega}{D}}L\right). \quad (10.120)$$

Again, by substituting this expression in (10.114), the FSW impedance expression is

$$Z_W(j\omega) = \frac{\sigma\sqrt{2}}{\sqrt{j\omega}} \coth\left(\sqrt{\frac{j\omega}{D}}L\right). \quad (10.121)$$

From the potential profile standpoint, the behavior is very similar to that of a capacitor.

10.2.9   Putting It All Together: The Randles Model

If we consider both redox and charge accumulation processes at the interface, we can develop the *Randles model* of the ionic/electronic interface as shown in Fig. 10.31 that takes into account most of the physical phenomena previously discussed.

The Randles model comprises a charge transfer resistance R_{CT}, the solution resistance R_S, the double-layer capacitance C_{DL} that could also be represented by a CPE as discussed earlier, and the Warburg impedance modeling the diffusion process. This model could be composed together with the solution model as illustrated in the same figure.

10.2.10   Model Analysis by Cole–Cole and Bode Plots

We have already shown Cole–Cole plots in the case of complex permittivity. The typical plot is a parametric frequency curve of the real part of the impedance or admittance versus the same entity's imaginary part with a changed sign. In this respect, Cole–Cole plots are nothing other than a Nyquist plot with a reversed axis, and it is a graphical representation of sinusoidal signal relationships (in amplitude and phase or real and imaginary parts) together with others such as the Bode plots. We will illustrate some simple examples of Cole–Cole plots to understand their main characteristics better.

In the case of simple parallel and series RC circuits, we have the plots of Fig. 10.32, where the plots start with $\omega = 0$ and end with $\omega = \infty$. Note that the shapes of the curves are antisymmetrical in the Z and Y plots between the two topologies. Impedance and admittance Cole–Cole plots are also referred to as *Z-arc* and *Y-arc* plots, respectively.

In the case where we have series and parallel of two RC circuits (thus two-time constants), as shown in Fig. 10.33, the semicircle is shifted on the right by the series resistance amount R_1. This example is suitable for the non-faradaic model illustrated in Fig. 10.22.

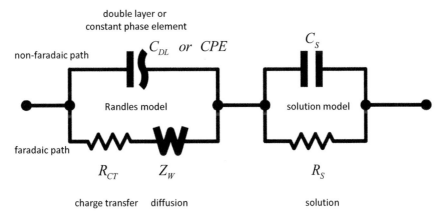

Figure 10.31   Randles model of the ionic/electronic interface and the solution model.

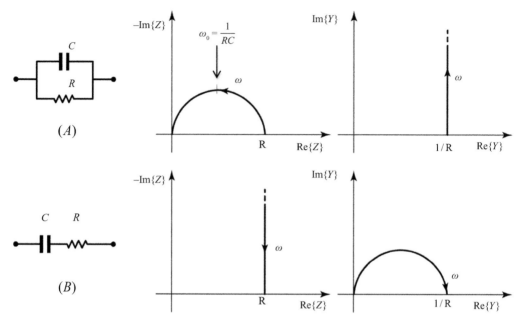

Figure 10.32 Cole and Cole Z and Y plots of a parallel RC circuit (A) and series RC circuit (B).

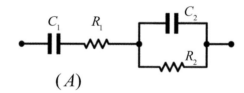

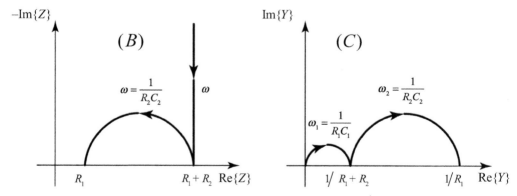

Figure 10.33 Cole–Cole plot of RC parallel circuit with a resistance in series. (A) Lumped model circuit. (B) Impedance Cole–Cole plot (Z-arc). (C) Admittance Cole–Cole plot (Y-arc).

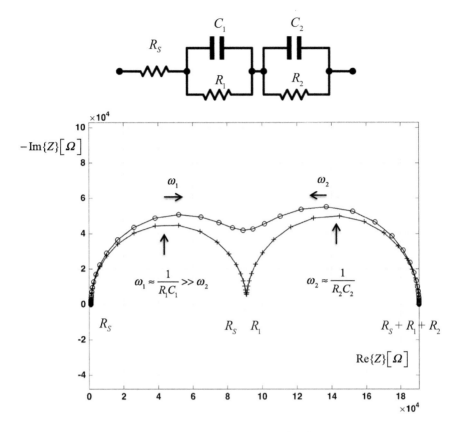

Figure 10.34 Impedance Cole–Cole plot of atwo-RC-parallel time-constant system. In one example, the poles are spaced apart 20 times and in the second 1,000 times.

As illustrated in Fig. 10.34, the impedance Cole-Cole plot is characterized by two semicircle bumps in the case of two parallel RC time-constants. First is shown the case where the two poles are highly separated so that the semicircles are well defined. Then is shown what happens whenever the two-time constants are fairly close together, where the two semicircles intersect.

An example of normalized Cole–Cole plots of a non-faradaic process such as that of Fig. 10.22 is shown in Fig. 10.35 for different conductivities of the solution. The semicircle is defined by the RC of the solution, while the interface defines the straight lines. The lower the slope of the lines, the lower the CPE's exponent. It is easy to show that for $\beta = 1$ the line is vertical (acting as a capacitor) and for $\beta = 1/2$ the slope of the line is 45° for axes of equal units. Most of the characteristics at low frequencies lie on the line, while only at very high frequencies we have points on the semicircle because the time constant of the solution is very small.

In the case of diffusion, it is useful to use the Randles model together with the solution model (Fig. 10.36A) previously introduced and shown in Fig. 10.31. In semi-infinite bulk, we have two semicircles, one for the solution (at very high frequencies) and another

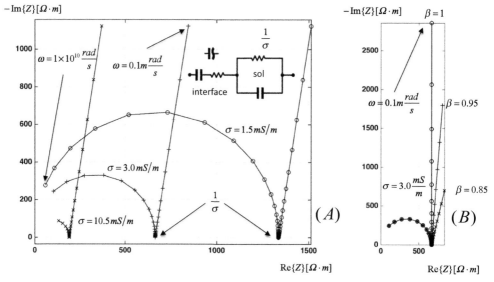

Figure 10.35 Normalized Cole–Cole plots of the non-faradaic model (blocking electrodes) of Fig. 10.22 using different KCl solutions (1E-4, 2E-4, and 7E-4 mol/L). (A) Variation of the solution conduction. (B) Variation of the CPE exponent. It should be pointed out that the maximum of the semicircle is achieved for very high frequencies (e.g., 1.9 GHz for 1.5 ms/m).

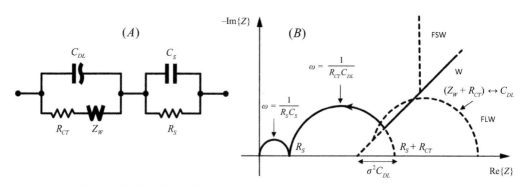

Figure 10.36 Cole–Cole plot of the Randles model.

for the time constant related to the double-layer capacitance and the charge transfer resistance. The boundaries of the semicircles with the real axis are 0, R_S, and $R_S + R_{CT}$, respectively, as in previous examples. At very low frequencies, diffusion comes into play, and the plot follows a straight line with a 45° slope since the Warburg element is a CPE impedance with $\beta = 1/2$. It is also easy to show that the intersect of the asymptotic line with the real axis is $\sigma^2 C_{DL}$ before the $R_S + R_{CT}$ point. If the diffusion process is intense, the straight line could be so shifted to the left to hide the semicircle completely.

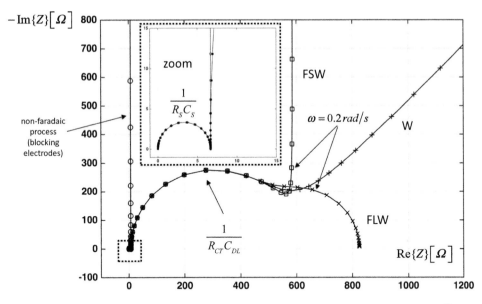

Figure 10.37 Simulated of the Randles model for a water solution with: $A = 1\,\text{cm}^2$, $L = 1\,\text{mm}$, $C = 20E-4$ mol/L, $C_{DL} = 70E-3$ F/m^2 $D = 2E-9$ m^2/s giving a value of Warburg constant $\sigma = 101.5\ \Omega \cdot \text{s}^{1/2}$. Then, to show the effects of finite-length conditions, the length has been restricted to 100 μm . The example is based on the aforementioned values for didactical purposes and could be far from experimental examples.

In the case of finite-length diffusion (FLW), the resistive behavior of the Warburg impedance couples with the charge transfer resistance to create with the double layer capacitance a time constant that could be identified in the Cole–Cole plot with a third semicircle at very low frequencies.

Finally, in the case of finite-space diffusion (FSW), the behavior is like a non-faradaic process (blocking electrodes), and at very low frequencies, the plot follows a straight line as in the capacitor case.

> **Warning** When the diffusion process is included in the model, we cannot normalize impedances with respect to the length and areas (as in Eqs. (10.82) and (10.83)), since Warburg impedances do not scale with respect to them.

A simulation example of the Cole–Cole plot for a Randles model is illustrated in Fig. 10.37, where all three cases are included. It is interesting to see that the FLW and FSW behaviors start when the diffusion length $\Delta x = \sqrt{D/\omega}$ is greater than that of the bulk (100 μm). The effects start at $\omega = 0.2$ rad/s when the diffusion length is $\Delta x \sim 100\ \mu$m.

Note that the semicircle due to the solution is very small compared to the semicircles due to other effects, which has been shown in an enlarged view.

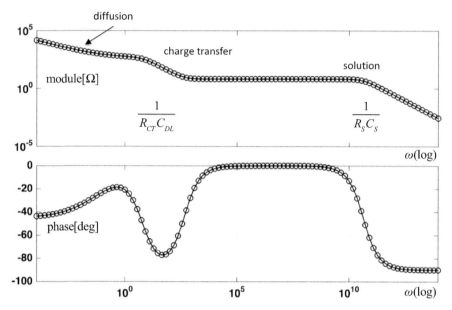

Figure 10.38 Bode plots of module and phase of the Randles model whose parameters have been used in Fig. 10.37.

It is interesting to see the same model by using Bode plots. Bode plots show the module (or magnitude) $|Z| = \sqrt{Re\{Z\}^2 + Im\{Z\}^2}$ and the impedance phase $\sphericalangle Z = \arctan\left(Im\{Z\}/Re\{Z\}\right)$ versus frequency.

Fig. 10.38 shows the Bode plot of the Randles model, whose Cole–Cole plot is illustrated in Fig. 10.37. For low frequencies, where diffusion dominates, the slope of the magnitude is $-1/2[\text{dec/dec}]$ (this means a half-decade of magnitude versus one decade of frequency) and starts from a phase set at $-45°$. Then, the magnitude falls with a slope of $-1[\text{dec/dec}]$ with the pole of the charge transfer. Finally, at very high frequencies, we have a unitary slope again due to the constant pole's solution time. The phase changes could be easily checked in the Cole–Cole plot.

A similar plot could illustrate the FSW and FLW behaviors, as shown in Fig. 10.39. As clearly shown, the main differences are at low frequencies where the FLW and FSW phases go to $-90°$ or $0°$, respectively. In this case, Cole–Cole plots are a better representation to show where the information resides. However, there is no general rule, and depending on the application, we can devise the best representation of the complex impedance.

The impedance spectroscopy could be very powerful to understand or detect even nanoscale processes. For example, a surface's nanometric porosity could dramatically change the diffusion process and could be easily detected by impedance spectroscopy.

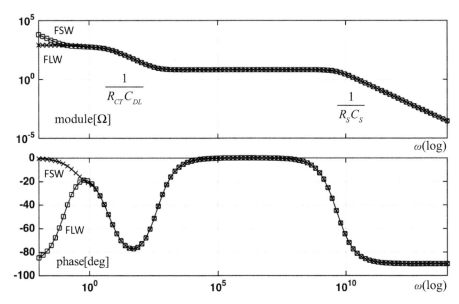

Figure 10.39 Bode plots of the finite length diffusion cases.

10.3 Biochemical Sensing

10.3.1 Basic Principles

Biosensing is one of the largest areas in sensors. Widely speaking, biosensors relate to the sensor technology aimed at measuring signals coming from life activity or products, and it is strictly related to biochemistry. Biosensors' stimuli could be physical properties of bodies and tissues, biochemical quantities and concentrations, cell or biomolecule counting, and other variables.

However, biosensors are transducers that provide quantitative information of biochemical analytes in a more restricted definition. Following our sensor taxonomy, based more on the interface architecture of sensing instead of the application, this section will be focused on electrochemical interfaces while other biosensing approaches, employing optical and mechanical transduction, will not be covered.

As discussed, electrical signals in biology are represented by ion transport because life is based on and has evolved from the water, where only ions could transport charges. Therefore, in electrochemical biosensors, the *electrode* will play a fundamental role as an interface between the ionic signaling of the biology environment with the solid-state, electrons-based electrical engineering and information technology platform illustrated in Fig. 10.40.

Signals are either variation of electric charges in electrodes given by a displacement of ions in the biophysical side and/or current to voltage relationship given by a faradaic process. We have two main cases: in the first one, the electronic interface just registers the ionic displacements by changes of current, voltages, or charges, as evidenced by

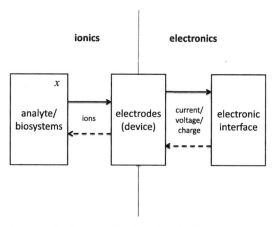

Figure 10.40 Concept of a (bio)chemical interface.

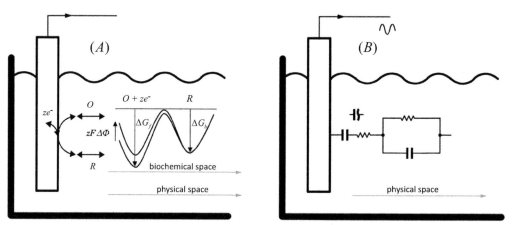

Figure 10.41 Schematic concepts of biosensing in the presence of faradaic (A) and non-faradaic processes (B).

the solid arrows of Fig. 10.40. For example, a biosensor that registers biopotentials (EEG, ECG) is based on this approach.

In the second case, the electronic interface induces electrical variation (represented by dashed arrows of Fig. 10.40) of the electrode versus time and detects the induced variations of voltages, currents, or charges. For example, chemical or biosensors based on impedance spectroscopy or voltammetry are based on this approach. The aforementioned cases could be applied for either faradaic or non-faradaic processes.

As discussed, the ion exchange could follow several schemes, as shown in Fig. 10.41. In the first case (Fig. 10.41A), a redox creates a direct link between the chemical space and the physical one. The relationship $\Delta G = -zF\Delta\Phi$ is the fundamental link between the two environments. This means that at any point of the space where the electric potential is defined, a change of the chemical variables is defined by means of the aforementioned relationship. In turn, this creates a variation of charge carriers'

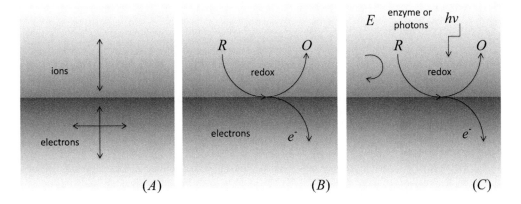

Figure 10.42 Ionic exchange processes at the (bio)chemical interface. (A) No exchange of charge between interface occurs. (B) A redox reaction realizes the exchange of charge between the interface. (C) Similar to (B); the redox is, however, activated by enzymes (E) or light energy (*hv*).

concentrations that are balanced in the drift-diffusion transport equations. There are two cases where we can implement (bio)sensing with the scheme of Fig. 10.41A:

- The redox is not spontaneous, but we can temporarily change the equilibrium by using an external potential. In this case (electrolytic cell), we should exert an external electric work to sense the properties of the reaction.
- Spontaneous redoxes set the electrode potential. In this case, we can monitor the properties of the reaction/process by an open circuit voltage sensing approach.

Of course, there are many possible combinations/variations of this sensing approach. When no redox is present, the interaction is essentially on the physical side, as shown in Fig. 10.41B. In this case, the interaction with the matter is essentially based on the polarization and conduction properties of the material.

In Fig. 10.42, we have schematized several processes that could be sensed at the interface:

- The stimulus (variable carrying the information) determines a displacement of ions in the ionic interface so that a corresponding displacement of electrons occurs on the electrode side. No exchange of charge exists between the two interfaces (Fig. 10.42A). The displacement of charges could be sensed as a change of capacitance or impedance. This approach could be extended in the case where the electrode is substituted by an electronic device (ISFET, nanowire, nanoribbon, nanotube) whose conductance (horizontal arrow of Fig. 10.42A) changes according to the number of ions in the proximity of the interface.

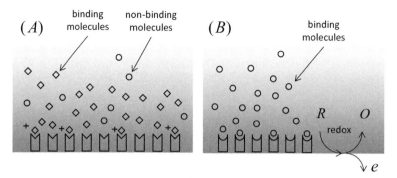

Figure 10.43 Surface functionalization approaches. (A) The electrode surface is functionalized by bioreceptors of target molecules. (B) The molecular site binding activates a redox reaction at the interface.

- The stimulus triggers a redox reaction at the interface, as shown in Fig. 10.42B. In this case, there is a charge exchange across the interface. Therefore, the stimulus linked to the reaction could be indirectly measured through electrodes.
- The redox is also regulated by a chain of several other chemical reactions, as shown in Fig. 10.42C. For example, the stimulus could trigger an enzyme-based redox. In other examples, the trigger of the redox is activated by external energy such as that coming from the light (photoelectrochemical sensing).

Another powerful approach in biosensing is based on the concept of *functionalization*. The surface of electrodes is covered by layers of molecules that are specific with respect to particular (bio)chemical molecular targets. As shown in Fig. 10.43A, such molecules (*binding sites* or *receptors*) have an *affinity* with respect to the target molecules of the stimulus. *Affinity* is a molecular characteristic in which the binding site has a complementary three-dimensional molecular structure with respect to the target, thus entering in chemical equilibrium with it. The molecular affinity is very important to achieve extremely high *specificity* (i.e., selectivity) in biosensing. The link of the target molecules with the binding sites induces charge displacements that the electronic interface could detect. In a more sophisticated approach (Fig. 10.43B), the electrode surface is functionalized as in the previous example; however, the molecular binding effect activates a redox reaction that the electronic interface could detect.

10.3.2   Electrode Polarization Approaches

As shown in Fig. 10.44A, the electrolyte–electrode interface could be described by a quasistatic characteristic, which could be nonlinear and described by the equation

$$F(V, I) = 0. \tag{10.122}$$

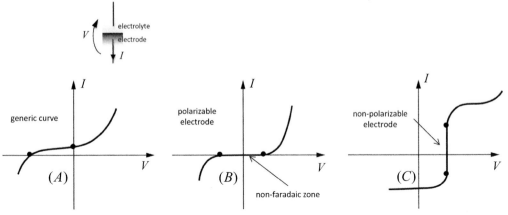

Figure 10.44 Quasistatic voltage–current relationship in the electrochemical interface versus a standard reference electrode. (A) Generic I–V relationship of the interface. The two points are called open-circuit and short-circuit bias points. (B) Characteristic of a polarizable electrode showing a non-faradaic behavior. (C) Example of a non polarizable electrode.

This is true only if we change the bias point using an extremely slow variation of the excitation versus time. For example, if we apply a very slowly varying voltage and measure the corresponding current, we get the plot Fig. 10.44A.

The evolution versus time of the bias point in the I–V plane is called *trajectory*. If the excitation is very slow, the trajectory follows the path indicated by the static characteristic (10.122) for both increasing and decreasing applied driving force variations. If the excitation is very rapid, the trajectory diverges from the static characteristic. If the excitation is cyclic, obtained using equal increasing and decreasing variations of the driving force, the trajectory follows a closed line called a limit cycle that is determined by the system's dynamic behavior. In a small-signal sinusoidal regime, the orbit is defined by time-dependent elements of the equivalent lumped circuit model.

The generic quasistatic characteristic of the interface shows two critical loci in the plot as shown in Fig. 10.44A: one is the *open-circuit point*, that is, the potential that the interface shows when no load is applied and thus no current is crossing the structure [Φ_{OC} of Eq. (10.28)]. The other point is the *short-circuit* one that could be located by the current flowing in the interface when it has been short-circuited.

An electrolytic cell curve could show a non-faradaic behavior for some parts of the curve, as shown in Fig. 10.44B. Therefore, in this zone, the electrode acts as a *polarizable electrode* where we need to fix the bias point using an external potential. In general, this region's size depends on the type of electrode material and the reaction. The region is due to the absence of a spontaneous reaction that could be activated only by operating on the electrode voltage. In the proximity of those points, the current starts an exponential behavior following the current-overpotential Eq. (10.99).

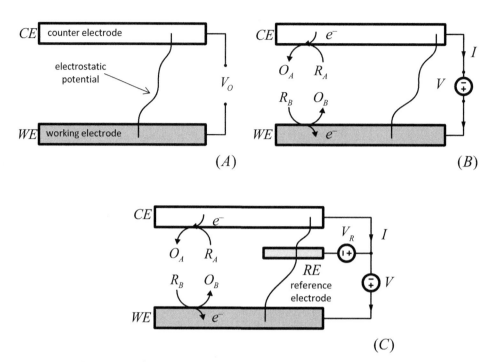

Figure 10.45 Electrode excitation. (A) Potentiometry. (B) Two-electrode voltammetry. (C) Three-electrode voltammetry.

There are cases where there is a spontaneous redox reaction at the electrode interface so that its potential is fixed by thermodynamic equilibrium. Such electrodes could be used as reference electrodes in biosensing, such as in the silver/chloride (Ag/AgCl) ones. Their typical characteristic is shown in Fig. 10.44C, where a part of the curve has a vertical section. This means that for any current included in the segment's range, the voltage of the electrode is fixed. This behavior is also referred to as a *nonpolarizable* electrode.

Now we will see possible electrode polarization approaches in (bio)chemical sensing.

In Fig. 10.45 are summarized three possible approaches:

- *Potentiometry* (Fig. 10.45A). The potential is sensed across the interface without any biasing. The change of the electrostatic potential induced by the stimulus (the variable we need to sense) is readout by a voltage amplifier. One electrode with known potential is called the *reference or counter electrode* (CE), while the electrode on which the reaction is observed is referred to as the *working electrode* (WE).
- *Two-electrode voltammetry* (Fig. 10.45B). In the case of a faradaic process, we are in the presence of redox reactions that fix the current regime. An external *driving potential E* is applied to the interface, and the corresponding interface current I is measured. The potential could be static or time variant. Again, the counter-electrode is characterized by a known potential using, for example, a nonpolarizable electrode.

- *Three-electrode voltammetry* (Fig. 10.45C). The problem with the two-electrode configuration is that the counter-electrode reaction also gives the current relationship. If the potential of the solution is not known, it could be problematic to reconstruct the characteristics of the WE. Thus, a *reference electrode* (RE) is introduced. The RE fixes the potential of the solution using a known redox potential. Therefore, the characteristic of the WE could be found disregarding the *I–V* relationship of the CE.

An electrical technique called potentiostat could better implement the three-electrode approach. Starting from the electrode scheme as shown in Fig. 10.45C, the reference electrode is inserted into a negative feedback loop of an operational amplifier (OPAMP), as shown in Fig. 10.45(B). An *RE* is usually a nonpolarizable electrode whose interface differential potential is well known, such as Ag/AgCl electrodes. The negative feedback acts so that the *RE* potential is equal to *E* using a virtual short-circuit approximation. Therefore, *V* changes the potential drop across *WE* irrespective of the potential drop across *CE*. Simultaneously, the current getting out of *CE* is equal to that entering the *WE*, since no current is flowing throughout *RE*.

The current variable could also be sensed at the *WE* by using a current-to-voltage converter (continuous-time approach) or a charge integrator (discrete-time approach), as shown in Fig. 10.46C. Finally, the *potentiostat* interface might be used to determine

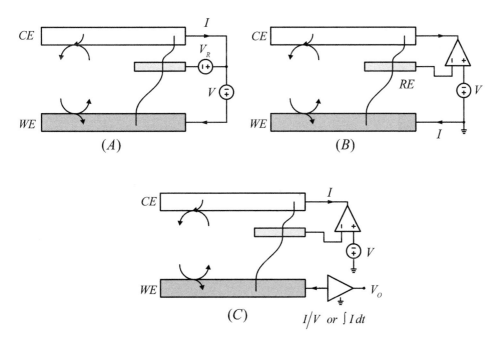

Figure 10.46 Potentiostat implementation of the three-electrode approach. (A) Base three-electrode scheme. (B) Reference electrode inserted into a negative feedback operational amplifier loop. (C) Potentiostat readout by current-to-voltage or current integration.

the quasistatic characteristic using a very slow variation of V or a time-response characteristic using a time-varying driving potential.

10.3.3 Applications of the Potentiostat

The *potentiostat* is a generic interface that could be used in several configurations and according to different working regimes to inspect (bio)chemical reactions.

The first application is for *impedance voltammetry*. The interface is excited around a bias point of the static characteristic using a sinusoidal voltage signal, and the current is monitored in the same regime as shown in Fig. 10.47. Of course, since the function could be nonlinear, the output is a distorted sinusoidal waveform, and the distortion depends on the size of the excitation and the bias point position on the characteristic. As shown in Fig. 10.47A, the signal is weakly distorted since the reference point lies in an almost linear part of the function. On the other side, Fig. 10.47B shows the case in which the waveform is highly distorted due to the input intensity and bias point position. The relationships could also be exchanged by using the current as excitation and sensing the potential drop; however, in this case, the potentiostat architecture should be modified.

A small harmonic excitation is used in the so-called small-signal regime according to the power series expansion approximation when it is desired to have a relationship free from harmonics due to distortion.

Using phasor representation, the current/voltage signals could be described as

$$V(t) = |V| \cos(\omega t + \varphi) = \text{Re}\{V(j\omega)e^{j\omega t}\}$$
$$I(t) = |I| \cos(\omega t) = \text{Re}\{I(j\omega)e^{j\omega t}\},$$

$$(10.123)$$

where ω is the angular frequency and φ is the phase of the sinusoidal waveform. The reference (in this case, the current) is usually assumed with zero phase, as shown in Fig. 10.48. Using the complex notation, the impedance is thus defined as

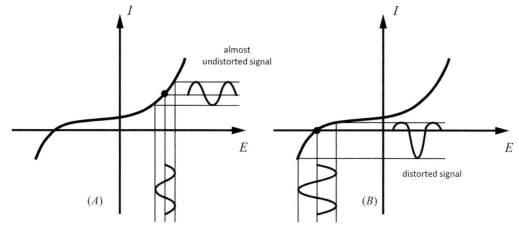

Figure 10.47 Sinusoidal regime for impedance spectroscopy. (A) Almost undistorted signal. (B) Distorted output signal.

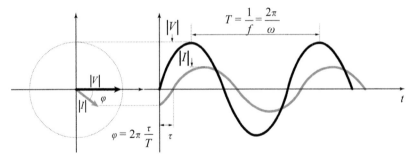

Figure 10.48 Sinusoidal relationships between voltage and current giving rise to the concept of impedance.

$$\left. \begin{array}{l} V(j\omega) = |V|e^{j\varphi} \\ I(j\omega) = |I|e^{j0} \end{array} \right\} \rightarrow Z(j\omega) = \frac{V}{I}(j\omega) = \frac{|V|}{|I|}e^{j\varphi}$$

$$= |Z|e^{j\sphericalangle Z} = \mathrm{Re}(Z) + j \cdot \mathrm{Im}(Z), \quad \text{and}$$

$$I(t) = |Z|\cos(\omega t + \sphericalangle Z), \tag{10.124}$$

where $\sphericalangle Z$ is the complex argument of the impedance.

With this setup, impedance voltammetry could be used to acquire the amplitude and phase versus frequency of sensed impedance that we have already referred to as *impedance spectroscopy* (IS). One of the scopes of IS is the characterization of the physical system with complex spectra and setup a possible electric circuit lumped model as already discussed. The circuit model could help determine the physical insight of the interface even at the nanometric level. The important point is that the magnitude and the phase relationships between current and voltage could be described by using topologically different but electrically equivalent models, as shown in Fig. 10.49D–F.

Another technique that could be used with a potentiostat is *cyclic voltammetry*, which is shown in Fig. 10.50. Different from impedance voltammetry, where a linear model of the interface is assumed, cyclic voltammetry is very powerful even for strongly nonlinear systems. It is based on determining the temporal characteristic of the current–voltage relationship by applying a voltage transient at the interface.

If the excitation is very slow, the *trajectory* of the I–V couple follows the "static characteristic" in both increasing and decreasing increments. However, if we apply a fast-changing variation, the I–V relationship's trajectory diverges from the static characteristic. For example, if we apply a voltage ramp to an RC model as shown in Fig. 10.50A, the corresponding current is shown in Fig. 10.50B. Note that the faster the current response, the steeper the voltage ramp in both the ascending and descending phase. If we collect the I and V response in a unique plot, as shown in Fig. 10.50C, we get several *orbits* characteristic of the physical process.

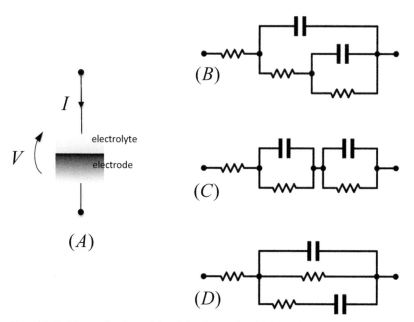

Figure 10.49 Linear circuit models of the electrochemical interface. (A) Physical representation of the interface. (B–D) Equivalent linear circuit models of the same current-voltage relationships.

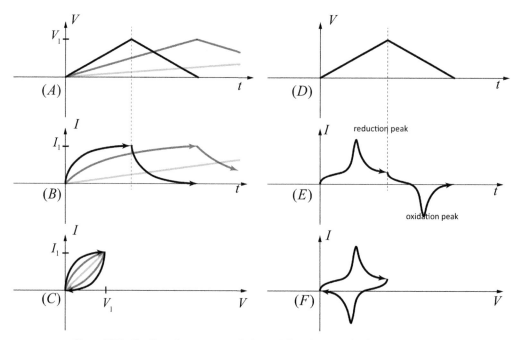

Figure 10.50 Cyclic voltammetry technique. (A) Voltage excitation versus time. (B) Example of the current evolution in the case of an RC circuit versus time. (C) Current–voltage cycle versus time in the same case of an RC circuit. (D–F) Typical voltage excitation, current response, and cyclic voltammetry of a faradaic process interface.

If we apply the same approach to a real electrochemical interface, we get the typical aspect ratio as in Fig. 10.50D–F, where the peaks are referred to as the *reduction peak* and *oxidation peak*.

To summarize, cyclic voltammetry is a very powerful technique because it considers even nonlinear effects due to large signal excitations that are not detectable with IS.

10.4 Biosensing for Electrophysiology

Electrophysiology is the branch of physiology studying the exchanges of ions in biological tissues and recording the corresponding signals. There are essentially two main electrophysiology techniques: the first is related to the in vivo recordings of biopotentials in living organisms such as the human body. The second is intracellular electric activity recordings in single cells or even single membrane ion channels performed with in vitro setups.

10.4.1 Biopotential Sensors

Bioelectric potentials are characterized by displacements of ions or charged molecules associated with biochemical reactions or with physiological phenomena of the body.

The transducing function is carried out by electrodes that consist of electric conductors in contact or in proximity to the body's tissues. There are two main electrode sensing approaches: the first is with skin coupling using *noninvasive electrodes* and the second with *invasive electrodes*. In noninvasive electrodes, there are several arrangements; among these is one with *wet contact* where the electrode is coupled with a gel to enhance the charge exchange between the interfaces. The other is employing a *dry contact* where the electrode interacts with the biological counterpart with capacitive coupling. On the other hand, the coupling is established using electrodes or needles implanted in the body to contact the tissue of interest directly.

The main bioelectric signals sensed by biopotential electrodes are

- *Electrocardiogram* (ECG), which is the procedure of recording the electric activity of the heart versus time using electrodes placed on a patient's body. The electrodes sense the tiny electrical changes on the skin arising from the heart muscle depolarizing during each heartbeat.
- *Electroencephalography* (EEG), which is characterized by monitoring the electrical activity of the brain. The EEG senses potential fluctuations resulting from ionic currents within the neurons of the brain. It is usually noninvasive, using electrodes placed along the scalp, although invasive electrodes are used in some applications.
- *Electromyography* (EMG) detects the electrical potential generated by muscle cells when they are electrically or neurologically activated. It is an electrodiagnostic approach for evaluating and recording the electrical activity produced by skeletal muscles.

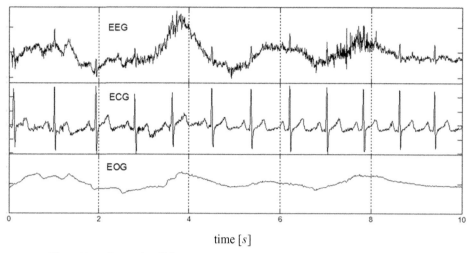

time [*s*]

Figure 10.51 Example of biopotential waveforms for EEG, ECG, and EOG signals . [From M. A. Garces and E. Laciar, Noise removal from EEG signals in polysomnographic records applying adaptive filters in cascade. In *Adaptive Filtering Applications*, 2011, pp. 173–196. Under CC license.]

- *Electrooculography* (EOG) is characterized by measuring the corneo-retinal potentials between the front and the back of the human eye. The applications are in the diagnosis of ophthalmological diseases and in recording eye movements.
- A *local field potential* (LFP) is a signal generated by the electric current flowing from neurons in a small nervous tissue volume. A potential is produced across the local extracellular space by action potentials varying due to synaptic activity. The potential is referred to a voltage recorded with a microelectrode embedded in vivo in a neuronal tissue or within brain tissue maintained in vitro.

Examples of biopotential waveforms are shown in Fig. 10.51, where the concurrent acquisition of signals from the same patient is taken.

The biopotential signal bandwidths and amplitude ranges are illustrated in Fig. 10.52 for both noninvasive and invasive electrodes. Note how the bandwidth ranges of EEG and ECG are mostly overlapped, and therefore, the ECG signal could interfere with EEG ones, as also clearly shown in Fig. 10.51. It is therefore imperative to remove ECG interference from an EEG. Similarly, interference of ECG over EMG could be found.

Another important problem in acquiring biopotentials is related to the interference with 50/60 Hz AC power lines due to capacitive coupling with the human body, as shown in Fig. 10.53. Even if the sought signal is the difference between V1 and V2, we can get a huge interference by AC coupling on both signals that could easily saturate the sensing amplifiers.

We can better see the problem using a common-mode/differential mode vector representation. Since the signals are taken at different points of the body, they could be

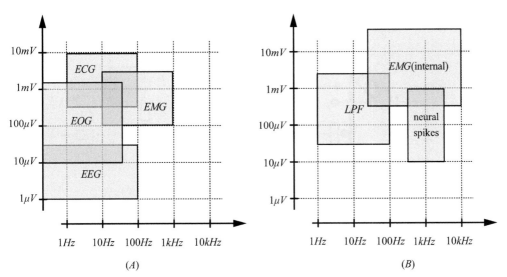

Figure 10.52 Biopotential signals bandwidth and strength for (A) noninvasive electrodes and (B) invasive electrodes.

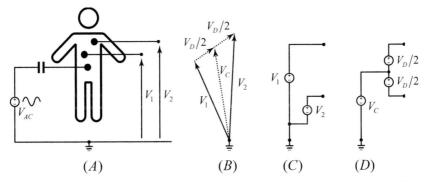

Figure 10.53 The problem of AC electrical power interference and common mode (A). Relationship between common and differential signals (B). Model of original signals (C) and their mapping into common and differential signals (D).

affected by different phase and amplitude with respect to each other, as shown in the vector representation of Fig. 10.53B. We can thus define the two signals:

$$\overline{V_D} = \overline{V_2} - \overline{V_1};$$
$$\overline{V_C} = \frac{\overline{V_1} + \overline{V_2}}{2},$$

(10.125)

where $\overline{V_C}$ is called *common-mode signal* and $\overline{V_D}$ is the *differential mode signal* that should be sensed. Since the common mode is huge due to the interference, the interface should have a high gain regarding the differential signal and be very insensitive to the common-mode one. A possible approach to reduce the common-mode interference is

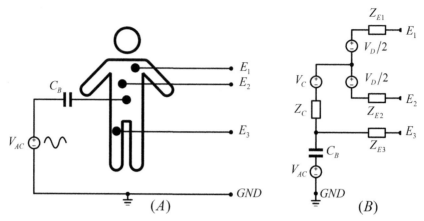

Figure 10.54 Electrode connection approach with "driven right leg" electrode (A) and related electric model.

shown in Fig. 10.54A, which consists of using an intermediate reference electrode E3 attached to the body called the *driven right leg (DRL) electrode*. Even if the reference ground is now the body, there could still be a common-mode signal due to biopotentials, such as the ECG signal interfering with the EEG (or EMG) one, as already discussed. The equivalent model is shown in Fig. 10.54B, where Z_{E1}, Z_{E2}, and Z_{E3} are the electric models of the skin/electrode contact. In other terms, even if the effects of the interference V_{AC} are reduced, there could still exist common-mode interferences from biopotentials that are modeled by the V_C generator of Fig. 10.54B.

To better understand the electric behavior of the skin/electrode contact, several electric lumped models are represented in Fig. 10.55. Figure 10.55A shows a cross-section of the skin where a parallel-connected RC configuration models each layer as in Fig. 10.55B. The deeper the layer, the lower C and R. The contact electrode is coupled with the tissue in several options. Historically, the most common connection is made by a faradaic interface made of an Ag/AgCl electrode, sometimes using a conductive gel to enhance the connection. Every time different materials are coupled together by a redox reaction, a *half-cell* potential V_{HC} is established across the interface, as shown in the electric model. Using a dominant time-constant approximation, represented by a single RC circuit, makes a useful simplification of the model, as shown in Fig. 10.55C. In this case, the behavior of the electrode is given by the impedance

$$Z_E(s) = R_B + \frac{R}{1 + sRC} \rightarrow z_1 = (R_B//R) \cdot C; \; p_1 = R \cdot C, \quad (10.126)$$

where z_1 and p_1 are the zero and the pole of the complex function. Therefore, the impedance module starts from $R_B + R$ and decreases down to R in the frequency range between $\omega_p = 1/|p_1|$ and $\omega_z = 1/|z_1|$.

Recent trends show the adoption of insulated or noncontact electrodes in places of contact electrodes where the model is illustrated in Fig. 10.55D, in which the electrode capacitor substitutes the half-cell potential.

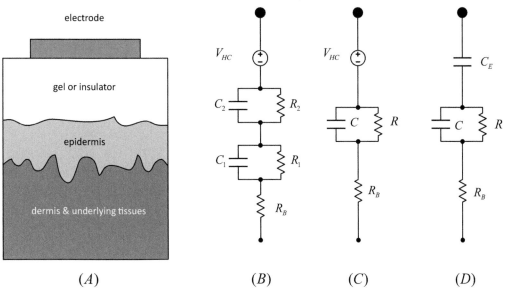

Figure 10.55   Models of the electrode/skin impedance. (A) Cross-section of the electrode/skin contact. (B) Model of the contact considering the different layers of the tissue. (C) Simplified model with dominant time constant and wet contact. (D) Model for a dry/insulated electrode.

The interaction of the body model with an instrumentation amplifier is illustrated in Fig. 10.56. The output of the symmetrical amplifier could be expressed in terms of differential and common-mode signals as

$$V_O = A_D V_D + A_C V_C, \tag{10.127}$$

so that we can define the *common mode rejection ratio*, which is a characteristic of the electronic interface, as

$$CMRR = \frac{A_D}{A_C}, \tag{10.128}$$

where the greater the CMRR, the better it is. However, suppose the input resistance of the differential amplifier is finite, as illustrated in Fig. 10.56. In that case, we have to deal with another source of imperfection called *imbalance CMRR,* not due to the intrinsic characteristics of the amplifier but to boundary operating conditions such as the difference of the impedances of the electrodes. It could be easily found that the imbalance CMRR is given by

$$\mathrm{CMRR_{IMB}} = \frac{2R_D}{\Delta Z_E}; \ \Delta Z_E = Z_{E1} - Z_{E2} \tag{10.129}$$

and that the total CMRR is given by

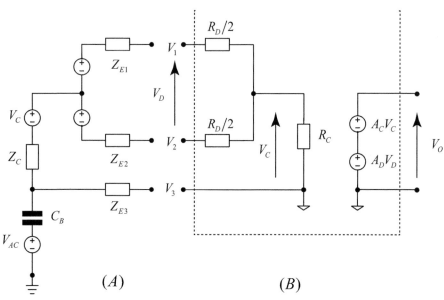

Figure 10.56 Interaction of the body model (A) with a differential instrumentation amplifier (B). The contact of the ground with the V3 point is called a "driven right leg" (DRL) electrode.

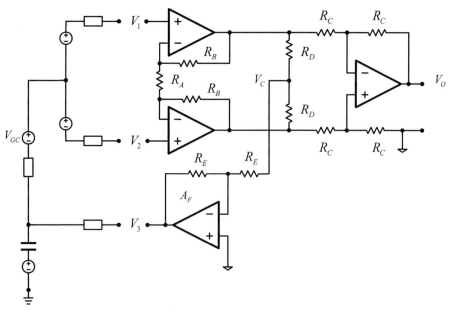

Figure 10.57 Implementation of the "driven right leg" electrode with an instrumentation amplifier.

$$\frac{1}{\text{CMRR}_{\text{TOT}}} = \frac{1}{\text{CMRR}} + \frac{1}{\text{CMRR}_{\text{IMB}}}, \qquad (10.130)$$

which shows how the prevailing CMRR is the lowest one between the two.

From Eq. (10.129), it is apparent that the imbalance CMRR could be greatly reduced by increasing input impedance as in CMOS technology, even if it offers a greater noise compared to BJT or JFET technology.

A common architecture of instrumentation amplifiers for biopotentials is shown in Fig. 10.57, as we have already discussed in Chapter 8. An important improvement of the CMRR could be obtained using an *actively decreasing common voltage technique* where an auxiliary amplifier of gain A_F realizes negative feedback for the common voltage. Using this technique, for high input impedance conditions, we have

$$-A_F V_C + V_{GC} = A_C \rightarrow V_C = \frac{V_{GC}}{1 + A_F}, \qquad (10.131)$$

thus greatly decreasing the effect of V_{GC} (for instance, ECG on EEG signals) on the common voltage effect.

The noise model of the electrode can be calculated as in Fig. 10.58, where $\overline{v_E^2}$ is the electrode contact noise (usually modeled by a $1/f^{\alpha}$ PSD originated from biochemical processes) $\overline{v_N^2}$ and $\overline{i_N^2}$ are the input-referred voltage and current noise of the amplifier, respectively. Therefore, the electrode-referred noise for an infinite input impedance of the amplifier is given by

$$\overline{v_{NE}^2}(\omega) = \overline{v_E^2} + \overline{v_N^2} + \left(\overline{i_N^2} + \frac{4kT}{R_E}\right)\frac{R_E^2}{1 + \omega^2 R_E^2 C_E^2}. \qquad (10.132)$$

It is interesting to note that, for a very low noise input amplifier,

$$\overline{v_{NE}^2}(\omega) \approx \frac{4kTR_E}{1 + \omega^2 R_E^2 C_E^2}, \qquad (10.133)$$

meaning that we can reduce the input noise as much as possible in the two extreme cases:

- In very low contact resistance, for contact sensing
- In very high contact capacitance, for contactless sensing

Unfortunately, the scheme shown in Fig. 10.58 is very much affected by external interferences. For this reason, an active shield could be implemented, as illustrated in Fig. 10.59. In this approach, the output drives with positive gain a shield so that the effect of perturbations on input by capacitive coupling is reduced. The mechanism acts as a negative Miller's capacitance effect on the input using a noninverting amplified by a gain A_0. Referring to the scheme of Fig. 10.59B, which is the circuital representation of Fig. 10.59A, the signal transfer function $H_S(j\omega)$ is

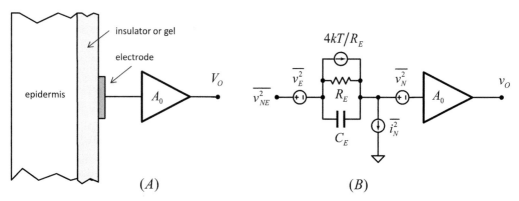

Figure 10.58 Active electrode configuration (A) and equivalent noise model (B).

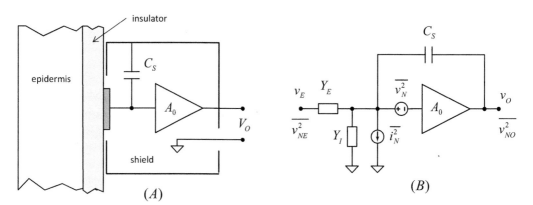

Figure 10.59 Actively shielded electrode configuration (A) and noise model (B).

$$H_S(j\omega) = \frac{v_O(j\omega)}{v_E(j\omega)} = A_0 \frac{Y_E(j\omega)}{Y_E(j\omega) + Y_I(j\omega) + j\omega(1 - A_0)C_S}$$

$$= A_0 \frac{G_E + j\omega C_C}{G_E + G_I + j\omega[C_E + C_I + (1 - A_0)C_S]}, \quad (10.134)$$

where $Y_E = G_E + j\omega C_E$ is the skin-electrode admittance and $Y_I = G_I + j\omega C_I$ is the amplifier input admittance. Note that the transfer function starts with a gain $G_E/(G_E + G_I)$ at low frequencies and rises to $C_E/(C_E + C_I + (1 - A_0)C_S)$ at high frequencies after one zero and one pole. Therefore, any changes of Y_E, which is the main effect of the interferences and offsets, could be neglected if

$$Y_I(j\omega) + j\omega(1 - A_0)C_S = 0 \rightarrow G_I + j\omega C_I = j\omega(A_0 - 1)C_S$$

$$\Leftrightarrow \begin{cases} G_I = 0 \\ C_I = (A_0 - 1)C_S. \end{cases} \quad (10.135)$$

The first condition of zero input conductance could be easily implemented with complementary metal–oxide–semiconductor (CMOS) technology, while the second condition could be realized under the knowledge of C_I which is not always ensured. If the condition is satisfied, the gain is real and equal to $G_C/(G_C + G_I)$. A possible approach consists of making a precise tuning of the gain to neglect the stray input capacitance. Another one is to set the $A_0 = 1$ (i.e., active shield with a buffer) and keep C_I as low as possible.

From the noise standpoint, we can see that, referring to Fig. 10.59B, the input-referred noise $\overline{v_{NE}^2}$ can be found as

$$
v_{NE} = \frac{Y_E(j\omega) + Y_I(j\omega) + j\omega C_S}{Y_E(j\omega)} v_N + \frac{1}{Y_E(j\omega)} i_N + \frac{1}{Y_I} i_{NR_I} + \frac{1}{Y_E} i_{NR_E}
$$

$$
= \frac{1/R_E + 1/R_I + j\omega(C_S + C_I + C_E)}{1/R_E + j\omega C_C} v_N + \frac{1}{1/R_E + j\omega C_E} i_N
$$

$$
+ \frac{1}{1/R_I + j\omega C_I} i_{NR_I} + \frac{1}{1/R_E + j\omega C_E} i_{NR_E}
$$

$$\tag{10.136}$$

$$
\rightarrow \overline{v_{NE}^2} = \frac{(1/R_E + 1/R_I)^2 + \omega^2(C_S + C_I + C_E)^2}{1/R_E^2 + \omega^2 C^2_E} \overline{v_N^2} + \frac{1}{1/R_E^2 + \omega^2 C^2_E} \overline{i_N^2}
$$

$$
+ \frac{4kT}{R_I} \frac{1}{1/R_I^2 + \omega^2 C_I^2} + \frac{4kT}{R_E} \frac{1}{1/R_E^2 + \omega^2 C_E^2},
$$

where we have not considered the noise of the source $\overline{v_E^2}$.

Now, we can make some conclusions for limiting cases. For very high input impedance (such as in CMOS technology) and contactless sensing, we have $R_I \gg 1$ and C_I includes the *stray capacitances* (i.e., the parasitic capacitances with respect to the reference of all the connections of the source to the input of the amplifier). We have

$$
\overline{v_{NE}^2} = \frac{1/R_E^2 + \omega^2(C_S + C_I + C_E)^2}{1/R_E^2 + \omega^2 C^2_E} \overline{v_N^2} + \frac{1}{1/R_E^2 + \omega^2 C^2_E} \overline{i_N^2}
$$

$$
+ \frac{4kT}{R_E} \frac{1}{1/R_E^2 + \omega^2 C_E^2}.
$$

$$\tag{10.137}$$

Therefore, for contactless high impedance electrodes, the voltage amplifier noise is amplified only by a $1 + (C_S + C_I)/C_E$ factor; thus, the most dominant noise contribution is given by current noise and noise from R_E. For CMOS low-impedance contact sensor, we have $R_E \ll 1$ and, as shown in (10.137), the noise is dominated by amplifier input noise. Finally, if the sensing is performed by a very low noise amplifier, (10.137) converges to (10.133) with the related conclusions.

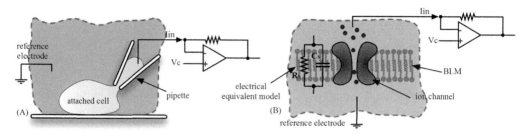

Figure 10.60 (a) Diagram showing the patch-clamp technique where a glass pipette is used to pull a patch of the cell membrane. A low-noise transimpedance amplifier measures channel currents. (B) The planar bilayer membrane (BLM) technique where a suspended lipid bilayer contains an ion channel. Again, the current is read by a low-noise transimpedance amplifier. The picture also shows the electric equivalent model of the BLM, consisting of a high-value resistor (of the order of GΩ or greater) in parallel with a capacitance C_S. [Adapted from M. Crescentini, M. Bennati, S. C. Saha, J. Ivica, M. R. R. de Planque, H. Morgan, and M. Tartagni, A low-noise transimpedance amplifier for BLM-based Ion Channel, *Sensors*, pp. 1–20, 2016]

10.4.2   Biosensing for Intracellular Recordings

Ion channels are nanoscale pores existing in a biological cell membrane, allowing communication between the inner and outer environments of the cell through *ionic currents*. The open/close behavior of the ion channel is modulated through different mechanisms, for example, voltage, ligand binding, pH change, or mechanical strain. Ion channels are crucial for the physiological control of living organisms, and their malfunctions are at the base of a variety of pathologies and diseases. Ion channel recording is an important component for drug discovery, DNA sequencing, and single-molecule detection. There are two main techniques for ion channel screening:

1. Patch-clamp, where a glass pipette or a microaperture in a solid-state device is used to pull a patch of the cell membrane (Fig. 10.60)
2. Planar bilayer lipid membranes (BLMs), where a single ion channel is inserted into a lipid bilayer suspended over a microaperture.

The patch-clamp technique is widely used for the study of membrane ion-channels. The advantages of patch-clamp are high fidelity, since the ion channels operate in their native physiological environment, together with a high level of automation and parallelization achieved by nowadays technology. This technique suffers from low specificity and high noise since several different ion channels are measured together, and the membrane provides a large capacitance. On the contrary, the BLM technique provides excellent electric sealing and high sensitivity of detection, down to a single molecule, with minimum noise and capacitance.

Specific requirements are mainly related to the kind of ion channel under investigation. For instance, potassium ion channels, such as KcsA, have fast responses (~100 μs) and zero-voltage conductivity lower than 100 pS, resulting in currents of the order of a few pA with applied voltages lower than 100 mV. In general, an ion channel has (1) very high output impedance (from 1 to 100 GΩ), (2) noise level smaller than 1 pA rms at

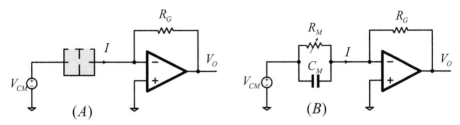

Figure 10.61 (A) Membrane current readout. (B) Equivalent circuit.

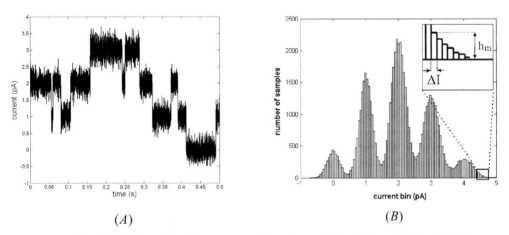

Figure 10.62 Examples of current recordings of gramicidin-A channels (A) and relative histograms (B).

1 kHz, (3) open/close events ranging from few milliseconds to hundreds of microseconds, and (4) capacitance of the order of tens of pF. Therefore, the main requirements for the electronic interfaces are noise floor lower than 10 fA/√Hz, high sensitivity (transresistance > 1 GΩ), and bandwidths higher than 10 kHz.

The natural or artificial membrane has ion channels that could be opened or closed to ion current according to several physiological or pathological causes. Therefore, the membrane could be modeled as in Fig. 10.61, where the capacitance represents the membrane capacitance while the variable resistance models the opening/closing of one or more ion channels. The main task is to record the current versus time to acquire data to be analyzed. As already discussed, for output voltages in the hundreds of mV and input full scale is in the order tens of pA, the feedback resistance should be on the order of GΩs. This is the reason why these amplifiers are frequently referred to as "GigaOhm current-to-voltage amplifies."

An example of signal recording is shown in Fig. 10.62. The level of current, as shown in Fig. 10.62A, depends on the number of single ion channels that are concurrently opened. The histogram of the states is shown in Fig. 10.63B.

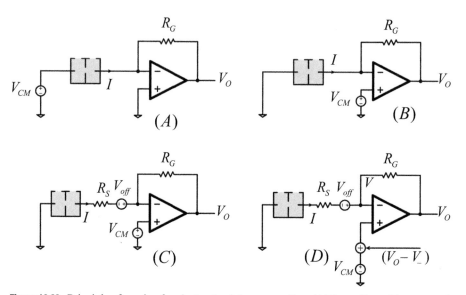

Figure 10.63 Principle of sensing for electrophysiology recording. (A) Recording with command potential on the cup. (B) Recording with command potential on the OPAMP using the virtual short-circuit of the OPAMP. (C) Problems arising from the offset of electrodes and resistance of the connection. (D) Remedies for the offset and resistive connections.

The principle of operation of the current sensing in intracellular recordings is shown in Fig. 10.63. The task is to record current signals that are correlated with variables of activation of ion channels. Figure 10.63A shows the starting point for the recording setup implemented with a current-to-voltage converter where the *command potential* V_{CM} is applied to one side of the two cell compartments. Thanks to the virtual short-circuit assumption, we can also apply the command potential to the noninverting input of the amplifier, as shown in Fig. 10.63B. Unfortunately, the system is affected by nonidealities such as a series resistance R_S and a variable offset due to variations of the built-in electrode potentials, as shown in Fig. 10.63C. In this case, the current that is readout does not correspond to an applied voltage V_{CM}

$$I = \frac{V_{CM} - R_S I - V_{off}}{R_G} \quad \text{instead of } I = \frac{V_{CM}}{R_G}. \quad (10.138)$$

Therefore, we need to apply a variable command voltage with respect to the input to counteract this effect:

$$I = \frac{V_{CM} - R_S I - V_{off} - \alpha(V_O - V_-)}{R_G} \quad \text{with } R_S I = \alpha(V_O - V_-) = \alpha R_G I, \quad (10.139)$$

as shown in Fig. 10.63D. This allows us to neglect the effects of both R_S and V_{off}.

Further Reading

Bard, A. J., and Faulkner, L. R., *Electrochemical Methods Fundamentals and Applications*. Hoboken, NJ: John Wiley & Sons, 2001.

Barsoukov, E. and Macdonald, J. R., *Impedance Spectroscopy: Theory, Experiment, and Applications*, 2nd ed. Hoboken, NJ: John Wiley & Sons, 2005.

Chi, Y. M., Jung, T., and Cauwenberghs, G., Dry-contact and noncontact biopotential electrodes: Methodological review. *IEEE Rev. Biomed. Eng.*, vol. 3, pp. 106–119, 2010.

Crescentini, M., Bennati, M., Carminati, M., and Tartagni, M., Noise limits of CMOS current interfaces for biosensors: A review. *IEEE Trans. Biomed. Circuits Syst.*, vol. 8, no. 2, pp. 278–292, April 2014.

Feynman, R. P., Robert, B. L., Sands M., and Gottlieb M. A., The Feynman Lectures on Physics. Reading, MA: Pearson/Addison-Wesley, 1963.

Haus, A. H. and Melcher, J. R., *Electromagnetic Fields and Energy*. Upper Saddle River, NJ: Prentice Hall, 1989.

Hille, B., *Ion Channels of Excitable Membranes*. Sunderland, MA: Sinauer Associates, 2001.

Morgan, H., and Green, N. G., AC *Electrokinetics: Colloids and Nanoparticles*. Research Studies Press, 2003.

Orazem, M. E. and Tribollet, B. *Electrochemical Impedance Spectroscopy*. Hoboken, NJ: John Wiley & Sons, 2008.

Webster, J. G., Clark, J. W., Neuman, M. R., Peura, R. A., Wheeler, L. A., and Olson, W. H., *Medical Instrumentation*, 4th ed. Hoboken, NJ: John Wiley & Sons, 2009.

11 Selected Topics on Mechanical and Thermal Transduction

This chapter focuses on mechanical and thermal transduction concepts related to the change of conductance and polarization in materials. Therefore, after introducing balogsic concepts, the transduction processes of piezoresistivity, piezoelectricity, and temperature effects on resistance are discussed. Finally, examples of applications of resistance sensors are given, focusing on some techniques to reduce errors due to influence variables.

> **Notation** The notation sometimes uses the same symbols among sections with different meanings. We have chosen not to unify the notation to be consistent with the literature currently used in the different contexts. We have summarized the notation in the Appendix of this chapter to help the reader.

11.1 Overview of Basic Concepts

11.1.1 Strain and Stress in One Dimension

Figure 11.1 shows an elastic, uniform, and isotropic material in mechanical equilibrium subject to deformation by symmetrical forces. After forces are exerted, the material is elongated by a factor δl and is restricted by a quantity δt with respect to the original length l and transversal size t, respectively.

The *strain* ε is the ratio between the variation of the length and its original length

$$\varepsilon = \frac{\delta l}{l} \equiv [\cdot] \equiv [\text{strain}]. \tag{11.1}$$

It is a dimensionless quantity, but we can use the auxiliary unit referred to as *strain* or ε. Furthermore, the ratio of the force over the section area of the slab is defined as the *stress*

$$\sigma = \frac{F}{A} \equiv [\text{Pa}], \tag{11.2}$$

which is expressed in units of pressure, that is, *pascals*. The relationship between stress and strain of materials is called a *strain–stress plot*.

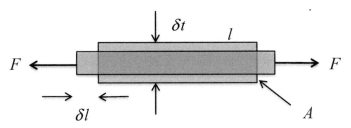

Figure 11.1 Deformation of a material subject to traction in one dimension.

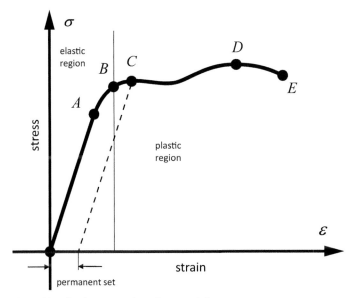

Figure 11.2 Strain–stress plot of a material.

A typical experimental strain–stress plot is shown in Fig. 11.2, where are evidenced characteristic points. The characteristic up to point A (*proportional limit*) is characterized by a linear relationship whose slope is called *Young's module* E_Y

$$E_Y = \frac{\sigma}{\varepsilon} \equiv [\text{Pa}], \tag{11.3}$$

which is also measured in pascals. The material in this region is said to behave in a *linear and elastic region*, following Hooke's law

$$\sigma = E_Y \cdot \varepsilon, \tag{11.4}$$

which states that the strain is proportional to the stress, as in a spring. From A to B (*elastic limit*), the relationship is no longer linear; however, we can always return to the original state with no variation in the form of the structure. This region is called the nonlinear *elastic region*. Up to point B, the solid body is deformed in the elastic region.

From point B onward, the external force action operates irreversible changes in the structure of the material (e.g., break of molecular bonds) so that we cannot return to the original state. Therefore, the region characterized by irreversible deformation (on the right of B) is referred to as the *plastic region*. From point C (*yield point*), the material has some elongation with no increase in load. Since we are in an irreversible state, we get an irreversible elongation called permanent set if we turn the external force back to zero. By increasing the forced elongation on the right of point C, the material offers an opposition again until point D (*ultimate strength*) is reached, which is the maximum value of the tensile strength that could be applied to the slab. Finally, a restriction neck appears until the *fracture point E,* where the device is irreversibly broken.

The illustrated plot is a typical example of a *ductile material*; however, there are other cases such as *brittle materials*, where the fracture is right after the elastic region, or *plastic materials* where the elastic region is very small compared to the elastic one.

11.1.2   Strain and Stress Applied to Orthogonal Axes

We have seen in Section 11.1.1 how the stress and the strain are operating in the simplest case of one dimension. However, forces act on a material in a complex manner, so we expand the concept of the relationship between stress and strain in a small elementary unit volume of material. We start considering an ideal experiment where forces are exerted on two orthogonal axes x and y even if the concept applies to any couple of the x, y, z triad. We use the Δ operator for infinitesimal quantities related to the *infinitesimal element* and the δ operator to indicate the *variation* effect on the material due to the applied stimulus.

Figure 11.3 shows the deformation of a finite elementary cubic volume of *uniform, isotropic,* and *perfectly elastic* material in mechanical equilibrium where all sides are equal to Δl with face areas denoted as ΔA_x, ΔA_y, and ΔA_z, where the notation means

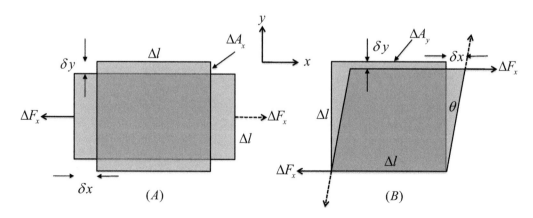

Figure 11.3 Normal deformation (A) and shear deformation (B) in the case in which the forces are acting in the x-direction.

that, for example, ΔA_x is the surface defined by a versor oriented along the x direction. Similarly, ΔF_x, ΔF_y, and ΔF_z, refer to forces exerting along the versors of the subscript. Therefore, we can define the *normal stress* and the *normal strain* on the x-direction as

$$\sigma_{xx} = \lim_{\Delta A_x \to 0} \frac{\Delta F_x}{\Delta A_x} \equiv [\text{Pa}]; \qquad \varepsilon_{xx} = \lim_{\Delta l \to 0} \frac{\delta x}{\Delta l} \equiv [\cdot], \tag{11.5}$$

where $\Delta F_x / \Delta A_x$ and $\delta x / \Delta l$ are the *mean* normal stress and normal strain defined on the elementary unit volume. Therefore, the relationships are intended as the limits of the mean values as ΔA_x and Δl become infinitesimal, defining functions of a point in the three-dimensional space of the continuum. We can extend the definition to other Cartesian variables.

On the other hand, a force could also be applied tangential to the surface A_y (Fig. 11.3B), inducing a torque on the material. Of course, since the elementary element should be in equilibrium, an equivalent, and opposite torque is necessary as boundary conditions on the body (dashed arrows). In this condition, we can define *shear stress* as (Fig. 11.3B)

$$\tau_{xy} = \lim_{\Delta A_y \to 0} \frac{\Delta F_x}{\Delta A_y} \equiv [\text{Pa}], \tag{11.6}$$

where $\Delta F_x / \Delta A_y$ is the *mean* shear stress of the unit volume. In the same manner, we can define the *(engineering) shear strain* as

$$\gamma_{xy} \hat{=} 2\varepsilon_{xy} = \lim_{\Delta l \to 0} \frac{\delta x}{\Delta l} = \tan(\theta) \approx \theta, \tag{11.7}$$

where again $\gamma_{xy} \equiv [\cdot]$.

Similarly to the one-dimensional case, we can define *Young's modulus* for a specific variable, for example, x, as

$$E_Y = \frac{\sigma_{xx}}{\varepsilon_{xx}} \equiv [\text{Pa}], \tag{11.8}$$

where, under the assumption of isotropic material, it is equivalent to the value calculated on other variables. However, conversely, with the one-dimensional case, we can also define the *shear modulus* as

$$G = \frac{\tau_{xy}}{\gamma_{xy}} \equiv [\text{Pa}], \tag{11.9}$$

which is another important characteristic of materials.

The characteristics of the material are also related to its volumetric properties against the deformation; for example, the change in the volume during the deformation fixes the relationship between δx and δy. For this reason, the *Poisson ratio* is defined as

$$v_{xy} = \frac{\delta y}{\delta x} = \frac{\varepsilon_{yy}}{\varepsilon_{xx}} \equiv [\cdot].$$

(11.10)

It can be easily shown that the Poisson ratio approaches $1/2$ when there is no change in volume with respect to a uniaxial strain. For materials having at least a plane of isotropy, it is

$$G = \frac{E_Y}{2(1+v)}.$$

(11.11)

11.1.3   The Stress Tensor

So far, we have discussed the concept of stress and strain using forces acting exclusively in tangential and orthogonal directions with respect to an aligned Cartesian reference.

We show how generic forces applied at the boundaries of a continuous body are decomposed in a generic point. The issue is that boundary conditions do not exert a "simple" force onto a generic point of the body, but the effect is "oriented" with respect to the coordination system. The best way to understand this is by performing a virtual experiment. Imagine carving an imaginary small cut in a continuous body in equilibrium, implying a negligible change in the original shape and no compromise the stability of the entire system. Now, we would like to calculate the necessary forces to recompose such fractures to the original continuum. This is summarized in the *Euler–Cauchy stress principle* stating that *upon any surface (real or imaginary) that divides the body, the action of one part of the body on the other is equivalent to the total of distributed forces acting on the surface dividing the body.* Thus, for a given cut, the stress vector with respect to such oriented surface considers the effects of all the boundary conditions on the body on that point.

As an example of the experiment, we refer to Fig. 11.4, where an elastic body is subject to compression. We could test the internal stresses (i.e., forces exerted on surfaces) by making an imaginary tiny cut of the area ΔA with a defined orientation and then estimating stresses by looking at the necessary forces to recompose the original body structure. Thus, we can use the strain effect on the body to understand the strength and orientation of the stress necessary to recompose the body.

We refer to the following vector quantity

$$\overline{T}^{(k)} = \lim_{\Delta A_k \to 0} \frac{\Delta \overline{F}}{\Delta A_k} \equiv [\text{Pa}],$$

(11.12)

where $\overline{T}^{(k)}$ is referred to as a *stress vector* (or *traction*) with respect to the surface ΔA_k defined by versor $\vec{k}$, which is the limit of the ratio of all forces $\Delta \overline{F}$ acting on an elementary surface ΔA_k. However, in the figure, we refer to a discrete area for clarity in the discussion. If we make the fracture normal to the longitudinal axis as in Fig. 11.4A, we can see that nothing is substantially changed since the two parts are pressed together by two equal stresses $T^{(b)}$, one against each other. However, if the cut is made on the longitudinal axis as in Fig. 11.4B, we observe a divergence of the two planes indicating two opposite tractions $\overline{T}^{(c)}$. Finally, if we make an oblique fracture as

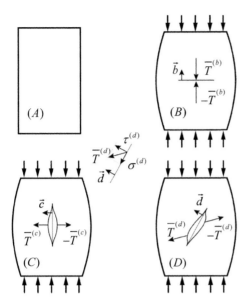

Figure 11.4 Mean surface traction with respect to a cut with different orientations in an elastic body subject to compression. (A) Undeformed shape. (B) Orthogonal cut with respect to the compression axis of the deformed shape. (C) Parallel cut with respect to the direction of the compression. (D) Oblique cut. The decomposition of the stress vector into the normal and shear stresses with respect to the (D) cut is shown at the center.

shown in Fig. 11.4D, the observed antisymmetric deformation reveals that the effect of the traction $\overline{T}^{(d)}$ exerted by the body is oriented in an oblique direction with respect to the surface. Therefore, depending on the orientation of the plane under consideration, the stress vector can be decomposed into two components: one normal to the plane $\sigma^{(d)}$ and the other $\tau^{(d)}$ tangential to the same, as shown in Fig. 11.4.

Thus, we can also define the two *scalar* quantities

$$\sigma^{(k)} = \lim_{\Delta A_k \to 0} \frac{\Delta F_n}{\Delta A_k} \equiv [\text{Pa}]; \ \tau^{(k)} = \lim_{\Delta A_k \to 0} \frac{\Delta F_s}{\Delta A_k} \equiv [\text{Pa}]; \text{ where}$$

$$|\overline{T}^{(k)}|^2 = \sigma^{(k)2} + \tau^{(k)2},$$

(11.13)

where σ is the *normal stress,* and it is the same ratio as before, considering only normal component forces ΔF_n acting on the same surface, and τ is the *shear stress* considering only tangential component forces ΔF_s.

Notation In this section, simple letters (i.e., V) as a scalar, letters with a bar $\overline{V}$ are vectors expressed in the components of a Cartesian coordinate system, and letters in square brackets $[V]$ as a matrix or a tensor in the same coordinate system. Furthermore, we use, depending on the context, the equivalences $x \leftrightarrow 1 \ y \leftrightarrow 2$, and $z \leftrightarrow 3$ for Cartesian axes.

To summarize, even if the boundary conditions of the continuous body are the same, the three stresses $T^{(b)}$ $T^{(c)}$ and $T^{(d)}$ are *not equal*. Therefore, for the given boundary conditions, the *traction vector* (or *stress vector*) of each point of the body *depends on the orientation of the virtual cut*.

What is the relationship between the cut orientation and the stress vector? Referring to a Cartesian coordinate system, (x, y, z) we have that the stress with respect to the surface defined by the versor[1] $\vec{k} = [k_x, k_y, k_z]^T$ is given by[2]

$$\overline{T}^{(k)} = [\sigma] \cdot \vec{k}, \tag{11.14}$$

where $[\sigma]$ is referred to as *Cauchy's stress tensor* with respect to that surface. Therefore, the stress of each point inside a material in the deformed state could be defined *not* by a single vector but by a *tensor*. Eq. (11.14) is the formal expression of *Cauchy's stress theorem*, stating that the stress status of a body in equilibrium is defined at any point by a stress tensor from which the exerted forces could be calculated for the surface identified by the surface versor $\vec{k}$. In expanded form (11.14) becomes

$$\begin{bmatrix} T_x^{(k)} \\ T_y^{(k)} \\ T_z^{(k)} \end{bmatrix} = \begin{bmatrix} \sigma_{xx} & \sigma_{xy} & \sigma_{xz} \\ \sigma_{yx} & \sigma_{yy} & \sigma_{yz} \\ \sigma_{zx} & \sigma_{zy} & \sigma_{zz} \end{bmatrix} \begin{bmatrix} k_x \\ k_y \\ k_z \end{bmatrix} = \begin{bmatrix} \sigma_{xx} & \tau_{xy} & \tau_{xz} \\ \tau_{yx} & \sigma_{yy} & \tau_{yz} \\ \tau_{zx} & \tau_{zy} & \sigma_{zz} \end{bmatrix} \begin{bmatrix} k_x \\ k_y \\ k_z \end{bmatrix}. \tag{11.15}$$

Note that the single scalar elements of the tensor $[\sigma]$ define the normal and shear stresses with respect to a surface oriented *according to the coordinate system* (x, y, z). For example, the normal and shear stresses with respect to the xy plane are $\sigma = \sigma_{zz}$ and $\tau = \sqrt{\tau_{zy}^2 + \tau_{zx}^2}$.

Equation (11.14) could be expressed in the two compact forms

$$T_i^{(k)} = \sum_j \sigma_{ij} k_j \text{ or } T_i^{(k)} = \sigma_{ij} k_j, \tag{11.16}$$

where it is understood that i represents either x, y, z and the sum is taken on $j = x, y, z$. σ_{ij} Which is a compact form of $[\sigma]$, a *second rank tensor* since it has two indexes and maps one vector into another one.

If we want now to calculate the total normal and shear components (scalar quantities), we could use dot product vector calculus and the Pythagorean theorem on (11.14):

$$\sigma^{(k)} = \overline{T}^{(k)} \cdot \vec{k} = \sigma_{ij} k_i k_j$$

$$\tau^{(k)} = \sqrt{\left|\overline{T}^{(k)}\right|^2 - \sigma^{(k)2}} \text{ where } \left|\overline{T}^{(k)}\right|^2 = T_x^{(k)2} + T_y^{(k)2} + T_x^{(k)2} = \sigma_{ij} \sigma_{ik} k_j k_k. \tag{11.17}$$

---

[1] Recall that a *versor* is a unit vector, where its components should relate to the equation $k_x^2 + k_y^2 + k_z^2 = 1$.
[2] The expression could be easily found by applying equilibrium of forces to an infinitesimal tetrahedron (called *Cauchy's tetrahedron*) with three faces orthogonal to the Cartesian coordinates and the fourth one oriented in an arbitrary direction specified by the given versor $\vec{k}$.

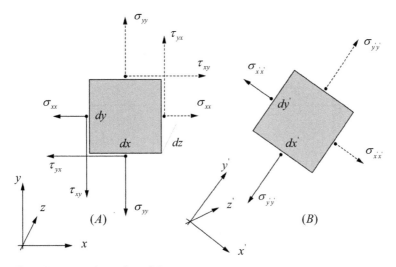

Figure 11.5 Transformation of the stress tensor.

An essential property of the stress tensor arises from the equilibrium conditions of the elementary cube, as shown in Fig. 11.5. According to the principle of conservation of linear momentum, if the continuum body is in static equilibrium, it can be demonstrated that the components of the Cauchy stress tensor satisfy the equilibrium equations. Thus, as shown in Fig. 11.5A, the normal stresses σ on opposite surfaces of the cube are equal. In the same way, according to the principle of conservation of angular momentum, equilibrium requires that the sum of moments with respect to an arbitrary point is zero. Therefore, as shown again in Fig. 11.5A, the shear stresses on adjacent surfaces are equal, that is, $\tau_{xy} = \tau_{yx}$ etc. This leads to the conclusion that the stress tensor is symmetric, thus having only six independent stress components instead of the originals nine.

Therefore, due to the symmetry of the tensor and using the coordinate notation, we have the following equivalent notations

$$[\sigma] \equiv \begin{bmatrix} \sigma_{xx} & \sigma_{xy} & \sigma_{xz} \\ \sigma_{yx} & \sigma_{yy} & \sigma_{yz} \\ \sigma_{zx} & \sigma_{zy} & \sigma_{zz} \end{bmatrix} = \begin{bmatrix} \sigma_1 & \tau_3 & \tau_2 \\ \tau_3 & \sigma_2 & \tau_1 \\ \tau_2 & \tau_1 & \sigma_3 \end{bmatrix}$$

$$= \begin{bmatrix} \sigma_{11} & \sigma_{12} & \sigma_{13} \\ \sigma_{21} & \sigma_{22} & \sigma_{23} \\ \sigma_{31} & \sigma_{32} & \sigma_{33} \end{bmatrix} = \begin{bmatrix} \sigma_1 & \sigma_6 & \sigma_5 \\ \sigma_6 & \sigma_2 & \sigma_4 \\ \sigma_5 & \sigma_4 & \sigma_3 \end{bmatrix}.$$

(11.18)

The last tensor representation is also referred to as *Voigt notation*. The third representation of the tensor of the (11.18) is also graphically represented in Fig. 11.6. Note that Voigt notation implies the following substitutions $11 \rightarrow 1$; $22 \rightarrow 2$; $33 \rightarrow 3$, and $12 \rightarrow 6$; $13 \rightarrow 5$; $23 \rightarrow 4$.

Since the stress tensor is *symmetric* due to the equilibrium constraints, then it (1) has only *real* eigenvalues, (2) could be *diagonalized*, and (3) and has *orthogonal*

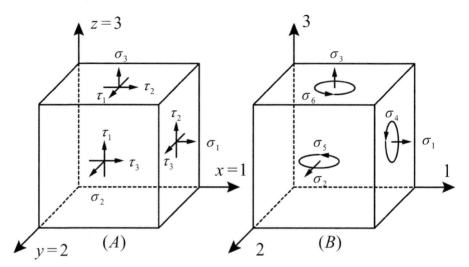

Figure 11.6 Compact notation of the stress tensor that benefits of its symmetry (A) and Voigt notation (B).

eigenvectors. This means that we could find a suitable new set of Cartesian coordinate axes, say (x', y', z') so that the stress is acting only in the normal direction of the surface[3], as shown in Fig. 11.5B. Therefore, for that kind of new coordinate system, we can write

$$[\sigma'] = \begin{bmatrix} \sigma'_1 & 0 & 0 \\ 0 & \sigma'_2 & 0 \\ 0 & 0 & \sigma'_3 \end{bmatrix},$$ (11.19)

where σ'_1, σ'_2, and σ'_3 are called *principal stresses,* and their direction vectors are the *principal directions.*[4] The special case where the principal stresses are equal is called a *hydrostatic tensor.*

The principal directions are those in which there is no shear stress because

$$T^{(n)} = [\sigma] \cdot \vec{n} = \lambda \cdot \vec{n}.$$ (11.20)

Thus, at every point of a stressed body, three planes are defined by versor $\vec{n}$ where the corresponding stress vector is perpendicular to the plane, and there are no shear stresses. The solutions in the λ variable are the eigenvalues, as indicated by the diagonalization process, and $\vec{n}$ is the versor of the principal direction.

To make an example using an alternative approach than using Eq. (11.20), we can use a rotation of the coordinate system (in a 2D coordinate system for simplicity) to find

---

[3] For the change of coordinate system, the linear algebra relationship $[\sigma'] = [A][\sigma][A^T]$ could be used, where $[A]$ is the coordinate transformation matrix where the superscript T is the transpose matrix.

[4] Principal stresses are *invariants* of the stress tensor, which are just the *eigenvalues* of the stress tensor, and the principal directions are *eigenvectors* of the tensor.

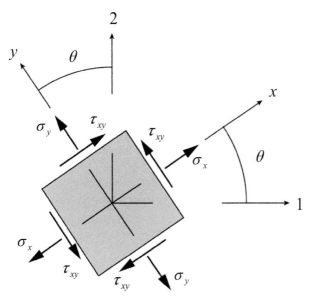

Figure 11.7 Coordinate transformations of the stress tensor.

principal directions. Referring to Fig. 11.7, and considering the rotation matrix $[R]$ and common trigonometric transformations, we have

$$[R] = \begin{bmatrix} \cos\theta & -\sin\theta \\ \sin\theta & \cos\theta \end{bmatrix} \rightarrow [\sigma_{(1,2)}] = [R][\sigma_{(x,y)}][R^T] \text{ from which}$$

$$\sigma_1 = \frac{\sigma_x + \sigma_y}{2} + \frac{\sigma_x - \sigma_y}{2}\cos(2\theta) + \tau_{xy}\sin(2\theta)$$

$$\sigma_2 = \frac{\sigma_x + \sigma_y}{2} - \frac{\sigma_x - \sigma_y}{2}\cos(2\theta) - \tau_{xy}\sin(2\theta) \qquad (11.21)$$

$$\tau_{12} = \tau_{21} = -\frac{\sigma_x - \sigma_y}{2}\sin(2\theta) + \tau_{xy}\cos(2\theta),$$

where σ_1, σ_2, τ_{12} are the elements of the stress tensor with respect to the new $(1,2)$ coordinate system and θ is the rotation angle of the new coordinate system. Note that σ_1 could also be found as the normal stress with respect to the versor $k_x = \cos(\theta)$ and $k_y = \sin(\theta)$ using the first expression of (11.17). Then σ_2 is found by adding $\pi/2$ to the angle.

By taking the derivative of the expressions with respect to θ and equating them to zero, we find the maximum normal stresses direction as

$$\tan(2\theta_{\sigma \max}) = \frac{2\tau_{yx}}{\sigma_x - \sigma_y}, \qquad (11.22)$$

where $\theta_{\sigma \max}$ is the angle of *principal direction*. Similarly, the maximum shear stress is given by

$$\tan(2\theta_{\tau \max}) = -\frac{\sigma_x - \sigma_y}{2\tau_{yx}}. \qquad (11.23)$$

The calculation of maximum stresses is important because it is the direction of the plane of each point of the body where cracks and fractures could arise.

11.1.4   Strain in Three Dimensions: The Strain Tensor

The change in the configuration of a body is given by two components: a rigid-body displacement and a deformation. The first consists of a concurrent translation and rotation of the body without changing its shape or size. The second implies the change in shape and/or size of the body from an initial or undeformed configuration.

We consider a two-dimensional, infinitesimal, rectangular material element with dimensions Δx and Δy, which, after deformation, takes the form of a rhombus, as shown in Fig. 11.8.

Looking at the figure and considering the displacement vector $\bar{u}(x, y)$, we can define the *normal strains* as

$$\varepsilon_{xx} = \frac{du_x}{dx}; \quad \varepsilon_{yy} = \frac{du_y}{dy}. \tag{11.24}$$

Furthermore, we have that the angle α could be approximated as

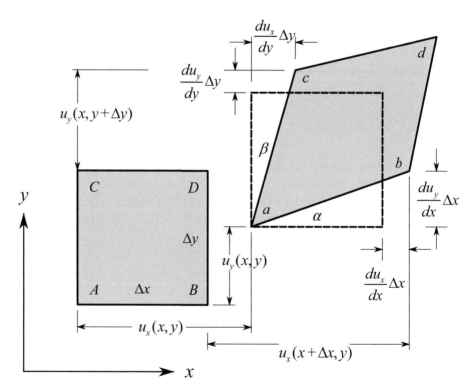

Figure 11.8 Deformation of an infinitesimal material element. [Adapted from Wikicommon]

$$\tan \alpha = \frac{\frac{du_y}{dx} \cdot \Delta x}{\Delta x + \frac{du_x}{dx} \cdot \Delta x} = \frac{\frac{du_y}{dx}}{1 + \frac{du_x}{dx}} \approx \alpha \approx \frac{du_y}{dx}. \qquad (11.25)$$

Therefore, the *shear strains* are

$$\frac{du_y}{dx} = \varepsilon_{yx} \approx \alpha;$$
$$\frac{du_x}{dy} = \varepsilon_{xy} \approx \beta; \qquad (11.26)$$

and the *engineering shear strain* is

$$\gamma_{xy} = \varepsilon_{xy} + \varepsilon_{yx} \approx \alpha + \beta. \qquad (11.27)$$

If we exclude rigid body rotation, we can *define* shear stress as an average of the two components so that it is symmetric by the interchange of indices

$$\varepsilon_{xy} \triangleq \frac{1}{2}\left[\frac{du_x}{dy} + \frac{du_y}{dx}\right] = \frac{1}{2}\left[\frac{du_y}{dx} + \frac{du_x}{dy}\right] = \varepsilon_{yx}$$

$$\rightarrow [\varepsilon] = \begin{bmatrix} \varepsilon_{xx} & \varepsilon_{xy} & \varepsilon_{xz} \\ \varepsilon_{yx} & \varepsilon_{yy} & \varepsilon_{yz} \\ \varepsilon_{zx} & \varepsilon_{zy} & \varepsilon_{zz} \end{bmatrix} = \begin{bmatrix} \varepsilon_{xx} & \gamma_{xy}/2 & \gamma_{xz}/2 \\ \gamma_{yx}/2 & \varepsilon_{yy} & \gamma_{yz}/2 \\ \gamma_{zx}/2 & \gamma_{zy}/2 & \varepsilon_{zz} \end{bmatrix}, \qquad (11.28)$$

where $[\varepsilon]$ is the *strain tensor* and, for definition, $\varepsilon_{xy} = \varepsilon_{yx}, \gamma_{xy} = \gamma_{yx},$ and $\gamma_{xy} = 2 \cdot \varepsilon_{xy}$ so that the strain tensor is symmetric. This definition models most of the practical cases because of some physical considerations; thus, we assume from now on strain tensor symmetric.

11.1.5 Relationships Between Shear and Stress in Three Dimensions

Similar to the case in one dimension, we can relate the strain and stress in three dimensions by means of a fourth-order tensor:

$$[\sigma] = [C] \cdot [\varepsilon] \text{ or}$$

$$\sigma_{ij} = \sum_{k,l} C_{ijkl}\varepsilon_{kl}; \quad C_{ijkl} \equiv [\text{Pa}] \text{ or simply } \sigma_{ij} = C_{ijkl}\varepsilon_{kl}, \qquad (11.29)$$

called *stiffness tensor* $[C]$, where each index could have (x, y, z) values. In practice, it is the inverted form of Hooke's law in three dimensions. Remembering that stress and strain tensors are composed of nine components, in principle, it consists of $3^2 \times 3^2 = 81$ elements[5]. However, the symmetry of the stress and strain tensors

---

[5] 3^2 means that there are nine permutations of two indexes (i, j) having three values (x, y, z).

could be used to ensure that the *stiffness tensor is also symmetric*, having 36 independent components, mapping the 6 normal and shear strains into the 6 corresponding strains. Moreover, by enforcing the thermodynamic properties of reversible elastic deformation, it could be shown that the tensor *might be reduced to just* 21 *independent elements* (6 diagonal plus 15 off-diagonal symmetric elements). Furthermore, for *anisotropic materials having three orthogonal planes of symmetry,*[6] the tensor is reduced to the following form:

$$
\begin{bmatrix} \sigma_1 \\ \sigma_2 \\ \sigma_3 \\ \tau_1 \\ \tau_2 \\ \tau_3 \end{bmatrix} = \begin{bmatrix} \sigma_1 \\ \sigma_2 \\ \sigma_3 \\ \sigma_4 \\ \sigma_5 \\ \sigma_6 \end{bmatrix} = \begin{bmatrix} C_{11} & C_{12} & C_{13} & 0 & 0 & 0 \\ C_{12} & C_{22} & C_{23} & 0 & 0 & 0 \\ C_{13} & C_{23} & C_{33} & 0 & 0 & 0 \\ 0 & 0 & 0 & C_{44} & 0 & 0 \\ 0 & 0 & 0 & 0 & C_{55} & 0 \\ 0 & 0 & 0 & 0 & 0 & C_{66} \end{bmatrix} \cdot \begin{bmatrix} \varepsilon_1 \\ \varepsilon_2 \\ \varepsilon_3 \\ \varepsilon_4 \\ \varepsilon_5 \\ \varepsilon_6 \end{bmatrix}, \qquad (11.30)
$$

where we have used the Voigt notation. Therefore, in the Voigt compact form, the tensor notation becomes

$$
\sigma_p = \sum_q C_{pq}\varepsilon_q; \text{ or simply } \sigma_p = C_{pq}\varepsilon_q. \qquad (11.31)
$$

Due to the redundancy of the physical system, now we use p, q running on six values each, that is, $6^1 \times 6^1 = 36$. This explains the reduction of the 81-element tensor size (not the rank) to 36 matrix elements.

For *cubic* crystals (such as silicon), we have $C_{12} = C_{13} = C_{23}$ and $C_{44} = C_{55} = C_{66}$. For *isotropic* materials, $C_{44} = C_{55} = C_{66}$ are functions of C_{11} and C_{12}. The aspect of (11.30) is valid concerning the *principal material coordinate system*, that is, aligned along the symmetry planes of crystals: in the presence of an arbitrary rotation, all the matrix terms are populated.

The inverted form of the previous relationship, which is a Hooke's law generalization, is given by

$$
\varepsilon_p = S_{pq}\sigma_q, \qquad (11.32)
$$

where $[S]$ is referred to as the *compliance tensor*.

All the symmetry relationships and the number of elements reduction of C are also valid for the S tensor.

---

[6] These are referred to as *orthotropic* materials and are a subset of *anisotropic* materials. It could be demonstrated that a material having two planes of symmetry must have a third one. *Trigonal, tetragonal, monoclinic,* and *triclinic* symmetries are *not* orthotropic. *Cubic* symmetry is a subclass of orthotropic materials. Finally, *isotropic* materials have an infinite number of planes of symmetry.

11.1.6   Elasticity for Isotropic Materials

In the case of *isotropic materials*, the relationships are further simplified. Isotropic materials are characterized by physical properties that are independent of the direction in space; therefore, the relationship should be independent of the coordinate system chosen to represent them. Starting from Hooke's law

$$\varepsilon_{xx} = \frac{1}{E_Y}[\sigma_{xx} - v(\sigma_{yy} + \sigma_{zz})]; \ x, y, z = \{1, 2, 3\} \ x \neq y \neq z$$

$$\varepsilon_{xy} = \frac{1}{2G}\sigma_{xy}; \ x, y = \{1, 2, 3\} \ x \neq y,$$

(11.33)

where $x \neq y \neq z$ means that we can exchange indices provided they are not equal. The second one of (11.33) is (11.9) reordered.

It is easy to show that the *compliance tensor* becomes

$$\begin{bmatrix} \varepsilon_{11} \\ \varepsilon_{22} \\ \varepsilon_{33} \\ \varepsilon_{12} \\ \varepsilon_{13} \\ \varepsilon_{23} \end{bmatrix} = \frac{1}{E_Y} \begin{bmatrix} 1 & -v & -v & 0 & 0 & 0 \\ -v & 1 & -v & 0 & 0 & 0 \\ -v & -v & 1 & 0 & 0 & 0 \\ 0 & 0 & 0 & (1+v) & 0 & 0 \\ 0 & 0 & 0 & 0 & (1+v) & 0 \\ 0 & 0 & 0 & 0 & 0 & (1+v) \end{bmatrix} \cdot \begin{bmatrix} \sigma_{11} \\ \sigma_{22} \\ \sigma_{33} \\ \sigma_{12} \\ \sigma_{13} \\ \sigma_{23} \end{bmatrix}.$$

(11.34)

By inverting the above matrix form, we get the *stiffness tensor* relationship

$$\begin{bmatrix} \sigma_{11} \\ \sigma_{22} \\ \sigma_{33} \\ \sigma_{12} \\ \sigma_{13} \\ \sigma_{23} \end{bmatrix} = k \cdot \begin{bmatrix} (1-v) & v & v & 0 & 0 & 0 \\ v & (1-v) & v & 0 & 0 & 0 \\ v & v & (1-v) & 0 & 0 & 0 \\ 0 & 0 & 0 & (1-2v) & 0 & 0 \\ 0 & 0 & 0 & 0 & (1-2v) & 0 \\ 0 & 0 & 0 & 0 & 0 & (1-2v) \end{bmatrix} \cdot \begin{bmatrix} \varepsilon_{11} \\ \varepsilon_{22} \\ \varepsilon_{33} \\ \varepsilon_{12} \\ \varepsilon_{13} \\ \varepsilon_{23} \end{bmatrix}.$$

$$k = \frac{E_Y}{(1+v)(1-2v)}.$$

(11.35)

A very useful case is where we have *plane stress*; that is, the stress is zero in one direction. An example could be tensions applied on the borders of a sheet of metal where no stress is applied in the direction perpendicular to the surface. In that case, using the "3" direction as the one on which no stress is applied ($\sigma_{13} = \sigma_{23} = \sigma_{33} = 0$), the (11.34) compliance tensor becomes

$$\begin{bmatrix} \varepsilon_{11} \\ \varepsilon_{22} \\ \varepsilon_{12} \end{bmatrix} = \frac{1}{E_Y} \begin{bmatrix} 1 & -v & 0 \\ -v & 1 & 0 \\ 0 & 0 & (1+v) \end{bmatrix} \cdot \begin{bmatrix} \sigma_{11} \\ \sigma_{22} \\ \sigma_{12} \end{bmatrix}.$$

(11.36)

Now, inverting Eq. (11.36), we get the stiffness tensor relationship

$$\begin{bmatrix} \sigma_{11} \\ \sigma_{22} \\ \sigma_{12} \end{bmatrix} = \frac{E_Y}{1 - v^2} \begin{bmatrix} 1 & v & 0 \\ v & 1 & 0 \\ 0 & 0 & (1 - v) \end{bmatrix} \cdot \begin{bmatrix} \varepsilon_{11} \\ \varepsilon_{22} \\ \varepsilon_{12} \end{bmatrix}. \tag{11.37}$$

Note that while the compliance tensor of the plane case is a subset of the general case, the stiffness tensor is *not*. This is because, on the definition of plane stress, the stress is a cause of deformation on which we could apply the superposition of effects. On the contrary, we should invert the matrix on a subset of the compliance tensor for the strain.

Plane stress will be useful for strain gauges application.

11.1.7   Deformation in Simple Structures

The relationships between the mechanical deformation and the applied forces in simple structures could be primarily beneficial for two reasons. On the one hand, they could be used to estimate the applied forces on physical structures on the base of deformations sensed by employing, for example, strain gauges. On the other hand, the relationships could be utilized at the design level of microelectromechanical systems (MEMs). The derivation of these relationships can be found in mechanical engineering textbooks. Here, we summarize the simplest relationships in common structures that are based on two assumptions: (1) small deformations of beam and plates and (2) thin plate in the case of planar structures; the thickness of the plate is small with respect to its size. More specifically, the assumptions above imply that the loads create a negligible contribution (with respect to the elastic constant) of in-plane axial stress on clamped beams and plates. If applied forces are high enough to violate this assumption, the deformation of the structures becomes nonlinear, and this condition is referred to as *large deflections* where analytical solutions could be found using energy methods. In any case, finite element method simulations are suggested with CAD tools to achieve the best estimation of the deformation.

The most elementary cases of structures used in sensor applications and MEMs are illustrated in Figs. 11.9 and 11.10.[7] The relative displacements and strains are tabulated in Tables 11.1 to 11.3.

In all expressions I is the *momentum of inertia* of the structure.

11.2   Piezoresistivity

Piezoresistivity is the property of materials to change their resistance under the effect of the strain induced by the stress. The conductance is proportional to the mobility that is strictly related to the mean free time between carrier collisions and to the equivalent mass (see Chapter 10):

$$J_{\text{drift}} = qn\mu E = \frac{1}{\rho} E; \ \mu = \frac{q \cdot \tau}{m^*}. \tag{11.38}$$

---

[7] Most of the values are taken from Kaajakari (2009), Liu (2012), Senturia (2000), and other sources.

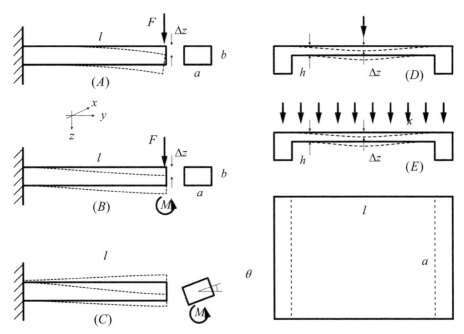

Figure 11.9 Deformations of simple mechanical structures. (A) Bending beam. (B) Bending beam with the clamped end. (C) Torsional beam. (D) Dual-fixed slab with a point load. (E) Dual-fixed slab with a uniform distributed load.

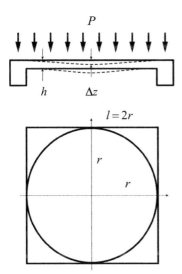

Figure 11.10 Deformations of circular and square plates under differential pressure.

Therefore, any change of the material at the dimension level of particles induces a change of both τ and m^*, thus, a macroscopic conductance variation for a constant electric field (i.e., applied potential). For a generic material or crystal, the

Table 11.1 Displacements and strains of free, clamped, and torsional beams.

	Bending beam (A)	Bending clamped beam (B)	Torsional beam (C)
Displacement	$\Delta z = \dfrac{l^3}{3EI}F$	$\Delta z = \dfrac{l^3}{12EI}F$	$\theta = M \Big/ \left[\dfrac{G \cdot ab^3}{l}\left[\dfrac{1}{3} - 0.21\dfrac{b}{a}\left(1 - \dfrac{b^4}{12a^4}\right)\right]\right];$ $a > b$
Displacement (rectangular profile $I = \dfrac{ab^3}{12}$)	$\Delta z = \dfrac{4l^3}{Eab^3}F$	$\Delta z = \dfrac{3l^3}{8Eab^3}F$	
Strain at the fixed end (maximum)	$\varepsilon = \dfrac{lb}{2EI}F$	$\varepsilon = \dfrac{lb}{4EI}F$	

I is the second-order moment of inertia, and E is Young's module. (Refer to Fig. 11.9.)

Table 11.2 Displacements and strains of free and clamped slabs.

	Slab with punctual load (D)	Slab with uniform distributed load (E) $F = P \cdot A$
Displacement	$\Delta z = \dfrac{l^3}{192EI}F$	$\Delta z = \dfrac{l^3}{384EI}F$
Displacement (rectangular profile $I = \dfrac{ah^3}{12}$)	$\Delta z = \dfrac{l^3}{16Eah^3}F$	$\Delta z = \dfrac{l^3}{32Eah^3}F$
Strain at fixed ends (maximum)	$\varepsilon = \dfrac{3}{4}\dfrac{l}{Eh^2}F$	$\varepsilon = \dfrac{1}{2}\dfrac{l}{Eh^2}F$
Strain at center	$\varepsilon = -\dfrac{3}{4}\dfrac{l}{Eh^2}F$	$\varepsilon = -\dfrac{1}{4}\dfrac{l}{Eh^2}F$

I is the second-order moment of inertia, and E is Young's module. (Refer to Fig. 11.9.)

Table 11.3 Displacements and strains of circular and square plates.

	Circular plate	Square plate
Displacement	$\Delta z = \dfrac{3(1 - v^2)r^4}{16Eh^3}P$	$\Delta z = \dfrac{12(1 - v^2)r^4}{47Eh^3}P$
Strain at fixed ends (maximum)	$\varepsilon = \dfrac{3}{4}\dfrac{r^2}{Eh^2}P$	$\varepsilon = \dfrac{48}{47}\dfrac{r^2}{Eh^2}P$

E is Young's module. P is the pressure. (Refer to Fig. 11.10.)

relationship above could also be extended into the three dimensions by using a resistivity tensor $[\rho]$

$$\overline{E} = [\rho] \cdot \overline{J} \text{ or}$$

$$\begin{bmatrix} E_1 \\ E_2 \\ E_3 \end{bmatrix} = \begin{bmatrix} \rho_1 & \rho_6 & \rho_5 \\ \rho_6 & \rho_2 & \rho_4 \\ \rho_5 & \rho_4 & \rho_3 \end{bmatrix} \cdot \begin{bmatrix} J_1 \\ J_2 \\ J_3 \end{bmatrix}, \tag{11.39}$$

where physical and energetic properties are used to show its symmetry in most cases. Note that off-diagonal elements imply that we should expect a contribution to the currents orthogonal to the applied electric field. For *amorphous isotropic* materials and cubic crystals (such as most semiconductors) along with the principal coordinate systems, we also have $\rho_4 = \rho_5 = \rho_6 = 0$.

In the presence of a strain of the material, using Voigt notation, we can refer to a change with respect to the resistivity of unstressed material ρ_0, which is the same in all directions

$$[\rho_0]\left(1 + \left[\frac{\delta\rho}{\rho_0}\right]\right) = [\rho_0](1 + [\pi] \cdot \overline{\sigma}), \tag{11.40}$$

where $[\pi]$ is called *piezoresistivity tensor* and σ is the strain components. It is a fourth-rank tensor and considers the variation of the resistivity induced by the material's forces. The elements $[\pi]$ are called *piezoresistance coefficients,* and they are measured in the following units:

$$\pi_{ij} = \frac{\delta\rho_i/\rho}{\sigma_j} \equiv [\text{Pa}^{-1}]. \tag{11.41}$$

However, due to the symmetry of both resistivity and stress tensors, under the same assumptions of isotropic materials and cubic crystals is

$$\overline{\sigma} = \begin{bmatrix} \sigma_1 \\ \sigma_2 \\ \sigma_3 \\ \sigma_4 \\ \sigma_5 \\ \sigma_6 \end{bmatrix}; \quad \rho_0 = \begin{bmatrix} \rho_0 \\ \rho_0 \\ \rho_0 \\ 0 \\ 0 \\ 0 \end{bmatrix}. \tag{11.42}$$

We analyze this behavior first for one-dimensional stresses in metals and alloys. Then, we will show the properties of piezoresistivity in crystals using the tensor concept.

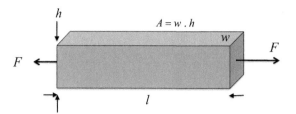

Figure 11.11 Effect of piezoresistivity for a conductor in one dimension.

11.2.1   Piezoresistivity in Metals and Alloys

For a conductor having the geometry as shown in Fig. 11.11, we have that its resistance at rest R_0 is proportional to the length L and inversely proportional to the area A

$$R_0 = \rho \frac{L}{A}, \tag{11.43}$$

where the proportionality coefficient ρ is referred to as *resistivity*.

Now, if we apply stress along the principal axis through a force F, we could write the relative change in the resistivity, using a first-order approximation, as decomposed in relative changes of its variables as[8]

$$\frac{\Delta R}{R_0} = \frac{\delta l}{l} - \frac{\delta A}{A} + \frac{\delta \rho}{\rho}. \tag{11.44}$$

Since

$$\frac{\delta A}{A} = -2v \frac{\delta l}{l}, \tag{11.45}$$

the total relative variation of resistance is

$$\frac{\Delta R}{R} = (1 + 2v)\frac{\delta l}{l} + \frac{\delta \rho}{\rho} = (1 + 2v + \pi E_Y)\frac{\delta l}{l} = G_F \varepsilon = G_F \frac{\sigma}{E_Y} = \frac{G_F}{AE_Y} \Delta F \equiv [\cdot], \tag{11.46}$$

where G_F is called the *gauge factor* of the stressed system. Equation (11.46) is at the base of *strain gauges* that are force resistance sensors.

We can also write

$$G_F = 1 + 2v + \pi E_Y \equiv [\cdot], \tag{11.47}$$

where π is again the *piezoresistive coefficient*.

Note that the first part of the G_F factor $(1 + 2v)$ is related to the resistance change due to the macroscopic variation of the volume, called the *Poisson effect*, while the second factor πE_Y is related to the fractional change in resistivity due to the

---

[8] From the relationship: $R = R(l, A, \rho) \rightarrow \Delta R = dR/dl \cdot \delta l + dR/dA \cdot \delta A + dR/dp \cdot \delta p.$

material strain. Depending on the material, we can have a prevalence of one effect over the other. As far as the first term is concerned, the gauge factor's expected value range would be $1 \leq G_F \leq 2$ corresponding to the theoretically allowable range for the Poisson ratio $0 \leq v \leq 1/2$. It should also be pointed out that the relationships above are referred to under linearity assumptions; however, for large strains, the *gauge factor could show nonlinear behavior.*

11.2.2 Piezoresistivity in Crystals

Piezoresistivity of semiconductor crystals is frequently used in microelectromechanical systems (MEMs), especially for silicon. It could be shown that in semiconductor crystals, the gauge factor is mostly due to the microscopic variation of the lattice instead of the macroscopic volume variations; thus

$$G_F = 1 + 2v + \frac{\delta \rho}{\rho} = 1 + 2v + \pi E_Y \approx \pi E_Y. \tag{11.48}$$

Referring to the diamond-cubic structure of silicon, a commonly used material in MEMs technology, due to the stiffness tensor structure of (11.30), it could be shown that, referring to the symmetry axis of a crystal, the tensor of piezoresistance coefficients is

$$\frac{1}{\rho}\begin{bmatrix} \delta\rho_1 \\ \delta\rho_2 \\ \delta\rho_3 \\ \delta\rho_4 \\ \delta\rho_5 \\ \delta\rho_6 \end{bmatrix} = \begin{bmatrix} \pi_{11} & \pi_{12} & \pi_{12} & 0 & 0 & 0 \\ \pi_{12} & \pi_{11} & \pi_{12} & 0 & 0 & 0 \\ \pi_{12} & \pi_{12} & \pi_{11} & 0 & 0 & 0 \\ 0 & 0 & 0 & \pi_{44} & 0 & 0 \\ 0 & 0 & 0 & 0 & \pi_{44} & 0 \\ 0 & 0 & 0 & 0 & 0 & \pi_{44} \end{bmatrix} \cdot \begin{bmatrix} \sigma_1 \\ \sigma_2 \\ \sigma_3 \\ \tau_1 \\ \tau_2 \\ \tau_3 \end{bmatrix}, \tag{11.49}$$

where the coefficients are dependent on both the doping and on the temperature.

Usually, the cut of silicon is classified according to the lattice's orientation, called Miller indexes, as shown in Fig. 11.12.

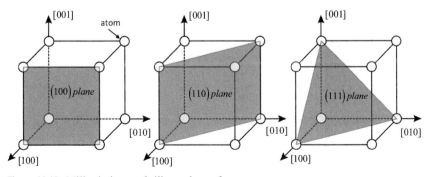

Figure 11.12 Miller indexes of silicon plane of cuts.

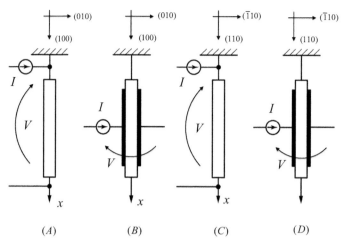

Figure 11.13 Some common cases of piezoresistivity in silicon crystals. Note that that the $(\bar{1}10)$ plane is orthogonal to the (110). (Adapted from Smith, 1954)

Note that plane names (hkl) denote the family of planes orthogonal to $hn_1 + kn_2 + ln_3$, where n_i are the components of the lattice versors, while $[hkl]$ denotes the direction in the corresponding basis.

Therefore, for the most common silicon crystal bulk technologies, the cases are reduced in those illustrated in Fig. 11.13. For a given fixed current I, applied along the direction illustrated in the cases, the voltage is sensed according to the directions indicated. Therefore, using the coordinate transformation[9] of the piezoresistance tensor, it can be shown that we get the following relationships

$$
\begin{aligned}
\text{case (A)}: \quad & \frac{\delta\rho/\rho}{\sigma} = \pi_{11} \\
\text{case (B)}: \quad & \frac{\delta\rho/\rho}{\sigma} = \pi_{12} \\
\text{case (C)}: \quad & \frac{\delta\rho/\rho}{\sigma} = \frac{1}{2}\left(\pi_{11} + \pi_{12} + \pi_{44}\right) \\
\text{case (D)}: \quad & \frac{\delta\rho/\rho}{\sigma} = \frac{1}{2}\left(\pi_{11} + \pi_{12} - \pi_{44}\right).
\end{aligned}
\tag{11.50}
$$

11.3   Piezoelectricity

We have already discussed in Chapter 10 the property of materials to have a polarization induced by an external electric field, referred to as *dielectric polarization*. The ideal relationship between the polarization density and the electric field of dielectric material is shown in Fig. 11.14A and is characterized by a linear

---

[9] $[\pi'] = A[\pi]A^T$, where $[\pi']$ is the tensor in the new coordinates and A is the *rotation matrix*.

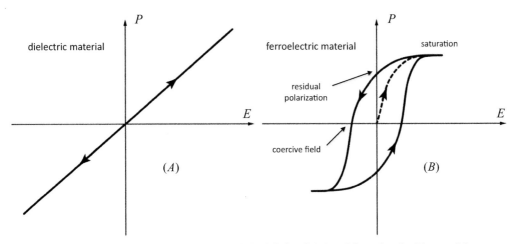

Figure 11.14 Effect of polarization of ideal dielectric(A) and ferroelectric (B) materials.

characteristic that could be run in both directions. The characteristic is called "ideal" since we expect that the polarization density follows the external excitation indefinitely. However, in real materials, we have that, due to the limited size of molecules and crystal lattices, the centers of polarization have a limited displacement, resulting in a saturation effect of the plot. In any case, when the external excitation is removed, no electric polarization density is present. The P(E) plot's saturation effect with no residual polarization density is referred to as *paraelectricity,* and the materials showing this effect are called paraelectric materials.

However, some materials, referred to as *ferroelectric materials*, show the property of spontaneous electric polarization; that is, they have electric polarization even in the absence of an external electric field. Furthermore, spontaneous polarization could be changed by applying an external electric force. Therefore, the typical experimental P(E) plot characteristic of ferroelectric materials appears in Fig. 11.14B. The behavior could be explained as follows. In the beginning, the material is characterized by spontaneous single electric dipoles (or subdomains of material with aligned dipoles) that are randomly displaced within the volume. Therefore, the total polarization of material is negligible due to dipoles' average effects. As we start increasing the electric field, the dipoles begin to orient toward the external excitation, thus increasing the overall polarization density. Once they have reached a joint alignment, the polarization increase is slower due to the saturation effects of polarization due to the limited displacement in that direction. If we decrease the electric field down to zero, we observe that residual polarization is present. This means that dipoles have moved to new resting states and retain memories of past excitation.

Residual polarization implies that some dipoles have moved to a new configuration, thus requires that some work is exerted from the external field to the matter. We will see that this has implications in the energetic balance of the material requiring heat dissipation. If we keep decreasing the electric field, we get a value, referred to as a *coercive field,* so that the polarization is neutralized.

Table 11.4 Polarization properties of the matter according to crystal classes.

32 classes	21 noncentro symmetric	20 piezoelectric	10 pyroelectric	ferroelectric	PbZr/TiO3, BaTiO3, PbTiO3
				nonferroelectric	ZnO, AlN
			10 nonpyroelectric		Quartz
	11 centrosymmetric				

Finally, if we keep varying the external excitation from lower to higher values, the P–E curve follows a hysteresis cycle, as shown in Fig. 11.14B.

An important class of materials regarding their electric polarization characteristic is referred to as *piezoelectric materials*. *Piezoelectricity* is the property of materials to create/displace electrical dipoles under deformation induced by mechanical stress. Piezoelectricity could be seen in two different ways on the same material. On the one hand, the creation of electrical dipoles under mechanical deformation is called the *direct piezoelectric effect*. On the other hand, the property of an external electric field to induce material deformation (e.g., contract or expand) is referred to as the *converse piezoelectric effect*. Piezoelectric materials could be ferroelectric (with spontaneous polarization) on nonferroelectric (without spontaneous polarization).

Another group of materials is the pyroelectric materials characterized by displaying variable polarization upon temperature change. For these materials, the variation of temperature forces the atoms of the crystal structure into a new configuration so as to enhance or reduce the effect of polarization.

As far as crystalline materials are concerned, since polarization properties are strictly linked to the atomic structure, the classification could be done according to the symmetry classification of crystals. It has been shown that any kind of crystal could be classified according to its lattice property of symmetry (*symmetry point groups*). Among the total of 32 classes of the point group, only 20 *noncentrosymmetric groups* (i.e., having a central point of symmetry) have a piezoelectric effect. Thus, noncentrosymmetry is a *necessary* requirement for piezoelectricity. Among them, 10 classes have spontaneous polarization, and 10 have nonspontaneous polarization. This classification, together with examples of materials, is shown in Table 10.4.

As an example of the noncentrosymmetry structure condition on piezoelectricity, refer to Fig. 11.15. In Fig. 11.15A is shown a cubic centrosymmetric lattice (e.g., NaCl) under stress. As clearly shown, the deformation does not induce polarization. Conversely, in Fig. 11.15B, a noncentrosymmetric hexagonal lattice (such as in quartz) under pressure shows dipole creation. Additionally, in Fig. 11.15C, another axis of deformation of the same lattice shows how the dipole is created along an *orthogonal* direction with respect to those related to the stress.

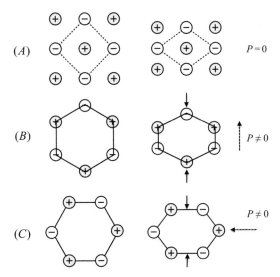

Figure 11.15 Examples of the piezoelectric effect in crystals. (A) A cubic lattice (centrosymmetric) does not show polarization under deformation. (B) A hexagonal lattice (noncentrosymmetric) showing polarization in the same direction. (C) A hexagonal lattice is showing polarization along the orthogonal direction.

Most piezoelectric materials are categorized into two classes:

- Crystals, where the creation of dipoles is due to the asymmetric deformation of charge centers in the lattice
- Polycrystals, such as ceramics, where piezoelectricity is due to the orientation of existing material subdomains having electrical dipole properties

Crystals showing piezoelectric effects are quartz (SiO_2), topaz, tourmaline, berlinite ($AlPO_4$), gallium orthophosphate ($GaPO_4$), lead magnesium niobate-lead titanate (PMN-PT), and aluminum nitride (AlN).

The second class of piezoelectric materials is based on polycrystalline materials composed of many crystal subdomains with dielectric polarization, called *ferroelectric domains*. A dielectric domain is a region of the polycrystalline material in which the molecular polarization is uniformly oriented, as shown in Fig. 11.16A. This means that the individual dipolar moments of the lattice structure are aligned with one another, and they point in the same direction, as shown in Fig. 11.16B. The polarization within each domain points to a unique direction, but the polarization distribution among different domains is almost uniform. The process of ferroelectric domain formation is as follows. Above a temperature called the *Curie temperature* (usually around 300°C for ceramics), the material could both assume a centrosymmetric structure and exhibit a completely random dipole displacement. The crystal structure below the Curie temperature acquires spontaneous polarization orienting dipoles in the same direction within ferroelectric clusters as in Fig. 11.16B. For the reason mentioned above, usually, materials become ferroelectric only under Curie temperature. The size and the

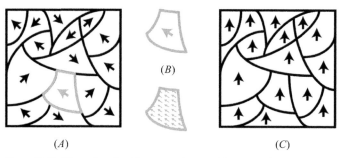

Figure 11.16 Piezoelectric effect on polycrystalline materials. (A, B) Ferroelectric domains. (C) The material after poling process.

orientation of the domains are determined by the minimization of the total energy of the material: often, the domains are separated by boundaries where dipoles are oriented about 180° or 90° each other on opposite sides. In any case, the net average dipole moment of the material is close to zero, as in Fig. 11.16A.

To enhance the piezoelectric effect, a technological production process called *poling* is applied to the material, which is subject to a high electric field under high temperature. When the material is cooled down, the ferroelectric dipole moments are kept almost uniformly and permanently oriented along the poling axis, as shown in Fig. 11.16C.

Among this class of materials, we can mention lead zirconate titanate ($Pb(ZrTi)O_3$), also known as PZT, one of the most used piezoelectric materials, barium titanate ($BaTiO_3$), lead titanate ($PbTiO_3$), potassium niobate ($KNbO_3$), lithium niobate ($LiNbO_3$), and lithium tantalate ($LiTaO_3$). Some of these materials could also be used in crystalline forms for piezoelectric applications.

Once the material is poled, it could be used as a piezoelectric sensor or actuator, inducing mechanical stress or external electric field. However, this class of materials shows a higher heat loss in displacing dipoles, thus reducing their Q-factor with respect to crystals.

11.3.1 Direct and Converse Piezoelectric Effect

The direct piezoelectric effect implies the generation of an electric field upon the material's stress. Before mathematical relationships, we should look better at the concept using Fig. 11.17.

In Fig. 11.17A, we have a poled piezoelectric material realizing a dielectric between two plates. We assume that the total free charge in the plates is set to zero by short-circuiting the structure for simplicity. The arrow also shows the direction of poling. What is important in calculating current and voltage signs is not the dipole moment itself but the *differential dipole moment* induced by the variation. As shown, a stressed dipole could be associated with the sum of the original dipole summed up to a differential moment with the *same* sign. Conversely, a compressed dipole could be modeled by the original dipole summed up with a differential dipole with an *opposite*

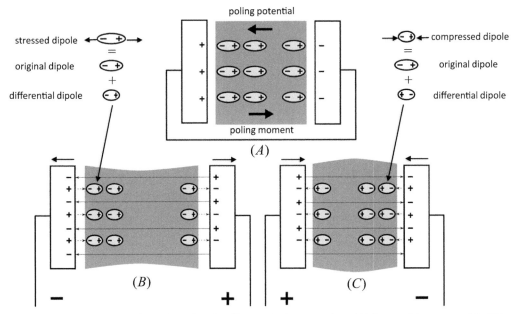

Figure 11.17 Direct piezoelectric effect and voltage sensing. (A) Reset of the system. (B) Effect of expansion. (C) Effect of the contraction.

sign. In Fig. 11.17B is shown a stressed structure where only differential dipoles are shown for clarity. The orientation of the dipoles induces the attraction of opposite charges on the plates. However, since the electrodes should be neutral (open circuit condition $D = 0$), an equal number of opposite charges creates an electric field between them as a charged structure with the signs shown. It is important to remind that what determines an electrostatic potential between two plates is an electric field E *between* them. Conversely, the compression of the structure induces differential dipole moments with opposite orientation with respect to the case before, as shown in Fig. 11.17C, inducing an opposite orientation of the electric field. To summarize, the *compression along the direction of polarization generates the same polarity voltage as the poling voltage*. Note also that the generated electric field acts in a way *to counteract the differential dipole formation*. Thus, referring to Fig. 11.17C, where we do not have a variation of charges during the compression, it is $\Delta D = 0$, $\Delta E \neq 0$.

Therefore, we can sense the electric field to evaluate the applied force. This is called *voltage sensing*[10] of the piezoelectric transducer. Voltage sensing is not the only way to sense the piezoelectric material. If we connect terminals in a short circuit fashion, as shown in Fig. 11.18, we get a reflux of charges from one side to the other to sense the deformation by current displacement. This is called *charge sensing*. Of course, we could get the same result by keeping the short-circuit from the beginning, obtaining the same result.

---

[10] Voltage sensing is usually critical due to leakage currents that affect piezoelectric materials and quickly reset the accumulated charge.

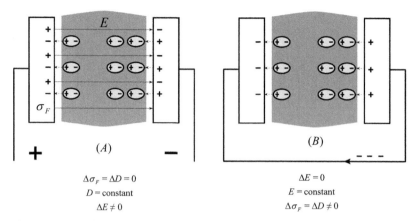

$$\Delta\sigma_F = \Delta D = 0 \qquad\qquad \Delta E = 0$$
$$D = \text{constant} \qquad\qquad\quad E = \text{constant}$$
$$\Delta E \neq 0 \qquad\qquad\quad \Delta\sigma_F = \Delta D \neq 0$$

Figure 11.18 Sensing approaches of the direct piezoelectric effect. (A) Voltage sensing in the open-circuit configuration. (B) Charge sensing in short-circuit configuration.

In general, the relationship between the stress T and the electric field E in one dimension and in *open circuit condition* is

$$E = -g \cdot T; \; g = \underbrace{\left.\frac{dE}{dT}\right|_{D=k}}_{\text{open circuit}} \equiv \left[\frac{\text{V} \cdot \text{m}}{\text{N}}\right], \tag{11.51}$$

where g is called *piezoelectric voltage constant*. Since this relationship is made using an open circuit condition, it is customary to use for g the notation with D in the superscript g^D, meaning that the free charges are set to a constant or zero during the deformation process.

The aforementioned condition is obtained with the physical boundary conditions $\Delta D \neq 0$, $\Delta E = 0$. It should be noted that since we get rid of the electric field that counteracts the differential dipole moments, the piezoelectric material shows *a lower stiffness* in charge sensing or whenever it operates in short-circuit conditions.

Conversely, in short-circuit conditions, the *direct piezoelectric effect* could be modeled as

$$D = d \cdot T; \; d = \underbrace{\left.\frac{dD}{dT}\right|_{E=k}}_{\text{short circuit}} \equiv \left[\frac{\text{C}}{\text{N}}\right], \tag{11.52}$$

where d is the *piezoelectric charge constant,* which is given by the variation of charges upon the applied stress for constant (such as zero) electric field.

The *converse effect* is shown in Fig. 11.19. First, the electrodes are reset by a short circuit, as in Fig. 11.19A. If we apply a potential in the same direction of the poling one, we have an expansion (Fig. 11.19B), while we have a contraction if we have a potential opposite to the poling one (Fig. 11.19C).

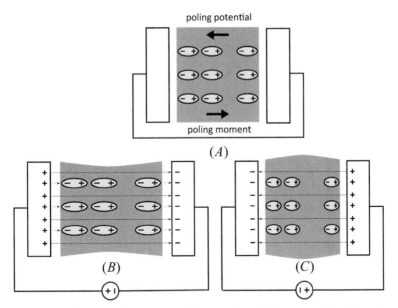

Figure 11.19 Converse piezoelectric effect. (A) Reset of the system. (B) Induced expansion. (C) Induced contraction.

The converse piezoelectric effect is modeled as

$$S = d \cdot E; \quad d = \frac{dS}{dE}\bigg|_{T=k} \equiv \left[\frac{\text{m}}{\text{V}}\right], \tag{11.53}$$

where S is the strain[11] and E is the electric field and d is again the *piezoelectric charge constant*. Note that to derive (11.53), we have kept the force exerted constant (e.g., zero).

One could wonder if d is the same as in (11.52). First, it could be easily shown from the standpoint of units $[\text{C/N}] \equiv [\text{m/V}]$; second, it could be demonstrated from the energy balance that the constants are the same.

Now recalling that a piezoelectric material also acts as an elastic stressed material obeying $S = s \cdot T$, where $s \equiv [\text{Pa}^{-1}]$ is the *elastic compliance constant*[12] and behaves as a dielectric: $D = \varepsilon \cdot E$, where ε is the *dielectric constant*, assuming linear material and applying superposition principle, we can get the *constitutive piezoelectric equations* in one dimension

$$\begin{cases} S = s^E \cdot T + d \cdot E \\ D = d \cdot T + \varepsilon^T \cdot E' \end{cases} \tag{11.54}$$

where the E and T superscripts mean that the differential quantities are calculated at constant electric field and constant traction, respectively. From these equations, it could be easily shown that $g = d/\varepsilon$ as used in (11.51).

---

[11] In piezoelectricity S is used for strain instead of ε, where it is assumed to be the dielectric constant.

[12] It is the reciprocal of the Young's modulus.

The constitutive equations could be read in this way: the first equation states that the application of a stress gives the strain of a piezoelectric material at constant electric field (Hooke's law) or by application of electric field at constant stress (reverse piezoelectric effect). The second relationship states that free charges on electrodes are given by exertion of stress at a constant electric field (direct piezoelectric effect) or by application of an electric field across a dielectric at constant stress.

These equations relate S and D to T and E; however, there are other three alternate forms of constitutive equations, i.e., relating T and D to S and E, and so forth, where all the coefficients of one form could be expressed in terms of coefficients of other forms, but we omit them for simplicity.[13]

All the constitutive forms could also be derived by applying the conservation of energy of piezoelectric material by means of the laws of thermodynamics

$$\frac{dU}{dt} = T \cdot \frac{dS}{dt} + E \cdot \frac{dD}{dt}$$

$$\rightarrow U = \int_0^T T \delta S + \int_0^D E \delta D, \tag{11.55}$$

where U is the stored energy density (per unit volume). This means that the variation of stored energy per volume is given by both mechanical work (by strain) and by electrical work (by dielectric polarization). An important insight of Eq. (11.55) is that energy transformations could be graphically identified in the (T, S) plots for mechanical work and in the (E, D) plots for electrical work, as we have discussed in

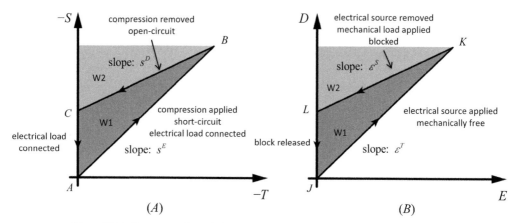

Figure 11.20 Representation of electromechanical conversion of a piezoelectric material. (A) Conversion of energy from mechanical to electrical. (B) Conversion of energy from electrical to mechanical. [Adapted from 176–1987 – ANSI/IEEE Standard on Piezoelectricity]

---

[13] Refer to 176–1987 – ANSI/IEEE Standard on Piezoelectricity, IEEE 1988, for further details.

Chapter 10. To better understand this, we can graphically see energy transformations examples using the plots of Fig. 11.20.

In Fig. 11.20A is shown the (T, S) plot (it is a strain–stress plot with reversed axis) of a piezoelectric material showing a possible conversion from mechanical to electrical work. First, the system is short-circuited, and compressive stress is applied so that the work done is W1 + W2 following a slope s^E, then the compressive stress is removed, and the circuit is opened so that the work done is W2. As shown in the figure, the material keeps a residual strain. This is because, as said before, the stiffness of material in short circuit conditions is smaller than that in open circuit ones. Finally, the electrical load is applied to return to the original state. Therefore, the energy conversion could be estimated by the *coupling factor k*

$$k^2 = \frac{\text{mechanical energy delivered to piezosystem}}{\text{total mechanical energy supplied}}$$

$$= \frac{W_1}{W_1 + W_2} = \frac{s^E - s^D}{s^E} = \frac{d^2}{s^E \varepsilon^T},$$

(11.56)

where the last equivalence could be found using the constitutive equations. In this case, the electrical energy is supplied in two steps: in the A–B step where we could reset the electric field by applying the electrical load itself, and in the C–A step, where the electrical load is used to release the strain. However, other cycles could be used, for example, compressing in open-circuit conditions and applying the load afterward, where conversion coupling factors might be calculated using the same graphical scheme.

In Fig. 11.20B is shown the (E, D) plot of a piezoelectric material showing the conversion from electrical to mechanical work. First, a potential is applied to the system while it is left mechanically free so that the work done is W1 + W2 following a slope ε^T; then the electrical circuit is opened while it is kept mechanically clamped without deformation and then, when $E = 0$, the block is removed so that mechanical work is done. Therefore, the energy conversion could be estimated by the *coupling factor k*

$$k^2 = \frac{\text{electrical energy delivered to piezosystem}}{\text{total electrical energy supplied}}$$

$$= \frac{W_1}{W_1 + W_2} = \frac{\varepsilon^T - \varepsilon^S}{\varepsilon^T} = \frac{d^2}{s^E \varepsilon^T},$$

(11.57)

where the last equivalence could be found using the constitutive equations, showing that the coupling factor is the same as before.

Even if a high k^2 is usually sought for effective transduction, it is *not a measure of efficiency*, since it indicates the net energy that is supplied to the piezosystem. However, we do not know if this energy is totally converted into electrical or mechanical ones or any heat loss. For example, referring to Fig. 11.20B, we have already seen in Chapter 10 that the $E \cdot \Delta D$ factor contributing to the area comprehends *both* energy storage and heat loss depending on the

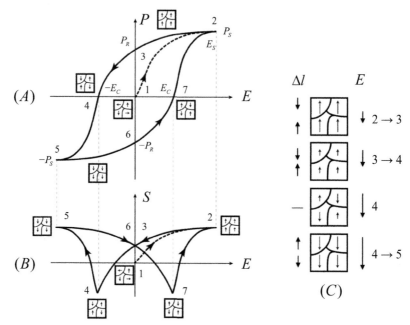

Figure 11.21 (A) Dielectric hysteresis loop. (B) Butterfly hysteresis loop. (C) Effect of the change of direction in strain for increasing electric field. For typical PZT "soft" materials P_S and P_R are around 0.3–0.5 C/m^2, E_S is about 1.5 kV/mm, and E_C is about 0.5 kV/mm.

phase lag of the displacement field. Furthermore, unconverted energy is not necessarily lost (dissipated into heat) and can, in many cases, be recovered. For example, we could take a material with both a very small W1 area and low losses, such as quartz. This means that it is acting as an almost perfect reversible elastic material but poorly converts the energy impressed from external sources. Therefore, coupling factors are useful for understanding if a piezoelectric material is effective for making systems convert mechanical/electrical energy (such as in energy harvesting), but they are not saying anything about efficiency as the Q factor. In other words, high coupling factors values are a necessary but not sufficient condition for energy transduction. For example, quartz has very low coupling factors but very high Q factors. Therefore, quartz is excellent for implementation in resonators but not in energy harvesters.

One of the best ways to evaluate the real heat loss is to characterize the *hysteretic loops*, as described in the text that follows. Figure 11.21 shows *dielectric* and *butterfly hysteresis loops* of the same material. The *dielectric hysteresis loop* shows a closed cycle in the (E, P) space, as shown in Fig. 11.21A. On the other side, the *butterfly hysteresis loop* is a closed curve in the (E, S) space, as shown in Fig. 11.21B. Starting from a random displacement of subdomains in position 1 in an unpoled material, the electric field's increase orients a more significant number of domains along the direction of the field. However, similarly to magnetic materials, it reaches a point where the number of oriented domains saturates P_S for an arbitrary strength of the electric field. This point is shown as 2 in the two plots. The butterfly loop also shows a saturation of the strain, as shown in Fig. 11.21B. The electric field where this happens is called *saturation electric field* E_S. If we now decrease the electric field down to zero, we see that not all the dipoles return to their original state, and some of them keep an orientation induced by the previous external electric field, resulting in a *residual polarization* P_R corresponding to point 3 of the loops. This is an index of a nonreversible physical effect. To reset the residual polarization, we have to apply a reverse electric field down to $-E_C$ as indicated in point 4, called *coercive field*. If we keep lowering the electric field in the negative direction, we reach a negative saturation polarization corresponding to point 5 $(-E_S, -P_R)$ in the loops. If we start increasing the electric field due to the dipoles' memory, we do not follow the same trajectory but a hysteresis curve as shown in Fig. 11.21A.

We can follow the corresponding history points in the butterfly loop of the (E, S) plane as illustrated in Fig. 11.21B. We have to note that in the (E, S) plot, the inversion of the strain from contraction to expansion in points 4 and 7. This effect is explained in Fig. 11.21C for the transition from points 3 to 5. If the dipoles have residual polarization, the effect of the increase of the reversed field acts to contract (thus reduce) the size of the material. If we keep increasing the reverse electric field, they might change the dipole direction in the opposite direction for some dipoles. When the dipole reversing effect equates to the reduction one, we are at point 4. Now, if we keep increasing the reverse electric field, the dipole starts inverting direction so that the material expands again until we reach the saturation point 5.

The (E, D) dielectric loop encloses an area proportional to the power dissipated by the material. This area is very small for crystalline materials such as quartz, while it is relevant for polycrystalline ceramics. To understand this, consider that the (E, P) plot is performed when no load is applied, either mechanical or electrical. Therefore, if we had no loss, we should have the same energy exchanged between the system in forward and backward mode, and thus, we should have the same area subtended by the $E \cdot \Delta P$ increment in both directions. Instead, the different areas subtended by the hysteretic loop indicate a loss that could be nothing other than heat. Another way to see this is to consider the discussion of Chapter 10, where a phase lag between electric field and polarization density contributes to the heat loss.[14] In fact, the closed-loop of Fig. 11.21A clearly indicates a phase lag between the input and the output. Finally, we could observe that the considerations above could be found even in the (E, D) plot other than that in the (E, P) one by virtue of the P and D relationship.

The butterfly loop shows that we cannot always get the same strain for equal potential. For the aforementioned reasons, in sensor and actuator applications, it is preferable to limit the hysteretic range in a unipolar or semipolar range, as shown in Fig. 11.22.

---

[14] $P_D = \overline{E} \cdot j\omega\varepsilon_0 \overline{E} + \overline{E} \cdot j\omega\overline{P} = \overline{E} \cdot j\omega\overline{D}$

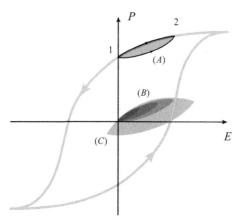

Figure 11.22 Hysteresis of piezoelectric materials.

Therefore, in a poled material, starting from point 1, we get a reduced unipolar hysteresis, as shown in Fig. 11.22A. The different amount of external electric field induces several loops, as shown in Fig. 11.22B, where they are shifted in the origin. Finally, we can also use a semipolar cycle, as shown in Fig. 11.22C.

Another important figure of merit for heat dissipation, especially for resonators, is the Q factor, as already discussed in Chapter 6. More specifically, in that case, also important as a figure of merit is the Q factor product with the coupling factor $Q \cdot k$.

11.3.2   Piezoelectricity in Three Dimensions

The constitutive equations of the piezoelectricity could be extended in three dimensions by using the concept of a tensor, as already seen in previous sections. Therefore, the equations could be written in tensor notation

$$\begin{cases} S_{ij} = s^E_{ijkl} \cdot \sigma_{kl} + d_{kij} \cdot E_k \\ D_i = d_{ikl} \cdot \sigma_{kl} + \varepsilon^T_{ik} \cdot E_k \end{cases}, \tag{11.58}$$

where we have used here σ in place of T. Similar to other contexts, symmetry could be found for piezoelectric systems to justify the Voigt simplified notation

$$\begin{cases} S_p = s^E_{pq} \cdot \sigma_q + d_{qp} \cdot E_q \\ D_p = d_{pq} \cdot \sigma_q + \varepsilon^T_{pq} \cdot E_q \end{cases}. \tag{11.59}$$

As examples, here is the expression for the PZT material

$$
\left\{
\begin{aligned}
\begin{bmatrix} S_1 \\ S_2 \\ S_3 \\ S_4 \\ S_5 \\ S_6 \end{bmatrix} &=
\begin{bmatrix}
s_{11} & s_{12} & s_{13} & 0 & 0 & 0 \\
s_{12} & s_{11} & s_{13} & 0 & 0 & 0 \\
s_{13} & s_{13} & s_{11} & 0 & 0 & 0 \\
0 & 0 & 0 & s_{44} & 0 & 0 \\
0 & 0 & 0 & 0 & s_{44} & 0 \\
0 & 0 & 0 & 0 & 0 & 2(s_{11}-s_{12})
\end{bmatrix}
\begin{bmatrix} \sigma_1 \\ \sigma_2 \\ \sigma_3 \\ \sigma_4 \\ \sigma_5 \\ \sigma_6 \end{bmatrix}
+
\begin{bmatrix}
0 & 0 & d_{31} \\
0 & 0 & d_{32} \\
0 & 0 & d_{33} \\
0 & d_{24} & 0 \\
d_{15} & 0 & 0 \\
0 & 0 & 0
\end{bmatrix}
\begin{bmatrix} E_1 \\ E_2 \\ E_3 \end{bmatrix} \\[2ex]
\begin{bmatrix} D_1 \\ D_2 \\ D_3 \end{bmatrix} &=
\begin{bmatrix}
0 & 0 & 0 & 0 & d_{15} & 0 \\
0 & 0 & 0 & d_{24} & 0 & 0 \\
d_{31} & d_{32} & d_{33} & 0 & 0 & 0
\end{bmatrix}
\begin{bmatrix} \sigma_1 \\ \sigma_2 \\ \sigma_3 \\ \sigma_4 \\ \sigma_5 \\ \sigma_6 \end{bmatrix}
+
\begin{bmatrix}
\varepsilon_{11} & 0 & 0 \\
0 & \varepsilon_{22} & 0 \\
0 & 0 & \varepsilon_{33}
\end{bmatrix}
\begin{bmatrix} E_1 \\ E_2 \\ E_3 \end{bmatrix}.
\end{aligned}
\right.
$$

$$(11.60)$$

and for quartz crystals

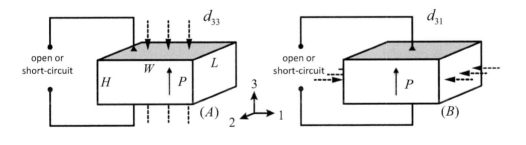

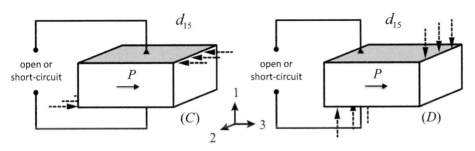

Figure 11.23 Experimental characterization and usage of the piezoelectric coefficients. P is the direction of poling.

$$
\left\{
\begin{aligned}
\begin{bmatrix} S_1 \\ S_2 \\ S_3 \\ S_4 \\ S_5 \\ S_6 \end{bmatrix}
&=
\begin{bmatrix}
s_{11} & s_{12} & s_{13} & s_{14} & 0 & 0 \\
s_{12} & s_{11} & s_{13} & -s_{14} & 0 & 0 \\
s_{13} & s_{13} & s_{11} & 0 & 0 & 0 \\
s_{14} & -s_{14} & 0 & s_{44} & 0 & 0 \\
0 & 0 & 0 & 0 & s_{44} & 0 \\
0 & 0 & 0 & 0 & 0 & 2(s_{11} - s_{12})
\end{bmatrix}
\begin{bmatrix} \sigma_1 \\ \sigma_2 \\ \sigma_3 \\ \sigma_4 \\ \sigma_5 \\ \sigma_6 \end{bmatrix}
+
\begin{bmatrix}
d_{11} & 0 & 0 \\
-d_{11} & 0 & 0 \\
0 & 0 & 0 \\
d_{14} & d_{24} & 0 \\
0 & -d_{14} & 0 \\
0 & -2d_{11} & 0
\end{bmatrix}
\begin{bmatrix} E_1 \\ E_2 \\ E_3 \end{bmatrix} \\[2em]
\begin{bmatrix} D_1 \\ D_2 \\ D_3 \end{bmatrix}
&=
\begin{bmatrix}
d_{11} & -d_{11} & 0 & d_{14} & 0 & 0 \\
0 & 0 & 0 & d_{24} & -d_{14} & -2d_{11} \\
0 & 0 & 0 & 0 & 0 & 0
\end{bmatrix}
\begin{bmatrix} \sigma_1 \\ \sigma_2 \\ \sigma_3 \\ \sigma_4 \\ \sigma_5 \\ \sigma_6 \end{bmatrix}
+
\begin{bmatrix}
\varepsilon_{11} & 0 & 0 \\
0 & \varepsilon_{22} & 0 \\
0 & 0 & \varepsilon_{33}
\end{bmatrix}
\begin{bmatrix} E_1 \\ E_2 \\ E_3 \end{bmatrix}. \quad (11.61)
\end{aligned}
\right.
$$

Note that usually, 3 is the *direction of polarization* (poling). Therefore, ε_{33} is the permittivity of the material in the polarization direction while ε_{11} is the perpendicular one.

An example of an application of relationships (11.60) is illustrated in Fig. 11.23, where H, W, L are the dimensions of the piezoelectric material. For instance, we would like to use a piezoelectric material as a pressure sensor: in this case, we could sense the system using electrodes on the same axis of the poling direction. Therefore, we can sense in two ways: charge-mode or voltage-mode. In the first case, using the second of (11.60) in the system shown in Fig. 11.23A and sensing in charge mode ($E = 0$), we have

$$
D_3 = d_{33} \cdot \sigma_3 \rightarrow Q = d_{33} \cdot F, \qquad (11.62)
$$

Table 11.5 Piezoelectric characteristics of some materials.

Material	d_{33} [pC/N]	d_{31} [pC/N]	d_{15} [pC/N]	g_{33} $\left[\dfrac{\text{mV} \cdot \text{m}}{\text{N}}\right]$	g_{31} $\left[\dfrac{\text{mV} \cdot \text{m}}{\text{N}}\right]$	$\varepsilon_{33}/\varepsilon_0$	$\sqrt{k^2}$	Q
Soft PZT[15]	400	−180	550	25	−11.3	1750	0.69	80
Hard PZT[16]	290	−130	475	29	−13.1	1500	0.66	1500
AlN	5.4	−2.78					0.47	
PMN-PT (33%) multi	2500	−1300	146	38	−18	8200	0.90	80
Quartz	−2.3 (d_{11})	−0.67 (d_{14})		−50		4.5	0.09	10^4–10^6

Note: These factors could greatly vary according to the implementation and should be taken only as a reference of comparison between materials.

[15] Data refer to the PI255 material by PI ceramic GmbH.
[16] Data refer to the PI141 material by PI ceramic GmbH.

which is the relationship between the force and the charge. Conversely, if we use the same system voltage-mode we have to leave electrodes in high impedance ($D = 0$); using the second of (11.60), we have

$$D_3 = 0 = d_{33} \cdot \sigma_3 + \varepsilon_{33} \cdot E_3 \rightarrow E_3 = g_{33} \cdot \sigma_3$$
$$\rightarrow V = g_{33} \frac{H}{W \cdot L} \cdot F, \tag{11.63}$$

which is the relationship between the force and the voltage at the electrodes.

If we use the configurations of Fig. 11.23B, C, and D, according to (11.60), we have to substitute in place of d_{33} in (11.62): d_{31}, d_{15}, and again d_{15}, respectively.

Conversely, if we use the system of Fig. 11.23A as an actuator and leave the material unclamped, that is, free to move ($\sigma = 0$), the displacement could be calculated as

$$S_3 = \frac{\Delta H}{H} = d_{33} \cdot E_3 \rightarrow \Delta H = d_{33} \cdot V, \tag{11.64}$$

where V is the applied voltage. Similar relationships could be found for the (B), (C), and (D) configurations using d_{31} and d_{15}.

Table 11.5 provides a summary of the principal characteristics of some piezoelectric materials. Note again how quartz has a very low coupling factor in the presence of high Q.

11.4 Resistance Temperature Sensing

The resistance of a conductor could also be used to sense temperature. The underlying phenomenon is quite complex because it involves concepts of quantum physics and energy bandgap theory and is beyond the objectives of this section. However, a quick explanation of principles could be summarized as follows.

The relationship between electrical conduction and temperature depends on the material, and in particular, it is different between metals and semiconductors. In general, metals or alloys' conduction is primarily due to the interactions between electrons and phonons. Phonon is a quasiparticle describing the vibrational modes of an elementary motion of a lattice of atoms or molecules. Therefore, it is related to the kinetic energy of the lattice and thus to the temperature of the material. The conduction in metals is due to the contributions of (1) electron scattering by phonons, (2) phonon-assisted interband electron scattering, and (3) electro-electron interaction. The relative incidence of these causes greatly depends on the kind of metal and on the temperature range. However, in general, we can state that the *electrical resistivity of metals increases with temperature*, especially due to the higher number of collisions between electrons and phonons. In classical physics approach terms, we could say that the higher is the vibrational energy of the lattice, the higher the number of electron collisions, thus its resistance. In general, the dependence follows a fairly linear behavior.

On the other side, the relationship between conductance and temperature in *semiconductors* should be modeled differently. Conduction in semiconductors is characterized by the relationship between the conduction and the valence bands that are separated by an energy gap. The bandgap allows for just a few electrons to jump to the conduction band and allow current to some extent. However, with the increase in temperature, they get sufficient energy (by phonons scattering) to overcome the energy barrier and jump into the conduction band to increase their conductivity. Therefore, the *electrical resistivity of semiconductors decreases with temperature*. Unfortunately, the dependence of this behavior is usually nonlinear.

11.4.1   Resistance Temperature Detectors

Resistance temperature detectors (RTDs) are resistive sensors made of metals or alloys, showing an increasing resistance versus temperature with a fairly linear behavior. As we have seen in Chapters 2 and 8, using a first-order approximation, we have

$$R(T) = R_0 + \Delta R = R_0 \left(1 + \frac{\Delta R}{R_0}\right) = R_0(1 + \alpha \cdot \Delta T);$$
$$\Delta T = T - T_0;$$
$$\alpha = \frac{1}{R_0}\frac{\Delta R}{\Delta T} \equiv \left[\frac{\Omega}{\Omega \cdot {}^0 C}\right] \equiv \left[\frac{\Omega}{\Omega \cdot K}\right], \tag{11.65}$$

where T_0 is the reference temperature and α is called *temperature coefficient of resistance* (TCR), which is also a *relative sensitivity* of the device. In most cases, the reference temperature is $T_0 = 0°C$ so that the TCR is calculated at the 100°C point as

$$\alpha = \frac{1}{R_0}\frac{R_{100} - R_0}{100}, \tag{11.66}$$

where $R_{100} = R(100°C)$. The typical TCR for metals and alloys ranges from 0.003 to 0.005 $\Omega/\Omega/°C$ and thus is quite low since we have a relative variation of 0.5% for 100° of variation. Using the linear relationship, the temperature could be deducted from the resistance from the equation

$$T = \frac{1}{\alpha} \cdot \left(\frac{R(T)}{R_0} - 1\right). \tag{11.67}$$

Platinum RTDs, also called *platinum resistance thermometers* (PRTs), are frequently used as precision temperature instruments, due to the stability properties of the noble metal and for its superior linearity with respect to other materials. However, to cope with the precision requirements of the PRTs, an interpolating function is used instead of (11.65). Therefore, the nonlinearity could be approximated by the following polynomial expression for $T_0 = 0°C$:

Table 11.6 Example of constants of some RTD materials.

Material	R_0	A	B	C	α	β	δ
	[Ω]	[$^\circ C^{-1}$]	[$^\circ C^{-2}$]	[$^\circ C^{-4}$]	[$\Omega/\Omega/^\circ C$]	[$\Omega/\Omega \cdot ^\circ C$]	[$\Omega/\Omega \cdot ^\circ C$]
Pt	100	3.908 $\times 10^{-3}$	-5.775 $\times 10^{-7}$	-4.183 $\times 10^{-12}$	0.00385	0.10863	1.4999
Ni	120				0.00672		
NiFe	604		polynomial form, see datasheets		0.00518	Polynomial form; see	
Cu	10				0.00427	datasheets	

$$R(T) = R_0[1 + A \cdot T + B \cdot T^2]. \tag{11.68}$$

Then, for experimental characterization reasons, it has been proposed to express the temperature according to the polynomial equation

$$T = \frac{1}{\alpha} \cdot \left(\frac{R(T)}{R_0} - 1\right) + \delta\left(\frac{T}{100} - 1\right)\frac{T}{100}, \tag{11.69}$$

where the first term is (11.67). The reason for this form is that the polynomial function is forced to intersect R_{100} for $T = 100^\circ C$ independently on the δ parameter value. For the determination of δ it is used a higher temperature, typically $T_R = 260^\circ C$ so that

$$\delta = T_R - \frac{R(T_R) - R_0}{R_0 \cdot \alpha} \bigg/ \left[\left(\frac{T_R}{100} - 1\right)\frac{T_R}{100}\right]. \tag{11.70}$$

Therefore, it is easy to show that

$$A = \alpha + \frac{\alpha\delta}{100} \equiv [^\circ C^{-1}]; \quad B = -\frac{\alpha\delta}{100^2} \equiv [^\circ C^{-2}]. \tag{11.71}$$

A further refinement of the interpolation is suggested when (and only if)$T < 0$, and it is given by

$$R(T) = R_0[1 + A \cdot T + B \cdot T^2 + C \cdot (T - 100) \cdot T^3], \tag{11.72}$$

referred to as the *Callendar–Van Dusen equation* with the temperature dependence as

$$T = \frac{1}{\alpha} \cdot \left(\frac{R(T)}{R_0} - 1\right) + \delta\left(\frac{T}{100} - 1\right)\frac{T}{100} + \beta\left(\frac{T}{100} - 1\right)\frac{T^3}{100^3} \tag{11.73}$$

so that

$$C = -\frac{\alpha\beta}{100^4}. \tag{11.74}$$

Equation (11.74) is useful because most PRTs and RTDs are characterized by A, B, and C constants that are tabulated and other metals or alloys.

Table 11.6 shows the interpolating constants of the most used RTD materials.

11.4.2   Thermistors

Thermistors differ from RTDs, as the sensing material is not a metal or alloys but is made of complex compounds of semiconductor materials. More specifically, thermistors could be classified according to two classes related to the following characteristics:

- *Negative temperature coefficient* (NTC) thermistors, where the resistance decreases with the temperature. They are made mainly of a sintered semiconductor material such as ferric oxide doped with titanium or nickel oxide doped with lithium.
- *Positive temperature coefficient* (PTC) thermistors, where the resistance increases with the temperature. They are made of doped polycrystalline ceramic containing barium titanate and/or other compounds.

The thermal behavior of PTC is complex and strictly dependent on the material, so it is not treated here. On the other hand, as discussed earlier, NTCs follow a decreasing characteristic of conductance, typical of semiconductors, that can be modeled with a high degree of precision by the *Steinhart–Hart equation*

$$\frac{1}{T} = a + b\ln(R) + c[\ln(R)]^3, \tag{11.75}$$

where a, b, and c are constant coefficients.

For practical uses, the cubic factor is neglected, and the following is used

$$a = \frac{1}{T_0} - \frac{1}{B}\ln(R_0); \ b = \frac{1}{B}, \tag{11.76}$$

where T_0 and $R_0 = R(T_0)$ are the reference temperature and resistances. Thus, the resistance versus temperature relationship becomes

$$R = R_0 e^{-B\left(\frac{1}{T_0} - \frac{1}{T}\right)}; \ B \equiv [°C] \equiv [K], \tag{11.77}$$

where B is called the *B parameter* of the thermistor. This expression is quite useful since the B parameter is related to the normalized sensitivity of the device

$$S' = \alpha = \frac{1}{R_0}\frac{dR}{dT}\bigg|_{T_0} = -\frac{B}{T^2}. \tag{11.78}$$

To evaluate the differences between RTDs and NTCs, the resistance versus temperature plot is shown in Fig. 11.24. From the plot, we can draw the following considerations:

- Platinum RTDs (PRTs) offer the best linearity but the worst sensitivity.

Table 11.7 Summary of mechanical and thermal properties of conductive materials.

Material	Composition	G_F (small deformations)	Resistance	TCR=α
		[·]	[$\Omega \cdot m$]	[$\Omega/\Omega/^{\circ}C$]
Constantan	45% NI, 55% Cu	2.0	4.89E-07	8.0E-06
Isoeleastic	36% Ni, 8% Cr, 0.5% Mo, 55.5% Fe	3.5	11.4E-07	
Manganin	84% Cu, 12% Mn, 4% Ni	0.5	4.82E-07	2.0E-06
Karma	74% Ni, 20% Cr, 3% Al, 3% Fe	2.4	13.51E-07	1E-06
Monel	67% Ni, 33% Cu	1.9	4.82E-07	600E-06
Nichrome	80% Ni, 20% Cu	2.0	11.80E-07	130E-06
Iridium-Platinum	95% Pt, 5% Ir	5.1	2.28E-07	400E-06
Nickel	Ni	−12.0	0.68E-07	6,000E-06

Note: All the data related to alloys could greatly vary due to variation of composition and manufacturing process.

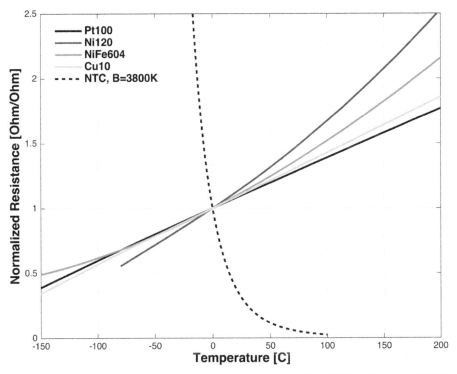

Figure 11.24 Comparison between characteristics of metal and alloy RTDs and a typical NTC. The RTD curves are referred to a pure platinum 100 Ω, pure nickel 120 Ω, nickel–iron alloy 604 Ω, and pure copper 10 Ω devices, respectively. Resistances are referred to 0°C. The NTC characteristic is related to a 1 kΩ device referred to as 25°C (The curve is shifted to the origin for clarity).

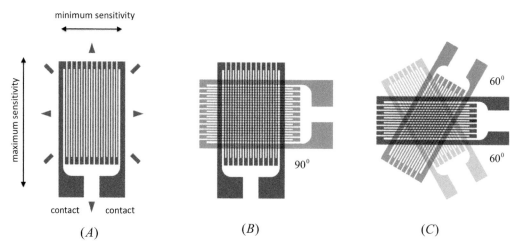

Figure 11.25 Implementation of planar strain gauges. (A) Planar strain gauge with maximum and minimum force sensitivity. (B) Orthogonal direction overlapping planar strain gauges. (C) 60° directions overlapping planar strain gauges.

- Metal RTDs offer better sensitivity but worst linearity with respect to PRTs.
- The NTC characteristic is strongly nonlinear but offers higher sensitivity.

11.5   Application of Force and Temperature Resistance Sensors

This section will introduce some applications where resistive temperature and force resistive sensors could be employed. Furthermore, we will discuss the case where mixed inputs (i.e., temperature variation in force measurement) and other errors should be considered. Before going into detail, in Table 11.7, we list the main thermal and mechanical coefficients of materials used for either thermal and/or force sensors (strain gauges).

11.5.1   Implementation and Readout Techniques of Resistive Sensors

Most implementations of resistive sensors (either force or temperature) in real component devices are based on the deposition or patterning of a conductive layer on an insulating substrate that is flexible in the case of strain gauges. A long line of metal is patterned onto the substrate in a zig-zag shape to achieve a higher resistance, as shown in Fig. 11.25. In strain gauges, this particular disposition offers a higher sensitivity versus stress on the wire along the longitudinal axis instead of the transversal one, as shown in Fig. 11.25A, because in that direction, there is a greater elongation of the entire conductive path. However, even the perpendicular elongation should be considered, as will be shown later. Several strain gauges could be placed on the same

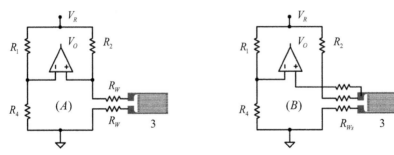

Figure 11.26 Resistance sensing by Wheatstone's bridge. (A) Two-wire quarter bridge configuration. (B) Three-wire quarter bridge configuration.

device substrate in overlapping layers, such as to sense orthogonal forces as shown in Fig. 11.25B or in 60° directions as in Fig. 11.25C.

The most common technique to sense a resistive thermal or force sensor is the use of a Wheatstone bridge, as shown in Fig. 11.26. As seen in Chapter 8, the quarter bridge configuration of Fig. 11.26A provides an output given by the relationship

$$\Delta V_O \approx V_R \frac{1}{4} \frac{\Delta R}{R_0}, \tag{11.79}$$

where V_R is the reference voltage, ΔR is the change of resistance of the sensor, and R_0 is the resistance at rest of the base leg of the bridge. More specifically, referring to (11.46) and (11.65)

$$\frac{\Delta R}{R_0} = \begin{cases} G_F \varepsilon; & \text{for force sensors (strain gauges)} \\ \alpha \Delta T; & \text{for temperature sensors.} \end{cases} \tag{11.80}$$

For example, in the simplest strain gauge application using the quarter bridge implementation, for the one-dimensional structure of section area A, characterized by a material with Young's modulus E_Y and subject to axial traction by force F, the output of the bridge is

$$\Delta V_O \approx V_R \frac{1}{4} \frac{\Delta R}{R_0} = \frac{V_R}{4} \frac{G_F}{AE_Y} F = \frac{V_R}{4} G_F \varepsilon = \frac{V_R}{4} \frac{G_F}{E_Y} \sigma. \tag{11.81}$$

Therefore, the output of the bridge is directly proportional to the force applied to the structure. It should be pointed out that the aforementioned relationship is for one-dimensional structure and *does not take into account* the effect of Poisson's ratio, as discussed in Section 11.5.3.

However, the quarter bridge approach has several drawbacks, especially considering the wire resistances R_W, which could be high to connect the sensor at long distances. This is because the monitoring of mechanical parts is usually realized by placing the interface in a remote location with respect to the sensing spot to reduce the size occupancy. The first drawback is related to the fact that

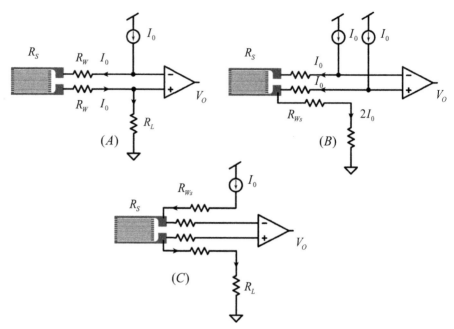

Figure 11.27 Resistance sensing by current sources. (A) Two-wire single current excitation. (B) Three-wire dual-matched current excitation. (C) Four-wire single current excitation (Kelvin sensing).

wire resistances contribute to R_0 thus reducing the sensitivity of the device. Second, any change of R_W due to temperature variations induces an error in the readout. A three-wire quarter bridge technique is used to reduce these drawbacks, as shown in Fig. 11.26B. By using this technique, the input of the instrumentation amplifier is connected to the "hot" terminal of the sensor by a third wire whose resistance does not affect the measurement due to the high impedance of the amplifier. Furthermore, the two original sensor wire resistances are placed on opposite voltage sensing points, gaining two advantages. First, the temperature variations of the wires are compensated; second, the sensitivity is increased with respect to the two-wire version since just one of them contributes to R_0.

Force and temperature resistive sensors could also be interfaced using current excitations, as shown in Fig. 11.27. The simplest approach is based on the two-wire connection, as shown in Fig. 11.27A, where a current reference I_0 sinks into the sensor, and an instrument amplifier senses the voltage drop. However, once again, the wire resistances induce errors since they are summed to the resistance of the sensor R_S: $V_O = I_0(R_S + 2R_W)$.

A second approach using a three-wire dual-matched current excitation is proposed to cope with this problem, as shown in Fig. 11.27B. In this approach, the voltage drop on the wire resistances has the same sign and therefore is canceled out in the differential sensing. A similar approach implements only a single current source but employs four wires, as illustrated in Fig. 11.27C. This is also referred to as *Kelvin sensing,* and thanks to the high impedance of the amplifier, no voltage drop is sensed on the wires in the readout of R_S.

11.5.2   Errors and Influence Variables in Strain Gauges

Several sources of errors in strain gauge sensors should be taken into account, especially for accuracy requirements.

The first is the influence of the transverse sensitivity, which is the sensitivity of the sensor on the direction which is normal to the direction for which it was built, as shown in Fig. 11.25A. Therefore, the true relative variation of resistance is given by

$$\frac{\Delta R}{R_0} = G_a \varepsilon_a + G_t \varepsilon_t = f(G_a, G_t) \tag{11.82}$$

where G_a is the *longitudinal gauge factor* for the longitudinal strain ε_a, and G_t is the *transverse gauge factor* for the transverse strain ε_t.[17] Equation (11.82) is experimentally characterized using the strain gauge on a material with known Poisson's ratio $v_0 = -\varepsilon_t/\varepsilon_a$ (usually $v_0 = 0.285$) to provide a nominal gauge factor G_F as

$$\frac{\Delta R}{R_0} = G_a(1 - v_0 K_t)\varepsilon_a = G_F \varepsilon_a$$
$$K_t = \frac{G_t}{G_a}, \tag{11.83}$$

where K_t is referred to as the *transverse sensitivity coefficient.* Therefore, if we apply the sensor to a material with different characteristics, we have an error

$$\frac{f(G_a, G_t) - G_F \varepsilon_a}{f(G_a, G_t)} = \frac{K_t \left(\dfrac{\varepsilon_t}{\varepsilon_a} + v_0 \right)}{1 - v_0 K_t} \approx K_t \cdot \frac{\varepsilon_t}{\varepsilon_a} \equiv [\cdot], \tag{11.84}$$

where ε_a and ε_t are referred to the material on which it is applied the strain gauge, while v_0 is provided by datasheets and referred to the reference material. The last relationship is a good approximation if the Poisson's ratio of the material on which we sense the strain is different from v_0.

Another important source of error is the effects induced by thermal effects. First, the resistivity of the strain gauge alloy changes versus temperature. Additionally, resistance changes because the thermal expansion coefficient of the alloy is usually different from that of the test material to which it is bonded. To summarize, the two effects could be modeled by the first-order approximation

---

[17] With the notation ε we implicitly refer to a *variation,* since it is defined as the ratio between the length and its value at rest.

$$\frac{\Delta R}{R_0} = [\alpha - G_F(\beta_S - \beta_G)]\Delta T = G_F \varepsilon_{ET}$$

$$\beta \equiv [°C^{-1}] \equiv [K^{-1}], \tag{11.85}$$

where again α is the temperature coefficient (TCR) of the sensor conductive material and β_S, β_G are the thermal expansion coefficients of the specimen and of the strain gauge, respectively, and ε_{ET} is called *temperature-induced apparent strain* or *thermal output*

$$\varepsilon_{ET} = \left[\frac{\alpha}{G_F} - (\beta_S - \beta_G)\right]\Delta T, \tag{11.86}$$

which is a fictional strain due to temperature variations (systematic error) that should be added to the real stimulus readout value.

Equation (11.85) is a linearization of a highly nonlinear function and should be better to refer to datasheet tables. Typical values of thermal expansion for metals (such as steel or aluminum) range from 10 to 20 ppm/°C, and similar ranges occur for the strain gauge alloys. Therefore, a difference of tens of ppm per degree should be considered. This generates errors that could be as high as hundreds of microstrains for hundreds of centigrade variations.

11.5.3   Sensing Techniques for Force Sensors and Error Compensation Techniques

Several techniques could be used to sense forces on the one hand and to reduce the errors early referred. More specifically, as long as errors compensation are concerned, *differential sensing* and/or *dummy sensor* techniques could be effectively employed, as already explained in Chapter 3. Among them, Wheatstone's half-bridge (Chapter 8) and the full-bridge interfaces are widely used, as illustrated in Fig. 11.28.

In general, a first-order approximation of Wheatstone's bridge interface gives the following output:

$$\Delta V_O = \frac{V_R}{4}\left[\frac{\Delta R_3}{R_3} - \frac{\Delta R_2}{R_2} + \frac{\Delta R_1}{R_1} - \frac{\Delta R_4}{R_4}\right]. \tag{11.87}$$

Applying strain gauges in the four points of the bridge, as in Fig. 11.28B, we have

$$\Delta V_O = \frac{V_R}{4}G_F[\varepsilon_3 - \varepsilon_2 + \varepsilon_1 - \varepsilon_4], \tag{11.88}$$

where ε_i are the strains sensed by gauges in the four points of the bridge. This condition could be useful to reduce errors. For instance, using strain gauges in the half-bridge configuration as in Fig. 11.28A, we could use one of them as a dummy sensor, that is, a strain gauge that is not subject to force but at the same temperature to reduce the *thermal output*

$$\Delta V_O = \frac{V_R}{4}G_F[(\varepsilon_3 + \varepsilon_{ET}) - (\varepsilon_{ET})] = \frac{V_R}{4}G_F \varepsilon_3. \tag{11.89}$$

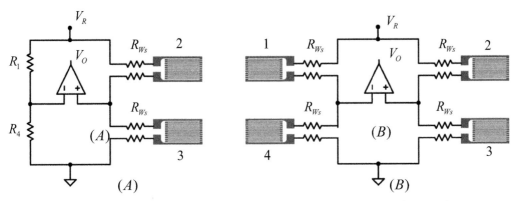

Figure 11.28 (A) half-bridge sensing technique and (B) full-bridge sensing techniques.

The same trick could be used to sense forces and reject undesired effects using the strain gauge dispositions, as illustrated in Fig. 11.29. For instance, using the full-bridge approach for bending strain as shown in Fig. 11.29A, we have that the upper strains are subject to tensile strain and the lower ones to compressive strain. Furthermore, at first-order approximation, we have that the absolute values of the strains are about the same, $\varepsilon_1 = \varepsilon_3 = \varepsilon$, $\varepsilon_2 = \varepsilon_4 = -\varepsilon$ and therefore

$$\Delta V_O = \frac{V_R}{4} G_F[\varepsilon_3 - (-\varepsilon_2) + \varepsilon_1 - (-\varepsilon_4)] = V_R G_F \varepsilon. \tag{11.90}$$

Note that, as before, temperature errors are self-compensated by the differential effect. Instead, for the half-bridge, the relationship is

$$\Delta V_O = \frac{V_R}{4} G_F[\varepsilon_3 - (-\varepsilon_2)] = \frac{V_R}{2} G_F \varepsilon. \tag{11.91}$$

An interesting application is that one illustrated in Fig. 11.29D for measuring axial forces considering Poisson's factor. We sense again antisymmetric stresses $\varepsilon_1 \leftrightarrow \varepsilon_4$; $\varepsilon_2 \leftrightarrow \varepsilon_3$ related to Poisson's factor

$$\Delta V_O = \frac{V_R}{4} G_F[\varepsilon_3 - (-v\varepsilon_2) + \varepsilon_1 - (-v\varepsilon_4)]. \tag{11.92}$$

The main advantages are insensitive to thermal effect and bending forces due to the differential approach. Then, to estimate the original force, we have first to calculate the stress by using the *plane stress* (because the dimension orthogonal to the strain gauge is free of forces) relationships by using (11.37)

$$\sigma_a = \frac{E_Y}{1 - v^2}(\varepsilon_a + v\varepsilon_t), \tag{11.93}$$

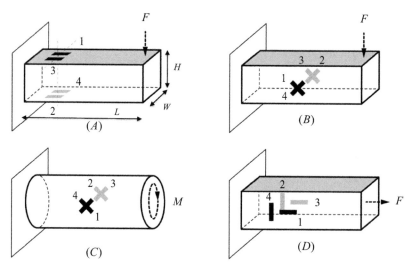

Figure 11.29 Main strain gauges dispositions for full-bridge approach. (A) Bending strain. (B) Shear strain. (C) Torsional strain. (D) Axial strain.

where ε_a and ε_t are the axial strain and the transverse one referred to the gauges, respectively. Therefore, using $\varepsilon_a = \varepsilon_1, \varepsilon_3$ and $\varepsilon_t = \varepsilon_2, \varepsilon_4$ in(11.92) and $\sigma_a = F/A$ we have

$$\Delta V_O = \frac{V_R}{2} G_F [\varepsilon_a + v\varepsilon_t] = \frac{V_R}{2} G_F \frac{\sigma_a}{E_Y}(1 - v^2) = \frac{V_R}{2} \frac{G_F}{AE_Y}(1 - v^2)F, \qquad (11.94)$$

which is similar to (11.81) but with the effect of Poisson's ratio taken into account.

If the strain gauges are not collinear with principal axes, we can estimate the stress in arbitrary directions with respect to those of the sensors, using (11.21) for conversion.

For bending structures (Fig. 11.29A), with a punctual force applied to the end, we could place strain gauges in the place of maximum strain and evaluate the force applied using (11.92) and

$$F = \frac{\sigma_a \cdot Z}{L}, \qquad (11.95)$$

where L is the length of the bar and Z is the *sectional modulus* which is dependent on the cross-sectional shape ($Z = W \cdot H^2/6$ for a rectangular shape, where W and H are the width and height of cross-sectional area).

Referring to Fig. 11.29B and C, we could also use the full bridge to determine either the bending force[18] or the torque using classical mechanics relationships using strain gauges at 45° of direction. In that case, we use the strain differences sensed at the gauges

---

[18] Using the approximations of the Zhuravskii theory.

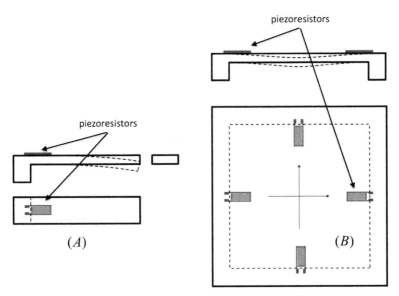

Figure 11.30 Application of piezoresistors in MEMS for cantilevers (A) or square planes (B).

$$\varepsilon_1 - \varepsilon_4 = \varepsilon_2 - \varepsilon_3, \tag{11.96}$$

11.5.4 Application to Microelectromechanical Systems

Piezoresistors (or even piezoelectric material) are employed in MEMs using some simple deformation structures as indicated in Section (11.5.3). Two very common implementations are illustrated in Fig. 11.30 for a cantilever and for a pressure sensor. We can use differential sensing and bridge sensing in previous sections in the latter case.

Piezoresistors could be implemented by patterning proper material using lithographic techniques on flexible surfaces. Many different implementations and approaches could be employed. For further reading, refer to the most common textbooks of MEMs technology.

As a general remark, we summarize some alternative and common implementation of MEMs technology for mechanical sensors:

- The flexible substrate is conductive, and the displacement is measured by capacitive sensing.
- The flexible structure is an insulator, and piezoresistors are patterned on the surface to measure displacement by resistance sensing.
- The flexible substrate is made of piezoelectric material, and the displacement is sensed by measuring the generated free charges on the conductive plates of the piezoelectric material.

Table 11.8 Notations of physical quantities among different sections.

Notation	Mechanical section	Piezoresistivity section	Piezoelectricity section	Units
Stress	T, σ, τ	σ, τ	T, σ, τ	[Pa]
Strain	ε		S	[·]
Stiffness coefficients (Hooke's law)	E_Y, G, C	E_Y		[Pa]
Dielectric permittivity			ε	$[\mathrm{F} \cdot \mathrm{m}^{-1}]$
Compliance coefficients	S		s	$[\mathrm{Pa}^{-1}]$
Electric field		E	E	[V/m]
Resistivity		ρ		$[\Omega \cdot \mathrm{m}]$

- The flexible structure is placed in electromechanical resonance, and the mechanical interactions are read out by variation of the operating frequency, such as in quartz crystal microbalance (QCM) devices.

11.5.5   Errors in Resistance Temperature Sensors

When we measure the temperature by using a resistor, we should consider that it dissipates power as heat and determines a temperature change. More specifically, from Fourier's heat law, we have

$$P = K\left(T(R) - T_0\right)$$
$$K \equiv [\mathrm{W}/^{\circ}\mathrm{C}], \tag{11.97}$$

where P is the power dissipated by the resistance, $T(R)$ is the temperature of the sensor that is sensed by means of the resistance, and T_0 is the temperature of the surroundings. The factor K is called *dissipation constant* and depends on the packaging and its disposition in the environment. If we consider the electrical power dissipated by the resistance, we have

$$T_0 = T(R) - \frac{I^2 R}{K}, \tag{11.98}$$

where I is the sensing current. Therefore, there is an error between the true temperature T_0 and the temperature interpolated by the sensor curve $T(R)$. This phenomenon is referred to as the *self-heating effect*. In general, the dissipation constant value is provided in datasheets; however, its variance is very difficult to calculate since it also depends on the ambient. Therefore, it should be better to design sensor interfaces to keep the self-heating effect below the precision requirements. Typical values range from 5 to 50 mW/°C.

11.6   Appendix: Notations Used in the Chapter

In the literature and books there are many overlapping notations for physical quantities, especially in thermal and mechanical transduction. The attempt to uniform the notation would be more confusing than using individual notation for the different sections of the Chapter. Table 11.8 provides a summary of the symbols used.

Further Reading

Caspari, E., and Merz, W. J., The electromechanical behavior of BaTiO$_3$ single-domain crystals, *Phys. Rev.*, vol. 80, no. 6, pp. 1082–1089, Dec. 1950.

Feynman, R. P., Robert, B. L., Sands, M., and Gottlieb, M. A., *The Feynman Lectures on Physics*. Reading, MA: Pearson/Addison-Wesley, 1963.

Institute of Electrical and Electronics Engineers, 176–1987 – ANSI/IEEE Standard on Piezoelectricity, 1988.

Kaajakari, V., *Practical MEMs*. Small Gear Publishing, 2009.

Kanda, Y., A graphical representation of the piezoresistance coefficients in silicon," *IEEE Trans. Electron Devices*, vol. 29, no. 1, pp. 64–70, 1982.

Kao, K. C., *Dielectric Phenomena in Solids*. Philadelphia: Elsevier Science, 2004.

Liu, C., *Foundations of MEMs*. Upper Saddle River, NJ: Pearson, 2012.

Kerr, D. R., and Milnes, A. G., Piezoresistance of diffused layers in cubic semiconductors, *J. Appl. Phys.*, vol. 34, no. 4, pp. 727–731, 1963.

Madou, M., *Fundamentals of Microfabrication: The Science of Miniaturization*, 2nd ed. Boca Raton, FL: CRC Press, 2002.

Nemeth, M. P., An in-depth tutorial on constitutive equations for elastic anisotropic materials," NASA Report NASA/TM–2011–217314, 2011.

Senturia, S. D., *Microsystem Design*. New York: Springer Science+Business Media, 2000.

Smith, C. S., Piezoresistance effect in germanium and silicon, *Phys. Rev.*, vol. 94, no. 1, pp. 42–49, 1954.

Vishay Precision Group, Strain gage thermal output and gage factor variation with temperature," 2012.

Warren Young, A. S., and Budynas, R., *Roark's Formulas for Stress and Strain*, 8th ed. New York: McGraw-Hill, 2011.

www.acromag.com, Criteria for temperature sensor selection of T/C and RTD Sensor types, 2011.

Part IV

Problems and Solutions

12 Problems and Solutions

Marco Crescentini[1]

12.1 Problems

1. A thermometer measures the temperature in the operating range [−40°C, 80°C] and provides an output voltage ranging from 0 V to 4 V. Assuming a purely linear input–output relationship, calculate the full-scale (*FS*), the full-scale output (*FSO*), the output offset, the input offset, and the slope of the static characteristic for this thermometer.

2. Calculate the relative sensitivity at room temperature of 25° C for the thermometer described in Problem 1.

3. A supermarket automatic weight meter needs to measure up to 50 kg of merchandise, and it is able to distinguish 5 g of variation to check the item barcode with its weight. What are the required resolution and the dynamic range? If I had to interface with an A/D converter, what would its effective number of bits (ENOB) be?

4. Calculate the sensitivity and the relative sensitivity at room temperature of 25° C for a PT1000 platinum resistance thermistor. Assume the following temperature–resistance relationship

$$R(T) = R_0(1 + AT + BT^2),$$

where $A = 3.91 \times 10^{-3}$° C^{-1} and $B = -5.77 \times 10^{-7}$° C^{-2}.

5. The following relationship rules a capacitive displacement sensor

$$C(d) = \frac{C_0}{\left(1 + \dfrac{d}{h_0}\right)},$$

where d is the displacement from the nominal position, C_0 is the nominal capacitance when the sensor is in the steady-state position, and h_0 is the nominal distance between the plates. Calculate the sensitivity and relative sensitivity around the zero-displacement point.

6. A Rogowski coil is used to sense a very fast variation of the magnetic field in particle accelerators, with a time constant of about 100 ns. Assuming the Rogowski coil's dynamic response can be described by a dominant pole approximation, calculate the minimum bandwidth of the sensor to get a settling error of less than 0.1%.

[1] Adjunct Professor, Alma Mater Studiorum, University of Bologna

7. A single-sided white spectrum describes additive thermal noise with a $10\ nV/\sqrt{Hz}$ noise density. Calculate the noise power in 1 MHz bandwidth and the related rms noise value.

8. An analog accelerometer with a sensitivity of $S = 0.3\ V/g$ has the following output noise characteristics: white noise $v_N = 400\ nV/\sqrt{Hz}$ and pink noise corner frequency $f_C = 2.3\ Hz$. What is the total noise if the system is limited to $f_0 = 10\ Hz$, and what is its minimum detectable signal?

9. A conductimeter is used to estimate the conductivity of a water sample. The measurement is repeated 10 times at time intervals of 1 second. The results of the measurement are reported in the table below. Assuming the conductivity meter has random errors and no systematic errors, estimate the measurement's uncertainty.

Experiment number	1	2	3	4	5	6	7	8	9	10
Measured conductivity C [mS/cm]	13.4	15.4	8.9	13.9	13.0	10.4	11.8	13.0	18.2	16.9

10. Evaluate the noise-free resolution of the conductivity meter described in the previous problem.

11. To improve the resolution of the conductivity meter described in Problem 9, we decided to perform post-processing averaging over M lectures (or experiments, as stated in the previous table). Define the number of lectures M to be averaged so as to get an rms error due to the noise of less than 1.5 mS/cm. Assuming each readout is taken at a time step of 1 second, what are the sensor throughput and bandwidth after the averaging process?

12. An amplifier with gain equal to $S = 0.1\ V/\mu V$ has $\sigma_N(y) = 350\ \mu V$ rms output noise in a given bandwidth and is interfaced with a 12b ENOB A/D amplifier with $V_R = 3.3\ V$. What is the input-referred resolution?

13. A conductimeter changes its conductance $G \equiv [mS]$ with respect to the measured conductivity $C \equiv [mS/cm]$, and it is characterized in temperature by the characteristic

$$G(C, T) = F(C, T) = \frac{1}{1 - k(T - 20)} \cdot \frac{\alpha C}{1 - \beta C}$$

with respect to the reference temperature $T_0 = 20°C$ where $\alpha = 0.8992$ cm, $\beta = 6.4 \times 10^{-5}$ cm/mS, and $k = 0.03°$ C. The input operating range is $C = 10 \div 100$ mS $\cdot$ cm^{-1}. Sketch a graph showing the ideal static characteristic and the boundary safety lines defining the maximum deviations of the characteristic due to temperature variations. Use a coverage factor $p = 2$.

The conductimeter gives an output value $G_0 = 50$ mS at the reference temperature. What are the estimated input value and the corresponding measurement uncertainty considering ambient temperature variations having Gaussian distribution with $\sigma(T) = 3°$ C?

14. With reference to Problem 13, calculate the average nonlinearity error with respect to the ideal characteristic at the reference temperature

$$G(C, T_0) = \alpha C$$

and with respect to the regression line. Finally, characterize the overall integral nonlinearity (INL).

15. A conductivity meter is used in an acquisition chain together with a resistance-to-voltage converter that suffers from saturation when the magnitude of the output voltage is greater than 2 V. The system is employed to measure the conductivity of biological fluids in the operating range 10 ms/cm to 20 ms/cm. The output of the conductivity meter is expressed in Ω, and its gain G_I is equal to 625,000 Ω^2 cm. The gain plot of the measurement chain is reported in Figure 12.1. Calculate the gain and offset of the resistance-to-voltage (R/V) converter that maximizes the *FSO* without causing saturation.

16. A Hall-effect sensor is a magnetic sensor that generates a differential voltage at the output when it is affected by a magnetic field along the normal direction with respect to the Hall plane. The following static characteristic rules the sensor

$$V_H = S_I I_b B_z,$$

where V_H is the output Hall voltage, S_I is a constant gain factor, I_b is the bias current flowing through the sensor, and B_z is the input magnetic field along the z-axis. The

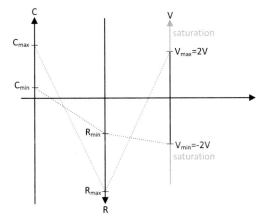

Figure 12.1 Gain plot of the acquisition chain composed of a conductivity meter (C to R) and a resistance-to-voltage converter (R to V).

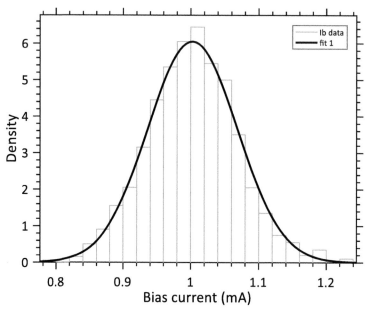

Figure 12.2 Statistical distribution of the bias current over 1000 samples.

bias current is generated by a noisy current source that can be statistically characterized by a mean value of 1 mA and a standard deviation of 66 µA, as shown by its statistical distribution reported in Fig. 12.2.

Assuming $S_I = 1000$ V/AT, evaluate the uncertainty on the output voltage due to the randomness of the bias current and graphically visualize the result on the static characteristic over the input range ±1 mT.

17. The Hall-effect sensors are always characterized by an additive thermal noise due to the agitation of the electrons in the Hall plate and an additive offset due to asymmetries in the physical realization of the sensor. Assuming the thermal noise described by an equivalent noise resistance of 3 kΩ at the temperature of 300 K and bandwidth of 100 kHz, and assuming an additive offset voltage described by a uniform distribution in the range ±0.3 mV, compute the input-referred noise and the input-referred offset by using the parameters of Problem 16.

18. For the Hall-effect sensor described in Problems 16 and 17 and taking into account a thermal noise of 2.2 µV rms, an output-offset uniformly distributed between ±120 µV, and random variations of the bias current with a standard deviation of 12 µA, calculate

 – The input-referred resolution with a coverage factor $p = 3$
 – The accuracy-related effective number of bits

Assume the magnetic field is typically a sine wave that ranges between −1 mT and 1 mT.

19. The output voltage of the system described in Problem 18 is time-averaged over 10 consecutive samples through a moving averages algorithm to improve the system's accuracy. Calculate the new accuracy-related ENOB.

20. A galvanically isolated current sensor is composed of three stages:

   1. The noiseless current to be sensed flows through a copper strip and generates a magnetic field according to Biot-Savart's law $B = \dfrac{\mu_0 I}{2\pi r}$, where μ_0 is the air permeability $(12.56\mathrm{E}^{-7}\ \mathrm{TmA}^{-1})$, and r is the distance of the magnetic sensing element. No noticeable errors or fluctuations are added by this stage.
   2. The generated magnetic field is sensed by a Hall-effect sensor with sensitivity $S_I = 1000$ V/AT and 1 mA of bias current. This stage adds its own thermal noise of 20 µV rms at the output.
   3. The Hall voltage is then amplified by a factor G_3 by an electronic amplifier that saturates for output voltage levels outside the [0, 1] V interval. This stage adds its own thermal noise of 20 µV rms at the output.

   The entire acquisition chain is depicted in Fig. 12.3. Note that the distance r between the copper strip and the Hall sensor cannot be arbitrarily reduced, since, for $r < 100$ µm, the thickness of the strip must be shrunk, and thus the maximum current is reduced according to
   $$I_{max} = \alpha r,$$

   where $\alpha = 200$ mA/µm. Assuming the application requires an input full-scale of 16 A (from 0 A to 16 A), calculate the distance r and the gain G_3 that maximize the FSO without causing saturation and maximizes the $\mathrm{SNR_p}$ for a DC input signal. Write the achieved input-referred $\mathrm{ENOB_p}$.

21. With reference to the acquisition chain described in Problem 20, calculate the minimum detectable signal and the DR for an input sine wave.

22. Given a power consumption of 22 mW and assuming a bandwidth of 500 kHz, calculate the sensor figure of merit (FoM) of the acquisition chain described in Problem 20.

23. A current readout circuit (i.e., current-to-voltage converter) has been characterized in the laboratory by applying known current values (we neglect the uncertainty on these reference values) and recording the output voltages after averaging over 10 seconds. The results are reported in the table below. Calculate

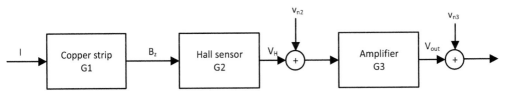

Figure 12.3 Acquisition chain realizing a galvanically isolated current sensor. Noise sources are modeled as additive noises.

– Offset and gain with respect to the best fitting straight line
– INL expressed as a percentage of the *FSO*
– Input-referred uncertainty due to nonlinearity error
– The accuracy-related ENOB for a DC input and neglecting gain and phase error

Reference input current (nA)	−20	−15.5	−11.1	−6.6	−2.2	2.2	6.6	11.1	15.5	20
Measured output voltage (V)	0.136	0.415	0.792	1.15	1.488	1.808	2.111	2.4	2.674	2.936

24. After a calibration process, it has been found that the characteristic of a microphone working in the *FS* range of 0.1 mPa can be described by

$$V_{OUT} = V_{off} + GP_{IN} + \alpha P_{IN}^2,$$

where G = 72 kV/Pa and α = 1 MV/Pa². Implement the equation on a numerical simulator (e.g., MATLAB) and evaluate the total harmonic distortion (THD) introduced by the microphone on an ideal input pressure sine wave of 5 µPa amplitude and 13 kHz frequency. To properly implement the equation in the numerical simulator, assume the output voltage ideally sampled at 200 kHz.

25. A laser meter has to measure distances from 100 m to 10 cm with 1% of accuracy. What should be the ENOB characteristics of the time-to-digital converter?

26. Thermal noise of 0.8 µV rms is added to the input of the A/D converter of Problem 23. Calculate the $ENOB_{SNDR}$ assuming a sine wave input.

27. Calculate the ENOB of the optimal A/D converter to be coupled with the microphone described in Problem 24 so as to not degrade the accuracy of the acquisition chain. Assume offset and gain error of the microphone completely compensated during calibration, an output-referred noise of the microphone of 330 µV rms and calculate the ENOB with respect to sinusoidal input signals.

28. Assuming the ADC selected in Problem 27 implemented in an architecture and technology at the limit of Walden's figure of merit of 1 fJ/*NL*, estimate the minimum power consumption of the ADC for a sampling frequency of 96 kHz.

29. The PT1000 thermistor described in Problem 4 is arranged as a Wheatstone bridge together with 3 precision resistors of value 1 kΩ (see Fig. 12.4). The bridge is supplied by V_B = 5 V. Calculate the total sensitivity of the whole sensor composed of thermistor and Wheatstone's bridge. Sketch the gain plots.

30. An instrumentation amplifier with gain G = 2 is connected at the output of the Wheatstone's bridge of Problem 29. The amplifier is powered by a double-supply voltage of ±15 V and has a perfectly linear input–output characteristic with saturation levels of 14 V and −14 V. Sketch the gain plot for this new sensing chain and calculate both the *FS* and the *FSO* that maximize the DR.

31. The acquisition chain of Problem 30 has a very narrow *FS*; modify the bridge to increase the *FS* by a factor of 5 without changing the *FSO*. The same PT1000 sensor must be used.

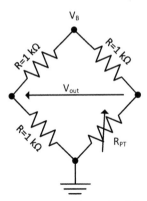

V_B

R=1 kΩ   R=1 kΩ

V_{out}

R=1 kΩ R_{PT}

Figure 12.4   Schematic of the Wheatstone bridge used with the PT1000 thermistor.

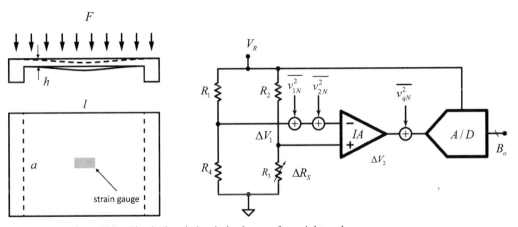

Figure 12.5   Physical and circuital scheme of a weight scale.

32. A weight scale is implemented, as illustrated in Fig. 12.5, where a strain gauge is placed underneath the middle point of a flat glass slab receiving the weight to be measured. The strain gauge is placed in a Wheatstone bridge which is interfaced to an A/D converter by an instrumentation amplifier (IA).

We want to measure up to 100 kg with a resolution of 10 g in less than 1 second. The data are the following: $h = 5$ mm, $l = a = 50$ cm, $R_x = 120\ \Omega$, $V_R = 3.3$ V, $G_F = 2.1$. What should be the input-referred noise of the IA and the characteristic of the A/D to have a stable (noise-free) readout of the measure?

12.2   Solutions

1. The full-scale span is the difference between the maximum and minimum value of the signal (in this case, the signal corresponds to the information). The *FSO* has the same definition.

$$FS = x_{max} - x_{min} = 120 \text{ °C}$$
$$FSO = y_{max} - y_{min} = 4 \text{ V}$$

A linear relationship means the output y can be related to the input x; thus, the slope S, the output offset y_{off}, and the input offset x_{off} can be computed as

$$S = \frac{FSO}{FS} = 33 \text{ m}\frac{V}{°C}$$

$$y_{off} = y_{min} - Sx_{min} = 1.32 \text{ V}$$

$$x_{off} = x_{min} - y_{min}/S = -40 \text{ °C}.$$

2. We first have to calculate the bias point y_0 for $x_0 = 25° \text{ C}$:

$$y_0 = Sx_0 + y_{off} = 2.155 \text{ V}$$

$$S' = \frac{S}{y_0} = 0.015 \, \frac{1}{°C}$$

3. The resolution (minimum detectable signal) is 5 g, and the dynamic range and the required effective number of bits are

$$DR = 20\log\frac{50}{5 \times 10^{-3}} = 80 \text{ dB}; \quad ENOB = \frac{80 - 1.76}{6.02} = 12.9 \text{ b.}$$

We should use an A/D converter with at least ENOB = 14 b. Note that we used $q=1$ because here, it is a DC measurement.

4. Note that a PT1000 sensor is a platinum thermistor with a nominal resistance value of 1 kΩ. Thus

$$S = \frac{dR}{dT} = R_0A + 2R_0BT$$

$$S|_{T=25} = 3.91 \, \frac{\Omega}{°C}$$

$$R(25°C) = R_0(1 + AT + BT^2) = 1097.4 \, \Omega$$

$$S' = \frac{S|_{T=25}}{R(25°C)} = 3.56E^{-3} \, \frac{1}{°C}.$$

5. The concept of sensitivity of can be extended to a multi-variate function by using partial derivatives and expressing a sensitivity to each variable. Thus, the sensitivity of the sensor to displacement can be calculated as

$$S = \frac{dC}{dd}\bigg|_{d=0} = \frac{-C_0}{h_0}\frac{1}{\left(1 + \dfrac{d}{h_0}\right)^2} \, \left[\frac{F}{m}\right]$$

$$S' = \frac{S|_{d=0}}{C_0} = \frac{-1}{h_0} \, \left[\frac{1}{m}\right].$$

6. A first-order system settles to the response value with an error of 0.1 % after 6.9 time constants.

$$BW \geq \frac{1}{2\pi(6.9\tau)} = 230 \text{ MHz}.$$

7. The noise power could be computed as

$$P_n = \int_0^{1MHz} (10E^{-9})^2 df = 100 \text{ pV}^2$$

$$n_{rms} = \sqrt{P_n} = 10 \text{ μV}.$$

8. We first calculate the coefficient of pink noise as

$$k_P = k_W \sqrt{f_C} = 606 \text{ nV}.$$

Then we assume a minimum integration frequency corresponding to 1 year of observation time

$$f_L = \frac{1}{365} = 3.2 \times 10^{-8} \text{ Hz}$$

therefore, the total output noise could be computed as

$$\sigma_N(y)@10 \text{ Hz} = \sqrt{k_W^2 \left(\frac{\pi}{2}f_0 + f_C \ln\frac{f_0}{f_L}\right)} = 3.11 \text{ μV}.$$

The input-referred noise and thus the minimum detectable signal is

$$MDS = \sigma_N(x) = \frac{\sigma_N(y)}{S} = 10.3 \text{ μg}.$$

9. The variance of the random process quantifies the power of a noise process. It is possible to estimate the variance of the noise process by using the experimental variance calculated over the measurement results (Fig. 12.6)

$$\overline{C} = \frac{1}{N}\sum_i C^{(i)} = 13.48 \frac{mS}{cm}$$

$$s_n^2(C) = \frac{1}{N-1}\sum_i \left(C^{(i)} - \overline{C}\right)^2 = 7.96 \left[\frac{mS}{cm}\right]^2$$

$$u(C) = \frac{S_n}{\sqrt{N}} = 2.51 \left[\frac{mS}{cm}\right],$$

10. The noise-free resolution is defined as the input-referred noise standard variation multiplied by 6.6 to take a confidence level of 99.9 % into care. Estimating the

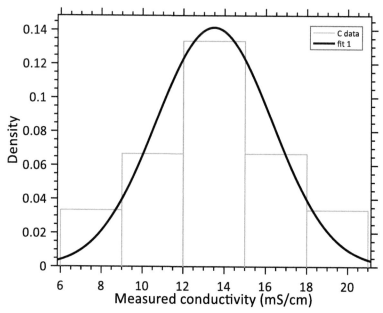

Figure 12.6 Probability density functions fitted over the measured values.

noise standard deviation by the experimental standard deviation $s_n(C)$ computed in the previous problem:

$$\sigma_n(C) \leftarrow s_n(C)$$

then the noise-free resolution is

$$\text{noise-free resolution} = 6.6 \cdot \sigma_n(C) = 18.62 \; \frac{\text{mS}}{\text{cm}}.$$

11. According to the central limit theorem, the averaging process reduces the noise variance (see Section 2.8.3).

$$s'_n \leq \frac{s_n}{\sqrt{M}} \Rightarrow M \geq \frac{s_n^2}{s'^2_n} = \frac{7.96}{2.25} = 3.5.$$

Thus, it is required to average over at least four lectures to get the desired resolution. Note the averaging process reduces the bandwidth by a factor M (i.e., 4) while the throughput is unaffected as long as the moving average is implemented.

12. We calculate first the quantization noise and then the overall noise at the input of the A/D converter (output of the amplifier):

$$\sigma_Q = \frac{3.3}{2^{12}\sqrt{12}} = 233 \; \mu V; \quad \sigma(v) = \sqrt{(350 \times 10^{-6})^2 + (233 \times 10^{-6})^2} = 420 \; \mu V.$$

Therefore, the input-referred noise (minimum detectable signal) is

$$\sigma(x) = \frac{350 \times 10^{-6}}{1 \times 10^5} = 4.2 \text{ nV}.$$

13. The characteristic for different influence parameters is plotted in Fig. 12.7, showing the nominal static characteristic of the sensor between two safety lines defined by a coverage factor $p = 2$.
    For the input operating range $FS = 90 \text{ mS} \times \text{cm}^{-1}$ and with respect to the reference temperature, the output operating range is $G = 8.99 \div 89.92 \text{ mS}$ so that $FSO = 80.93 \text{ mS}$.
    The input is estimated from the output value $G_0 = 50 \text{ mS}$ using the inverse characteristic function at reference temperature $T_0 = 20°C$ so that

$$C_0 = F^{-1}(G_0, T_0) = \frac{G_0}{(\alpha + \beta G_0)} = 55.41 \text{ mS/cm}.$$

For the uncertainty, the law of propagation of uncertainty states that the variations propagate along the sensor chain by means of the square of the sensitivities computed around the bias point or the estimated point

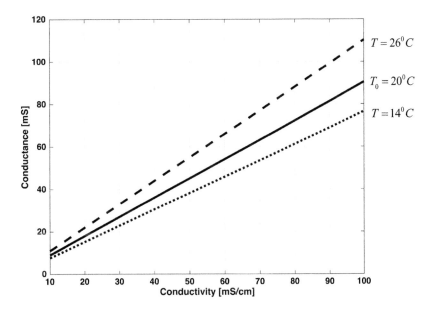

Figure 12.7 Static characteristic and safety lines describing a coverage factor of 2.

$$u_T^2(G) = \left(\frac{dG}{dT}\bigg|_{C_0,T_0}\right)^2 u^2(T) = [k \cdot G(C_0, T_0)]^2 u^2(T); \text{ where } G(C_0, T_0) = \frac{\alpha C_0}{1 - \beta C_0},$$

from which

$$u_T(C) = [k \cdot G(C_0, T_0)]\sigma(T) = 4.5 \ \frac{mS}{cm}.$$

The output uncertainty is now referred to the input by the derivative of the function

$$F'(C_0, T_0) = \frac{dG}{dC}\bigg|_{T_0,C_0} = \frac{\alpha}{(1 - \beta C_0)^2} = 0.90 \ \frac{mS \times cm}{mS}.$$

Therefore, the input-referred uncertainty is

$$u_T(G) = \frac{u_T(C)}{F'(C_0, T_0)} = \frac{u_T(C)}{0.90} = 4.97 \ \frac{mS}{cm}$$

so that we can say that the measurement is $C_0 = 55.41 \pm 14.9$ mS/cm with 99.7% (± 3 sigmas) of confidence that the true value is inside the interval.

14. We can derive the solution to the problem numerically using MATLAB. We first characterize the error with respect to the reference line(s) as

$$\Delta y_{(D)ppm}(x) = \frac{y_{reference}(x) - y_{real}(x)}{FSO} \times 10^6 \equiv [\text{ppm}].$$

The systematic errors are shown in Fig. 12.8 for the two reference lines, respectively. Then, we calculate the average error and the variance. For the ideal characteristic reference, we have

$$\overline{\Delta y_{(D)ppm}} = \frac{1}{FS}\int_{FS} \Delta y_{(D)}(x)dx = 2646 \text{ ppm}$$

$$\sigma^2_{(D)ppm}(y) = \frac{1}{FS}\int_{FS} \left(\Delta y_{(D)}(x) - \overline{\Delta y_{(D)}}\right)^2 dx \rightarrow \sigma_{(D)}(y) = 2097 \text{ ppm}$$

We can also find the interpolating linear regression function numerically as

$$G = 0.906 \cdot C - 0.136$$

Thus, the gain and offset errors are

$$\Delta y_{off} = -0.136$$
$$\Delta y_{gain(max)} = (0.906 - 0.899) \cdot C = 0.636 \ (\text{max for } C = 100)$$
$$\max(\Delta y_{off} + \Delta y_{gain}) = 0.636 - 0.136 = 0.500 \ mS \rightarrow \frac{0.500}{80.93} = 6178 \text{ ppm}.$$

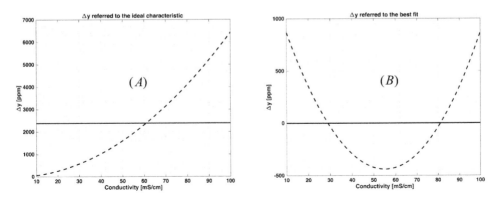

Figure 12.8 Deviation of the real characteristic with respect to the ideal characteristic (A) and with respect to the best fit line (B). All values are normalized to *FSO*.

With respect to that function, the average is, of course, zero (sum of residuals of the regression interpolation should be zero), while the standard deviation is

$$\sigma^2_{(D)ppm}(y) = \frac{1}{FS} \int_{FS} \left(\Delta y_{(D)}(x) - \overline{\Delta y_{(D)}} \right)^2 dx \ \rightarrow \ \sigma_{(D)}(y) = 436 \text{ ppm}$$

We can always refer the error to absolute values using

$$\sigma_{(D)}(y) = \frac{\sigma_{(D)ppm}(y) \cdot FSO}{10^6}.$$

In the case of regression line reference, the INL is

$$\text{INL} = \frac{\max\{\Delta y_{(D)}\}}{FSO} = 969 \text{ ppm}.$$

However, in the case of distortion with respect to the regression line, we have always to consider the gain and offset errors to be added (or canceled out by calibration).

Now, we can refer the nonlinearity error to the input, that in the case of ideal reference characteristic is

$$\sigma_{(D)}(x) = \frac{\sigma_{(D)}(y)}{\alpha} = 0.1887 \text{ mS/cm}.$$

Therefore, referring to the previous problem result, we can say that we have an uncertainty (only due to nonlinearities) of the readout: $C_0 = 45.10 \pm 0.1887$ mS/cm(± 1 sigmas)

Using this value, the accuracy (DC measurement) due only to nonlinearity errors is:

$$\text{SNR}_a = 20\log\frac{FSO}{\sigma_{(D)}(y)} = 123 \text{ dB} \rightarrow \text{ENOB}_a = \frac{123 - 1.76}{6.02} = 20.1 \text{ b}$$

On the other hand, assuming offset and gain errors compensated, the nonlinearity error with respect to the regression line could be even improved as

$$\sigma_{(D)}(x) = 0.0393 \text{ mS/cm}.$$

15. First, we must calculate the full-scale of the resistance variation

$$R_{min} = G_1 \cdot C_{min} = 3125 \text{ } \Omega$$
$$R_{max} = G_1 \cdot C_{max} = 6250 \text{ } \Omega.$$

The gain that maximizes the FSO can be calculated as

$$G_2 = \frac{FSO}{R_{max} - R_{min}} = 1.28 \frac{\text{mV}}{\Omega}.$$

Since the output voltage changes of sign, contrarily to the input conductivity, then an additive offset voltage must be added to prevent saturation of the R/V converter. This voltage is simply defined as the voltage to be added so that the minimum output voltage is equal to the negative saturation level:

$$V_{min} = -2 \text{ V}$$
$$V_{min} = R_{min} \cdot G_2 + V_{os} = -2$$
$$V_{os} = -2 - R_{min} \cdot G_2 = -6 \text{ V}.$$

16. The problem can be solved similarly to Problem 13 since the dispersion on the bias current can be treated as a dispersion on an influencing parameter. Thus, the sensitivity of the output voltage to the bias current must be evaluated, and uncertainty must be propagated by using the law of propagation of uncertainty

$$u_{I_b}^2(V_H) = \left(\frac{dV_H}{dI_b}\right)^2 \cdot u^2(I_b) = (S_I B_z)^2 \cdot u^2(I_b).$$

Note that the output-referred uncertainty depends on the value of the input signal B_z and a single value cannot be estimated without knowing the input magnetic field. Conversely, it is possible to plot the characteristic function graphically, showing the effect of bias current dispersion.

17. The description of noise and offset can be added to the model as follows

$$V_H = S_I I_b B_z + V_N + V_{off},$$

where V_n is the thermal noise and V_{off} is the additive offset.
The noise is described as a thermal noise generated by an equivalent resistance of value 3 kΩ; thus, its variance can be evaluated as

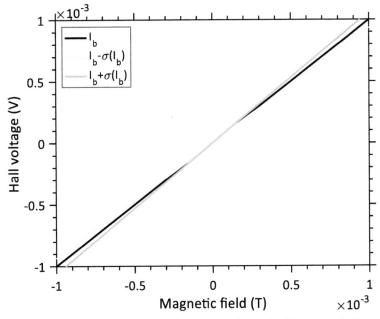

Figure 12.9   Variation of the static characteristic due to the bias current's dispersion.

$$u_N^2(V_H) = \sigma_N^2 = 4KTR_{eq}B = 4.97E^{-12} \text{ V}^2.$$

To refer these uncertainties back to the input, we have to express the input quantity B as a function of the output Hall voltage.

$$V_H = S_I I_b B_z + V_N + V_{\text{off}}$$

$$B_z = \frac{V_H - V_N - V_{\text{off}}}{S_I I_b}$$

$$u_N^2(B_z) = \left(\frac{dB_z}{dV_H}\right)^2 u_N^2(V_H) = \frac{u_N^2(V_H)}{(S_I I_b)^2} = 4.97E^{-12} \text{ T}^2$$

$$u_N(B_z) = 2.23 \ \mu\text{T}$$

$$u_{\text{off}}^2(V_H) = \frac{0.002^2}{12} = 3E^{-8} \text{ V}^2$$

$$u_{\text{off}}^2(B_z) = \left(\frac{dB_z}{dV_H}\right)^2 u_{\text{off}}^2(V_H) = \frac{u_{\text{off}}^2(V_H)}{(S_I I_b)^2} = 3E^{-8} \text{ T}^2$$

$$u_{\text{off}}(B_z) = 173 \ \mu\text{T}.$$

18. The concept of resolution is related to random variation of the output. These variations can be due to thermal noise and other sources of uncertainty. Therefore, only thermal noise and random variations of the bias current must be accounted for in calculating the input-referred noise. The input-

referred resolution is computed by referring to the input, by means of the sensitivity, the output-referred resolution.

$$u_N^2(V_H) = \sigma_N^2(V_H) = u_{I_b}^2(V_H) + u_N^2(V_H) = \left(\frac{dV_H}{dI_b}\right)^2 u^2(I_b) + u_N^2(V_H)$$

$$\sigma_N^2(V_H) = (S_I B_z)^2 \cdot u^2(I_b) + u_N^2(V_H)$$

$$\Delta x = \frac{2 \cdot p \cdot \sigma_N(V_H)}{S_I I_b} = \frac{2p}{S_I I_b} \sqrt{(S_I B_z)^2 \cdot u^2(I_b) + u_N^2(V_H)}.$$

Since the random dispersion of the bias current is not constant over the *FS*, the resolution is not constant over the *FS*. However, we can calculate a minimum and a maximum input-referred resolution (for a magnetic field of 1 mT)

$$\Delta x_{min} = \frac{2p}{S_I I_b} u_n(V_H) = 13.37 \, \mu\text{T}$$

$$\Delta x_{max} = \frac{2p}{S_I I_b} \sqrt{(S_I B_z)^2 \cdot u^2(I_b) + u_n^2(V_H)} = 181 \, \mu\text{T}.$$

Accuracy-related ENOB takes into account randomness and systematic errors; thus, it is not constant as well. Better accuracy is achieved for small-input magnetic fields. To be on the safe side, a minimum ENOB can be calculated on the basis of the maximum error given at the maximum input signal power. Total uncertainty (at 1 sigma) on the estimate of the magnetic field can be calculated by RSS all the uncertainty components as long as they are uncorrelated

$$u_{tot}(B_z) = \sqrt{u_{off}^2(B_z) + u_N^2(B_z) + \max[u_{I_b}^2(B_z)]} = 75 \, \mu\text{T}$$

$$\text{SNR}_a = 20\log\left(\frac{FS}{q \cdot u_{tot}(B_z)}\right) = 27.38 \, \text{dB}$$

$$\text{ENOB}_a = \frac{\text{SNR}_a - 1.76}{6.02} = 4.2 \, \text{bit}.$$

19. The average is performed over the time domain; thus, only sources of uncertainty that generate a dispersion of the output voltage over the time domain are reduced. This means that offset is unaffected while the uncertainties due to thermal noise and bias variation are divided by the square root of 10

$$u_{tot}(B_z) = \sqrt{u_{off}^2(B_z) + \frac{u_n^2(B_z)}{N} + \frac{\max[u_{I_b}^2(B_z)]}{N}} = 70 \, \mu\text{T}$$

$$\text{SNR}_a = 20\log\left(\frac{FS}{q \cdot u_{tot}(B_z)}\right) = 28 \, \text{dB}$$

$$\text{ENOB}_a = \frac{\text{SNR}_a - 1.76}{6.02} = 4.4 \, \text{bit}.$$

The accuracy improvement given by the averaging process is negligible as far as the offset limits the accuracy.

20. As discussed in Chapter 3, the maximization of the resolution in an acquisition chain is achieved by maximizing the operating ranges at each stage, according to all the limitations provided by the technology. In this case, the gain G1 can be theoretically increased at will by reducing the distance r. The technology poses a limit to the gain for r lower than 100 μm by relating the FS to the distance r. From this relationship, it is possible to find the minimum distance that allows the desired input FS

$$r = \frac{I_{max}}{\alpha} = 80 \ \mu m$$

$$G1 = \frac{\mu_0}{2\pi r} = 2.5 \ \frac{mT}{A}.$$

The gain of the third stage is now adjusted to map the FS to the maximum FSO without causing saturation

$$G3 = \frac{FSO}{FS \cdot G1 \cdot G2} = 25 \ \frac{V}{V}$$

We can now calculate the SNR_p and the $ENOB_p$ (at 1 sigma)

$$SNR_p = 20\log \left(\frac{FS}{\frac{v_{n2}}{G1 \cdot G2} + \frac{v_{n3}}{G1 \cdot G2 \cdot G3}} \right) = 66 \ dB$$

$$ENOB_p = \frac{SNR_p - 1.76}{6.02} = 10.7 \ bit.$$

21. According to

$$MDS = \sigma(x) = \frac{v_{n2}}{G1 \cdot G2} + \frac{v_{n3}}{G1 \cdot G2 \cdot G3} = 8.3 \ mA$$

$$DR = 20\log \left(\frac{FS}{q \cdot MDS} \right) = 56.9 \ dB.$$

22. The FoM$_B$ is given by

$$FoM_B = \frac{P \cdot \sigma^2(x)}{BW} = 3.05 \ pJ \cdot A^2.$$

23. The solution of the problem is very similar to the one proposed for Problem 12.12, thus we propose in here the numerical solution to the problem in MATLAB code.

```
ref = [..] %insert here the reference values
out = [..] %insert here the measured output values
[P,S] = polyfit(ref,out,1); % fitting with a straight line of
coefficients P
Vos_stima = P(2) % offset from linear fitting
G_stima = P(1) % gain from linear fitting
yfit = polyval(P,ref); %evaluate the fitted point
err = out - yfit; %out-referred errors
FSO = max(out)-min(out);
INL = max(err)*100/FSO %output-referred INL
sigma_out=std(err);
sigma_in=std(err)/G_stima %input-referred distortion-
related uncertainty
FS = max(ref) - min(ref)
SNRa = 20*log10(FS / sigma_in);
ENOBa= (SNRa-1.76)/6.02
```

Numerical results are the following

$$V_{off} = 1.58 \text{ V}$$
$$G = 72.8 \ \frac{\text{mV}}{\text{nA}}$$
$$\text{INL} = 2.4\%\text{FSO}$$
$$\sigma_D(I) = 0.9 \text{ nA}$$
$$\text{ENOB}_a = 5.16 \text{ bit.}$$

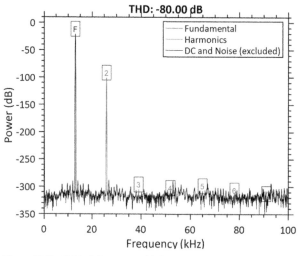

Figure 12.10 FFT of the output highlighting the THD.

24. The THD could be found in many numerical simulators where they have pre-compiled functions for its calculation. The following MATLAB code solves the problem and plots the FFT of the output voltage signal as shown in Fig. 12.10:

```
Vos = 440e-6; % offset voltage
G = 25e3; % gain
alpha =1e6; % second-order coefficient
Fs = 200e3; % sampling freq.
Ts = 1/Fs; % time step
Fin = 13e3; % signal freq.
Ampin = 5e-6; % amplitude of the pressure signal
t = linspace(0,0.05,0.05/Ts); % time axis from 0 to 500 ms
sampled at ts
in = Ampin.*sin(2*pi*Fin.*t);
out = Vos + G.*in + alpha.*in.^2;
%plot(t,in,t,out)
figure
thd(out,Fs,10) %thd computed by evaluating the first ten
harmonic components
```

25. The 1% of accuracy means that both random and systematic errors should be as low as to achieve the following signal-to-noise-and-distortion-ratio

$$SNDR = 20\log\left(\frac{1}{0.01}\right) = 40 \text{ dB}$$

Therefore, for the given operating range, the required covered dynamic range should be

$$DR = SNDR + 20\log\frac{L_{max}}{L_{min}} = 40 + 20\log\frac{100}{10 \times 10^{-2}} = 100 \text{ dB}$$

Now, the TDC should have an effective number of bits (including distortion) of

$$ENOB = \frac{100 - 1.76}{6.02} = 16.32b.$$

26. As discussed in Section 2.11, the quantization noise must be handled in different ways depending on the properties of the input signal. If the input signal is noise-free or affected by a very small noise, then the quantization noise is, actually, a systematic error in general correlated with the input signal. Conversely, if the input signal is affected by a large noise, then the quantization noise can be treated as a white noise uncorrelated with respect to the input noise. To discriminate between these two cases, we have to calculate the factor k:

$$k = \frac{V_{LSB}}{\sigma_N} = \frac{596}{800} = 0.745.$$

A factor $k < 2$ implies that the thermal noise is almost completely uncorrelated with respect to quantization noise; thus, they can be combined by means of the geometric mean

$$u_c(x) = \sqrt{\sigma_N^2 + \sigma_Q^2} = \sqrt{\sigma_N^2 + \frac{V_{LSB}^2}{12}} = 818 \text{ nV}.$$

The SNDR is given by

$$SNDR = 20\log \frac{FS}{q \cdot u_c(x)} = 132.7 \text{ dB}$$

$$ENOB_{SNDR} = \frac{SNDR - 1.76}{6.02} = 21.7 \text{ bit}$$

Due to the thermal noise, almost two least significant bits become "noisy."

27. As discussed in Section 3.4.3, the optimal A/D converter must have 1 bit more in accuracy so as to not degrade the accuracy of the overall acquisition chain with respect to the accuracy of the sensor (and its analog interface). Thus, we have to start with the computation of the accuracy of the analog acquisition chain.

   Since offset and gain error are assumed to be completely compensated, then noise and nonlinear distortion are the main sources of uncertainty. The rms noise is given and can be treated as the noise-related standard uncertainty. Uncertainty due to nonlinearity errors can be numerically computed as follows:

```
Pin = logspace (-8,-4,100); % span of the whole input FS
out = Vos + G.*Pin + alpha.*Pin.^2;
[P,S] = polyfit (Pin,out,1); % fitting with a straight line of
coefficients P to compensate for real offset and gain errors
outfit = polyval (P, Pin) ; % evaluate the fitted points
nl_error =out - outfit; % evaluate the nonlinearity errors
u_nl = std (nl_error) ; %output-related distortion-induced
uncertainty
```

Now, the accuracy of the acquisition chain (without the ADC) can be expressed in terms of "virtual bits." Note that all the variables are referred to the output of the microphone, i.e., the input of the ADC

$$u_c^2 = u_N^2 + u_d^2 = 647 \text{ } \mu V$$

$$SNDR = 20\log \left(\frac{FSO}{q \cdot u_c} \right) = 62.7 \text{ dB}$$

$$ENOB_{SNDR} = \frac{SNDR - 1.76}{6.02} = 10.1 \text{ bit.}$$

Thus, the optimal A/D converter should have more than 11 bits of ENOB.

28. The problem can be solved by looking at Murmann's plot (Fig. 3.35 of Chapter 3), where the Walden FOM for ADC is depicted. The given ADC should have more than 11 bits; thus, we will solve the problem for both an 11-bit ADC and a 12-bit ADC. Note that these are not nominal bits but represent the $\text{ENOB}_{\text{SNDR}}$ of the converter, which takes into care the quantization, the thermal noise, and all the other sources of inaccuracy. We have the extract the value of SNDR from the $\text{ENOB}_{\text{SNDR}}$ as follows:

$$\text{SNDR} = 1.76 + 6.02 \cdot \text{ENOB}_{\text{SNDR}}$$
$$11\text{-bit} \rightarrow \text{SNDR} = 68 \text{ dB}$$
$$12\text{-bit} \rightarrow \text{SNDR} = 74 \text{ dB}.$$

The limit of 1 fJ/NL shown in the Walden figure of merit allows to calculate the power-to-sampling-frequency ratio easily, and the minimum required power consumption

$$11\text{-bit}: 1 \text{ fJ} \cdot \text{NL} = 1 \text{ fJ} \cdot 2^{11} = 2 \text{ pJ} \quad \Rightarrow P = 2 \times 10^{-12} \cdot 96 \times 10^{3} = 192 \text{ nW}$$
$$12\text{-bit}: 1 \text{ fJ} \cdot \text{NL} = 1 \text{ fJ} \cdot 2^{12} = 4.1 \text{ pJ} \quad \Rightarrow P = 4.2 \times 10^{-12} \cdot 96 \times 10^{3} = 394 \text{ nW}.$$

29. First of all, we have to express the output voltage of the Wheatstone bridge as a function of the temperature

$$V_{\text{out}} = \frac{V_B}{2} - \frac{V_B R_{PT}(T)}{R + R_{PT}(T)}.$$

Our system's sensitivity can be defined as the derivative of the output voltage with respect to the temperature. Given the complexity of the relationship, we can split the problem as follows

$$S = \frac{dV_{\text{out}}}{dT} = \frac{dV_{\text{out}}}{dR_{PT}} \frac{dR_{PT}}{dT},$$

where the second ratio is the sensitivity of the PT1000 sensor we have already computed in Problem 4; thus, we have to compute the sensitivity of the Wheatstone bridge to the variable R_{PT}.

$$\frac{dV_{\text{out}}}{dR_{PT}} = -V_B \frac{R - 2R_{PT}}{(R + R_{PT})^2}.$$

The resistance of the platinum thermistor can be expressed as the nominal value plus a small change; thus

$$\frac{dV_{\text{out}}}{dR_{PT}} = -V_B \frac{R - 2(R + \Delta R)}{\left(R + (R + \Delta R)\right)^2} = V_B \frac{R + 2\Delta R}{(2R + \Delta R)^2} \xrightarrow{\Delta R \ll R} \frac{V_B}{4} = 1.25 \frac{\text{V}}{\Omega}.$$

As a result

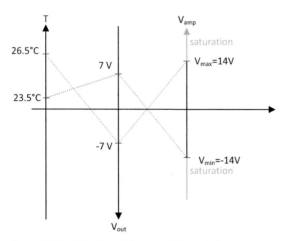

Figure 12.11 Gain plot of the acquisition chain.

$$S = \frac{dV_{out}}{dR_{PT}} \cdot S|_{T=25} = 4.89 \, \frac{V}{°C}.$$

30. To maximize the *DR*, the *FSO* should be as high as possible. In this case, the *FSO* is limited by the saturation levels of the instrumentation amplifier; thus the *FSO* = 28 V. The intermediated full-scale *FS*1, that is, the full scale at Wheatstone's bridge output, is simply related to the *FSO* by the gain of the amplifier

$$FS1 = \frac{FSO}{G} = 14 \text{ V}$$

and the input *FS* is related to *FS*1 by the bridge sensitivity to temperature variations

$$FS = \frac{FS1}{S} = 2.86 \, °C.$$

The sensing chain's gain plot is depicted in Fig. 12.11, where it has been taken into account that the PT1000 sensor has its nominal value of 1 kΩ at 25°C; thus, the 0 V output of the Wheatstone bridge corresponds to 25°C of temperature.

31. The only way to enlarge the *FS* without changing the *FSO* consists of reducing the gain. Since we have to work on the bridge, but we cannot change the thermistor, we can only lower the bias voltage V_B by a factor of 5 to get the desired result. Setting V_B = 1 V

$$V_B = 1 \text{ V}$$

$$S = \frac{V_B}{4} \cdot S|_{T=25} = 0.98 \, \frac{V}{°C}$$

$$FS = \frac{FSO}{S \cdot G} = 14 \, °C.$$

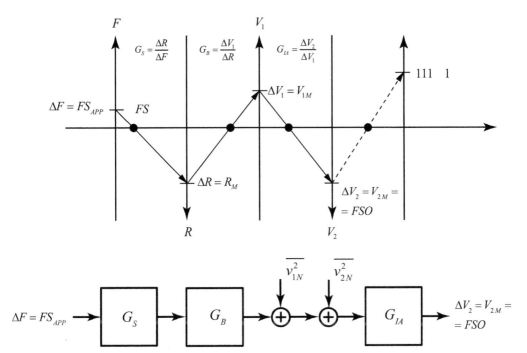

Figure 12.12 Gain plot of the exercise.

32. The first step is to draw the gain plot of the system, as illustrated in Fig. 12.12. The required *total gain* of the analog part is given by

$$S_T = \frac{FSO}{FS_{\text{APP}}} = \frac{V_R}{\Delta F} = \frac{3.3}{980} = 3.37 \times 10^{-3} \frac{V}{N} \rightarrow G_{T(\text{dB})} = 20\log S_T = -49.45 \text{ dB},$$

since 100 kg are 980 N.

The physical transduction can be calculated as (Chapter 11) where the maximum strain in the center of the slab is

$$\varepsilon_{\max} = \frac{1}{4} \frac{l}{Eah^2} F = 0.0109\%,$$

which is compatible with the limits of the material. From the preceding, we can get the relative variation of the resistance

$$\frac{\Delta R}{R_0} = G_F \cdot \varepsilon_{\max} = 0.02\% \rightarrow \Delta R = 0.027 \Omega.$$

Therefore, the gain of the first stage is

$$G_S = \frac{\Delta R}{\Delta F} = 28 \times 10^{-6} \frac{\Omega}{N} \rightarrow G_{S(\text{dB})} = -91.06 \text{ dB}$$

As far as the second stage is concerned, the gain is (Chapter 8)

$$G_1 = G_B = \frac{\Delta V_1}{\Delta R} = \frac{1}{4} \frac{V_R}{R_0} = 6.8 \times 10^{-3} \frac{V}{\Omega} \rightarrow G_{1(\text{dB})} = G_{B(\text{dB})} = -43.2 \text{ dB}$$

and

$$\Delta V_1 = G_B \Delta R = 189 \ \mu V$$

Now, we can calculate the gain of the last stage as

$$G_2 = G_{IA} = \frac{V_R}{\Delta V_1} = 17.5 \times 10^{-3} \rightarrow G_{2(\text{dB})} = G_{IA(\text{dB})} = 84.8 \text{ dB}$$

Note that $G_{AT(\text{dB})} = G_{S(\text{dB})} + G_{B(\text{dB})} + G_{IA(\text{dB})}$.

Now, if we want to reach 10 g of the resolution, we have to get a DR of the analog part of

$$\text{noise-free resolution} = \log_2 \frac{100}{0.01} = 13.2\text{b}.$$

This means that the effective resolution, that is, the ENOB, should be at least 2.7b more (Chapter 2):

$$\text{ENOB}_{AT} = \text{effective resolution} + 2.43 = 15.72\text{b}$$
$$\rightarrow DR_{AT(\text{dB})} = 6.02 \cdot \text{ENOB}_{AT} + 1.76 = 96.3 \text{ dB} \rightarrow DR_{AT} = 4.3 \times 10^9$$

where AT subscript stands for "analog interface." We have to calculate first the dynamic range of the Wheatstone bridge, since

$$\overline{v_{1N}^2} = 4kTR_0 \cdot BW = 1.97 \times 10^{-18} \rightarrow DR_{1(\text{dB})} = DR_{S(\text{dB})} = 20\log \frac{\Delta V_1}{v_{1Nrms}} = 102.5 \text{ dB}$$
$$\rightarrow \text{ENOB}_1 = \frac{102.5 - 1.76}{6.02} = 16.7\text{b}.$$

Therefore, to understand the DR of the instrumentation amplifier, we have to use the resolution rule of the acquisition chain:

$$\frac{1}{DR_{AT}} = \frac{1}{DR_0} + \frac{1}{DR_1} + \frac{1}{DR_2}.$$

and since and since we have no input noise so that $DR_0 = \infty$, we have

$$\frac{1}{DR_2} = \frac{1}{DR_{AT}} - \frac{1}{DR_1} \rightarrow DR_2 = 5.7 \times 10^9 \rightarrow DR_{2(\text{dB})} = 97.5 \text{ dB} \rightarrow \text{ENOB}_2 = 15.92\text{b}.$$

Therefore, the input-referred noise should be

$$\overline{v_{2N}^2} = \frac{\Delta V_1^2}{DR_2 \cdot BW} \rightarrow v_{2N}(f) = 2.5\text{nV}/\sqrt{\text{Hz}},$$

which is quite challenging.

Now, we have to choose the A/D converter whose DR should contribute so that the total dynamic range $DR_{T(\text{dB})}$ is as close as possible to $DR_{AT(\text{dB})} = 96.3\text{dB}$. So, if the use, for example, an A/D converter with ENOB=18b, we have

$$DR_{A/D} = 6.02 \cdot 18 + 1.76 = 110 \text{ dB}$$
$$\frac{1}{DR_T} = \frac{1}{DR_1} + \frac{1}{DR_2} + \frac{1}{DR_{A/D}} \rightarrow DR_{T(\text{dB})} = 96.20 \text{ dB}.$$

If we had used an ENOB = 17b A/D converter, we would get a total of $DR_{T(\text{dB})} = 95.70\text{dB}$, which is a bit low for the total requirement.

In conclusion, we should use an A/D converter with a noise-free number of bits 18b − 3b = 15b for the final requirements that can be achieved by truncating 3 bits in an ENOB=18b converter.

Index